AF522210

ENCYCLOPEDIA OF BIOCHEMISTRY

ENCYCLOPEDIA OF BIOCHEMISTRY

Vol. 3

By

Sananda Chatterjee

DAE, DCA, CMCNet

Scientific Consultant

Institute for Natural Sciences

Kolkata

(West Bengal)

DISCOVERY PUBLISHING HOUSE PVT. LTD.

NEW DELHI-110 002

Published by:
Namit Wasan
DISCOVERY PUBLISHING HOUSE PVT. LTD.
4383/4B, Ansari Road, Darya Ganj
New Delhi-110 002 (India)
Phone : +91-11-23279245; 23253475; 43596065
E-mail : discoverybooksindia@gmail.com
discoverypublishinghouse@gmail.com
namitwasan9@gmail.com
web : www.discoverypublishinggroup.com

Reprinted: **2020**
First Edition: **2012**

ISBN: 978-81-8356-876-0 (Set)

Encyclopedia of Biochemistry

Printed at:
Infinity Imaging Systems
Delhi

Preface

It is always an ultimate outlook to give students the best information in a good package. So I took up the job of writing a series of books in chemistry starting from a very introductory level to this advance level of biochemistry.

Biochemistry, to express something about it, in one word, would be difficult for me. It should be ones part of life; the reactions of protein, amino-acids, the DNA RNA—double helix pattern. The nynhydrin reactions are all the wonders of biochemistry—man's anato-biochemical linkages.

Students who want to become doctors or medical practitioners must be a good biochemist first because biochemistry is the theory behind and the medicine is the practical application.

The effects and side effects of the particular medicine, whether it is allergic or non-allergic, to the patients are well understood by the doctors but are taught in biochemistry.

The chemical functions of the different human organs and systems; the kidney, the respiratory system, the blood, the chemical functions of hormones, vitamins, the liver etc are all well covered by this book.

The subject of this book is therefore deeply intended for those upcoming doctors who keep themselves for the real service of the mankind.

The object of this book is not only meant for the doctors but it is the last bangle of that long garland which has started all along from the introduction chapter of The Architecture of Chemistry, The Mans Scientific Attitude.

Biochemistry also comes in use in the field of agriculture, the biochemical configuration of the soil. The biochemical configuration of the plants to be cultivated, the application of the pesticides, and the relation of the pesticides with the human population are well studied under biochemistry.

The Agricultural Scientist, the Pathological Scientist, Forensic Scientist, is definitely inter-joined with each other by a single buffer of the Biochemistry.

This book is dedicated to those promising scientific readers and interested students of biochemistry and advanced chemical science, who wants to make a fruitful attempt in the field of modern scientific world.

The author will be grateful and surely thankful to see the best utilization of this book in any of the field of application.

Sananda Chatterjee

Contents

Thyroid Hormone Transport—Transport in the Blood—Thyroxine-Binding Globulin—Transthyretin—Albumin—Lipoproteins—Consequences of Abnormal Binding Protein Concentrations—Transport Across the Cell Membrane—Organic Anion Transporters—Amino Acid Transporters—The Transport Mechanism of the Insulin Hormone in Blood—Insulin —Glucagon—Levels—Insulin action in Fat Cells—Resting Conditions—At Work —After a Good Meal—Tissue Distribution of Glucose after a Meal—Energy metabolism during a fast, in Starvation and Uncontrolled Diabetes—Starvation—Diabetes; Metabolism in the Absence of Insulin—Aimasthænia - The Blood diseases —Bone Marrow —Stroma—Stem Cells—Types of Stem Cells—Diseases Involving the Bone Marrow—Examination—Anaemia or Anæmia—Signs and Symptoms—Diagnosis—Classification—Cause—Thalassemia—Alpha (a) thalassemias—Beta (b) thalassemias—Delta (d) thalassemia—In Combination with Other Hemoglobinopathies—Thalassemia Major —Haemophilia —Causes of Haemophilia—Symptoms of Haemophilia—In vivo Hemolysis—In vitro Hemolysis—Hemolysis due to Mechanical Blood Processing during Surgery—Leukæmia—The Blood Cancer—Classification—Signs and Symptoms—Causes—Edema—Formation—Mechanism—Generalized Edema—Organ-specific Edema—The Subacute CO Poisoning—The Blood Grouping and Testing Methods—Technique—The Blood Grouping and Blood Cell Counting—Rare Blood Types—Determination of Blood Groups2—Blood Counting—Method A —Method B.

Genetic Methodology

CHROMOSOME AND MITOCHONDRIA

A **chromosome** is an organized structure of DNA and protein that is found in cells. A chromosome is a single piece of DNA that contains many genes, regulatory elements and other nucleotide sequences. Chromosomes also contain DNA-bound proteins, which serve to package the DNA and control its functions. The word *chromosome* comes from the Greek (*chroma*, colour) and *soma* (body) due to their property of being stained very strongly by some dyes. Chromosomes vary extensively between different organisms. The DNA molecule may be circular or linear, and can contain anything from 10,000 to 100,000,000 nucleotides in length. Typically eukaryotic cells (cells with nuclei) have large linear chromosomes and prokaryotic cells (cells without defined nuclei) have smaller circular chromosomes, although there are many exceptions to this rule. Furthermore, cells may contain more than one type of chromosome; for example, mitochondria in most eukaryotes and chloroplasts in plants have their own small chromosomes. In eukaryotes, nuclear chromosomes

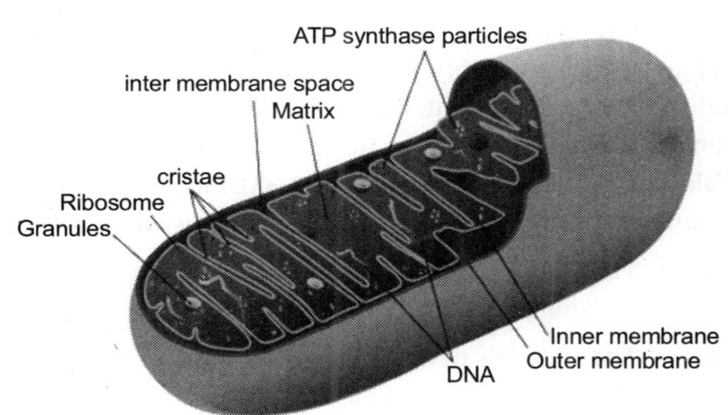

Fig. 6.1 : The Mitochondria

are packaged by proteins into a condensed structure called chromatin. This allows the very long DNA molecules to fit into the cell nucleus. The structure of chromosomes and chromatin varies through the cell cycle. Chromosomes may exist as either duplicated or unduplicated—unduplicated chromosomes are single linear strands, whereas duplicated chromosomes (copied during synthesis phase) contain two copies joined by a centromere. Compaction of the duplicated chromosomes during mitosis and meiosis results in the classic four-arm structure (pictured to the right). "Chromosome" is a rather loosely defined term. In prokaryotes, a small circular DNA molecule may be called either a plasmid or a small chromosome. These small circular genomes are also found in mitochondria and chloroplasts, reflecting their bacterial origins. The simplest chromosomes are found in viruses: these DNA or RNA molecules are short linear or circular chromosomes that often lack any structural proteins. His is a brief history of research in a complex field where each advance was hard won, and often hotly disputed at the time.

Visual Discovery of Chromosomes

Textbooks have often said that chromosomes were first observed in plant cells by a Swiss botanist named Karl Wilhelm von Nägeli in 1842. However, this opinion has been challenged, perhaps decisively, by Henry Harris, who has freshly reviewed the primary literature. In his opinion the claim of Nägeli to have seen spore mother cells divide is mistaken, as are some of his interpretations. Harris considers other candidates, especially Wilhelm Hofmeister, whose publications in 1848-9 include plates that definitely show mitotic events. Hofmeister was also the choice of Cyril Darlington.

The work of other cytologists such as Walther Flemming, Eduard Strasburger, Otto Bütschli, Oskar Hertwig and Carl Rabl should be acknowledged. The use of basophilic aniline dyes was a new technique for effectively staining the chromatin material in the nucleus. Their behavior in animal (salamander) cells was later described in detail by Walther Flemming, who in 1882 "provided a superb summary of the state of the field." The name *chromosome* was invented in 1888 by Heinrich von Waldeyer. However, van Beneden's monograph of 1883 on the fertilised eggs of the parasitic roundworm *Ascaris megalocephala* was the outstanding work of this period. His conclusions are classic:

- Thus there is no fusion between the male chromatin and the female chromatin at any stage of division....
- The elements of male origin and those of female origin are never fused together in a cleavage nucleus, and perhaps they remain distinct in all the nuclei derived from them. "It is not easy to identify who first discerned chromosomes during mitosis, but there is no doubt that those who first saw them had no idea of their significance... [but] with the work of Balbiani and van Beneden we move away from... the mechanism of cell division to a precise delineation of chromosomes and what they do during the division of the cell."

Van Beneden's master work was closely followed by that of Carl Rabl, who reached similar conclusions. This more or less concludes the first period, in which chromosomes were visually sighted and the morphological stages of mitosis were described. Coleman also gives a useful review of these discoveries.

Nucleus as the Seat of Heredity

The origin of this epoch-making idea lies in a few sentences tucked away in Ernst Haeckel's *Generelle Morphologie* of 1866. The evidence for this insight gradually accumulated until, after twenty or so years, two of the greatest in a line of great German scientists spelled out the concept. August Weismann proposed that the germ line is separate from the soma, and that the cell nucleus is the repository of the hereditary material, which, he proposed, is arranged along the chromosomes in a linear manner. Furthermore, he proposed that at fertilisation a new combination of chromosomes (and their hereditary material) would be formed. This was the explanation for the reduction division of meiosis (first described by van Beneden).

Chromosomes as Vectors of Heredity

In a series of outstanding experiments, Theodor Boveri gave the definitive demonstration that chromosomes are the vectors of heredity. His two principles were:

The continuity of chromosomes; and

The individuality of chromosomes.

It is the second of these principles that was so original. Boveri was able to test the proposal put forward by Wilhelm Roux, that each chromosome carries a different genetic load, and showed that Roux was right. Upon the rediscovery of Mendel, Boveri was able to point out the connection between the rules of inheritance and the behaviour of the chromosomes. It is interesting to see that Boveri influenced two generations of American cytologists: Edmund Beecher Wilson, Walter Sutton and Theophilus Painter were all influenced by Boveri (Wilson and Painter actually worked with him). In his famous textbook *The Cell*, Wilson linked Boveri and Sutton together by the Boveri-Sutton theory. Mayr remarks that the theory was hotly contested by some famous geneticists: William Bateson, Wilhelm Johannsen, Richard Goldschmidt and T.H. Morgan, all of a rather dogmatic turn-of-mind. Eventually complete proof came from chromosome maps – in Morgan's own lab

Chromosomes in Eukaryotes

Eukaryotes (cells with nuclei such as plants, yeast, and animals) possess multiple large linear chromosomes contained in the cell's nucleus. Each chromosome has one centromere, with one or two arms projecting from the centromere, although, under most circumstances, these arms are not visible as such. In addition, most eukaryotes have a small circular mitochondrial genome, and some eukaryotes may have additional small circular or linear cytoplasmic chromosomes.

In the nuclear chromosomes of eukaryotes, the uncondensed DNA exists in a semi-ordered structure, where it is wrapped around histones (structural proteins), forming a composite material called chromatin.

Chromatin

Chromatin is the complex of DNA and protein found in the eukaryotic nucleus which packages chromosomes. The structure of chromatin varies significantly between different stages of the cell cycle, according to the requirements of the DNA.

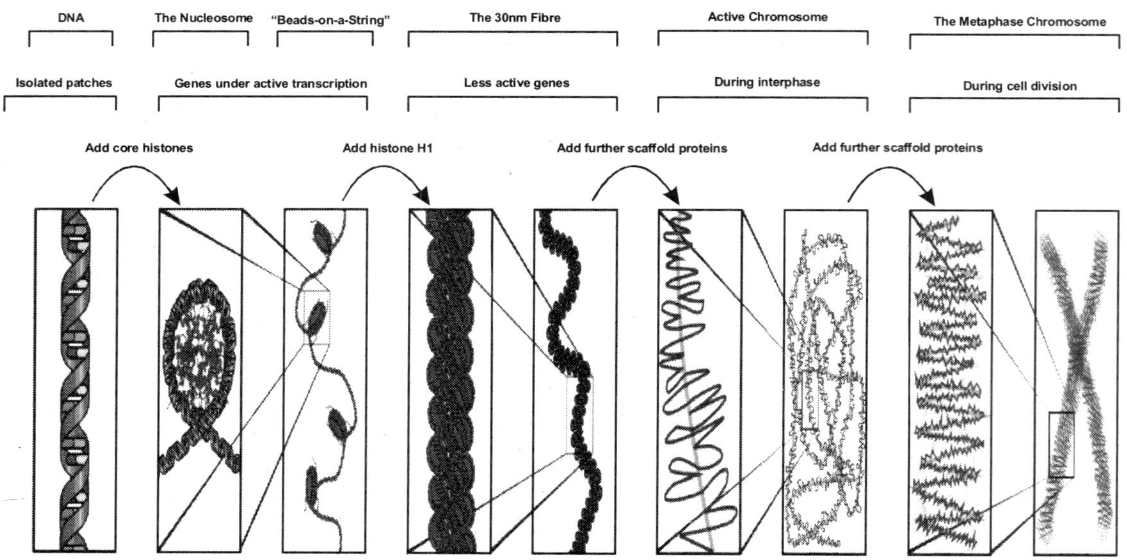

Fig. 6.2 : The major structures in DNA compaction; DNA, the nucleosome, the 10nm "beads-on-a-string" fibre, the 30nm fibre and the metaphase chromosome

Interphase Chromatin

During interphase (the period of the cell cycle where the cell is not dividing) two types of chromatin can be distinguished:

- Euchromatin, which consists of DNA that is active, e.g., expressed as protein.
- Heterochromatin, which consists of mostly inactive DNA. It seems to serve structural purposes during the chromosomal stages. Heterochromatin can be further distinguished into two types:
  - *Constitutive heterochromatin*, which is never expressed. It is located around the centromere and usually contains repetitive sequences.
  - *Facultative heterochromatin*, which is sometimes expressed.

Individual chromosomes cannot be distinguished at this stage - they appear in the nucleus as a homogeneous tangled mix of DNA and protein.

In the early stages of mitosis or meiosis (cell division), the chromatin strands become more and more condensed. They cease to function as accessible genetic material (transcription stops) and become a compact transportable form. This compact form makes the individual chromosomes visible, and they form the classic four arm structure, a pair of sister chromatids attached to each other at the centromere. The shorter arms are called *p arms* (from the French *petit*, small) and the longer arms are called *q arms* (*q* follows *p* in the Latin alphabet). This is the only natural context in which individual chromosomes are visible with an optical microscope.

During divisions long microtubules attach to the centromere and the two opposite ends of the cell. The microtubules then pull the chromatids apart, so that each daughter cell inherits one set of chromatids. Once the cells have divided, the chromatids are uncoiled and can function again as chromatin. In spite of their appearance, chromosomes are structurally highly condensed, which enables these giant DNA structures to be contained within a cell nucleus.

The self-assembled microtubules form the spindle, which attaches to chromosomes at specialized structures called kinetochores, one of which is present on each sister chromatid. A special DNA base sequence in the region of the kinetochores provides, along with special proteins, longer-lasting attachment in this region.

Chromosomes in Prokaryotes

The prokaryotes - bacteria and archaea - typically have a single circular chromosome, but many variations do exist. Most bacteria have a single circular chromosome that can range in size from only 160,000 base pairs in the endosymbiotic bacteria *Candidatus Carsonella ruddii*, to 12,200,000 base pairs in the soil-dwelling bacteria *Sorangium cellulosum*. Spirochaetes of the genus *Borrelia* are a notable exception to this arrangement, with bacteria such as *Borrelia burgdorferi*, the cause of Lyme disease, containing a single linear chromosome.

Structure in Sequences

Prokaryotic chromosomes have less sequence-based structure than eukaryotes. Bacteria typically have a single point (the origin of replication) from which replication starts, whereas some archaea contain multiple replication origins. The genes in prokaryotes are often organized in operons, and do not usually contain introns, unlike eukaryotes.

DNA Packaging

Prokaryotes do not possess nuclei. Instead, their DNA is organized into a structure called the nucleoid. The nucleoid is a distinct structure and occupies a defined region of the bacterial cell. This structure is, however, dynamic and is maintained and remodeled by the actions of a range of histone-like proteins, which associate with the bacterial chromosome. In archaea, the DNA in chromosomes is even more organized, with the DNA packaged within structures similar to eukaryotic nucleosomes.

Bacterial chromosomes tend to be tethered to the plasma membrane of the bacteria. In molecular biology application, this allows for its isolation from plasmid DNA by centrifugation of lysed bacteria and pelleting of the membranes (and the attached DNA).

Prokaryotic chromosomes and plasmids are, like eukaryotic DNA, generally supercoiled. The DNA must first be released into its relaxed state for access for transcription, regulation, and replication.

Number of Chromosomes in various Organisms Eukaryotes

These tables give the total number of chromosomes (including sex chromosomes) in a cell nucleus. For example, human cells are diploid and have 22 different types of autosomes, each present as two copies, and two sex chromosomes. This gives 46 chromosomes in total. Other organisms have more

Table 6.1 : Chromosome numbers in some plants

Plant Species	#
Arabidopsis thaliana (diploid)	10
Rye (diploid)	14
Maize (diploid)	20
Einkorn wheat (diploid)	14
Durum wheat (tetraploid)	28
Bread wheat (hexaploid)	42
Potato (tetraploid)	48
Cultivated tobacco (diploid)	48
Adder's Tongue Fern (diploid)	approx 1,440

than two copies of their chromosomes, such as Bread wheat, which is *hexaploid* and has six copies of seven different chromosomes - 42 chromosomes in total.

Table 6.2 : Chromosome numbers (2n) in some animals

Species	#	Species	#
Common fruit fly	8	Guinea Pig	64
Dove	16	Garden snail	54
Earthworm *Octodrilus complanatus*	36	Tibetan fox	36
Domestic cat	38	Domestic pig	38
Laboratory mouse	40	Laboratory rat	42
Rabbit	44	Syrian hamster	44
Hare	46	Human	46
Gorillas, Chimpanzees	48	Domestic sheep	54
Elephants	56	Cow	60
Donkey	62	Horse	64
Dog	78	Kingfisher	132
Goldfish	100-104	Silkworm	56

Chromosome numbers in other organisms

Species - Large

Chromosomes - Intermediate

Chromosomes - Small

Chromosomes

Trypanosoma brucei-11-6-~100

Chicken-8-2 sex chromosomes-60

Normal members of a particular eukaryotic species all have the same number of nuclear chromosomes (Table 6.1 and Table 6.2). Other eukaryotic chromosomes, i.e., mitochondrial and plasmid-like small chromosomes, are much more variable in number, and there may be thousands of copies per cell.

Asexually reproducing species have one set of chromosomes, which is the same in all body cells.Sexually reproducing species have somatic cells (body cells), which are diploid [2n] having two sets of chromosomes, one from the mother and one from the father. Gametes, reproductive cells, are haploid [n]: They have one set of chromosomes. Gametes are produced by meiosis of a diploid germ line cell. During meiosis, the matching chromosomes of father and mother can exchange small parts of themselves (crossover), and thus create new chromosomes that are not inherited solely from either parent. When a male and a female gamete merge (fertilization), a new diploid organism is formed.

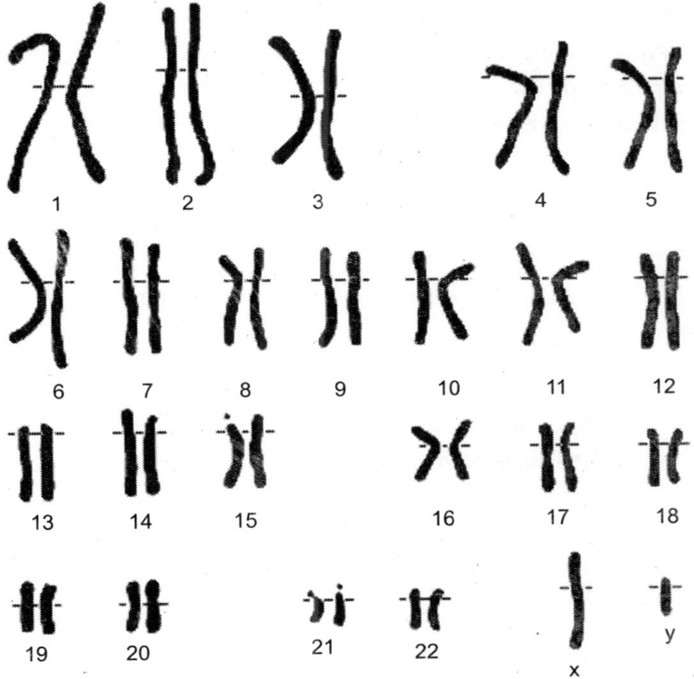

Fig. 6.3 : The 23 human chromosome territories

Some animal and plant species are polyploid [Xn]: They have more than two sets of homologous chromosomes. Plants important in agriculture such as tobacco or wheat are often polyploid, compared to their ancestral species. Wheat has a haploid number of seven chromosomes, still seen in some cultivars as well as the wild progenitors. The more-common pasta and bread wheats are polyploid, having 28 (tetraploid) and 42 (hexaploid) chromosomes, compared to the 14 (diploid) chromosomes in the wild wheat.

Prokaryotes

Prokaryote species generally have one copy of each major chromosome, but most cells can easily survive with multiple copies. For example, *Buchnera*, a symbiont of aphids has multiple copies of its chromosome, ranging from 10-400 copies per cell. However, in some large bacteria, such as *Epulopiscium fishelsoni* up to 100,000 copies of the chromosome can be present.[43] Plasmids and plasmid-like small chromosomes are, as in eukaryotes, very variable in copy number. The number of plasmids in the cell is almost entirely determined by the rate of division of the plasmid - fast division causes high copy number, and *vice versa*.

Karyotype

In general, the **karyotype** is the characteristic chromosome complement of a eukaryote species. The preparation and study of karyotypes is part of cytogenetics.

Although the replication and transcription of DNA is highly standardized in eukaryotes, *the same cannot be said for their karotypes*, which are often highly variable. There may be variation between species in chromosome number and in detailed organization. In some cases, there is significant variation within species. Often there is:

1. variation between the two sexes;
2. variation between the germ-line and soma (between gametes and the rest of the body);
3. variation between members of a population, due to balanced genetic polymorphism;
4. geographical variation between races;
5. mosaics or otherwise abnormal individuals. Also, variation in karyotype may occur during development from the fertilised egg.The technique of determining the karyotype is usually called *karyotyping*. Cells can be locked part-way through division (in metaphase) in vitro (in a reaction vial) with colchicine. These cells are then stained, photographed, and arranged into a *karyogram*, with the set of chromosomes arranged, autosomes in order of length, and sex chromosomes (here XY) at the end.

Like many sexually reproducing species, humans have special gonosomes (sex chromosomes, in contrast to autosomes). These are XX in females and XY in males.

Historical Note

Investigation into the human karyotype took many years to settle the most basic question. How many chromosomes does a normal diploid human cell contain? In 1912, Hans von Winiwarter reported 47 chromosomes in spermatogonia and 48 in oogonia, concluding an XX/XO sex determination mechanism. Painter in 1922 was not certain whether the diploid number of man is 46 or 48, at first favouring 46. He revised his opinion later from 46 to 48, and he correctly insisted on man's having an XX/XY system.

New techniques were needed to definitively solve the problem:

1. Using cells in culture
2. Pretreating cells in a hypotonic solution, which swells them and spreads the chromosomes
3. Arresting mitosis in metaphase by a solution of colchicine
4. Squashing the preparation on the slide forcing the chromosomes into a single plane
5. Cutting up a photomicrograph and arranging the result into an indisputable karyogram.

It took until the mid-1950s until it became generally accepted that the human karyotype include only 46 chromosomes. Considering the techniques of Winiwarter and Painter, their results were quite remarkable. Chimpanzees (the closest living relatives to modern humans) have 48 chromosomes.

Chromosomal Aberrations

Chromosomal aberrations are disruptions in the normal chromosomal content of a cell, and are a major cause of genetic conditions in humans, such as Down syndrome. Some chromosome abnormalities do not cause disease in carriers, such as translocations, or chromosomal inversions, although they may lead to a higher chance of birthing a child with a chromosome disorder. Abnormal numbers of chromosomes or chromosome sets, aneuploidy, may be lethal or give rise to genetic disorders. Genetic counseling is offered for families that may carry a chromosome rearrangement.The gain or loss of chromosome material can lead to a variety of genetic disorders. Human examples include:

- Cri du chat, which is caused by the deletion of part of the short arm of chromosome 5. "Cri du chat" means "cry of the cat" in French, and the condition was so-named because affected babies make high-pitched cries that sound like those of a cat. Affected individuals have wide-set eyes, a small head and jaw, and are moderately to severely mentally retarded and very short.
- Wolf-Hirschhorn syndrome, which is caused by partial deletion of the short arm of chromosome 4. It is characterized by severe growth retardation and severe to profound mental retardation.
- Down's syndrome, usually is caused by an extra copy of chromosome 21 (trisomy 21). Characteristics include decreased muscle tone, stockier build, asymmetrical skull, slanting eyes and mild to moderate mental retardation.
- Edwards syndrome, which is the second-most-common trisomy; Down syndrome is the most common. It is a trisomy of chromosome 18. Symptoms include mental and motor retardation and numerous congenital anomalies causing serious health problems. Ninety percent die in infancy; however, those that live past their first birthday usually are quite healthy thereafter. They have a characteristic clenched hands and overlapping fingers.
- Patau Syndrome, also called D-Syndrome or trisomy-13. Symptoms are somewhat similar to those of trisomy-18, but they do not have the characteristic hand shape.
- Idic15, abbreviation for Isodicentric 15 on chromosome 15; also called the following names due to various researches, but they all mean the same; IDIC(15), Inverted dupliction 15, extra Marker, Inv dup 15, partial tetrasomy 15.

- Jacobsen syndrome, also called the terminal 11q deletion disorder. This is a very rare disorder. Those affected have normal intelligence or mild mental retardation, with poor expressive language skills. Most have a bleeding disorder called Paris-Trousseau syndrome.
- Klinefelter's syndrome (XXY). Men with Klinefelter syndrome are usually sterile, and tend to have longer arms and legs and to be taller than their peers. Boys with the syndrome are often shy and quiet, and have a higher incidence of speech delay and dyslexia. During puberty, without testosterone treatment, some of them may develop gynecomastia.
- Turner syndrome (X instead of XX or XY). In Turner syndrome, female sexual characteristics are present but underdeveloped. People with Turner syndrome often have a short stature, low hairline, abnormal eye features and bone development and a "caved-in" appearance to the chest.
- XYY syndrome. XYY boys are usually taller than their siblings. Like XXY boys and XXX girls, they are somewhat more likely to have learning difficulties.
- Triple-X syndrome (XXX). XXX girls tend to be tall and thin. They have a higher incidence of dyslexia.
- Small supernumerary marker chromosome. This means there is an extra, abnormal chromosome. Features depend on the origin of the extra genetic material. Cat-eye syndrome and isodicentric chromosome 15 syndrome (or Idic15) are both caused by a supernumerary marker chromosome, as is Pallister-Killian syndrome.

Chromosomal mutations produce changes in whole chromosomes (more than one gene) or in the number of chromosomes present.

- Deletion - loss of part of a chromosome
- Duplication - extra copies of a part of a chromosome
- Inversion - reverse the direction of a part of a chromosome
- Translocation - part of a chromosome breaks off and attaches to another chromosome.

Most mutations are neutral - have little or no effect.

A detailed graphical display of all human chromosomes and the diseases annotated at the correct spot may be found at.

Human Chromosomes

Chromosomes can be divided into two types-autosomes, and sex chromosomes. Certain genetic traits are linked to your sex, and are passed on through the sex chromosomes. The autosomes contain the rest of the genetic hereditary information. All act in the same way during cell division.

Human cells have 23 pairs of large linear nuclear chromosomes, giving a total of 46 per cell. In addition to these, human cells have many hundreds of copies of the mitochondrial genome. Sequencing of the human genome has provided a great deal of information about each of the chromosomes. Below is Table 6.3 compiling statistics for the chromosomes, based on the Sanger Institute's human genome information in the Vertebrate Genome Annotation (VEGA) database. Number of genes is an estimate as it is in part based on gene predictions. Total chromosome length is an estimate as well, based on the estimated size of unsequenced heterochromatin regions.

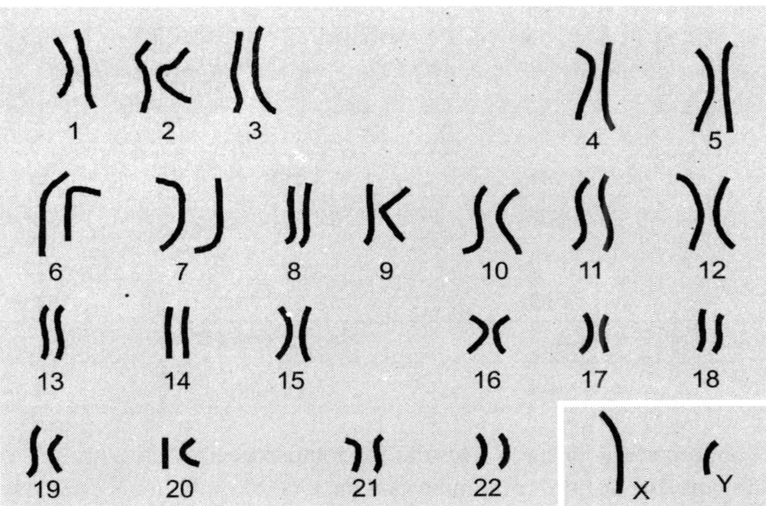

Fig. 6.4 : The 22 autosomes are numbered by size. The other two chromosomes, X and Y are the sex chromosomes. This picture of the human chromosomes lined up in pairs is called a karyotype.

Chromosome	Genes	Total bases	Sequenced bases
1	4,220	247,199,719	224,999,719
2	1,491	242,751,149	237,712,649
3	1,550	199,446,827	194,704,827
4	446	191,263,063	187,297,063
5	609	180,837,866	177,702,766
6	2,281	170,896,993	167,273,993
7	2,135	158,821,424	154,952,424
8	1,106	146,274,826	142,612,826
9	1,920	140,442,298	120,312,298
10	1,793	135,374,737	131,624,737
11	379	134,452,384	131,130,853
12	1,430	132,289,534	130,303,534
13	924	114,127,980	95,559,980

Chromosome	Genes	Total bases	Sequenced bases
14	1,347	106,360,585	88,290,585
15	921	100,338,915	81,341,915
16	909	88,822,254	78,884,754
17	1,672	78,654,742	77,800,220
18	519	76,117,153	74,656,155
19	1,555	63,806,651	55,785,651
20	1,008	62,435,965	59,505,254
21	578	46,944,323	34,171,998
22	1,092	49,528,953	34,893,953
X (sex chromosome)	1,846	154,913,754	151,058,754
Y (sex chromosome)	454	57,741,652	25,121,652

Sex chromosome abnormalities occur as a result of chromosome mutations brought on by mutagens (like radiation) or problems that occur during meiosis. One type of mutation is caused by chromosome breakage. The broken chromosome fragment may be deleted, duplicated, inverted, or translocated to a non-homologous chromosome. Another type of mutation occurs during meiosis and causes cells to have either too many or not enough chromosomes.

Normal Sex Chromosomes

In human sexual reproduction, two distinct gametes fuse to form a zygote. Gametes are reproductive cells produced by a type of cell division called meiosis. They contain only one set of chromosomes and are said to be haploid (one set of 22 autosomes and one sex chromosome). When the haploid male and female gametes unite in a process called fertilization, they form what is called a zygote. The zygote is diploid, meaning that it contains two sets of chromosomes (two sets of 22 autosomes and two sex chromosomes). The male gametes or sperm cells in humans and other mammals are heterogametic and contain one of two types of sex chromosomes. They are either X or Y. The female gametes or eggs however, contain only the X sex chromosome and are homogametic. The sperm cell determines the sex of an individual in this case. If a sperm cell containing an X chromosome fertilizes an egg, the resulting zygote will be XX or female. If the sperm cell contains a Y chromosome, then the resulting zygote will be XY or male.

Sex Chromosome Abnormalities

A chromosome abnormality that causes individuals to have an abnormal number of chromosomes is called aneuploidy. If the zygote has an additional chromosome (three instead of two) it is trisomic for that chromosome. If the zygote is missing a chromosome it is monosomic. Aneuploid cells occur as a result of chromosome breakage or nondisjunction errors that happen during meiosis. Two types of errors occur during nondisjunction: homologous chromosomes don't separate during anaphase I of meiosis I or sister chromatids don't separate during anaphase II of meiosis II. In sex chromosomes, nondisjunction results in a number of abnormalities. Klinefelter syndrome is a disorder in which males have an extra X chromosome. The genotype for males with this disorder is XXY. People with Klinefelter syndrome may also have more than one extra chromosome resulting in genotypes which include XXYY, XXXY, and XXXXY. Other mutations result in males that have an extra Y chromosome and a genotype of XYY. These males were once thought to be taller than average males and overly aggressive based on prison studies. Additional studies however have found XYY males to be normal. Tuner syndrome is a condition that affects females. Individuals with this syndrome, also called monosomy X, have a genotype of only one X chromosome (XO). Trisomy **X** females have an additional X chromosome and are also referred to as metafemales (XXX).Nondisjunction can occur in autosomal cells as well. Down syndrome is most commonly the result of nondisjunction affecting autosomal chromosome 21. Down syndrome is also referred to as trisomy 21 because of the extra chromosome. Table 6.3 includes information on sex chromosome abnormalities, resulting syndromes, and expressed physical traits.

Table 6.3 : Chromosome Abnormalities

Genotype	Gender	Syndrome	Physical Traits
XXY, XXYY, XXXY	male	Klinefelter syndrome	sterility, small testicles, breast enlargement
XYY	male	XYY syndrome	normal male traits
XO	female	Turner syndrome	sex organs don't mature at adolescence, sterility, short stature
XXX	female	Trisomy X	tall stature, learning disabilities, limited fertility

GENETIC METHODOLOGY AN INTRODUCTION

The segments of DNA which have been associated with specific features or functions of an organism are called ***genes. Genetics*** *studies how living organisms inherit features from their ancestors – for example, children often look like their parents. Genetics seeks to identify which features are inherited, and explain how these features are passed from generation to generation.*

In genetics, a feature of an organism is called a "trait". Some traits are features of an organism's physical appearance, for example, a person's eye-color, height or weight. There are many other trait types, and these range from aspects of behavior to resistance to disease. Traits are often inherited, for example tall and thin people tend to have tall and thin children. Other traits come from the interaction between inherited features and the environment. For example a child might inherit the tendency to be tall, but if there is very little food where they live and they are poorly nourished, they will still be short. The way genetics and environment interact to produce a trait can be complicated: for example, the chances of somebody dying of cancer or heart disease seem to depend on both their family history and their lifestyle.

Genetic information is carried by a long molecule called DNA which is copied and inherited across generations. Traits are carried in DNA as instructions for constructing and operating an organism. These instructions are contained in segments of DNA called genes. DNA is made of a sequence of simple units, with the order of these units spelling out instructions in the genetic code. This is similar to the order of letters spelling out words. The organism "reads" the sequence of these units and decodes the instruction. Not all the genes for a particular instruction are exactly the same.

Different forms of one type of gene are called different alleles of that gene. As an example, one allele of a gene for hair color could carry the instruction to produce a lot of the pigment in black hair, while a different allele could give a garbled version of this instruction, so that no pigment is produced and the hair is white. Mutations are random events that change the sequence of a gene and therefore create a new allele. Mutations can produce a new trait, such as turning an allele for black hair into an allele for white hair. The appearance of new traits is important in evolution.

Genes are inherited as units, with parents dividing out their genes to their offspring. You can think of this process like mixing two hands of cards, shuffling them, and then dealing them out again. Humans have two copies of each of their genes (*i.e,* two alleles) and when people reproduce they make copies of their genes in eggs or sperm, but only put in one copy of each type of gene. An egg then joins with a sperm to give a child with a new set of genes. This child will have the same number of genes as its parents but for any gene one of their two copies will come from the father, and one from the mother. The effects of this mixing depends on the types (the alleles) of the gene you are interested in.

If the father has two alleles specifying green eyes, and the mother has two alleles specifying brown eyes, all their children will get two alleles giving different instructions, one for green eyes and one for brown. The eye colour of these children depends on how these alleles work together. If one allele overrides the instructions from another, it is called the *dominant* allele, and the allele that is overridden is called the *recessive* allele. In the case of a daughter with both green and brown alleles, brown is dominant and she ends up with brown eyes.

However, the green eye color allele is still there in this brown-eyed girl, it just doesn't show. This is a difference between what you see on the surface (the set of observable traits of an organism, also called its phenotype) and which genes are in this organism (its genotype). In this example you can call the brown allele "B" and the green allele "g". (It is normal to write dominant alleles with capital letters and recessive ones with lower-case letters.) The brown-eyed daughter has the "brown eye phenotype" but her genotype is Bg, with one copy of the B allele, and one of the g allele.

Now imagine that this woman grows up and has children with a brown-eyed man who also has a Bg genotype. Her eggs will be a mixture of two types, one sort containing the B allele, and one sort the g allele. Similarly, her partner will produce a mix of two types of sperm containing one or the other of the two alleles. Now, when the alleles are mixed up in the offspring, these children have a chance of getting either brown or green eyes, since they could get a genotype of BB = brown eyes, Bg = brown eyes or gg = green eyes. In this generation, there is therefore a chance of the recessive allele showing itself in the phenotype of the children - some of them may have green eyes like their grandfather.

Many traits are inherited in a more complicated way than the example above. This can happen when there are several genes involved, each contributing a small part to the end result. Tall people tend to have tall children because their children get a package of many alleles that each contribute a bit to how much they grow. However, there are not clear groups of "short people" and "tall people", like there are groups of people with brown or green eyes. This is because of the large number of genes involved; this makes the trait very variable and people are many different heights. Inheritance can also be complicated when the trait depends on the interaction between genetics and the environment. This is quite common, for example, if a child does not eat enough nutritious food this will not change traits like eye color, but it could stunt their growth.

Inherited Diseases

Some diseases are hereditary and run in families; others, such as infectious diseases, are caused by the environment. Other disorders are caused by a combination of hereditary and environmental factors.[5]

Diseases that are caused by a single allele of a gene and are inherited in families are called genetic disorders. These include diseases like Huntington's disease, Cystic fibrosis or Duchenne muscular dystrophy. Cystic fibrosis, for example, is caused by mutations in a single gene called *CFTR* and is inherited as a recessive trait. Other diseases are influenced by genetics, but which alleles a person gets from their parents only changes their risk of getting a disease. Most of these diseases are inherited in a complex way, with either multiple genes involved, or both genes and the environment being important.As an example, the risk of breast cancer is 50 times higher in the families most at risk, compared to the families least at risk. This variation is probably due to a large number of alleles. Each of them changes the risk a little bit. Several of the genes involved have been identified, such as *BRCA1* and *BRCA2*, but not all of them. However, although some of the risk is genetic, being overweight, drinking a lot of alcohol, or not exercising, all increase the risk of this cancer. A woman's risk of breast cancer is therefore the result of a large number of alleles and her environment, so it is very hard to predict.

How Genes Work

Genes make Proteins

The function of genes is to provide the information needed to make molecules called proteins in cells. Cells are the smallest independent parts of organisms: the human body contains about 100 trillion cells, while very small organisms like bacteria are just a single cell. A cell is like a miniature and very complex factory that can make all the parts needed to produce a copy of itself, which happens when cells divide. There is a simple division of labor in cells - genes give instructions and proteins carry out these instructions, tasks like building a new copy of a cell, or repairing damage. Each type of protein is a specialist that only does one job, so if a cell needs to do something new, it must make a new protein to do this job. Similarly, if a cell needs to do something faster or slower than before, it makes more or less of the protein responsible. Genes tell cells what to do by telling them which proteins to make and in what amounts.Genes are expressed by being transcribed into RNA, and this RNA then translated into protein.

Proteins are made of a chain of 20 different types of amino acids. This chain folds up into a compact shape, rather like an untidy ball of rope. The shape of the protein is determined by the sequence of amino acids along its chain and it is this shape that, in turn, determines what the protein will do. For example, some proteins have depressions in their surface that perfectly match another molecule, allowing the protein to bind to this molecule very tightly. Other proteins are enzymes, which are like tiny machines that can alter other molecules.

The information in DNA is held in the sequence of the repeating units along the DNA chain. These units are four types of nucleotides (A,T,G and C) and the sequence of nucleotides stores information in an alphabet called the genetic code. When a gene is read by a cell the DNA sequence is copied into a very similar molecule called RNA (this process is called transcription). Transcription is controlled by other DNA sequences (such as promoters), which show a cell where genes are, and control how often they are copied. The RNA copy made from a gene is then fed through a structure called a ribosome, which translates the sequence of nucleotides in the RNA into the correct sequence of amino acids and joins these amino acids together to make a complete protein chain. The new protein then folds up into its active form. The process of moving information from the language of DNA into the language of amino acids is called translation. DNA is unwound and nucleotides are matched to make two new strands.

If the sequence of the nucleotides in a gene changes, the sequence of the amino acids in the protein it produces may also change - if part of a gene is deleted, the protein produced will be shorter and may not work any more. This is the reason why different alleles of a gene can have different effects in an organism. As an example, hair color depends on how much of a dark substance called melanin is put into the hair as it grows. If a person has a normal set of the genes involved in making melanin, they make all the proteins needed and they grow dark hair. However, if the alleles for a particular protein have different sequences and produce proteins that do not do the job correctly, no melanin will be produced and the hair will be white. This condition is called albinism and the person with this condition is called an albino.

Genes are Copied

Genes are copied each time a cell divides into two new cells. The process that copies DNA is called DNA replication. It is through a similar process that a child inherits genes from its parents, when a copy from the mother is mixed with a copy from the father.

DNA can be copied very easily and accurately because each piece of DNA can direct the creation of a new copy of its information. This is because DNA is made of two strands that pair together like the two sides of a zipper. The nucleotides are in the center, like the teeth in the zipper, and pair up to hold the two strands together. Importantly, the four different sorts of nucleotides are different shapes, so in order for the strands to close up properly, an **A** nucleotide must go opposite a **T** nucleotide, and a **G** opposite a **C**. This exact pairing is called base pairing.

When DNA is copied, the two strands of the old DNA are pulled apart by enzymes which move along each of the two single strands pairing up new nucleotide units and then zipping the strands closed. This produces two new pieces of DNA, each containing one strand from the old DNA and one newly made strand. This process isn't perfect and sometimes the proteins will make mistakes and put the wrong nucleotide into the strand they are building. This causes a change in the sequence of that gene. These changes in DNA sequence are called mutations. Mutations produce new alleles of genes. Sometimes these changes stop the gene from working properly, like the melanin genes discussed above. In other cases these mutations can change what the gene does or even let it do its job a little better than before. These mutations and their effects on the traits of organisms are one of the causes of evolution.

Genes and Evolution

A population of organisms evolves when an inherited trait becomes more common or less common over time. For instance, all the mice living on an island would be a single population of mice. If over a few generations, white mice went from being rare, to being a large part of this population, then the coat color of these mice would be evolving. In terms of genetics, this is called a change in allele frequency—such as an increase in the frequency of the allele for white fur.

Alleles become more or less common either just by chance (in a process called genetic drift), or through natural selection. In natural selection, if an allele makes it more likely that an organism will survive and reproduce, then over time this allele will become more common. But if an allele is harmful, natural selection will make it less common. For example, if the island was getting colder each year and was covered with snow for much of the time, then the allele for white fur would become useful for the mice, since it would make them harder to see against the snow. Fewer of the white mice would be eaten by predators, so over time white mice would out-compete mice with dark fur. White fur alleles would become more common, and dark fur alleles would become more rare.

Mutations create new alleles. These alleles have new DNA sequences and can produce proteins with new properties. So if an island was populated entirely by black mice, mutations could happen creating alleles for white fur. The combination of mutations creating new alleles at random, and natural selection picking out those which are useful, causes adaptation. This is when organisms change in ways that help them to survive and reproduce.

Genetic Engineering

Since traits come from the genes in a cell, putting a new piece of DNA into a cell can produce a new trait. This is how genetic engineering works. For example, crop plants can be given a gene from an Arctic fish, so they produce an antifreeze protein in their leaves. This can help prevent frost damage. Other genes that can be put into crops include a natural insecticide from the bacteria *Bacillus Thuringiensis.* The insecticide kills insects that eat the plants, but is harmless to people. In these plants the new genes are put into the plant before it is grown, so the genes will be in every part of the plant, including its seeds. The plant's offspring will then inherit the new genes, something which has led to concern about the spread of new traits into wild plants.

The kind of technology used in genetic engineering is also being developed to treat people with genetic disorders in an experimental medical technique called gene therapy. However, here the new gene is put in after the person has grown up and become ill, so any new gene will not be inherited by their children. Gene therapy works by trying to replace the allele that causes the disease with an allele that will work properly. Genetic engineering is a laboratory technique used by scientists to change the *DNA of living organisms.*

DNA is the blueprint for the individuality of an organism. The organism relies upon the information stored in its DNA for the management of every biochemical process. The life, growth and unique features of the organism depend on its DNA. Molecular biologists have discovered many enzymes which change the structure of DNA in living organisms.

Some of these enzymes can cut and join strands of DNA. Using such enzymes, scientists learned to cut specific genes from DNA and to build customized DNA using these genes. They also learned about **vectors**, strands of DNA such as viruses, which can infect a cell and insert themselves into its DNA. With this knowledge, scientists started to build vectors which incorporated genes of their choosing and used the new vectors to insert these genes into the DNA of living organisms. Genetic engineers believe they can improve the foods we eat by doing this. For example, tomatoes are sensitive to frost. This shortens their growing season. Fish, on the other hand, survive in very cold water. Scientists identified a particular gene which enables a flounder to resist cold and used the technology of genetic engineering to insert this 'anti-freeze' gene into a tomato. This makes it possible to extend the growing season of the tomato.

GENE: MUTATION

DNA replication is for the most part extremely accurate, with an error rate per site of around 10^{-6} to 10^{-10} in eukaryotes. Rare, spontaneous alterations in the base sequence of a particular gene arise from a number of sources, such as errors in DNA replication and the aftermath of DNA damage. These errors are called mutations. The cell contains many DNA repair mechanisms for preventing mutations and maintaining the integrity of the genome; however, in some cases—such as breaks in both DNA strands of a chromosome — repairing the physical damage to the molecule is a higher priority than producing an exact copy. Due to the degeneracy of the genetic code, some mutations in protein-coding genes are *silent*, or produce no change in the amino acid sequence of the protein for which they code; for example, the codons UCU and UUC both code for serine, so the U↔C mutation has no effect on the protein. Mutations that do have phenotypic effects are most often neutral or deleterious to the organism, but sometimes they confer benefits to the organism's fitness.

Mutations propagated to the next generation lead to variations within a species' population. Variants of a single gene are known as alleles, and differences in alleles may give rise to differences in traits. Although it is rare for the variants in a single gene to have clearly distinguishable phenotypic effects, certain well-defined traits are in fact controlled by single genetic loci. A gene's most common allele is called the wild type allele, and rare alleles are called mutants. However, this does not imply that the wild-type allele is the ancestor from which the mutants are descended.

Mutations can be caused by copying errors in the genetic material during cell division, by exposure to ultraviolet or ionizing radiation, chemical mutagens, or viruses, or can be induced by the organism itself, by cellular processes such as hypermutation. In multicellular organisms with dedicated reproductive cells, mutations can be subdivided into **germ line mutations**, which can be passed on to descendants through the reproductive cells, and **somatic mutations**, which involve cells outside the dedicated reproductive group and which are not usually transmitted to descendants. If the organism can reproduce asexually through mechanisms such as cuttings or budding the distinction can become blurred. For example, plants can sometimes transmit somatic mutations to their descendants asexually or sexually where flower buds develop in somatically mutated parts of plants. A new mutation that was not inherited from either parent is called a ***de novo* mutation**. The source of the mutation is unrelated to the consequence, although the consequences are related to which cells were mutated.

Mutations create variation within the gene pool. Less favorable (or *deleterious*) mutations can be reduced in frequency in the gene pool by natural selection, while more favorable (*beneficial* or *advantageous*) mutations may accumulate and result in adaptive evolutionary changes. For example, a butterfly may produce offspring with new mutations. The majority of these mutations will have no effect; but one might change the color of one of the butterfly's offspring, making it harder (or easier) for predators to see. If this color change is advantageous, the chance of this butterfly surviving and producing its own offspring are a little better, and over time the number of butterflies with this mutation may form a larger percentage of the population.

Neutral mutations are defined as mutations whose effects do not influence the fitness of an individual. These can accumulate over time due to genetic drift. It is believed that the overwhelming majority of mutations have no significant effect on an organism's fitness. Also, DNA repair mechanisms are able to mend most changes before they become permanent mutations, and many organisms have mechanisms for eliminating otherwise permanently mutated somatic cells.

Mutation is generally accepted by the scientific community as the mechanism upon which natural selection acts, providing the advantageous new traits that survive and multiply in offspring or disadvantageous traits that die out with weaker organisms.

By Effect on Structure

The sequence of a gene can be altered in a number of ways. Gene mutations have varying effects on health depending on where they occur and whether they alter the function of essential proteins. Structurally, mutations can be classified as:

- Small-scale mutations, such as those affecting a small gene in one or a few nucleotides, including:
  - Point mutations, often caused by chemicals or malfunction of DNA replication, exchange a single nucleotide for another. These changes are classified as transitions or transversions. Most common is the transition that exchanges a purine for a purine (A ↔ G) or a pyrimidine for a pyrimidine, (C ↔ T). A transition can be caused by nitrous acid, base mis-pairing, or mutagenic base analogs such as 5-bromo-2-deoxyuridine (BrdU). Less common is a transversion, which exchanges a purine for a pyrimidine or a pyrimidine for a purine (C/T ↔ A/G). An example of a transversion is adenine (A) being converted into a cytosine (C). A point mutation can be reversed by another point mutation, in which

the nucleotide is changed back to its original state (true reversion) or by second-site reversion (a complementary mutation elsewhere that results in regained gene functionality). Point mutations that occur within the protein coding region of a gene may be classified into three kinds, depending upon what the erroneous codon codes for:

- Silent mutations: which code for the same amino acid.
- Missense mutations: which code for a different amino acid.
- Nonsense mutations: which code for a stop and can truncate the protein.

– Insertions add one or more extra nucleotides into the DNA. They are usually caused by transposable elements, or errors during replication of repeating elements (e.g. AT repeats). Insertions in the coding region of a gene may alter splicing of the mRNA (splice site mutation), or cause a shift in the reading frame (frameshift), both of which can significantly alter the gene product. Insertions can be reverted by excision of the transposable element.

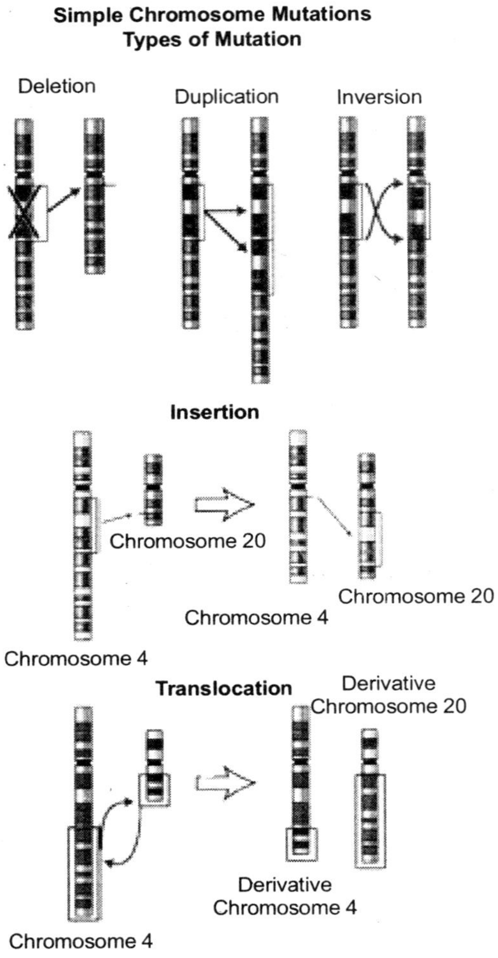

– Deletions remove one or more nucleotides from the DNA. Like insertions, these mutations can alter the reading frame of the gene. They are generally irreversible: though exactly the same sequence might theoretically be restored by an insertion, transposable elements able to revert a very short deletion (say 1–2 bases) in *any* location are either highly unlikely to exist or do not exist at all. Note that a deletion is not the exact opposite of an insertion: the former is quite random while the latter consists of a specific sequence inserting at locations that are not entirely random or even quite narrowly defined.

Fig. 6.5 : Illustrations of five types of chromosomal mutations

• Large-scale mutations in chromosomal structure, including:

– Amplifications (or gene duplications) leading to multiple copies of all chromosomal regions, increasing the dosage of the genes located within them.

- Deletions of large chromosomal regions, leading to loss of the genes within those regions.
- Mutations whose effect is to juxtapose previously separate pieces of DNA, potentially bringing together separate genes to form functionally distinct fusion genes (e.g. bcr-abl). These include:
  * Chromosomal translocations: interchange of genetic parts from nonhomologous chromosomes.
  * Interstitial deletions: an intra-chromosomal deletion that removes a segment of DNA from a single chromosome, thereby apposing previously distant genes. For example, cells isolated from a human astrocytoma, a type of brain tumor, were found to have a chromosomal deletion removing sequences between the "fused in glioblastoma" (fig) gene and the receptor tyrosine kinase "ros", producing a fusion protein (FIG-ROS). The abnormal FIG-ROS fusion protein has constitutively active kinase activity that causes oncogenic transformation (a transformation from normal cells to cancer cells).
  * Chromosomal inversions: reversing the orientation of a chromosomal segment.
- Loss of heterozygosity: loss of one allele, either by a deletion or recombination event, in an organism that previously had two different alleles.

By Effect on Function

- **Loss-of-function mutations** are the result of gene product having less or no function. When the allele has a complete loss of function (null allele) it is often called an amorphic mutation. Phenotypes associated with such mutations are most often recessive. Exceptions are when the organism is haploid, or when the reduced dosage of a normal gene product is not enough for a normal phenotype (this is called haploinsufficiency).
- **Gain-of-function mutations** change the gene product such that it gains a new and abnormal function. These mutations usually have dominant phenotypes. Often called a neomorphic mutation.
- **Dominant negative mutations** (also called antimorphic mutations) have an altered gene product that acts antagonistically to the wild-type allele. These mutations usually result in an altered molecular function (often inactive) and are characterised by a dominant or semi-dominant phenotype. In humans, Marfan syndrome is an example of a dominant negative mutation occurring in an autosomal dominant disease. In this condition, the defective glycoprotein product of the fibrillin gene (FBN1) antagonizes the product of the normal allele.
- **Lethal mutations are mutations** that lead to the death of the organisms which carry the mutations.
- **A back mutation** or **reversion** is a point mutation that restores the original sequence and hence the original phenotype.

By Effect on Fitness

In applied genetics it is usual to speak of mutations as either harmful or beneficial.

- A **harmful mutation** is a mutation that decreases the fitness of the organism.
- A **beneficial mutation** is a mutation that increases fitness of the organism, or which promotes traits that are desirable.

In theoretical population genetics, it is more usual to speak of such mutations as deleterious or advantageous. In the neutral theory of molecular evolution, genetic drift is the basis for most variation at the molecular level.

- A **neutral mutation** has no harmful or beneficial effect on the organism. Such mutations occur at a steady rate, forming the basis for the molecular clock.
- A **deleterious mutation** has a negative effect on the phenotype, and thus decreases the fitness of the organism.
- An **advantageous mutation** has a positive effect on the phenotype, and thus increases the fitness of the organism.
- A nearly **neutral mutation** is a mutation that may be slightly deleterious or advantageous, although most nearly neutral mutations are slightly deleterious.

By Inheritance

- inheritable generic in pro-generic tissue or cells on path to be changed to gametes.
- non inheritable somatic (eg, carcinogenic mutation)
- non inheritable post mortem aDNA mutation in decaying remains.

By pattern of Inheritance

The human genome contains two copies of each gene – a paternal and a maternal allele.

- A **heterozygous mutation** is a mutation of only one allele.
- A **homozygous mutation** is an identical mutation of both the paternal and maternal alleles.
- Compound **heterozygous** mutations or a genetic compound comprises two different mutations in the paternal and maternal alleles.
- A wildtype or homozygous non-mutated organism is one in which neither allele is mutated. (Just not a mutation)

By Impact on Protein Sequence

- A **frameshift mutation** is a mutation caused by insertion or deletion of a number of nucleotides that is not evenly divisible by three from a DNA sequence. Due to the triplet nature of gene expression by codons, the insertion or deletion can disrupt the reading frame, or the grouping of the codons, resulting in a completely different translation from the original. The earlier in the sequence the deletion or insertion occurs, the more altered the protein produced is.

- **Missense mutations** or *nonsynonymous mutations* are types of point mutations where a single nucleotide is changed to cause substitution of a different amino acid. This in turn can render the resulting protein nonfunctional. Such mutations are responsible for diseases such as Epidermolysis bullosa, sickle-cell disease, and SOD1 mediated ALS *(Boillée 2006, p. 39)*.
- A **neutral mutation** is a mutation that occurs in an amino acid codon which results in the use of a different, but chemically similar, amino acid. This is similar to a silent mutation, where a codon mutation may encode the same amino acid (see Wobble Hypothesis); for example, a change from AUU to AUC will still encode leucine, so no discernible change occurs (a silent mutation).
- A **nonsense mutation** is a point mutation in a sequence of DNA that results in a premature stop codon, or a *nonsense codon* in the transcribed mRNA, and possibly a truncated, and often nonfunctional protein product.
- **Silent mutations** are mutations that do not result in a change to the amino acid sequence of a protein. They may occur in a region that does not code for a protein, or they may occur within a codon in a manner that does not alter the final amino acid sequence. The phrase *silent mutation* is often used interchangeably with the phrase *synonymous mutation*; however, synonymous mutations are a subcategory of the former, occurring only within exons. The name silent could be a misnomer. For example, a silent mutation in the exon/intron border may lead to alternative splicing by changing the splice site (*see Splice site mutation*), thereby leading to a changed protein.

Special Classes

- **Conditional mutation** is a mutation that has wild-type (or less severe) phenotype under certain "permissive" environmental conditions and a mutant phenotype under certain "restrictive" conditions. For example, a temperature-sensitive mutation can cause cell death at high temperature (restrictive condition), but might have no deleterious consequences at a lower temperature (permissive condition).

Causes of Mutation

Two classes of mutations are spontaneous mutations (molecular decay) and induced mutations caused by mutagens.

Spontaneous mutations on the molecular level include:

- **Tautomerism** – A base is changed by the repositioning of a hydrogen atom, altering the hydrogen bonding pattern of that base resulting in incorrect base pairing during replication.
- **Depurination** – Loss of a purine base (A or G) to form an apurinic site (AP site).
- **Deamination** – Hydrolysis changes a normal base to an atypical base containing a keto group in place of the original amine group. Examples include C → U and A → HX (hypoxanthine), which can be corrected by DNA repair mechanisms; and 5MeC (5-methylcytosine) → T, which is less likely to be detected as a mutation because thymine is a normal DNA base.

- Transition – A purine changes to another purine, or a pyrimidine to a pyrimidine.
- Transversion – A purine becomes a pyrimidine, or vice versa.

Induced mutations on the molecular level can be caused by:

- Chemicals
  - Hydroxylamine NH_2OH
  - Base analogs (e.g. BrdU)
  - Alkylating agents (e.g. *N*-ethyl-*N*-nitrosourea) These agents can mutate both replicating and non-replicating DNA. In contrast, a base analog can only mutate the DNA when the analog is incorporated in replicating the DNA. Each of these classes of chemical mutagens has certain effects that then lead to transitions, transversions, or deletions.
  - Agents that form DNA adducts (e.g. ochratoxin A metabolites)
  - DNA intercalating agents (e.g. ethidium bromide)
  - DNA crosslinkers
  - Oxidative damage
  - Nitrous acid converts amine groups on A and C to diazo groups, altering their hydrogen bonding patterns which leads to incorrect base pairing during replication.
- Radiation
  - Ultraviolet radiation (nonionizing radiation). Two nucleotide bases in DNA – cytosine and thymine – are most vulnerable to radiation that can change their properties. UV light can induce adjacent thymine bases in a DNA strand to pair with each other, as a bulky dimer.
  - Ionizing radiation
- Viral infections

DNA has so-called hotspots, where mutations occur up to 100 times more frequently than the normal mutation rate. A hotspot can be at an unusual base, e.g., 5-methylcytosine.

Mutation rates also vary across species. Evolutionary biologists have theorized that higher mutation rates are beneficial in some situations, because they allow organisms to evolve and therefore adapt more quickly to their environments. For example, repeated exposure of bacteria to antibiotics, and selection of resistant mutants, can result in the selection of bacteria that have a much higher mutation rate than the original population (mutator strains).

Nomenclature

Nomenclature of mutations specify the type of mutation and base or amino acid changes.

- Nucleotide substitution (e.g. 76A>T) - The number is the position of the nucleotide from the 5' end, the first letter represents the wild type nucleotide, and the second letter represents the nucleotide which replaced the wild type. In the given example, the adenine at the 76th position was replaced by a thymine.

  - If it becomes necessary to differentiate between mutations in genomic DNA, mitochondrial DNA, and RNA, a simple convention is used. For example, if the 100th base of a nucleotide sequence mutated from G to C, then it would be written as g.100G>C if the mutation occurred in genomic DNA, m.100G>C if the mutation occurred in mitochondrial DNA, or r.100g>c if the mutation occurred in RNA. Note that for mutations in RNA, the nucleotide code is written in lower case.
- Amino acid substitution (e.g. Δ111E) – The first letter is the one letter code of the wild type amino acid, the number is the position of the amino acid from the N terminus, and the second letter is the one letter code of the amino acid present in the mutation. Nonsense mutations are represented with an X for the second amino acid (e.g. D111X).
- Amino acid deletion (e.g. ΔF508) – The Greek letter Δ (delta) indicates a deletion. The letter refers to the amino acid present in the wild type and the number is the position from the N terminus of the amino acid were it to be present as in the wild type.

Harmful Mutations

Changes in DNA caused by mutation can cause errors in protein sequence, creating partially or completely non-functional proteins. To function correctly, each cell depends on thousands of proteins to function in the right places at the right times. When a mutation alters a protein that plays a critical role in the body, a medical condition can result. A condition caused by mutations in one or more genes is called a genetic disorder. Some mutations alter a gene's DNA base sequence but do not change the function of the protein made by the gene. Studies in the fly *Drosophila melanogaster* suggest that if a mutation does change a protein, this will probably be harmful, with about 70 percent of these mutations having damaging effects, and the remainder being either neutral or weakly beneficial.

If a mutation is present in a germ cell, it can give rise to offspring that carries the mutation in all of its cells. This is the case in hereditary diseases. On the other hand, a mutation can occur in a somatic cell of an organism. Such mutations will be present in all descendants of this cell, and certain mutations can cause the cell to become malignant, and thus cause cancer.

Often, gene mutations that could cause a genetic disorder are repaired by the DNA repair system of the cell. Each cell has a number of pathways through which enzymes recognize and repair mistakes in DNA. Because DNA can be damaged or mutated in many ways, the process of DNA repair is an important way in which the body protects itself from disease.

Beneficial Mutations

Although many mutations are deleterious, mutations may have a positive effect given certain selective pressures in a population.

For example, a specific 32 base pair deletion in human CCR5 (CCR5-Δ32) confers HIV resistance to homozygotes and delays AIDS onset in heterozygotes. The CCR5 mutation is more common in those of European descent. One theory for the etiology of the relatively high frequency of CCR5-Δ32 in the European population is that it conferred resistance to the bubonic plague in mid-14th century Europe. People who had this mutation were able to survive infection; thus, its frequency in the population increased. It could also explain why this mutation is not found in Africa where the bubonic plague never reached. Newer theory says the selective pressure on the CCR5 Delta 32 mutation has been caused by smallpox instead of the bubonic plague.

GENETIC MAPPING

Gene Mapping of an organism also called a linkage map is the order of genes on a chromosome, an order that can be determined by the methods of classical genetics. Here we are using the bacterium *E coli* as an example.

In the classical genetic mapping for decoding it has been known about some bacteria's like E coli can transfer DNA from one cell to another by the method of **conjugation.** The cells are in physical contact as DNA is transferred from one to another. (See Fig. 6.6) The longer cells are in contact, the more DNA can be transferred. Genes close to each other on the bacterial genome" travel together in the course of conjugation. If the flask containing the cell is shaken the contacts are broken and less DNA transferred. The amount of DNA transferred in the process can be detected by the use of recipient cells that are missing some genes (usually ones that direct the biosynthesis of some important compound) and then testing of those genes and accepting of presence in the recombinants.

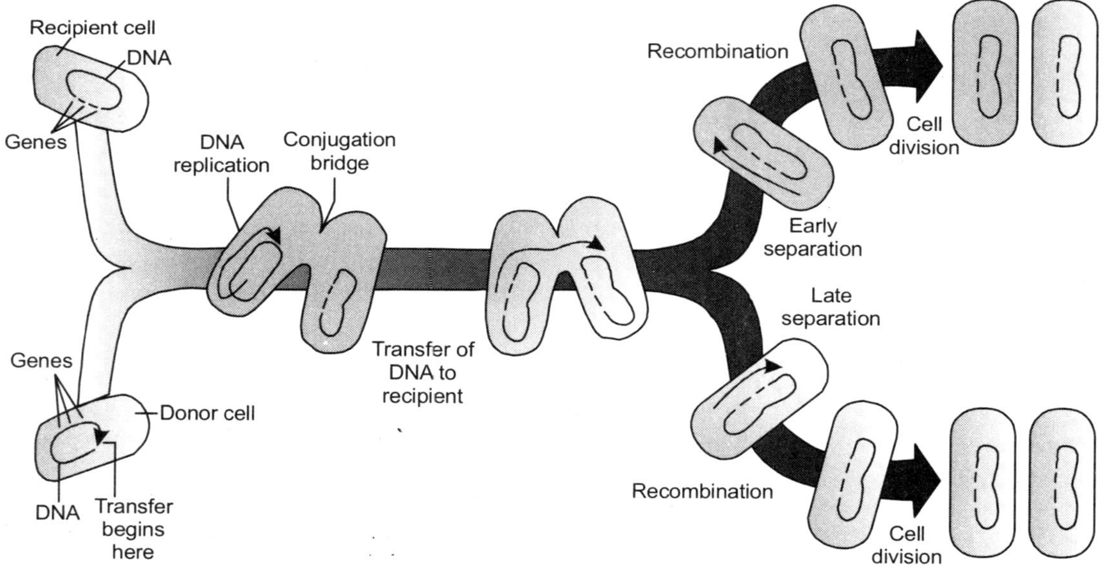

Fig. 6.6 : The Conjugation in Bacteria. The amount if DNA transferred depends on the length of time the bacteria remain in contact. Genes that lie far apart are not transferred together

The order in which genes are transferred can be determined by varying the amount of time allowed for conjugation. Genes that are closer together tend to be linked in the course of recombination and to be transferred together even during short recombination periods. Genes that are farther apart are much less likely to be transferred together. This method and variations on it produce the genetic map (See Fig. 6.7)

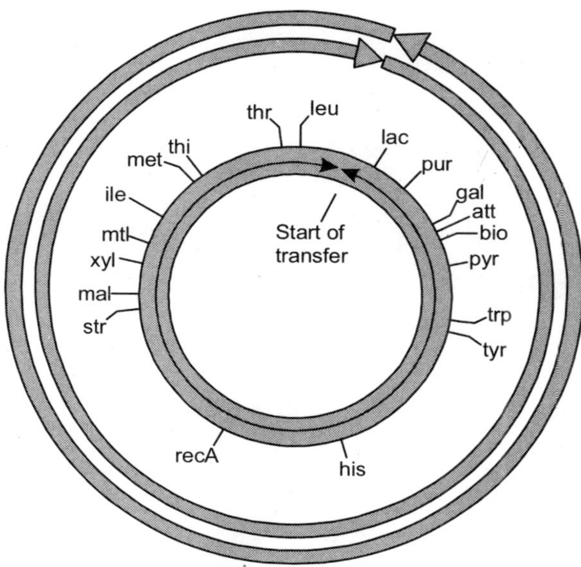

Fig. 6.7 : The Genetic map of E. Coli the three letter symbol represents genes the arrows represents the direction of transfer when two different donors are used in the process of conjugation, the genes lie close to each other n on the map are transferred together

The construction of the map is based entirely on genetic methods that rely on visible traits and does not deal directly with the nature of the molecules involved. It is also possible construct a physical or molecular map of the DNA sequence of chromosomes, using the methods disused in this chapter; we shall now see what the physical map has been in common with the classical genetic map. (see Fig. 6.8)

Construction of a physical map of chromosomes depends upon an important way on the organism involved. In a procaryote there is normally a single chromosome, not separated from the rest of the cell; a procaryote chromosome is circle of double – helical DNA. In a eucaryote there are multiple chromosomes sequestered* in the nucleus; each contains a single linear DNA double helix. In procaryote genes lie close to each other without intervening DNA. In eucaroytes genes a separated by long stretches of non – coding. DNA sequencing the chromosome. Dealing with all the complexities requires the powerful methods developed for the recombinant DNA technology. The physical map of E coli genome is digested with a restriction nuclease. A particular enzyme was chosen that gives rise to relatively few fragments, which can be separated by gel electrophoresis. The positions of the fragments on the circular genome were determined by comparison with the classical genetic map on which the positions of more than 1000 genes has already been determined at the time this work was done. Clone segments

*to hold (as metallic on) in solution usually by inclsion in appropriate co-ordination complex.

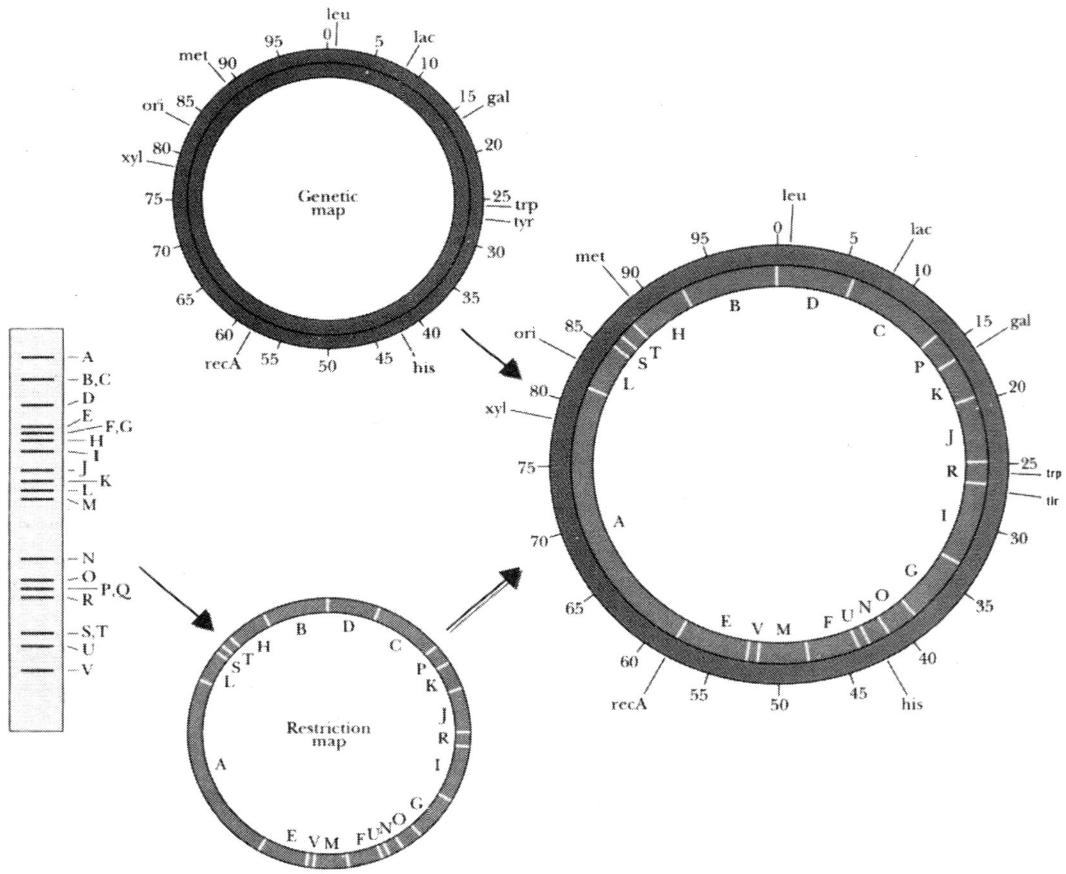

Fig. 6.8 : The Comparison of the physical and genetic map of E. Coli. The rectangle on the left represents the results of gel electrophoresis of a restriction enzyme digest of E. Coli genome. This information is used to construct a map of the physical fragments of the genome and to see how the physical map of the physical fragments "line up" with the genetic map

of these known genes were available as probes in DNA blotting to locate a given gene on a given fragment. Consequently it has been possible to co-relate the physical and genetic map of *E. coli*. Mapping genes on the larger and more complex chromosomes of eucaryotic cell represents a considerable challenge. Classical genetic studies on the fruit fly *Drosophila melanogaster* in the early 20th century determined linkage patterns of genes via methods similar to those just described for bacteria. This work produced detailed maps of genes on chromosomes of eucaryotic cells, but the positions of

Note that two kinds of maps parallel to each other. Any newly found gene can be placed accurately on a restriction fragment as well as on the genetic linkage map

many more genes were completely unknown, Even less information was available for other eucaroytes, and these situations remained unchanged until recently. In humans the first genes to be mapped where those X chromosomes. Females have X chromosomes and males have one X and one Y chromosomes. If the gene of one male X chromosome becomes non-functional then there is no chance of replacement by the corresponding gene on the other hand the X chromosome as can happen in the females. As a result some human males exhibit traits that rise from non – functional genes on the X chromosomes (as discussed earlier) the genes involved in the traits. (See Fig. 6.9).

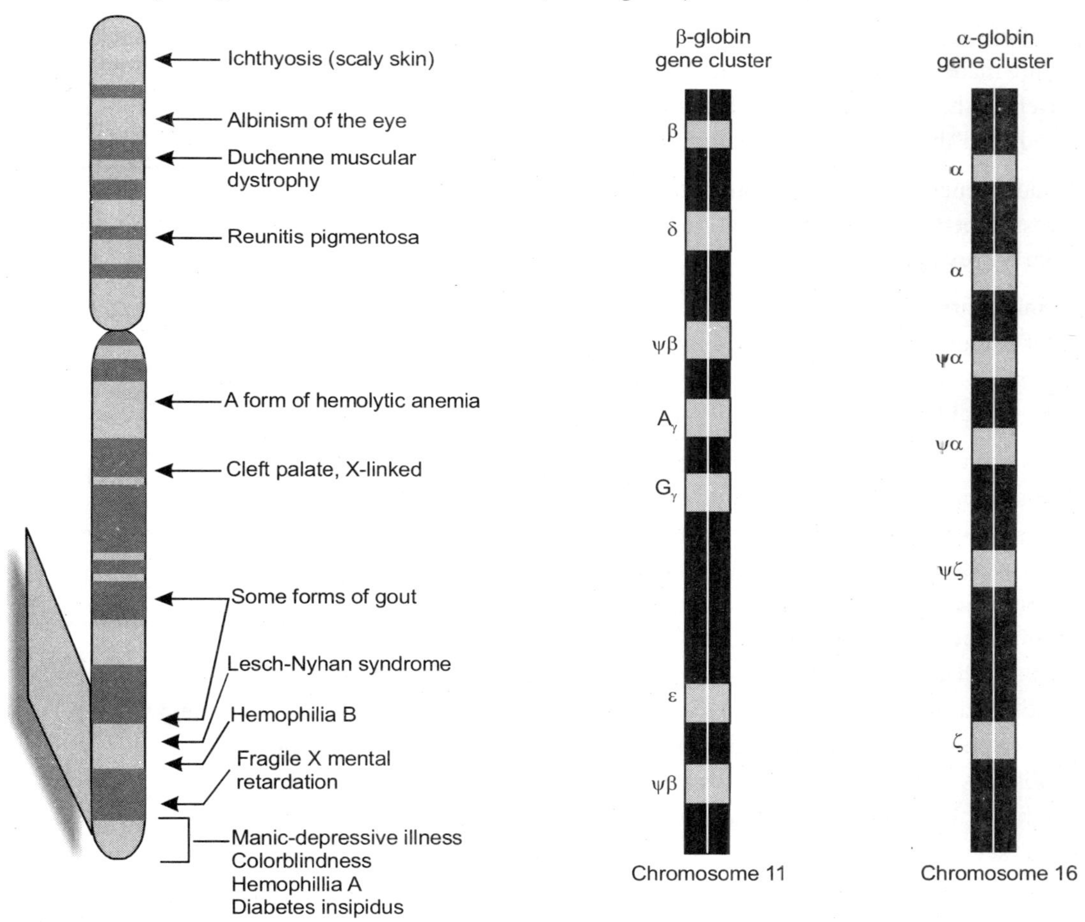

Fig. 6.9 : A schematic diagram of the human X-chromosome showing the location of genes associated with the number of diseases

Fig. 6.10 : The Maps of Gene clusters that code for the variant forms of α and β chains of human hemoglobin. Different α-type and β-type chains are made during embryonic development and after child birth. The Greek letter psi (ψ) indicates pseudo-genes, which are non- functional copies of genes)

Progress in locating human genes on chromosomes other than the sex chromosomes (X and Y) was slow until recombinant DNA become available.

Globin genes are an example of a cluster of genes on an autosomal (non-sex) chromosome. The family of genes that encodes the β-chain of hemoglobin consists of a cluster arranged along chromosome 11 in the order in which the genes are expressed in the course of development (see Fig. 6.10).

Five different β-type chains are made from the early embryonic stage to shortly after birth. The form known as ε-globin (epsilon globin) forms Gγ and A γ are made later in the life while the gene for the epsilon for is switched off. After birth, the genes for δ-globin (delta globin) and β-globin are expressed, and the genes for the gamma forms are switched off in turn.

> Note that the map contains two pseudo-genes designated yb(the Greek letter ψβ means pseudo genes)

Pseudo genes are non-functional copies of genes that almost normally be expressed because of some mutation. Pseudo genes occur to varying extents in eucaroytes and are particularly prevalent in mammals.

A similar arrangement appears in the α-globin gene cluster on chromosome 16 (see Fig. 6.10). The variant of α-globin that is produced in the early stage of the embryo is ξ-globin (zeta globin), this gene is switched off later in development, at which time the α-globin gene is expressed.

> Note the presence of pseudo genes for both the alpha and zeta forms (ψα and ψξ respectively). The mechanism of regulation of these two genes cluster on different chromo-somes is a topic of considerable interest to biologists.

It is not at all surprising that human globin gene is the first mammalian genes to be cloned using recombinant DNA technology. There was a considerable chance to develop gene therapies for the genetic diseases that arise from abnormal hemoglobin; it was also comparatively easy to carry out the actual cloning process. Protein synthesis in human red blood cells is largely decided to the production of hemoglobin, which accounts for about 90% of the protein in these cells. Most of the m RNA in the red blood cells is of two kinds, those for the α-globins and those for the β-globins. Using these two readily available m RNAs, it was possible to construct probes to identify the globin genes in cDNA libraries. Each of these genes was contained in a clone That contained stretches of DNA flanking the gene itself. The flanking sequences were cloned by themselves they could then be used to construct probes for segments of DNA farther along the **chromosome waiting.** It was repeated until cluster of the β-globin gene family was obtained. We shall now see some changes in the genetic code.

Important Points

- The information in the genes that are made up of DNA is in the form of a chemicalc ode, called the genetic code.
- There are small differences between every individual in their genetic information that mkes each of us unique.

- Some changes to the gneetic information do not significantly alter the gene message so the information is still understood by the cell. These changes are common and are called neutral gene variations (variants) or polynorphisms.
- Other changes to the gneetic information cause the message to be changed so that it is no longer understood by the cell
  - The gene is described as broken or faulty. Changes that mke the genes faulty are called mutations.
  - Language to describe mutations needs to be sensitive but also reflect the function of the gene change.
- Faulty genes (mutated genes) may
  - Cause a problem with the development and functioning of different body systems or organs and result in a genetic condition.
  - Be beenficial eg. having a faulty copy of the gene that tells the body to make haemoglobin makes a person more immune to malaria
- Some gene changes
  - Are inherited from our parents
  - Occur in the egg, sperm, during or shortly after conception and are described as 'new" or 'spontaneous" gene changes
  - Build up in our body's cells during our lifetime (not inherited)
- Some faulty genes directly or indirectly cause genetic conditions that run in families (inherited)
- Everyone is born with several faulty genes that usually cause no problem.

REGULATION OF GENE EXPRESSION

Regulation of gene expression (or **gene regulation**) includes the processes that cells and viruses use to turn the information in genes into gene products. Although a functional gene product may be an RNA or a protein, the majority of known mechanisms regulate protein coding genes. Any step of the gene's expression may be modulated, from DNA-RNA transcription to the post-translational modification of a protein.

Gene regulation is essential for viruses, prokaryotes and eukaryotes as it increases the versatility and adaptability of an organism by allowing the cell to express protein when needed. The first discovered example of a gene regulation system was the lac operon, discovered by Jacques Monod, in which protein involved in lactose metabolism are expressed by *E.coli* only in the presence of lactose and absence of glucose.

Furthermore, gene regulation drives the processes of cellular differentiation and morphogenesis, leading to the creation of different cell types in multicellular organisms where the different types of cells may possess different gene expression profiles though they all possess the same genome sequence.

Regulated Stages of Gene Expression

Any step of gene expression may be modulated, from the DNA-RNA transcription step to post-translational modification of a protein. The following is a list of stages where gene expression is regulated:

- chromatin domains
- Transcription
- Post-transcriptional modification
- RNA transport
- Translation
- mRNA degradation
- Post-translational modifications

Modification of DNA

In eukaryotes, the accessibility of large regions of DNA can depend on its chromatin structure which can be altered as a result of histone modifications which are directed by DNA methylation, ncRNA or DNA binding protein.

Chemical

Methylation of DNA is a common method of gene silencing. DNA is typically methylated by methyltransferase enzymes on cytosine nucleotides in a CpG dinucleotide sequence (also called "CpG islands" when densely clustered). Analysis of the pattern of methylation in a given region of DNA (which can be a promoter) can be achieved through a method called bisulfite mapping. Methylated cytosine residues are unchanged by the treatment, whereas unmethylated ones are changed to uracil. The differences are analyzed by DNA sequencing or by methods developed to quantify SNPs, such as Pyrosequencing (Biotage) or MassArray (Sequenom), measuring the relative amounts of C/T at the CG dinucleotide. Abnormal methylation patterns are thought to be involved in carcinogenesis.

Structural

Transcription of DNA is dictated by its structure. In general, the density of its packing is indicative of the frequency of transcription. Octameric protein complexes called nucleosomes are responsible for the amount of supercoiling of DNA, and these complexes can be temporarily modified by processes such as phosphorylation or more permanently modified by processes such as methylation. Such modifications are considered to be responsible for more or less permanent changes in gene expression levels.

Histone acetylation is also an important process in transcription. Histone acetyltransferase enzymes (HATs) such as CREB binding protein also dissociate the DNA from the histone complex, allowing transcription to proceed. Often, DNA methylation and histone deacetylation work together in gene silencing. The combination of the two seems to be a signal for DNA to be packed more densely, lowering gene expression.

Regulation of Transcription

Regulation of transcription controls when transcription occurs and how much RNA is created. Transcription of a gene by RNA polymerase can be regulated by at least five mechanisms:

- **Specificity factors** alter the specificity of RNA polymerase for a given promoter or set of promoters, making it more or less likely to bind to them (i.e. sigma factors used in prokaryotic transcription).
- **Repressors** bind to non-coding sequences on the DNA strand that are close to or overlapping the promoter region, impeding RNA polymerase's progress along the strand, thus impeding the expression of the gene.
- **General transcription factors** These transcription factors position RNA polymerase at the start of a protein-coding sequence and then release the polymerase to transcribe the mRNA.
- **Activators** enhance the interaction between RNA polymerase and a particular promoter, encouraging the expression of the gene. Activators do this by increasing the attraction of RNA polymerase for the promoter, through interactions with subunits of the RNA polymerase or indirectly by changing the structure of the DNA.
- **Enhancers** are sites on the DNA helix that are bound to by activators in order to loop the DNA bringing a specific promoter to the initiation complex.

Posttranscriptional Regulation

After the DNA is transcribed and mRNA is formed there must be some sort of regulation on how much the mRNA is translated into proteins. Cells do this by modulating the capping, splicing, addition of a Poly(A) Tail, the sequence-specific nuclear export rates and in several contexts sequestration of the RNA transcript. These processes occur in eukaryotes but not in prokaryotes. This modulation is a result of a protein or transcript which in turn is regulated and may have an affinity for certain sequences.

- **Capping** changes the five prime end of the mRNA to a three prime end by 5'-5' linkage, which protects the mRNA from 5' exonuclease, which degrades foreign RNA. The cap also helps in ribosomal binding.
- **Splicing** removes the introns, noncoding regions that are transcribed into RNA, in order to make the mRNA able to create proteins. Cells do this by spliceosome's binding on either side of an intron, looping the intron into a circle and then cleaving it off. The two ends of the exons are then joined together.
- **Addition of poly(A) tail** otherwise known as poly-adenylation. Junk RNA is added to the 3' end, and acts as a buffer to the 3' exonuclease in order to increase the half life of mRNA.

In both prokaryotes and eukaryotes a large number of RNA binding proteins exist, which often are directed to their target sequence by the secondary structure of the transcript, which may change depending on certain conditions, such as temperature or presence of a ligand (aptamer), some transcripts act as ribozymes and self-regulate their expression.

Examples of Gene Regulation

- Enzyme induction is a process in which a molecule (e.g. a drug) induces (i.e. initiates or enhances) the expression of an enzyme.
- The induction of heat shock proteins in the fruit fly *Drosophila melanogaster*.
- The Lac operon is an interesting example of how gene expression can be regulated.
- Viruses despite having only a few genes, possess mechanisms to regulate their gene expression, typically into an early and late phase, using collinear systems regulated by anti-terminators (lambda phage) or splicing modulators (HIV)
- Up-regulation occurs for example when a cell is deficient in some kind of receptor. In this case, more receptor protein is synthesized and transported to the membrane of the cell and thus the sensitivity of the cell is brought back to normal reestablishing homeostasis.
- Down-regulation occurs for example when a cell is overly stimulated by a neurotransmitter, hormone, or drug for a prolonged period of time and the expression of the receptor protein is decreased in order to protect the cell.

Inducible vs. Repressible Systems

Gene Regulation can be summarized as how they respond:

- Inducible systems - An inducible system is off unless there is the presence of some molecule (called an inducer) that allows for gene expression. The molecule is said to "induce expression". The manner in which this happens is dependent on the control mechanisms as well as differences between prokaryotic and eukaryotic cells.
- Repressible systems - A repressible system is on except in the presence of some molecule (called a corepressor) that suppresses gene expression. The molecule is said to "repress expression". The manner in which this happens is dependent on the control mechanisms as well as differences between prokaryotic and eukaryotic cells.

Developmental Biology

A large number of studied regulatory systems come from developmental biology. Examples include:

- The collinearity of the Hox gene cluster with their nested antero-posterior patterning
- It has been speculated that pattern generation of the hand (digits-interdigits) The gradient of Sonic hedgehog (secreted inducing factor) from the zone of polarizing activity in the limb which creates a gradient of active Gli3 which activates Gremlin which inhibits BMPs also secreted in the limb resulting in the formation of an alternating pattern of activity as a result of this reaction-diffusion system.
- Somatogenesis is the creation of segmatation (somites) form a uniform tissue (PSM) sequentially from anterior to posterior, this is achieved in amniotes possibly by means of two opposing

gradients, Retinoic acid in the anterior (wavefront) and an oscillating gradient in the posterior (clock) composed of FGF + Notch and Wnt in antiphase.

- Sex determination in the soma of a Drosophila requires the sensing of the ratio of autosomal genes to sex chromosome encoded genes, which results in the production of sexless splicing factor in females resulting in the female isoform of doublesex.

Theoretical Circuits

- Repressor/Inducer: an activation of a sensor results in the change of expression of a gene
- negative feedback: the gene product downregulates its own production directly or indirectly, which can result in
  - keeping transcript levels constant/proportional to a factor
  - inhibition of run-away reactions when coupled with a positive feedback loop
  - creating an oscillator by taking advantage in the time delay of transcription and translation, given that the mRNA and protein half-life is shorter
- positive feedback: the gene product upregulates its own production directly or indirectly, which can result in
  - signal amplification
  - bistable switches when two genes inhibit each other and have both positive feedback
  - pattern generation

Methods

Generally, most experiments investigating differential expression used whole cell extracts of RNA, called steady-state levels, to determined which genes changed and by how much they did, these are however not informative of where the regulation has occurred and may actually mask conflicting regulatory processess it is the most commonly analysed (QPCR and DNA microarray).

When studying gene expression there are several methods to look at the various stages. In eukaryotes these include:

- The chromatin conformation of the region can be determined by ChIP-chip analysis by pulling down RNA Polymerase II, Histone 3 modifications, Trithorax-group protein,Polycomb-group protein or any other DNA binding element to which a good antibody is available.
- Epistatic interactions can be investigated by synthetic genetic array analysis
- Due to post-transcriptional regulation, transcription rates and total RNA levels differ significantly, to measure the transcription rates nuclear run-on assays can be done and newer high-throughput methods are being developed, using thiol labelling instead of radioactivity.
- Only 5% of the RNA polymerised in the nucleus actually exists and not only introns, abortive products and non-sense transcripts are degradated therefore the differences in nuclear and

cytoplasmic levels can be see by separating the two fractions by gentle lysis. Alternative splicing can be analysed with a splicing array or with a tiling array All in vivo RNA is complexed as RNPs. The quantity of transcripts bound to specific protein can be also analysed by RIP-Chip, for example DCP2 will give an indication of sequestered protein, ribosome bound gives and indication of transcripts active in transcription (although it should be noted that a more dated method, called polysome fractionation, is still popular in some labs)

- Protein levels can be analysed by Mass spectrometry, which can only be compare to QPCR data as microarray data is relative and not absolute.
- RNA and protein degradation rates are measured by means of transcription inhibitors (actinomycin D or α-amanitin) or translation inhibitors (Cycloheximide) respectively.

LAC—OPERON HIS OPERON AND TRP—OPERON MODEL

The operon model of prokaryotic gene regulation was proposed by Fancois Jacob and Jacques Monod. Groups of genes coding for related proteins are arranged in units known as operons. An operon consists of an operator, promoter, regulator, and structural genes. The regulator gene codes for a repressor protein that binds to the operator, obstructing the promoter (thus, transcription) of the structural genes. The regulator does not have to be adjacent to other genes in the operon. If the repressor protein is removed, transcription may occur.

The ***lac* operon** is an operon required for the transport and metabolism of lactose in *Escherichia coli* and some other enteric bacteria. It consists of three adjacent structural genes, a promoter, a terminator, and an operator. The *lac* operon is regulated by several factors including the availability of glucose and of lactose. Gene regulation of the *lac* operon was the first genetic regulatory mechanism to be elucidated and is often used as the canonical example of prokaryotic gene regulation.

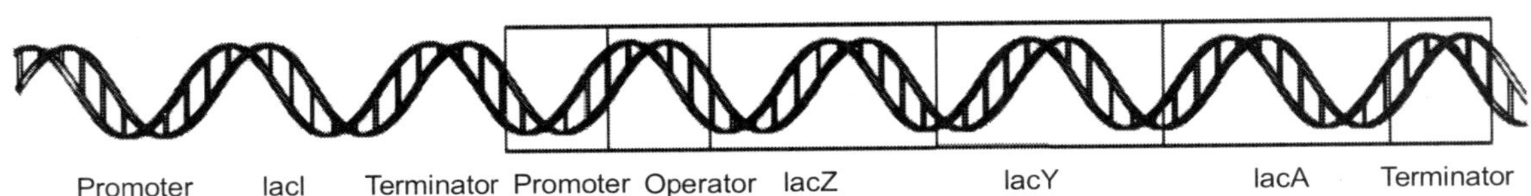

Fig. 6.11 : Showing the Lac Operon Gene

In its natural environment, *lac* operon is a complex mechanism to digest lactose efficiently. The cell can use lactose as an energy source, but it must produce the enzyme β-galactosidase to digest it into glucose. It would be inefficient to produce enzymes when there is no lactose available, or if there is a more readily-available energy source available (e.g. glucose). The *lac* operon uses a two-part control mechanism to ensure that the cell expends energy producing β-galactosidase, galactose permease and transacetylase only when necessary. It achieves this with the *lac* repressor, which halts production in the absence of lactose, and the Catabolite activator protein (CAP), which assists in production in the absence of glucose. This dual control mechanism causes the sequential utilization of glucose and lactose in two distinct growth phases, known as diauxie. Similar diauxic growth patterns have been observed in bacterial growth on mixtures of other sugars as well, such as glucose and xylose or glucose and arabinose, etc. The genetic control mechanisms underlying such diauxic growth patterns are known as xyl operon and ara operon, etc.

Structure of the *Lac* Operon

- The *lac* operon consists of three structural genes, and a promoter, a terminator, regulator, and an operator. The three structural genes are: *lacZ*, *lacY*, and *lacA*.
- *lacZ* encodes β-galactosidase (LacZ), an intracellular enzyme that cleaves the disaccharide lactose into glucose and galactose.
- *lacY* encodes β-galactoside permease (LacY), a membrane-bound transport protein that pumps lactose into the cell.
- *lacA* encodes β-galactoside transacetylase (LacA), an enzyme that transfers an acetyl group from acetyl-CoA to β-galactosides.

Only *lacZ* and *lacY* appear to be necessary for lactose catabolism.

Specific control of the *lac* genes depends on the availability of the substrate lactose to the bacterium. The proteins are not produced by the bacterium when lactose is unavailable as a carbon source. The *lac* genes are organized into an operon; that is, they are oriented in the same direction immediately adjacent on the chromosome and are co-transcribed into a single polycistronic mRNA molecule. Transcription of all genes starts with the binding of the enzyme RNA polymerase (RNAP), a DNA-binding protein, which binds to a specific DNA binding site, the *promoter*, immediately upstream of the genes. From this position RNAP proceeds to transcribe all three genes (*lacZYA*) into mRNA. The DNA sequence of the *E. coli* lac operon, the lacZYA mRNA, and the *lacI* genes are available from GenBank (view).

The first control mechanism is the regulatory response to lactose, which uses an intracellular *regulatory protein* called the *lactose repressor* to hinder production of β-galactosidase in the absence of lactose. The *lacI* gene coding for the repressor lies nearby the *lac* operon and is always expressed (*constitutive*). If lactose is missing from the growth medium, the repressor binds very tightly to a short DNA sequence just downstream of the promoter near the beginning of *lacZ* called the *lac operator*. The repressor binding to the operator interferes with binding of RNAP to the promoter, and therefore mRNA encoding LacZ and LacY is only made at very low levels. When cells are grown in the presence of lactose, however, a lactose metabolite called allolactose, which is a recombination of glucose and galactose, binds to the repressor, causing a change in its shape. Thus altered, the repressor is unable to bind to the operator, allowing RNAP to transcribe the *lac* genes and thereby leading to high levels of the encoded proteins.

The second control mechanism is a response to glucose, which uses the Catabolite activator protein (CAP) to greatly increase production of β-galactosidase in the absence of glucose. Cyclic adenosine monophosphate (cAMP) is a signal molecule whose prevalence is inversely proportional to that of glucose. It binds to the CAP, which in turn allows the CAP to bind to the CAP promoter (on the left in the diagram below), which assists the RNAP in binding to the DNA. In the absence of glucose, the prevalence of cAMP and binding of the CAP to the DNA significantly increases the production of β-galactosidase, enabling the cell to digest the lactose needed to produce glucose.

The Figure 6.12 below summarizes these statements:

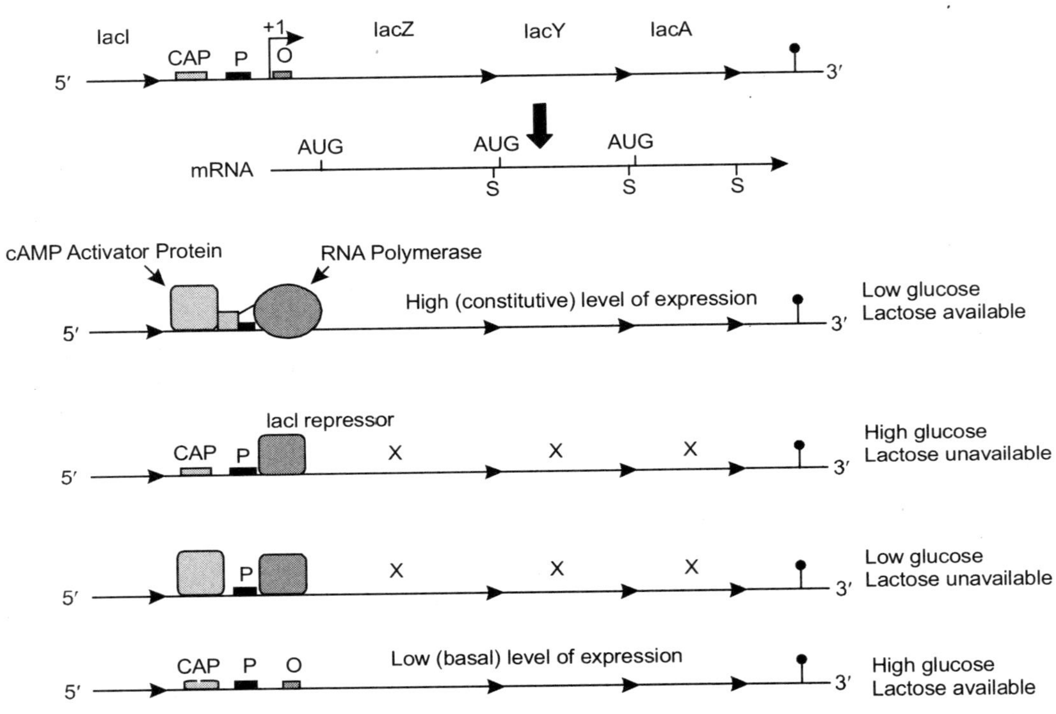

Fig. 6.12 : The Lac Operon in Detail

Genetic Nomenclature

Three-letter abbreviations are used to describe phenotypes in bacteria including *E. coli*.

Examples include:

- Lac (the ability to use lactose),
- His (the ability to synthesize the amino acid histidine)
- Mot (swimming motility)
- Sm^R (response to the antibiotic streptomycin)

In the case of Lac, wild type cells are Lac^+ and are able to use lactose as a carbon and energy source, while Lac^- mutant derivatives cannot use lactose. The same three letters are typically used (lower-case, italicized) to label the genes involved in a particular phenotype, where each different gene is additionally distinguished by an extra letter. The *lac* genes encoding enzymes are *lacZ*, *lacY*, and *lacA*. The fourth *lac* gene is *lacI*, encoding the lactose repressor—"I" stands for *inducibility*.

One may distinguish between *structural* genes encoding enzymes, and regulatory genes encoding proteins that affect gene expression. Current usage expands the phenotypic nomenclature to apply to proteins: thus, LacZ is the protein product of the *lacZ* gene, β-galactosidase. Various short sequences that are not genes also affect gene expression, including the *lac* promoter, *lac p*, and the *lac* operator, *lac o*. Although it is not strictly standard usage, mutations affecting *lac o* are referred to as *lac* o^c, for historical reasons.

A number of lactose derivatives or analogs have been described that are useful for work with the lac operon. These compounds are mainly substituted galactosides, where the glucose moiety of lactose is replaced by another chemical group.

- **Isopropyl-β-D-thio-galactoside (IPTG)** is frequently used as an inducer of the *lac* operon for physiological work.[1] IPTG binds to repressor and inactivates it, but is not a substrate for β-galactosidase. One advantage of IPTG for *in vivo* studies is that since it cannot be metabolized by *E. coli* its concentration remains constant and the rate of expression of *lac p/o*-controlled genes, is not a variable in the experiment. IPTG intake is dependent on the action of lactose permease.
- **Phenyl-β-D-galactose (phenyl-Gal)** is a substrate for β-galactosidase, but does not inactivate repressor and so is not an inducer. Since wild type cells produce very little β-galactosidase, they cannot grow on phenyl-Gal as a carbon and energy source. Mutants lacking repressor are able to grow on phenyl-Gal. Thus, minimal medium containing only phenyl-Gal as a source of carbon and energy is selective for repressor mutants and operator mutants. If 10^8 cells of a wild type strain are plated on agar plates containing phenyl-Gal, the rare colonies which grow are mainly spontaneous mutants affecting the repressor. The relative distribution of repressor and operator mutants is affected by the target size. Since the *lacI gene* encoding repressor is about 50 times larger than the operator, repressor mutants predominate in the selection.
- Other compounds serve as colourful indicators of β-galactosidase activity.
  - ONPG is cleaved to produce the intensely yellow compound, orthonitrophenol, and is commonly used as a substrate for assay of β-galactosidase *in vitro*.
  - Colonies that produce β-galactosidase are turned blue by X-gal (5-bromo-4-chloro-3-indolyl-β-D-galactoside).
- Allolactose is an isomer of lactose and is the inducer of the lac operon. Lactose is galactose-(β1->4)-glucose, whereas allolactose is galactose-(β1->6)-glucose. Lactose is converted to allolactose by â-galactosidase in an alternative reaction to the hydrolytic one. A physiological experiment which demonstrates the role of LacZ in production of the "true" inducer in *E. coli* cells is the observation that a null mutant of *lacZ* can still produce LacY permease when grown with IPTG but not when grown with lactose. The explanation is that processing of lactose to allolactose (catalyzed by β-galactosidase) is needed to produce the inducer inside the cell.

A conceptual breakthrough of Jacob and Monod was to recognize the distinction between regulatory substances and sites where they act to change gene expression. A former soldier, Jacob used the analogy of a bomber that would release its lethal cargo upon receipt of a special radio transmission or signal. A working system requires both a ground transmitter and a receiver in the airplane. Now, suppose that the usual transmitter is broken. This system can be made to work by introduction of a second, functional transmitter. In contrast, he said, consider a bomber with a defective receiver. The behavior of *this* bomber cannot be changed by introduction of a second, functional aeroplane.

To analyze regulatory mutants of the *lac* operon, Jacob developed a system by which a second copy of the *lac* genes (*lacI* with its promoter, and *lacZYA* with promoter and operator) could be introduced into a single cell. A culture of such bacteria, which are diploid for the *lac* genes but otherwise normal, is then tested for the regulatory phenotype. In particular, it is determined whether LacZ and LacY are made even in the absence of IPTG. This experiment, in which genes or gene clusters are tested pairwise, is called a *complementation test.*

This test is illustrated in the Fig. 6.13 (*lacA* is omitted for simplicity). First, certain haploid states are shown (i.e. the cell carries only a single copy of the *lac* genes). Panel (a) shows repression, (b) shows induction by IPTG, and (c) and (d) show the effect of a mutation to the *lacI* gene or to the

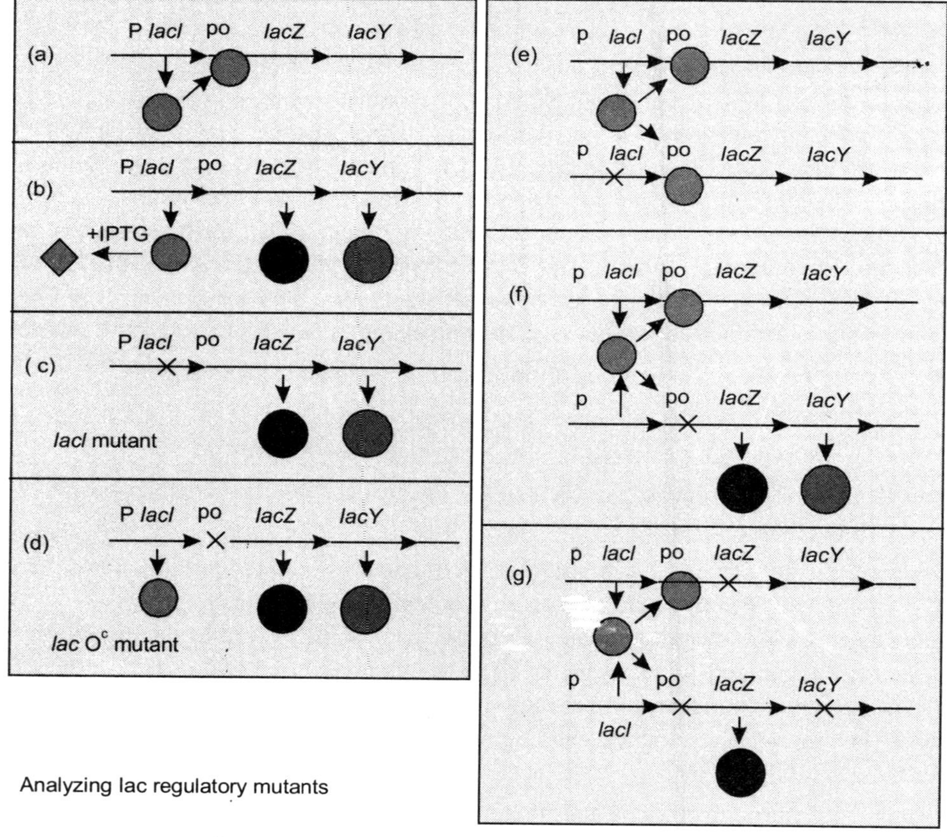

Fig. 6.13 : Analysing the Lac Regulatory Mutants

operator, respectively. In panel (e) the complementation test for repressor is shown. If one copy of the *lac* genes carries a mutation in *lacI*, but the second copy is wild type for *lacI*, the resulting phenotype is normal—but lacZ is expressed when exposed to inducer IPTG. Mutations affecting repressor are said to be *recessive* to wild type (and that wild type is *dominant*), and this is explained by the fact that repressor is a small protein which can diffuse in the cell. The copy of the *lac* operon adjacent to the defective *lacI* gene is effectively shut off by protein produced from the second copy of *lacI*.

If the same experiment is carried out using an operator mutation, a different result is obtained (panel (f)). The phenotype of a cell carrying one mutant and one wild type operator site is that LacZ and LacY are produced even in the absence of the inducer IPTG; because the damaged operator site, does not permit binding of the repressor to inhibit transcription of the structural genes. The operator mutation is dominant. When the operator site where repressor must bind is damaged by mutation, the presence of a second functional site in the same cell makes no difference to expression of genes controlled by the mutant site.

A more sophisticated version of this experiment uses *marked* operons to distinguish between the two copies of the *lac* genes and show that the unregulated gene(s) are the ones next to the mutant operator (panel (g). For example, suppose that one copy is marked by a mutation inactivating *lacZ* so that it can only produce the LacY protein, while the second copy carries a mutation affecting *lacY* and can only produce LacZ. In this version, only the copy of the *lac* operon that is adjacent to the mutant operator is expressed without IPTG. We say that the operator mutation is *cis-dominant*, it is dominant to wild type but affects only the copy of the operon which is immediately adjacent to it.

This explanation is misleading in an important sense, because it proceeds from a description of the experiment and then explains the results in terms of a model. But in fact, it is often true that the model comes first, and an experiment is fashioned specifically to test the model. Jacob and Monod first imagined that there must be a *site* in DNA with the properties of the operator, and then designed their complementation tests to show this.The dominance of operator mutants also suggests a procedure to select them specifically. If regulatory mutants are selected from a culture of wild type using phenyl-Gal, as described above, operator mutations are rare compared to repressor mutants because the target-size is so small. But if instead we start with a strain which carries two copies of the whole *lac* region (that is diploid for *lac*), the repressor mutations (which still occur) are not recovered because complementation by the second, wild type *lacI* gene confers a wild type phenotype. In contrast, mutation of one copy of the operator confers a mutant phenotype because it is dominant to the second, wild type copy.

The experimental microorganism used by François Jacob and Jacques Monod was the common laboratory bacterium, *E. coli*, but many of the basic regulatory concepts that were discovered by Jacob and Monod are fundamental to cellular regulation in organisms. The key idea is that proteins are not synthesized when they are not needed— *E. coli* conserves cellular resources and energy by not making the three Lac proteins when there is no need to metabolize lactose, such as when other sugars like glucose are available. The following section discusses how *E. coli* controls certain genes in response to metabolic needs.

During World War II, Monod was testing the effects of combinations of sugars as nutrient sources for *E. coli*. He found that bacteria grown with two different sugars often displayed two phases of growth. For example, if glucose and lactose were both provided, glucose would be metabolized first (growth phase I, and then lactose (growth phase II). This phenomenon is called diauxis.

Metabolism of lactose does not occur during the first part of the diauxic growth curve because β-galactosidase is not made when both glucose and lactose are present in the medium.

Explanation of this depended on the characterization of additional mutations affecting the *lac* genes other than those explained by the classical model. Two other genes, *cya* and *crp*, were known that map far away from *lac*, and when mutant result in a decreased level of expression in the *presence* of IPTG and even in a strain which is mutant for repressor or operator. The discovery of cyclic AMP in 1957 (in eukaryotic cells) and a decade later in *E. coli* led to the demonstration that mutants defective in one of these genes could be restored to full activity by the addition of cyclic AMP to the medium.

The *cya* gene encodes adenylate cyclase, which produces cyclic AMP. In a *cya* mutant, the absence of cyclic AMP makes the expression of the *lacZYA* genes more than ten times lower than normal. Addition of cyclic AMP corrects the low Lac expression characteristic of *cya* mutants. The second gene, *crp*, encodes a protein called catabolite activator protein (CAP) or cAMP receptor protein (CRP).

This dual regulation causes the lactose metabolism enzymes to be made in small quantities in the presence of both glucose and lactose (sometimes called leaky expression) due to lactose inhibiting LacI from binding to the operator, but at high cAMP concentrations and in the presence of lactose there are high levels of expression. Leaky expression is necessary in order to allow for metabolism of some lactose after the glucose source is expended, but before *lac* expression is fully activated.

In summary:

- When lactose is absent then there is very little Lac enzyme production (the operator has LacI bound to it).
- When lactose is present but a preferred carbon source (like glucose) is also present then a small amount of enzyme is produced (LacI is not bound to the operator).
- When lactose is the favoured carbon source (for example in the absence of glucose) cAMP-CAP binds upstream of the promoter at a specific site. This bends the DNA around the protein which creates tension, and allows the RNA polymerase to bind to the promoter and Lac enzyme production is maximised. The DNA is not easily unwound under normal conditions, without the bound CAP, as the DNA contains a large number of the nucleotides which have 3 hydrogen bonds between them, needing more energy to part them

The delay between growth phases reflects the time needed to produce sufficient quantities of lactose-metabolizing enzymes. First, the CAP regulatory protein has to assemble on the *lac* promotor, resulting in an increase in the production of *lac* mRNA. More available copies of the *lac* mRNA results in the production (see translation) of significantly more copies of LacZ (β-galactosidase, for lactose metabolism) and LacY (lactose permease to transport lactose into the cell). After a delay needed to increase the level of the lactose metabolizing enzymes, the bacteria enter into a new rapid phase of cell growth.

Two puzzles of catabolite repression relate to how cAMP level is actually coupled to the presence of glucose, and secondly, why the cells should even bother. After lactose is cleaved it actually forms glucose and galactose (easily converted to glucose). In metabolic terms, lactose is just as *good* a carbon and energy source as glucose. The cAMP level is related not to intracellular glucose concentration but to the rate of glucose transport, which influences the activity of adenylate cyclase. (In addition, glucose transport also leads to direct inhibition of the lactose permease.) As to why *E. coli* works this way, one can only speculate. All enteric bacteria ferment glucose, which suggests they encounter it frequently. It is possible that a small difference in efficiency of transport or metabolism of glucose v. lactose makes it advantageous for cells to regulate the *lac* operon in this way.

The lac repressor is a tetramer of identical subunits. Each subunit contains a helix-turn-helix (HTH) motif capable of binding to DNA. The operator site where repressor binds is a DNA sequence with inverted repeat symmetry. The two DNA half-sites of the operator together bind to two of the subunits of the tetrameric repressor. Although the other two subunits of repressor are not doing anything in this model, this property was not understood for many years.

Eventually it was discovered that two additional (minor) operators are involved in *lac* regulation. One (O_3) lies in the end of the *lacI* gene and the other (O_2) is about 400 bp downstream in the early part of *lacZ*. These two sites were not found in the early work because they have redundant functions and individual mutations do not affect repression very much. Single mutations to either O_2 or O_3 have only 2 to 3-fold effects. However, their importance is demonstrated by the fact that a double mutant defective in both O_2 and O_3 is dramatically de-repressed (by about 70-fold).

In the current model, repressor is bound simultaneously to both the main operator O_1 and to either O_2 or O_3. The intervening DNA loops out from the complex. The redundant nature of the two minor operators suggests that it is not a specific looped complex that is important. One idea is that the system works through tethering. If bound repressor releases from O_1 momentarily, binding to a minor operator keeps it in the vicinity, so that it may rebind quickly. This would increase the affinity of repressor for O_1.

The repressor is an allosteric protein, i.e. it can assume either one of two slightly different shapes, which are in equilibrium with each other. In one form the repressor is capable of binding to the operator DNA, and in the other form it cannot bind to the operator. According to the classical model of induction, binding of the inducer, either allolactose or IPTG, to the repressor affects the distribution of repressor between the two shapes. Thus, repressor with inducer bound is stabilized in the non-DNA-binding conformation. However, this simple model cannot be the whole story, because repressor is bound quite stably to DNA, yet it is released rapidly by addition of inducer. Therefore it seems clear that repressor can also bind inducer while still bound to DNA. It is still not entirely known what the exact mechanism of binding is.

The *lac* gene and its derivatives are amenable to use as a reporter gene in a number of bacterial-based selection techniques such as two hybrid analysis, in which the successful binding of a transcriptional activator to a specific promoter sequence must be determined. In LB plates containing X-gal, the colour change from white colonies to a shade of blue, corresponds to about 20-100 β-galactosidase units, while tetrazolium lactose and MacConkey lactose media have a range of 100-1000 units, being most sensitive in the high and low parts of this range respectively. Since MacConkey lactose and tetrazolium lactose media both rely on the products of lactose breakdown, they require the presence of both *lacZ* and *lacY* genes. The many *lac* fusion techniques which include only the *lacZ* gene are thus suited to the X-gal plates or ONPG liquid broths

Trp operon is an operon - a group of genes that are used, or transcribed, together - that codes for the components for production of tryptophan. The Trp operon is present in many bacteria, but was first characterized in *Escherichia coli*. This is the specific example we will discuss here. It is regulated so that when tryptophan is present in the environment, it is not used. It was an important experimental system for learning about gene regulation, and is commonly used to teach gene regulation.

Discovered in 1953 by Jacques Monod and colleagues, the trp operon in *E. coli* was the first repressible operon to be discovered. While the lac operon can be activated by a chemical (allolactose), the tryptophan (Trp) operon is inhibited by a chemical (tryptophan). This operon contains five structural genes: trp E, trp D, trp C, trp B, and trp A, which encodes tryptophan synthetase. It also contains a promoter which binds to RNA polymerase and an operator which blocks transcription when bound tc the protein synthesized by the repressor gene (trp R) that binds to the operator. In the lac operon, lactose binds to the repressor protein, allowing gene transcription, while in the trp operon, tryptophan binds to the repressor protein effectively blocking gene transcription. In both situations, repression is that of RNA polymerase transcribing the genes in the operon. Also unlike the lac operon, the trp operon contains a leader peptide and an attenuator sequence which allows for graded regulation.

It is an example of negative regulation of gene expression. Within the operon's regulatory sequence, the operator is blocked by the repressor protein in the presence of tryptophan (thereby preventing transcription) and is liberated in tryptophan's absence (thereby allowing transcription). The process of attenuation complements this regulatory action.

RepressionThis is a negative repressive feedback mechanism. The repressor for the trp operon is produced upstream by the trpR gene, which is continually expressed at a low level. It creates monomers, which associate into tetramers. These tetramers are inactive and "floating" around within the cell. When tryptophan is present, it binds to the tryptophan repressor tetramers causing a change in conformation, which allows the repressor to bind the operator. This prevents RNA polymerase from binding to and transcribing the operon, so tryptophan is not produced from its precursor. When tryptophan is not present, the repressor is in its native conformation and cannot bind the operator region, so transcription is not inhibited by the repressor.Attenuation is a second mechanism of negative feedback in the trp operon. While the TrpR repressor decreases transcription by a factor of 70, attenuation can further decrease it by a factor of 10, thus allowing accumulated repression of about 700-fold. Attenuation is made possible by the fact that in prokaryotes (which have no nucleus), the ribosomes begin translating the mRNA while RNA polymerase is still transcribing the DNA sequence. This allows the process of translation to directly affect transcription of the operon.

At the beginning of the transcribed genes of the trp operon is a sequence of 140 nucleotides termed the leader transcript (trpL). This transcript includes four short sequences designated 1-4. Sequence 1 is partially complementary to sequence 2, which is partially complementary to sequence 3, which is partially complementary to sequence 4. Thus, three distinct secondary structures (hairpins) can form:

1-2, 2-3 or 3-4. The hybridization of strands 1 and 2 to form the 1-2 structure prevents the formation of the 2-3 structure, while the formation of 2-3 prevents the formation of 3-4. The 3-4 structure is a transcription termination sequence, once it forms RNA polymerase will disassociate from the DNA and transcription of the structural genes of the operon will not occur.

Part of the leader transcript codes for a short polypeptide of 14 amino acids, termed the leader peptide. This peptide contains two adjacent tryptophan residues, which is unusual, since tryptophan is a fairly uncommon amino acid (about one in a hundred residues in a typical *E. coli* protein is tryptophan). If the ribosome attempts to translate this peptide while tryptophan levels in the cell are low, it will stall at either of the two trp codons. While it is stalled, the ribosome physically shields sequence 1 of the transcript, thus preventing it from forming the 1-2 secondary structure. Sequence 2 is then free to hybridize with sequence 3 to form the 2-3 structure, which then prevents the formation of the 3-4 termination hairpin. RNA polymerase is free to continue transcribing the entire operon. If tryptophan levels in the cell are high, the ribosome will translate the entire leader peptide without interruption and will only stall during translation termination at the stop codon. At this point the ribosome physically shields both sequences 1 and 2. Sequences 3 and 4 are thus free to form the 3-4 structure which terminates transcription. The end result is that the Operon will be transcribed only when tryptophan is unavailable for the ribosome, while the trpL transcript is constitutively expressed.

To ensure that the ribosome binds and begins translation of the leader transcript immediately following its synthesis, a pause site exists in the trpL sequence. Upon reaching this site, RNA polymerase pauses transcription and apparently waits for translation to begin. This mechanism allows for synchronization of transcription and translation, a key element in attenuation.

A similar attenuation mechanism regulates the synthesis of histidine, phenylalanine and threonine.

ROLE OF HISTONE AND NON-HISTONE PROTEINS

In biology, histones are the chief protein components of chromatin. They act as spools around which DNA winds, and they play a role in gene regulation. Without histones, the unwound DNA in chromosomes would be very long. For example, each human cell has about 1.8 meters of DNA, but wound on the histones it has about 90 millimeters of chromatin, which, when duplicated and condensed during mitosis, result in about 120 micrometers of chromosomes.

There are a total of six classes of histones (H1, H2A, H2B, H3, H4, and H5) organized into two super classes as follows:

- core histones – H2A, H2B, H3 and H4
- linker histones – H1 and H5

Two of each of the core histones assemble to form one octameric nucleosome core particle by wrapping 146 base pairs of DNA around the protein spool in 1.65 left-handed super-helical turn. The linker histone H1 binds the nucleosome and the entry and exit sites of the DNA, thus locking the DNA into place and allowing the formation of higher order structure. The most basic such formation is the 10 nm fiber or beads on a string conformation. This involves the wrapping of DNA around nucleosomes with approximately 50 base pairs of DNA spaced between each nucleosome (also referred to as linker

DNA). The assembled histones and DNA is called chromatin. Higher order structures include the 30 nm fiber (forming an irregular zigzag) and 100 nm fiber, these being the structures found in normal cells. During mitosis and meiosis, the condensed chromosomes are assembled through interactions between nucleosomes and other regulatory proteins.

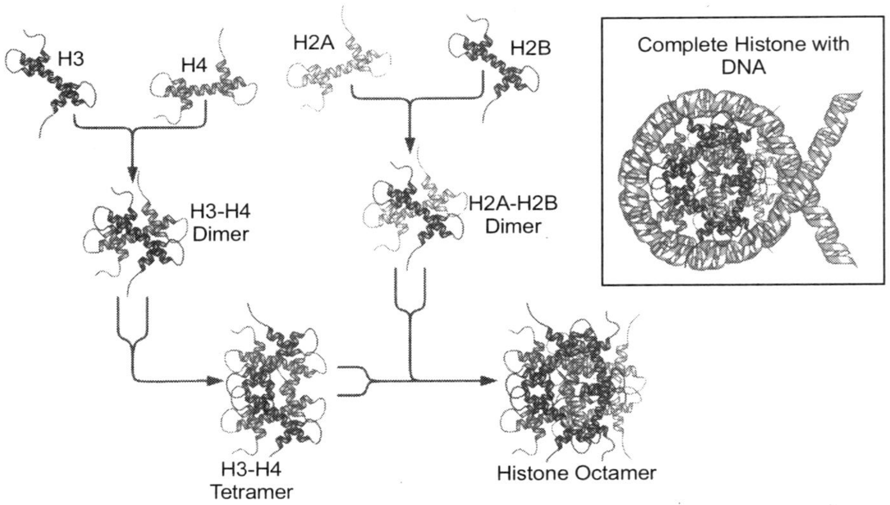

Fig. 6.14 : the Structure of Nucleosome with Histone Protein

The following is a list of human histoneproteins:

Super family	Family	Subfamily	Members
1	2	3	4
Linker	H1	H1F	H1F0, H1FNT, H1FOO, H1FX
		H1H1	HIST1H1A, HIST1H1B, HIST1H1C, HIST1H1D, HIST1H1E, HIST1H1T
Core	H2A	H2AF	H2AFB1, H2AFB2, H2AFB3, H2AFJ, H2AFV, H2AFX, H2AFY, H2AFY2, H2AFZ
		H2A1	HIST1H2AA, HIST1H2AB, HIST1H2AC, HIST1H2AD, HIST1H2AE, HIST1H2AG, HIST1H2AI, HIST1H2AJ, HIST1H2AK, HIST1H2AL, HIST1H2AM
		H2A2	HIST2H2AA3, HIST2H2AC
	H2B	H2BF	H2BFM, H2BFO, H2BFS, H2BFWT

...(Contd.)

1	2	3	4
		H2B1	HIST1H2BA, HIST1H2BB, HIST1H2BC, HIST1H2BD, HIST1H2BE, HIST1H2BF, HIST1H2BG, HIST1H2BH, HIST1H2BI, HIST1H2BJ, HIST1H2BK, HIST1H2BL, HIST1H2BM, HIST1H2BN, HIST1H2BO
		H2B2	HIST2H2BE
	H3	H3A1	HIST1H3A, HIST1H3B, HIST1H3C, HIST1H3D, HIST1H3E, HIST1H3F, HIST1H3G, HIST1H3H, HIST1H3I, HIST1H3J
		H3A2	HIST2H3C
		H3A3	HIST3H3
	H4	H41	HIST1H4A, HIST1H4B, HIST1H4C, HIST1H4D, HIST1H4E, HIST1H4F, HIST1H4G, HIST1H4H, HIST1H4I, HIST1H4J, HIST1H4K, HIST1H4L
		H44	HIST4H4

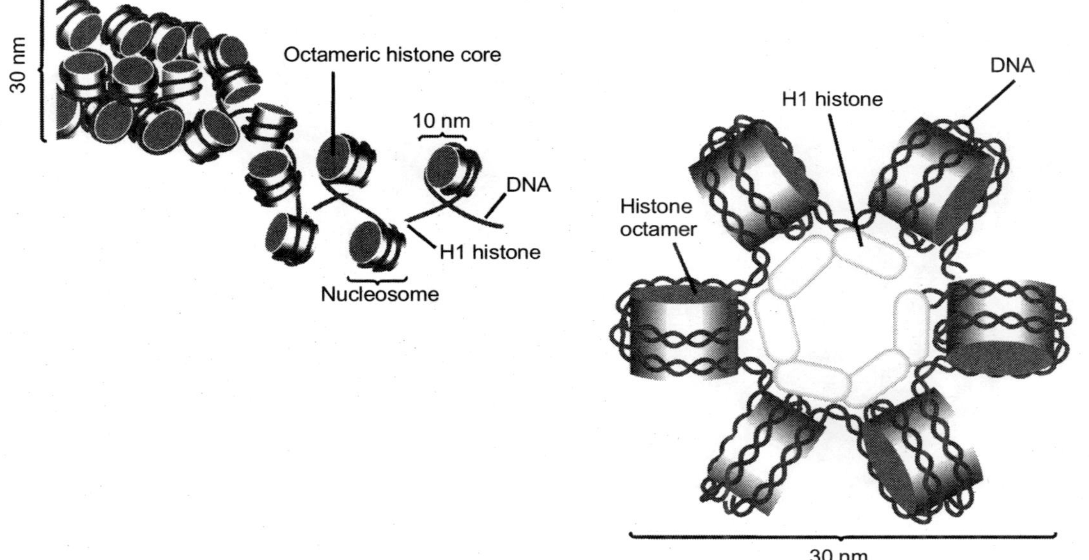

Fig. 6.15 : The Structure of Histone Protein

EUCARYOTE AND PROCARYOTE CELLS AND THEIR DIFFERENCE

A cell that contains membrane-bound compartments in which specific metabolic activities take place. Most important among these compartments is the nucleus, which houses the eukaryotic cell's DNA. It is this nucleus that gives the eukaryote – literally, "true nucleus" – its name. Eukaryotic organisms also have other specialized, membrane-bounded structures, called organelles, which are small structures within cells that perform dedicated functions. Eukaryotic cells are typically 10 to 100 micrometers across, or about 10 times the size of prokaryotic cells. The set of ideas most favored by biologists to explain how eukaryotic cells first came about is called the endosymbiotic theory. This theory is able to account well for the fact that two of the organelles found in eukaryotic cells, mitochondria and chloroplasts, have their own DNA that is completely distinct from the DNA housed in the nucleus. According to the endosymbiotic theory, the eukaryotic cell evolved from a symbiotic community of prokaryotic cells. Specifically, the mitochondria and the chloroplasts are what remains of ancient symbiotic oxygen-breathing bacteria and cyanobacteria, respectively, whereas the rest of the cell seems

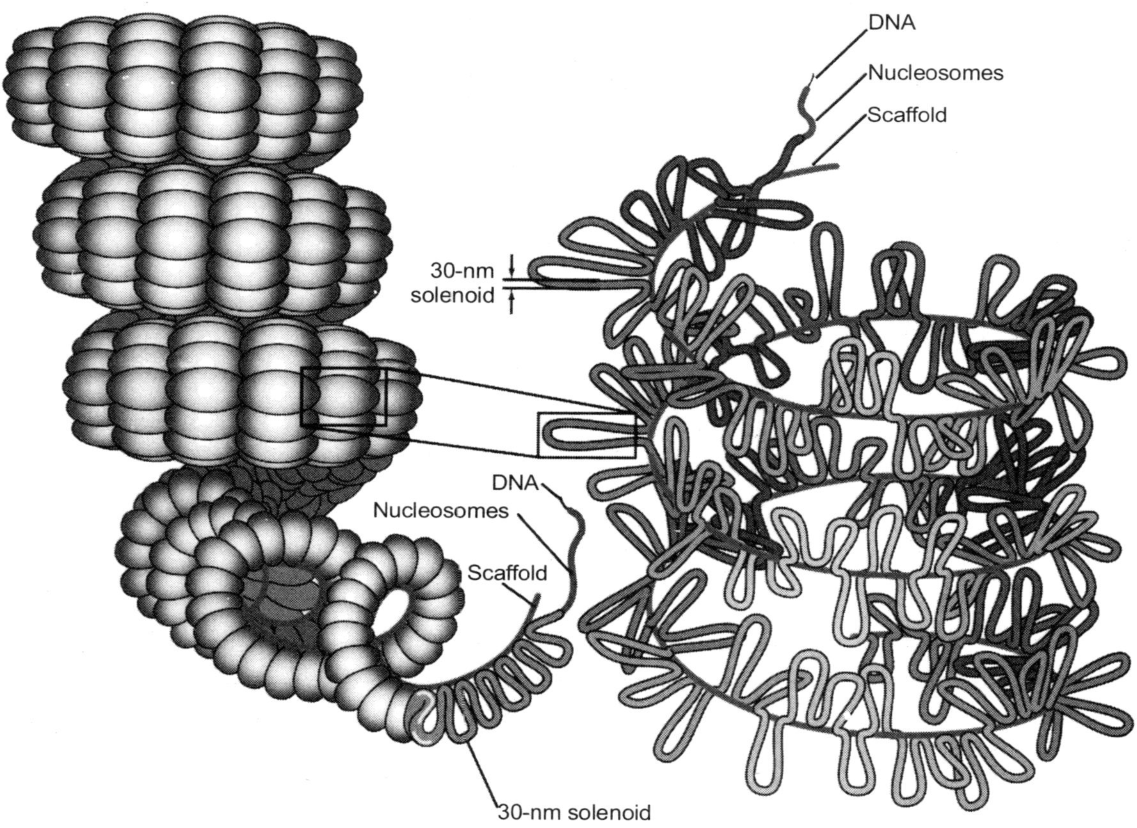

Fig. 6.16 : The Eucaryote and the Procaryote

to be derived from an ancestral archaean prokaryote cell. The origin of the eukaryotic cell was a milestone in the evolution of life. Although eukaryotes use the same genetic code and metabolic processes as prokaryotes, their higher level of organizational complexity has permitted the development of truly multicellular organisms. Without eukaryotes, the world would lack mammals, birds, fish, invertebrates, mushrooms, plants, and complex single-celled organisms.

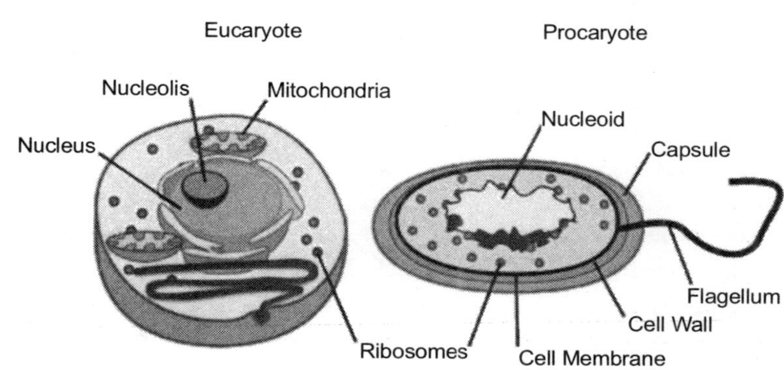

Fig. 6.17 : The Eucaryotic and the Procaryote

All eukaryotic cells have a number of features in common. These include:

- nucleus
- cell (plasma) membrane
- endoplasmic reticulum
- ribosomes
- centriole
- cytoplasm
- mitochondria
- Golgi apparatus
- cytoskeleton

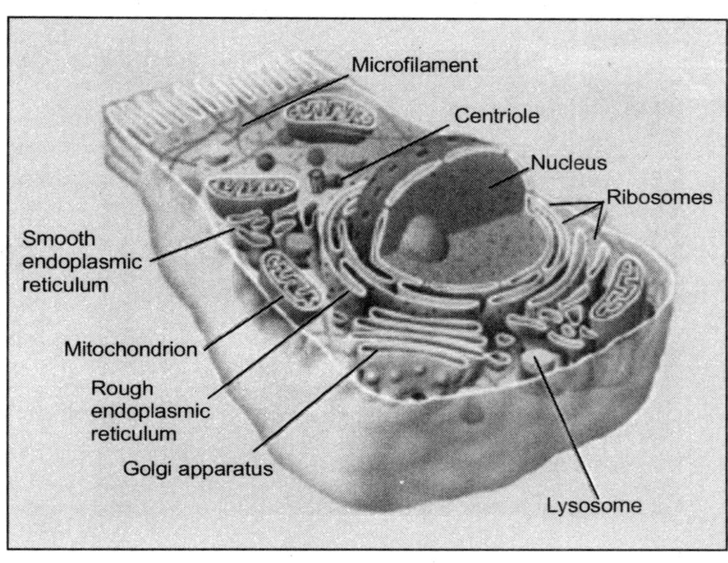

Fig. 6.18 : The Eucaryotic Cell

However, there are also some important differences between animal cells and plant cells.

Comparison of Eukaryotic and Prokaryotic Cells

feature	eukaryotic cells	prokaryotic cells
1	2	3
types of organism	Found in "complex" organisms, including all plants and animals	Found in "simple" organisms, including bacteria and cyanobacteria
specialization	Can specialize for certain functions, such as absorbing nutrients from food or transmitting nerve impulses; groups cells can form large, multicellular organs and organisms	Usually exist as single, virtually identical cells
size	Most animal cells are 10-30 micrometers across; most plant cells are 10-100 micrometers across	Most are 1-10 micrometers across
nucleus	Contain a nucleus and many other organelles, each surrounded by a membrane (the nucleus and mitochondrion have two membranes)	Lack a nucleus and other membrane-encased organelles
nucleolus	One (or more) present in each nucleus	Absent
DNA	DNA always in combination with histone proteins	Simple duplex not associated with histones (i.e. basic proteins) Absent
spindle	Present temporarily during mitosis and meiosis	Absent
sexual system	Complete nuclear fusion between gametes, with equal contributions from both genomes	Unidirectional transfer of genes from donor to recipient
cell wall	Present in plant cells, but never contain muramic acid	Present, but chemically different in many respects from eukaryotes (e.g. presence of muramic acid)
internal membranes	Complex compartmentalization into endoplasmic reticulum, Golgi bodies, lysosomes, etc	Usually simple and often transient, if present at all
ribosomes	80 S with subunits (60 S + 40 S)	70 S* with subunits (30 S + 50 S)
photosynthesis	Complex chloroplasts (membrane-bounded organelles)	Simple chromatophores

...(Contd.)

1	2	3
respiration	Virtually all forms are aerobic, though a few are faculatively anaerobic (e.g. yeasts); and, uniquely, the trichomonads are obligate anerobes	Simple chromatophores
electron transport system & ATP synthesis	Found on the inner membrane of special membrane-bound organelles: mitochondria (oxidative phosphorylation) and chloroplasts (photophosphorylation). Virtually all forms are aerobic, though a few are faculatively anaerobic (e.g. yeasts); and, uniquely, the trichomonads are obligate anerobes	Localized on the cell membrane

REPLICATION OF DNA

Naturally occurring DNA exist in many forms single and double stranded DNA's are known, and both can exist in linear and both can exist in linear and circular forms. As a result it is difficult to generalize about all possible cases of DNA replication. Since many DNAs are double stranded DNA, DNA feature that we shall discuss here are first investigated in prokaryotes, particularly Escherichia coli, we shall use information obtained by experiments on this organization throughout our discussion of the topic.

The process by which one double helical DNA molecule is duplicated to produce two such double stranded molecules is a pretty complicated one. The very complexity allows for a high degree of fine tuning which in turn ensures considerable fidelity in replication. The cell faces three important challenges in carrying out the necessary steps. The first challenge is how to separate the two DNA strands. In addition to achieve continuous unwinding of the double helix, the cell also has to protect the unwound portions of DNA form the action nucleases that preferentially attack single – stranded DNA. Two anti parallel stands must be synthesize in the same direction on anti-parallel templates. In other words the templates have one 5' → 3' strand and one 3' → 5' strand as does the newly synthesized DNA. The third task is how to guard against errors in replication, ensuring that the correct base is added to the growing polynucleotide chain.

SEMI CONSERVATIVE REPLICATION

The DNA replication, involves separation of the two original ones as templates. Each new DNA molecule contains one strand from the original DNA and one newly synthesized strand. This situation is called semi-conservative replication (see Fig. 6.19)

The details of the process differ in procaryotes and eucaryotes, but the semi conservative nature of replication is observed in all organisms.

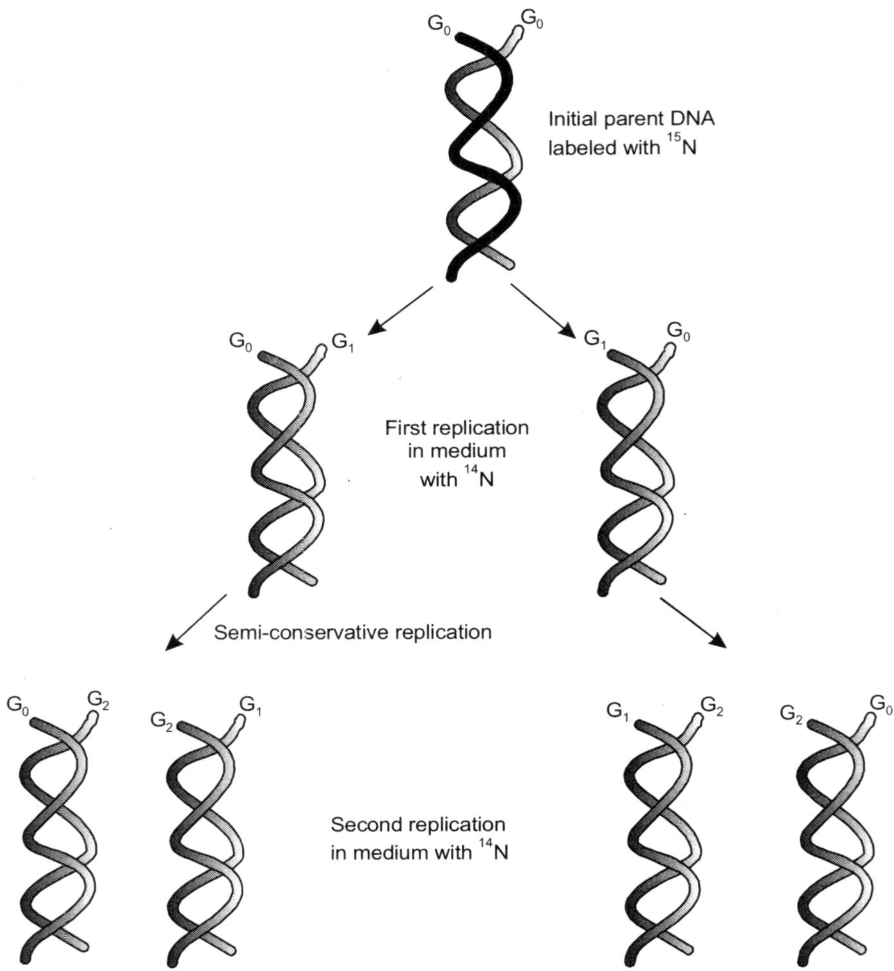

Fig. 6.19 : The Labelling Pattern of ^{15}N strands in semi-conservative replication (G_0 is original strands G_1 in new strands after first generation G_2 is new strands after second generation)

Semi-conservative replication of DNA was established in 1950s by experiments performed by Nelson and Stahl. *E. coli* bacteria were grown in the presence of sole nitrogen source that is ^{15}ammonium chloride, where ^{15}N is a heavy isotope of nitrogen. The usual of nitrogen is ^{14}N. In such a medium all newly formed nitrogen compounds including purine and pyrimidine nulceo-bases become labelled with ^{15}N. The ^{15}N labelled DNA has a higher density than unlabelled DNA which contains the usual isotope, ^{14}N. In this experiment the ^{15}N labelled where then transferred in a medium consisting only ^{14}N. The cells continued to grow in the new medium, which every new generation of growth, a sample of DNA was extracted and analyzed by the technique of density gradiant centrifugation (see Fig. 6.20).

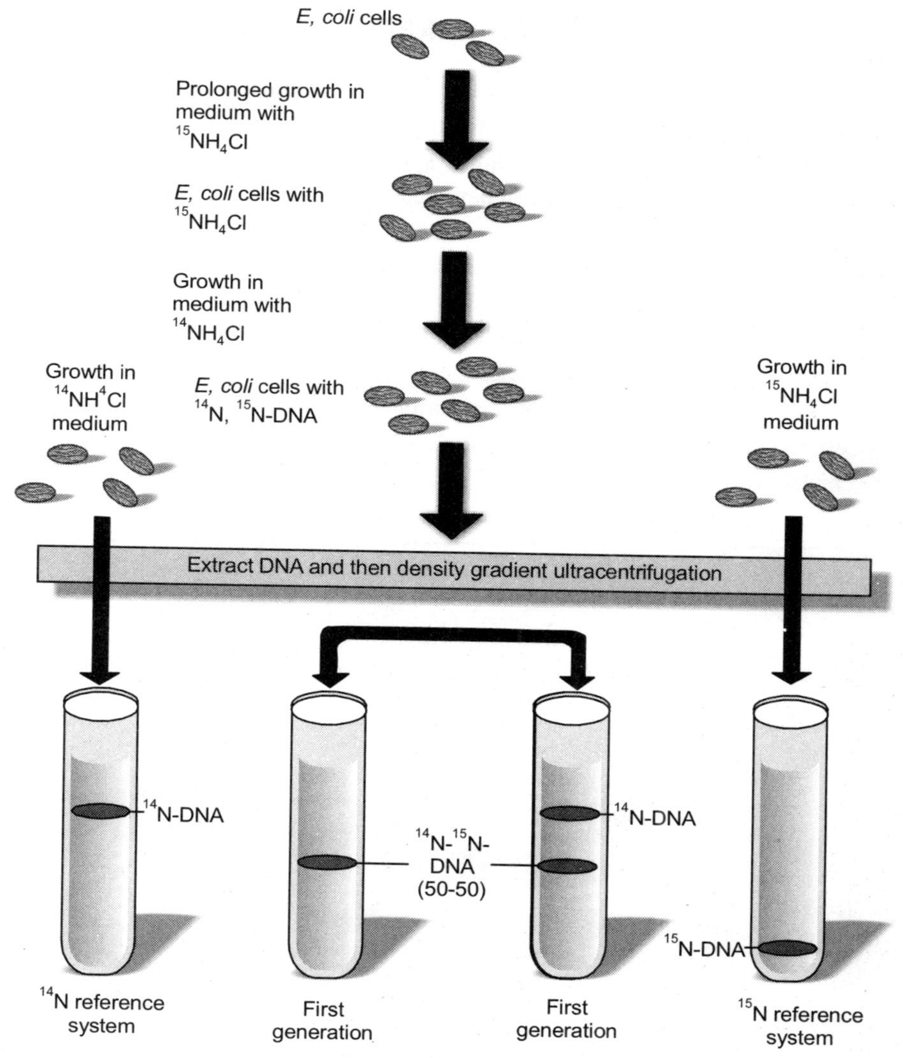

Fig. 6.20 : The Experimental Evidence for Semiconservative Replication. Heavy DNA labelled ^{15}N forms a band at the bottom of the tube and light DNA with ^{14}N forms a band at an intermediate position has one heavy strand and one light strand

This technique depends on the fact that heavy ^{15}N DNA (DNA that contains ^{15}N alone) will appear from the band at the bottom of the tube, Light ^{14}N DNA (^{14}N alone) will appear at the top of the tube the DNA containing 50-50 mixture will appear half way between the bands. In the actual experiment this 50-50 hybrids DNA was observed after one generation a result to be expected with semi conservative replication. After two generations in the lighter medium, half should be lighter ^{15}DNA. This prediction of the kind amount of DNA observed is confirmed by the experiment.

BI DIRECTIONAL REPLICATION

During replication the DNA double the helix unwinds at a specific point called the **'origin of replication'** (Ori). New polynucleotide chains are synthesized using each of the exposed strands as template. Two possibilities exist for the growth of the new strands; synthesis can take place in both directions from the origin of replication or in only one direction. It has been established that DNA synthesis is bidirectional in most organism, with the exception of a few viruses. For each origin of replication there are two points (replication of forks) at which new polynucleotide chains are formed. A **bubble** of newly synthesized DNA between regions of the original DNA is a manifestation of the advance of the replication forks in the opposite directions. There is one such bubble (and one origin of replication) in the circular DNA of procaryotes in Fig. 6.21A in eucaryotes several origins of replication and thus several bubbles exist (Fig. 6.21b). The bubbles grow larger and eventually coalesce, giving rising two complete daughter DNAs. This directional growth of both new polynucleotide chains represents **net chain growth**. Both new nucleotide chains are synthesized in the 5'-3'-direction.

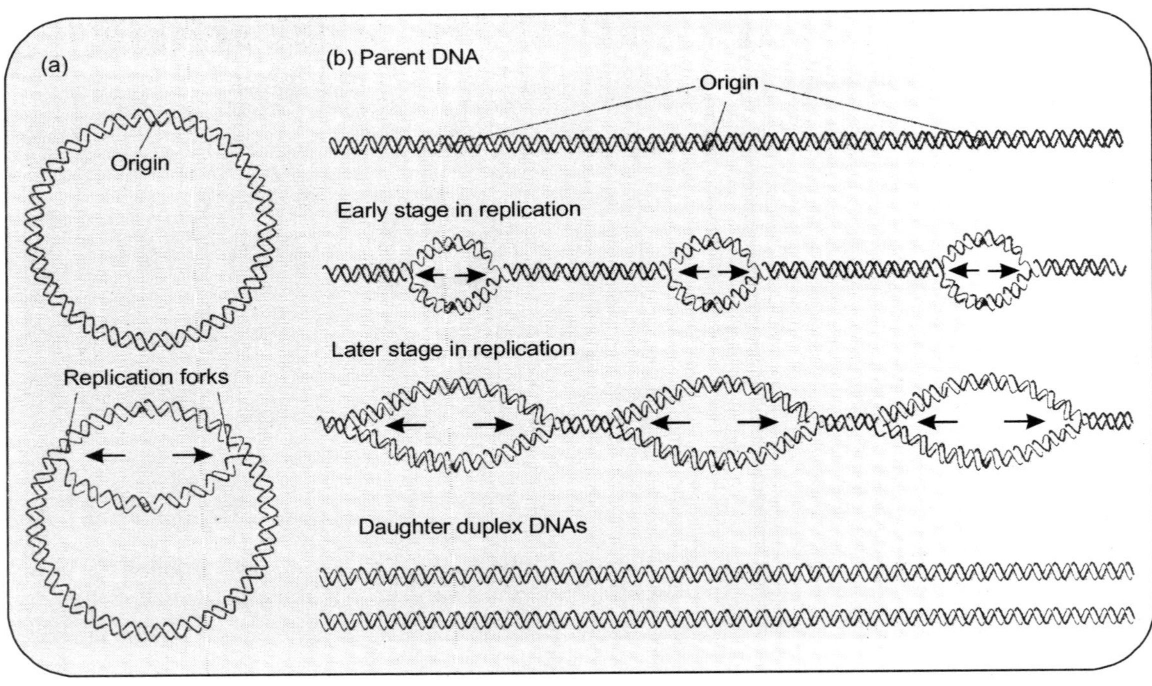

Fig. 6.21 : The Bidirectional Replication of DNA in Procaryotes (one origin of replication) and in eucaryotes (several origins). Bidirectional replication refers to overall synthesis.

DNA REPLICATION REQUIRES THE COMBINED ACTIONS OF SEVERAL ENZYMES

Regarding the process of separating the two strands of the original DNA so that it can be replicated, two things can be asked, that is how to achieve the continuous un winding the double helix, this is true in the case of the procaryotic DNA which exists in super coiled position and secondly how to protect simple stranded stretched of DNA that are exposed to intra cellular nucleases as a result of the unwinding

An Enzyme called DNA gyrase catalyzes the conversion of relaxed circular DNA with the nick in one strand to the super coiled form with the nick sealed (see Fig. 6.22).

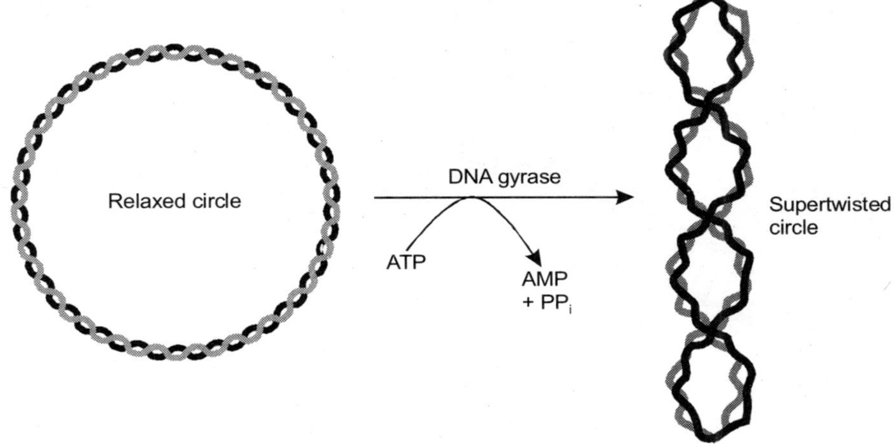

Fig. 6.22 : The Introduction of DNA Gyrase and the Twisting of the in Circular DNA

A slight unwinding of the helix before the nick is sealed introduces the super coiling. The energy required for the process is supplied by the hydrolysis of ATP. Some evidence exists that DNA gyrase causes a double strand break in DNA in the process of converting the relaxed circular form to the super coiled from (a mode of action, typical of Type II topoisomers, the class of enzymes of which DNA gyrase belongs). In replication, the role of the gyrase is somewhat different. It introduces a nick in super coiled DNA; the reverse of the reaction that produces super coiling gives rise to a **swivel point** in the DNA at the site of the nick. The gyrase opens and reseals the swivel point in advance of the replication fork (see Fig. 6.23)

The newly formed DNA automatically assumes the super coiled form since it does not have the nick at the swivel point; A helix destabilizing protein also called a helicase, promotes in winding by binding at the replication fork. This protein is also called the **rep protein.** Another protein called SSB (single strand binding) protein stabilizes the single stranded regions by binding tightly to these portions of the molecule. The presence of this DNA binding protein protects the single – stranded regions from hydrolysis.

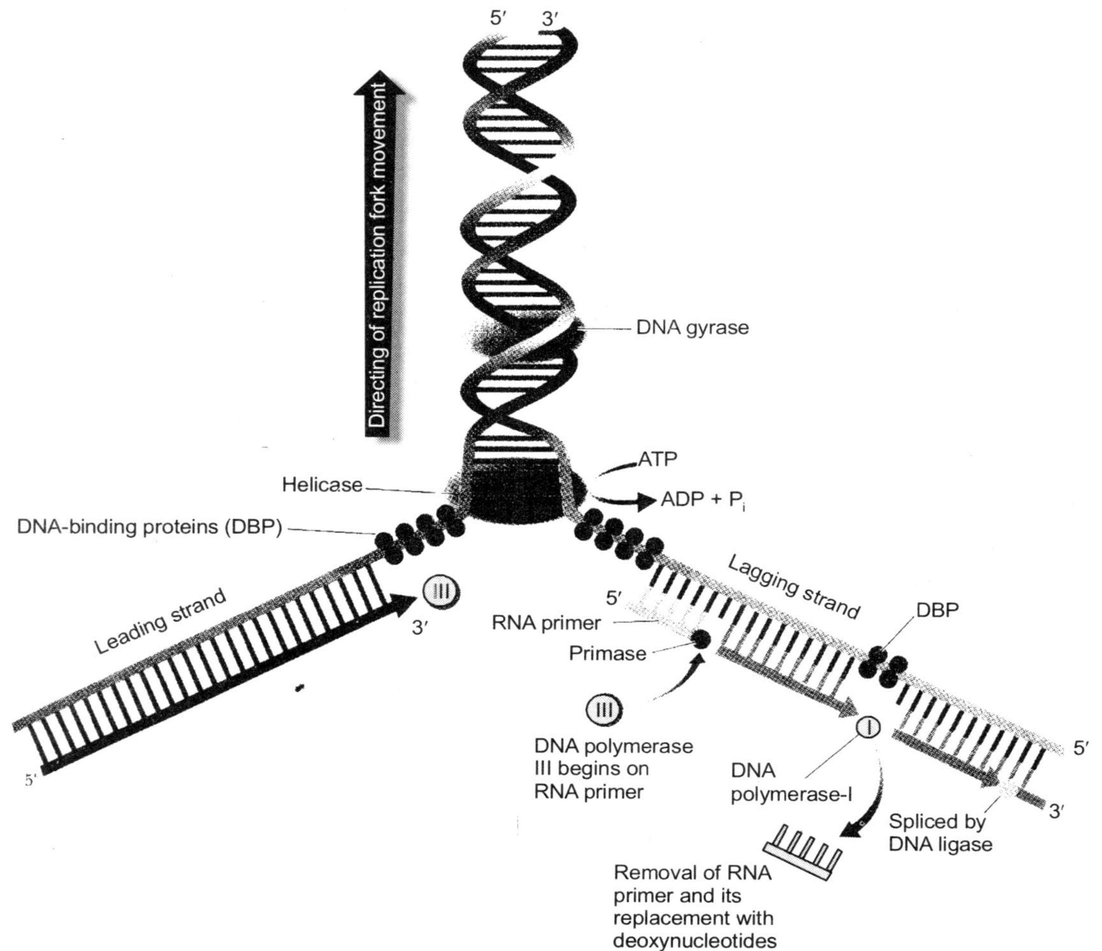

Fig. 6.23 : The Steps in DNA replication. The exact site of the swivel point for unwinding of parental DNA and the site of action of DNA gyrase are not known with certainty. The DNA binding protein is also called a single – strand binding protein

DIFFERENT ASPECTS OF DNA

The DNA Polymerase Reaction

A major challenge for the cell in the DNA replication is how to achieve 5' → 3' polymerization in the opposite direction from the template strand, which is itself exposed from the 5' → 3' direction. (There is no problem with the other strand, which is exposed by unwinding from 3' to 5' end.

The problem is solved by different modes of polymerization for the two growing strands. One newly formed strand (the leading strand) is formed continuously from its 3' end at the replication fork on the exposed 3' to 5' template strand. The other strand (the lagging strand) is formed discontinuously in small fragments (typically 100 to 2000 nucleotides long) sometimes called **Okazaki fragments** (See Fig. 6.24). The fragments of the lagging strand are then linked enzymatically by an enzyme called DNA ligase).

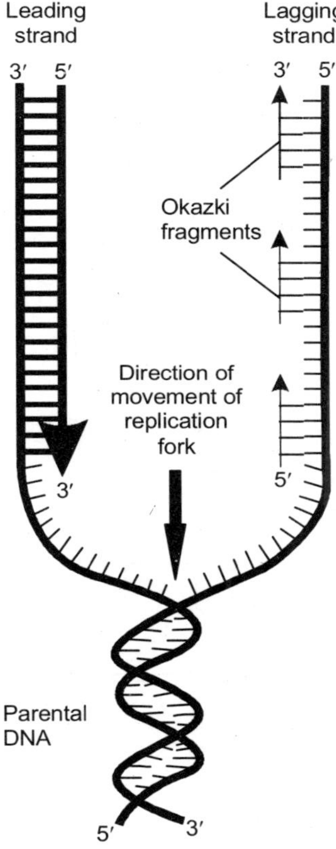

Fig. 6.24 : The Discontinuous and continuous synthesis of DNA strands at a replication fork. The leading strand of the DNA is synthesized continuously in the 5' ↑ 3' direction. In the lagging strand, short fragments (Okazaki fragments) are synthesized, also 5' ↑ 3' direction

The fragments of the lagging strand are then linked enzymatically by an enzyme called DNA ligase).

The first DNA polymerase was located in the *E. coli*, A universal feature of DNA replication is that the nascent chain grows from the 5' to the 3' end there is an 5'-phosphate on the sugar at one end and free 3'-hydroxyl on the sugar at the other end DNA, polymerase catalyses the successive addition of each new nucleotide to the growing chain. The 3'-hydroxyl group at the end of the growing chains a nucleotide. It attacks the phosphorous adjacent to the sugar in the nucleotide to be added to the growing chain, leading to the elimination of the pyrophosphate and the formation of anew phosphodieaster bond.

We discussed nucleophilic attack by a hydroxyl group at length in the case of serine protease; here we see another instance of this kind of mechanism.

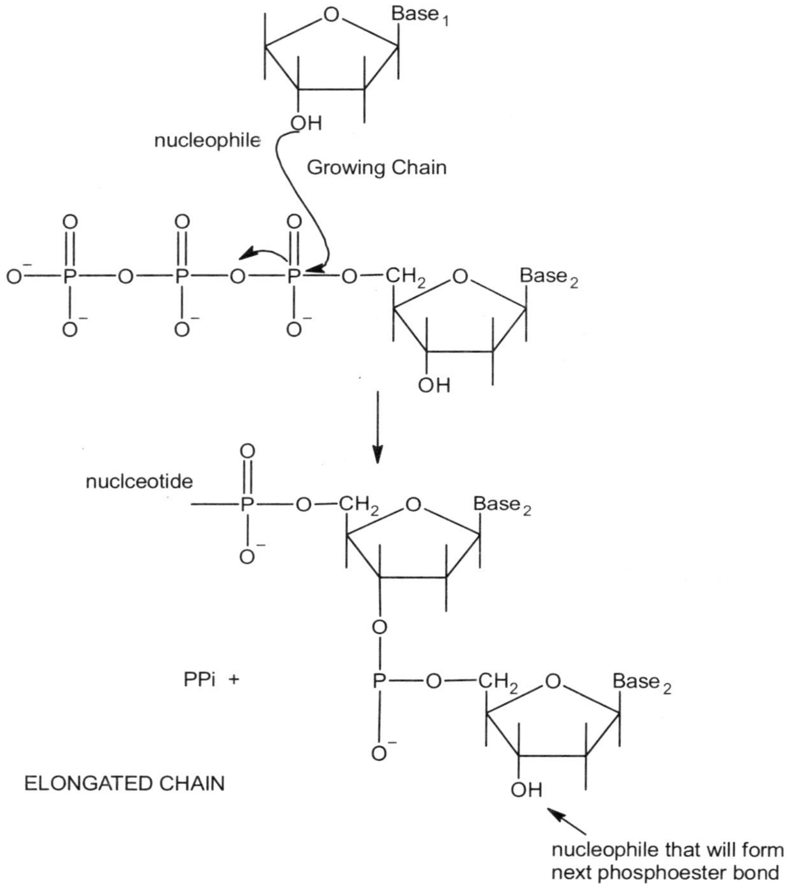

Base 1

OH

nucleophile

Growing Chain

O — P — O — P — O — P — O — CH_2 O Base 2

OH

nuclceotide

— P — O — CH_2 O Base 2

PPi +

O

P — O — CH_2 O Base 2

ELONGATED CHAIN

OH

nucleophile that will form next phosphoester bond

Fig. 6.25

Functions	Polymerase I	Polymerase	Polymerase III
Polymerization 5' → 3'	+	+	+
Exonuclease 3' →5'	+	+	+
Exonuclease 5' →3'	+	+	+
Molecular Weight	109,000	120,000	140,000
Molecules/cells	400	30	10-20
Number of nucleotides polymerized/ min at 37° by one enzyme	600		9000

There are actually three polymerases in *E coli*, some of their properties are listed in the table above, DNA polymerase I(Pol I) was discovered first with the subsequent discoveries of polymers II (Pol II) and III (Pol III), Polymerase I and II consist of single polypeptide chain, but polymerase that is s multi submit protein (Fig. 6.26). All these enzymes require the presence of **primer**, a short strand of RNA to which the growing polymerized chain is covalently attached in the early stages of replication.

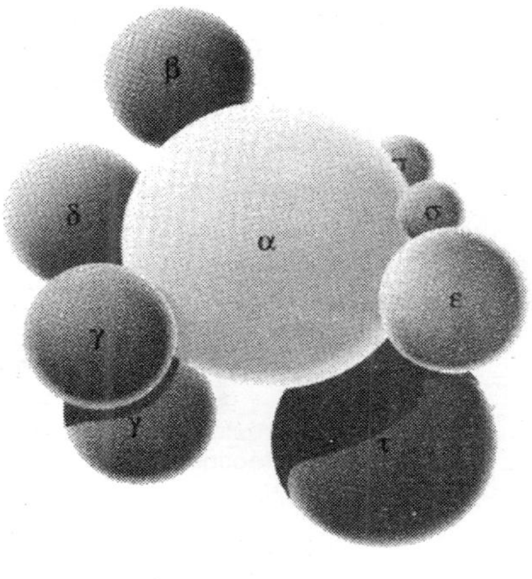

The DNA polymerase reaction requires all four deoxyribonulceoside triphosphate dTTP, dATP, dGTP and dCTP Mg^{2+} and DNA itself are also necessary. Because of the requirement of an RNA primer, all four ribocucleoside triphosphate ATP, UTP FTP and CTP are needed as well; they are incorporated into the primer. The primer (RNA) is hydrogen bonded

$$\text{dTTP, dATP, dGTP, dCTP} \xrightarrow[\text{DNA Apolymerase}]{\text{DNA dependent}} \text{DNA (polymer)} + \text{PPi}$$

is the template the primer provides a stable framework which the nascent chain can start to grow. The newly synthesized DNA strands begin to grow by forming a solvent linkage to the free 3' hydroxyl group of the primer. It is now known that DNA polymerase I has a specialized function in replication, that of repair and patching of DNA and that DNA polymerase III is the enzyme primarily responsible for the polymerization of the newly formed DNA strand. The major function of DNA polymerase II is not known. The exonuclease activity listed in the table above is the part of the proof reading and repair functions of DNA polymerase; it is a process by which incorrect nucleotides are removed from the polynucleotide so that the correct nucleotides can be incorporated. The 3' → 5' exonuclease activity, which all these polymerases process, is a part of the **proof reading function** where the incorrect nucleotides are removed in the course of replication and replaced by the correct ones. Proof reading is done by one nucleotide at a time.

Fig. 6.26 : The subunit structure of DNA polymerase III of *E coli*. The a subunit is the polymerase itself (the protein of molecular weight 140,000 listed in the table) the whole complex, called the DNA polymerase holoenzyme, has a molecular weight of 550,000. The β subunit is required for recognition of the parental DNA. Once the complex is bound to the DNA to be copied, the β subunit dissociates, leaving an active polymerase of molecular weight of 400,000.

DNA REPAIR MECHANISM

On the average, DNA replication takes place only once each generation in each cell, unlike other process such as DNA and proteins synthesis, which occur many times. It is essential that the fidelity of the replication process be as high as possible to prevent mutations which are errors in replication. Most mutations are harmful even fatal to organisms. Nature has devised several ways to ensure that the base sequence of DNA is copied faithfully. Errors in replication occur spontaneously only once in every 10^9 to 10^{10} base pairs.

Proof reading removes incorrect nucleotides immediately after they are added to the growing DNA chain once in every 10^4 to 10^5 base pairs (See Fig. 6.27).

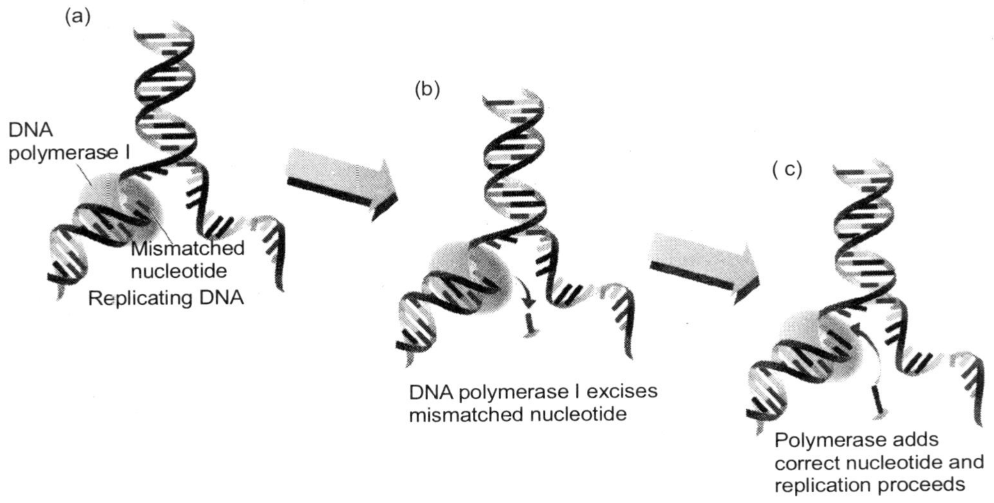

Fig. 6.27 : The Proof-reading in DNA replication (*a*) When an incorrect nucleotide has been incorporated into a growing polynucleotide chain, (*b*) DNA polymerase I uses its exo-nuclease activity to remove the mismatched nucleotide. (*c*) The polymerase then adds the correct nucleotide, and replication proceeds

DNA polymerase I uses its 3' exonuclease activity to remove the incorrect nucleotide (Fig. 6.27b). Replication resumes when the correct nucleotides are added, also by DNA polymerase I (Fig. 6.27c). The specificity of hydrogen bonded base pairing accounts for one error in every 10^4 to 10^5 base pairs, the proof reading function of DNA polymerase improves the fidelity of replication to one error in every 10^9 to 10^{10} base pairs.

During replication a cut-and-patch process catalyzed by polymerase I take place. The cutting is the removal of the RNA primer by the 5' exonuclease function of the polymerase, and the required deoxynucleotides, done by the polymerase function of the same enzyme. Existing DNA can also be repaired by polymerase I, using the cut-and-patch method, if one or more bases have been damaged by an external agent. In addition to experiencing those spontaneous mutations cause by misreading of the genetic code, the organisms are frequently exposed to mutagen agents that produce mutations. Common mutagens include ultra-violet light, ionizing radiation (radioactivity) and various chemical agents, all of which lead to changes in DNA over and above those produced by spontaneous mutation. In the repair of existing DNA the process of cutting and patching is called excision repair; the damaged portion of the DNA is removed, and the new, correct portion is substituted.

Deflects in DNA repair mechanism can have drastic consequences. One of the remarkable examples is the disease **xeroderma pigmentosum**; affected individuals develop innumerous skin cancers are an early age because they do not have the repair system to correct damage caused by UV probability the missing enzyme. The carcinogenic lesions eventually spread throughout the body causing death.

DNA is sometimes required to be modified after replication; the most important among them is methylation. Specific adenine and thymine residues are affected in eucaryotes. The products include N^6-methyl adenine and 5-methyl cytosine. S-Adenosylmethiane is the methyl group donor. The residues modified are frequently found in sequences that would otherwise be attached by restriction endonucleases. These enzymes, which occur mainly in bacteria, produce breaks in specific places in both strands of DNA.

The methylation protects the DNA of the bacterium from the restriction endonucleases (see Fig. 6.28) but DNA of an infected bacteriophage which does not have the methyl groups and is cleaved. The cleavage of the bacteriophage DNA restricts the growth of the bacteriophage in the infected cell giving rise to the name *restriction nuclease.*

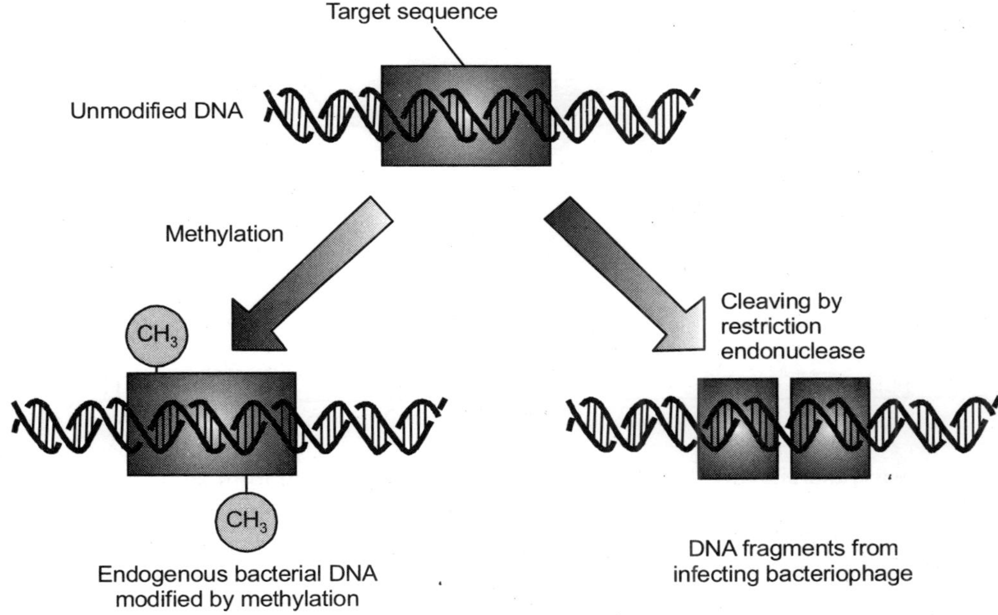

Fig. 6.28 : Showing the Methylation of Endogenous DNA Protects from its own Restriction Nuclease

RNA (Ribonucleic Acid)

Ribonucleic acid (RNA) is a biologically important type of molecule that consists of a long chain of nucleotide units. Each nucleotide consists of a nitrogenous base, a ribose sugar, and a phosphate. RNA is very similar to DNA, but differs in a few important structural details: in the cell, RNA is usually single-stranded, while DNA is usually double-stranded; RNA nucleotides contain ribose while DNA contains deoxyribose (a type of ribose that lacks one oxygen atom); and RNA has the base uracil rather than thymine that is present in DNA.

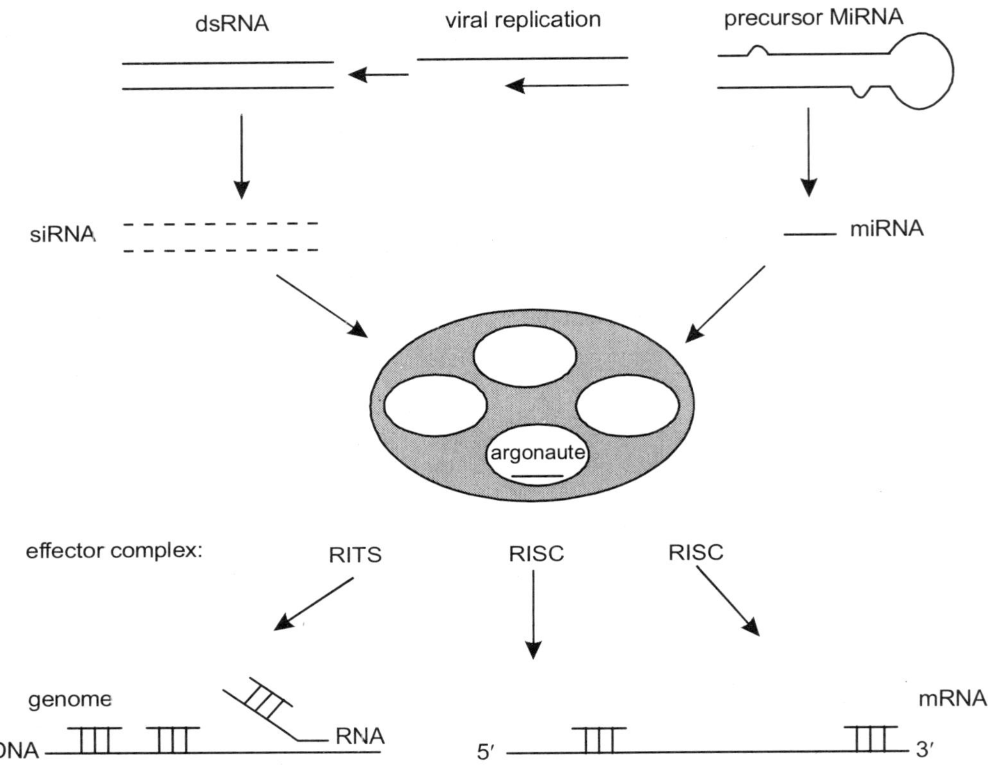

Fig. 6.29 : Showing the enzyme dicer trims double stranded RNA, to form small interfering RNA or microRNA. These processed RNAs are incorporated into the RNA-induced silencing complex (RISC), which targets messenger RNA to prevent translation.

RNA is transcribed from DNA by enzymes called RNA polymerases and is generally further processed by other enzymes. RNA is central to the synthesis of proteins. Here, a type of RNA called messenger RNA carries information from DNA to structures called ribosomes. These ribosomes are made from proteins and ribosomal RNAs, which come together to form a molecular machine that can read messenger RNAs and translate the information they carry into proteins. There are many RNAs with other roles – in particular regulating which genes are expressed, but also as the genomes of most viruses.

TYPES OF RNA

Messenger RNA

Messenger RNA (mRNA) is synthesized from a gene segment of DNA which ultimately contains the information on the primary sequence of amino acids in a protein to be synthesized. The genetic code as translated is for m-RNA not DNA. The **messenger RNA carries the code** into the cytoplasm where protein synthesis occurs.

Genetic Code

Each gene (or distinct segment) on DNA contains instructions for making one specific protein with order of amino acids coded by the precise sequence of heterocyclic amines on the nucleotides. Since proteins have a variety of functions including those of enzymes mistakes in the primary sequence of amino acids in proteins may have lethal effects.

How can a polymeric nucleotide with only four different heterocyclic amines specify the sequence of 20 or more different amino acids? If each nucleotide coded for a single amino acid, then obviously only 4 of the 20 amino acids could be accommodated. If the nucleotides were used in groups of two, there are 16 different combinations possible which is still inadequate.

It has been determined that the genetic code is actually based upon **triplets** of nucleotides which provide 64 different codes using the 4 nucleotides. During the 1960's, a tremendous effort was devoted to proving that the code was read as triplets, and also to solving the genetic code. The genetic code was originally translated for the bacteria *E. coli*, but its universality has since been established. The genetic code is "read" from a type of RNA called **messenger RNA (mRNA)**. **Each nucleotide triplet, called a codon**, can be "read" and translated into an amino acid to be incorporated into a protein being synthesized. Several distinctive features of the genetic code are clearly evident. First, all of the 64 codons or triplets have a known function, with 61 coding for amino acids and the other 3 serving as a stop or termination signal for protein synthesis. Secondly, the code is degenerate, meaning that there are usually several codons for each amino acid. Only methionine and tryptophan have a single codon. More specifics on the importance of the degeneracy of the genetic code will be discussed in a later section.

Ribosomal RNA

In the cytoplasm, ribsomal RNA (rRNA) and protein combine to form a nucleoprotein called a ribosome. The ribosome serves as the site and carries the enzymes necessary for protein synthesis. In the graphic on the left, the ribosome is shown as made from two sub units, 50S and 30 S. There are about equal parts rRNA and protein. The far left graphic shows the complete ribosome with three tRNA attached.

The ribosome attaches itself to m-RNA and provides the stabilizing structure to hold all substances in position as the protein is synthesized. Several ribosomes may be attached to a single RNA at any time. In upper right corner is the 30S sub unit with mRNA and tRNA attached.

Note: The coordinates used in this display have only the alpha carbons of the proteins (CA) and the DNA backbone atoms.

Transfer RNA

The existence of tRNA was first hypothesized by Francis Crick, based on the assumption that there must exist an adapter molecule capable of mediating the translation of the RNA alphabet into the protein alphabet. Significant research on structure was conducted in the early 1960s by Alex Rich and Don Caspar, two researchers in Boston, the Jacques Fresco group in Princeton University and a United Kingdom group at King's College London. A later publication reported the primary structure in 1965 by Robert W. Holley. The secondary and tertiary structures were derived from X-ray crystallography studies reported independently in 1974 by American and British research groups headed, respectively, by Alexander Rich and Aaron Klug.

Transfer RNA (tRNA) contains about 75 nucleotides, three of which are called anticodons, and one amino acid. The tRNA reads the code and carries the amino acid to be incorporated into the developing protein.

There are at least 20 different tRNA's - one for each amino acid. The basic structure of a tRNA is shown in the left graphic. Part of the tRNA doubles back upon itself to form several double helical sections. On one end, the amino acid, phenylalanine, is attached. On the opposite end, a specific base triplet, called the anticodon, is used to actually "read" the codons on the mRNA.

The tRNA "reads" the mRNA codon by using its own anticodon. The actual "reading" is done by matching the base pairs through hydrogen bonding following the base pairing principle. Each codon is "read" by various tRNA's until the appropriate match of the anticodon with the codon occurs.

In this example, the tRNA anticodon (AAG) reads the codon (UUC) on the mRNA. The UUC codon codes for phenylalanine which is attached to the tRNA. Remember that the codons read from the mRNA make up the genetic code as read by humans.

tRNA has primary structuresecondary structure (usually visualized as the *cloverleaf structure*), and tertiary structure (all tRNAs have a similar L-shaped 3D structure that allows them to fit into the P and A sites of the ribosome).

1. The 5'-terminal phosphate group.
2. The acceptor stem is a 7-bp stem made by the base pairing of the 5'-terminal nucleotide with the 3'-terminal nucleotide (which contains the CCA 3'-terminal group used to attach the amino acid). The acceptor stem may contain non-Watson-Crick base pairs.
3. The CCA tail is a CCA sequence at the 3' end of the tRNA molecule. This sequence is important for the recognition of tRNA by enzymes critical in translation. In prokaryotes, the CCA sequence is transcribed. In eukaryotes, the CCA sequence is added during processing and therefore does not appear in the tRNA gene.
4. The D arm is a 4 bp stem ending in a loop that often contains dihydrouridine.
5. The anticodon arm is a 5-bp stem whose loop contains the anticodon.
6. The T arm is a 5 bp stem containing the sequence TΨC where Ψ is a pseudouridine.
7. Bases that have been modified, especially by methylation, occur in several positions outside the anticodon. The first anticodon base is sometimes modified to inosine (derived from adenine) or pseudouridine (derived from uracil).

An **anticodon** is a unit made up of three nucleotides that correspond to the three bases of the codon on the mRNA. Each tRNA contains a specific anticodon triplet sequence that can base-pair to one or more codons for an amino acid. For example, one codon for lysine is AAA; the anticodon of a lysine tRNA might be UUU. Some anticodons can pair with more than one codon due to a phenomenon known as wobble base pairing. Frequently, the first nucleotide of the anticodon is one of two not found on mRNA: inosine and pseudouridine, which can hydrogen bond to more than one base in the corresponding codon position. In the genetic code, it is common for a single amino acid to be specified by all four third-position possibilities; for example, the amino acid glycine is coded for by the codon sequences GGU, GGC, GGA, and GGG.

To provide a one-to-one correspondence between tRNA molecules and codons that specify amino acids, 61 tRNA molecules would be required per cell. However, many cells contain fewer than 61 types of tRNAs because the wobble base is capable of binding to several, though not necessarily all, of the codons that specify a particular amino acid.

Aminoacylation is the process of adding an aminoacyl group to a compound. It produces tRNA molecules with their CCA 3' ends covalently linked to an amino acid.

Each tRNA is aminoacylated (or *charged*) with a specific amino acid by an aminoacyl tRNA synthetase. There is normally a single aminoacyl tRNA synthetase for each amino acid, despite the fact that there can be more than one tRNA, and more than one anticodon, for an amino acid. Recognition of the appropriate tRNA by the synthetases is not mediated solely by the anticodon, and the acceptor stem often plays a prominent role.

Reaction:

1. amino acid + ATP → aminoacyl-AMP + PPi
2. aminoacyl-AMP + tRNA → aminoacyl-tRNA + AMP

The ribosome has three binding sites for tRNA molecules: the A, P, and E sites. During translation the A site binds an incoming aminoacyl-tRNA as directed by the codon currently occupying this site. This codon specifies the next amino acid to be added to the growing peptide chain. The A site only works after the first aminoacyl-tRNA has attached to the P site. The P-site codon is occupied by peptidyl-tRNA that is a tRNA with multiple amino acids attached as a long chain. The P site is actually the first to bind to aminoacyl tRNA. This tRNA in the P site carries the chain of amino acids that has already been synthesized. The E site is occupied by the empty tRNA as it is about to exit the ribosome.

Organisms vary in the number of tRNA genes in their genome. The nematode worm *C. elegans*, a commonly used model organism in genetics studies, has 29,647 genes in its nuclear genome, of which 620 code for tRNA. The budding yeast *Saccharomyces cerevisiae* has 275 tRNA genes in its genome. In the human genome, which according to current estimates has about 27,161 genes in total, there are about 4,421 non-coding RNA genes, which include tRNA genes. There are 22 mitochondrial tRNA genes; 497 nuclear genes encoding cytoplasmic tRNA molecules and there are 324 tRNA-derived putative pseudogenes.

Cytoplasmic tRNA genes can be grouped into 49 families according to their anticodon features. These genes are found on all chromosomes, except 22 and Y chromosome. High clustering on 6p is observed (140 tRNA genes), as well on 1 chromosome.

tRNA molecules are transcribed (in eukaryotic cells) by RNA polymerase III, unlike messenger RNA which is transcribed by RNA polymerase II. pre-tRNAs contain introns; in bacteria these self-splice, whereas in eukaryotes and archaea they are removed by tRNA splicing endonuclease.

Precursor mRNA (pre-mRNA), also termed **heterogeneous nuclear RNA (hnRNA)**, is an immature single strand of messenger ribonucleic acid (mRNA). pre-mRNA is synthesized from a DNA template in the cell nucleus by transcription.

Once pre-mRNA has been completely processed, it is termed "mature messenger RNA", "mature mRNA", or simply "mRNA".

TRANSCRIPTION PROCESS RNA

The detail of RNA transcription differs somewhat in procaryotes, and eucaryotes. Most of the research of the subject has been done in procaryotes, especially E. coli but some general features are found in all organisms except RNA viruses; the most extensively studied RNA polymerase is that isolated formed E. Coli. The molecular weight of this enzyme is about 500,000 and it has a multisubunit structure. Four different types of subunits designatedα, β, β' and σ have been identified. The actual composition of the enzyme is $\alpha_2\beta\beta'\sigma$. The σ subunit is rather loosely bound to the rest of the enzyme (the $\alpha_2\beta\beta'$ portion) which is called the core enzyme. The whole enzyme consists of all the subunits including the σ subunit. The essential role of the σ subunit is the recognition of the promoter locus* the loosely bound σ subunit is released after transcription.

The promoter region to which RNA polymerase binds is closer to the 3' end of the DNA than is the actual gene for the RNA to be synthesized. (The RNA is formed) from the 5' to the 3' end, so the polymerase moves along the DNA from the 3' to the 5' end). The binding site for the polymerase is said to lie upstream of the start of transcription. The base sequence of promoter regions has been determined for a number of procaryote genes, and a striking feature is that they contain many sequences in common (consensus sequences) Promoter regions are A-T-rich, which two hydrogen bonds per base pair and thus more easily unwound than G-C rich regions, with three hydrogen bonds per base pair. In procaryotes these consensus** regions recently occur.

35 base pairs (bp) and 10 bp upstream from the start of transcription, these two regions are called the -35 box and the -10box (also called) ***the pribnow box***. The sequences for a procaryotic promoter is

−35			−10
TTGACA	17	bp	TATAAT

In the eucaryotes the similar sequences is called TATA box lies about 25 bp upstream from the start of the transcription.

*A DNA sequence that signals the start of the RNA transcription.

**Strongly grounded or allied

Termination of RNA transcription also involves specific sequences downstream of the actual size gene for the RNA to be transcribed and in procaryotes a protein denoted as ρ (rho) is involved in the termination process. This protein, which is tetramer with a molecular weight of approximate 200,000 binds of DNA, RNA and RNA polymerase the ρ-factor triggers the release of the completed RNA from the DNA – RNA polymerase complex (see Fig. 6.30). Non-ρ-protein has been detected in the eucaryotic system.

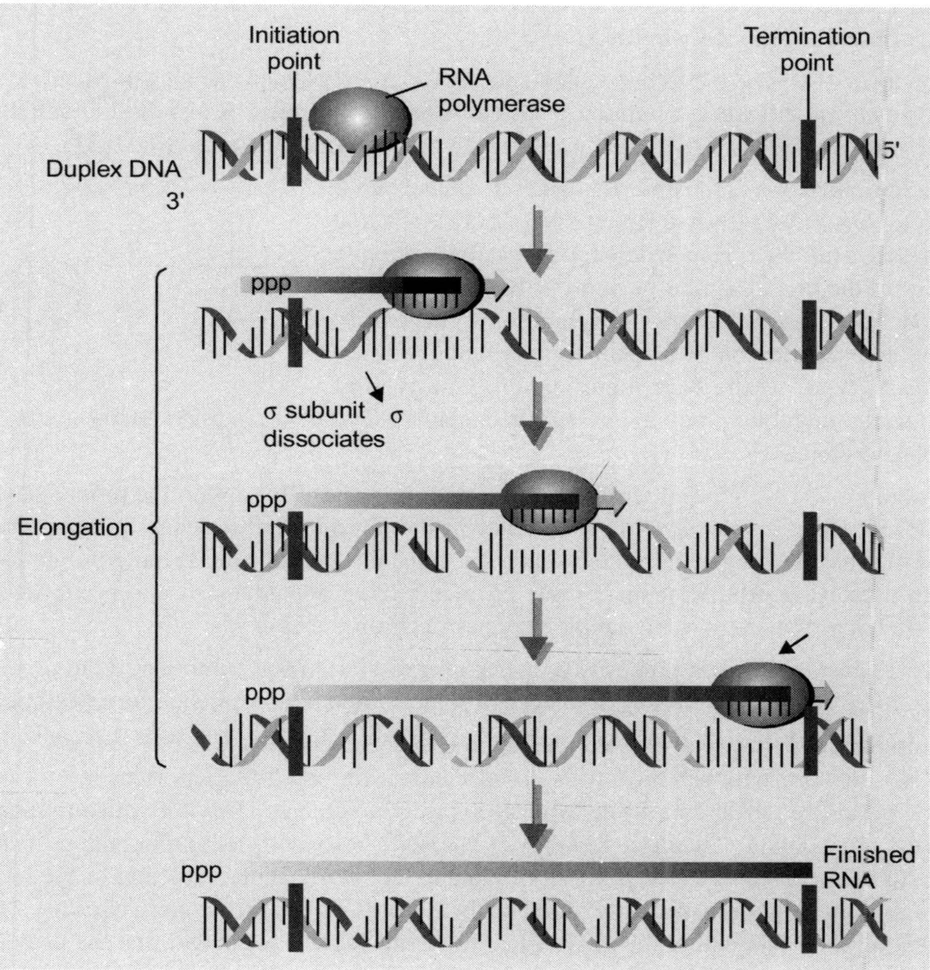

Fig. 6.30 : The Transcription of RNA on a DNA template. In E. Coli the transcription begins when the RNA polymerize holoenzyme ($\alpha_2\beta\beta'\sigma$) binds to DNA. The σ subunit is required for binding but dissociates form the complex after initiation of transcription. The synthesis of RNA proceeds from 5' to 3' end. The termination of chain growth needs subunit.

POST TRANSCRIPTIONAL MODIFICATION OF RNA

The three kinds of RNA the transfer RNA ribosomal RNA and messanger RNA are all basically modified enzymatically after transcription to give rise the functional form of the RNA. The type of processing is totally different between procaryotes and eucaryotes, specially in the case of mRNA. The initial size of the RNA transcripts is greater than the final size because of the leader sequence at the 5' end and trailer sequence at the 3' end. The leader and the tailor sequences must be removed and the other forms of trimming are possible as well. Terminal sequences can be added after transcription and base modification is frequently observed especially in RNA.

The precursor of several tRNA molecules is frequently transcribed in one long polypeptide sequence. All three types of modification trimming addition of terminal sequences also base modification take place in the transformation of the initial transcript to the mature tRNAs (see Fig. 6.31).

The enzyme that is responsible for generating that the 5 ends of the E Coli tRNA; RNAse P, consists of both RNA and protein. The RNA middle is responsible for the catalytic activity. This was one of the first examples of catalytic RNA. Some base modifications take place before trimming and some occur after; Methylation and substitution of sulphur for oxygen are two of the more usual types of the base modification. One type of methylated nucleotide found only in eucaryotes contains a 2' – O methylribosyl group.

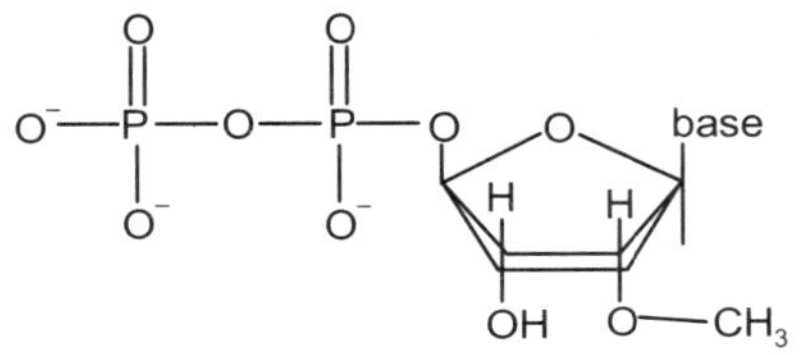

2′–O Methylribosyl link

The trimming and addition of terminal nucleotides produce tRNAs with the proper size and base sequence. Every tRNA contains a CCA sequence at the 3' end. The presence of this portion of the molecule is of great importance in protein synthesis, since the 3' end is the accepter for amino acids to be added to a growing protein chain. Trimming of the large precursors of eucaryotic tRNAs takes place in the nucleus, but most methylating enzymes occur in the cytosol.

The processing of rRNA is primarily a matter of methylation and trimming to the proper size. In procaryotes, there are three rRNAs in an intact ribosome, which has sedimentation co-efficient of 70S.

In the similar subunit, which has sedimentation co-efficient of 16S. The 50S subunit contains two kinds of RNA subunit, with sedimentation co-efficient of 5S and 23S. The ribosome of eucaryotes have sedimentations co-efficient of 80S with 40S and 60S subunits, The 40S subunit consist a 18S RNA, and the 60S subunit contains a 5S RNA, a 5.8 RNA base modification in both procaryotic and eucaryotic rRNA are accomplished primarily by methylation and sometimes that of the next nearest neighbor is as well. The polyadenylate "tail" at the 3' ends of the messenger (typically 100 to 200 nucleotides long) is added before the mRNA leaves the nucleus. It is thought that the presence of the tail served to protect the mRNA from nucleases and phophatases, which would degrade it, According to this point of view the adenylate residues would be cleaved off before the portion of the molecule that contains the actual message is attached. The presence of the 5 cap also protects the mRNA from exonuclease degradation. The genes of procaryotes are continuous in every base pair; in a continuous procaryotic gene it is reflected in the base sequence of mRNA. The genes of eucaryotes are not

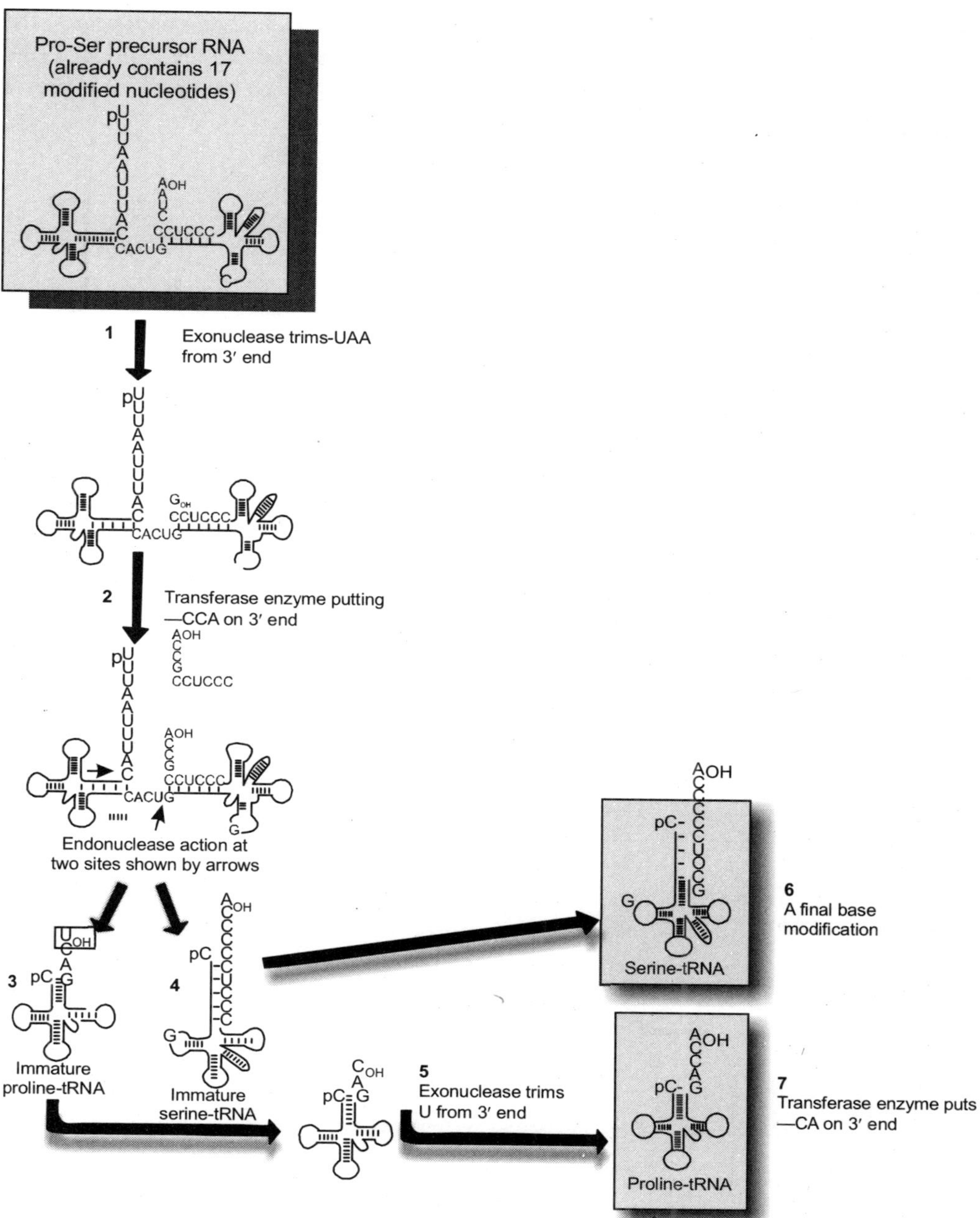

Fig. 6.31 : The Post-transcriptional Modification of tRNA precursor. Dots represent the hydrogen-bonded base pairs symbols G_{OH} C_{OH} A_{OH} and U_{OH} refer to a free 3' end without a phosphate group G_m^2 is methylated Guanine

necessarily continuous rather frequently contain intervening sequences that do not appear in the final base sequence of the mRNA for that gene product. The DNA sequences that are expressed (the ones actually retained in the final product) are called ***exones***[1] and the ones non exposed or expressed are called **introns.**

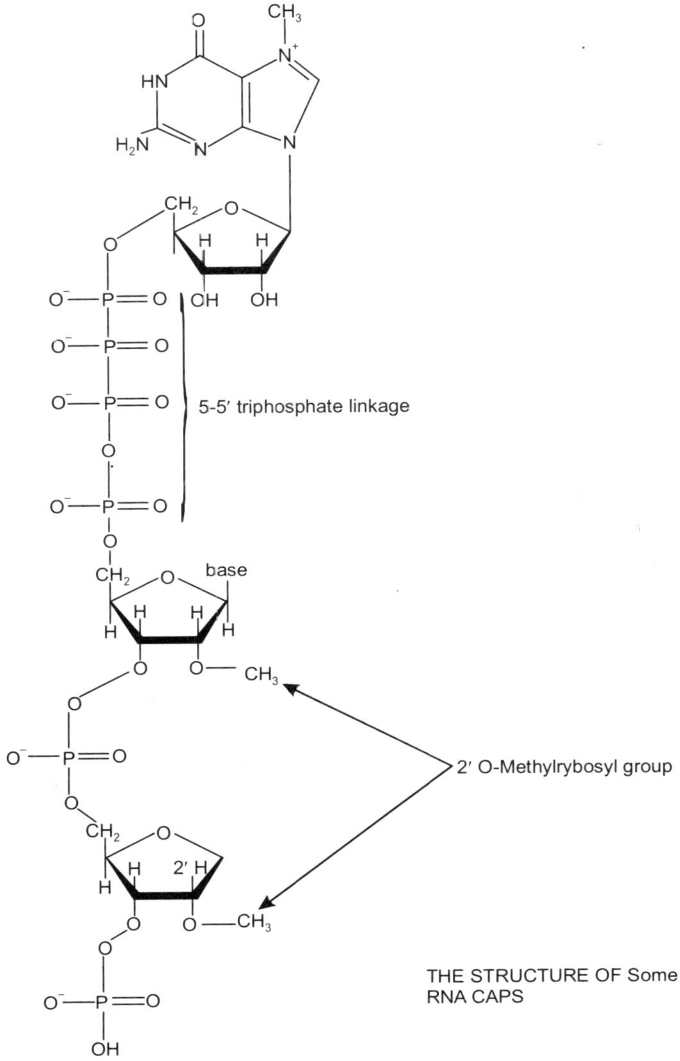

5-5′ triphosphate linkage

2′ O-Methylrybosyl group

THE STRUCTURE OF Some m RNA CAPS

[1]An **exon** is a nucleic acid sequence that is represented in the mature form of an RNAmolecule after either portions of a precursor RNA (introns) have been removed by cis-splicing or by two or more precursor RNA molecules have been ligated by trans-splicing. The mature RNA molecule can be a messenger RNA or a functional form of a non-coding RNA such as rRNA or tRNA. Depending on the context, exon can refer to the sequence in the DNA or its RNA transcript.

...(Contd.)

The β-globin gene of mouse which codes for the β - chain of hemoglobin well known example. The actual gene is split into three part with two intervening sequence (see Fig. 6.32)

...(Contd.)

The term *exon* was coined by American biochemist Walter Gilbert in 1978:

The notion of the cistron..must be replaced by that of a transcription unit containing regions which will be lost from the mature messenger—which I suggest we call introns (for intragenic regions)—alternating with regions which will be expressed— exons.

This definition was originally made for protein-coding transcipts that are spliced before being translated. The term later came to include sequences removed from rRNA and tRNA, and it also was used later for RNA molecules originating from different parts of the genome that are then ligated by trans-splicing

Function

In many genes, each exon contains part of the open reading frame (ORF) that codes for a specific portion of the complete protein. However, the term *exon* is often misused to refer only to coding sequences for the final protein. This is incorrect, since many noncoding exons are known in human genes (Zhang 1998).

To the right is a diagram of an heterogeneous nuclear RNA (hnRNA), which is an unedited mRNA transcript, or pre-mRNAs. Exons can include both sequences that code for amino acids (red) and untranslated sequences (grey). Stretches of unused sequence called introns (blue) are removed, and the exons are joined together to form the final functional mRNA. The notation 5' and 3' refer to the direction of the DNA template in the chromosome and is used to distinguish between the two untranslated regions (grey). and the ones non exposed or expressed are called ***introns.***

[2] An **intron** is a DNA region within a gene that is not translated into protein. These non-coding sections are transcribed to precursor mRNA (pre-mRNA) and some other RNAs (such as long noncoding RNAs), and subsequently removed by a process called splicing during the processing to mature RNA. After intron splicing (ie. removal), the mRNA consists only of exon derived sequences, which are translated into a protein.The word *intron* is derived from the term *intragenic region* and also called *intervening sequence* (IVS) Introns are common in eukaryotic pre-mRNA, but in prokaryotes they are only found in tRNA and rRNA; introns have variable length and alternate with exons in intron-containing genes.The number and length of introns varies widely among species, and among genes within the same species. Some eukaryotes, e.g. sac fungi, have evolved genomes with few introns, while the genomes of many other eukaryote groups are rich in introns (several per gene).

Simple illustration of a pre-mRNA, with introns (top). After the introns have been removed via splicing, the mature mRNA sequence is ready for translation (bottom).

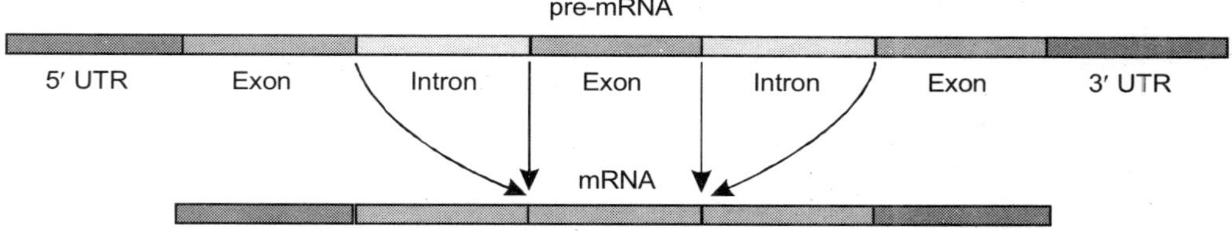

Alternative splicing of introns within a gene may introduce greater variability of protein sequences translated from a single gene. The control of mRNA splicing is performed by a wide variety of signaling molecules.

In eukaryotes the entire DNA sequence, both Introns and exones, is transcribed to produce a precursor of the mature mRNA. In the proceeding of mRNA, the non coding sequences, the excised

---

Introns may also contain "old code", or sections of a gene that were once translated into a protein, but have since become inactive. It was generally assumed that the sequence of any given intron is junk DNA with no biological function. More recently, however, this is being disputed. For example, a point mutation in intron 7 of the human gene *TPH1* is highly correlated to the development of the psychiatric disorder schizophrenia.Introns contain several short sequences that are important for efficient splicing, such as acceptor and donor sites at either end of the intron as well as a branch point site, which are required for proper splicing by the spliceosome. Some introns are known to enhance the expression of the gene that they are contained in by a process known as intron-mediated enhancement (IME).One of the most important roles of introns currently under investigation is the transcription of the introns to small regulatory RNA, such as a type of RNAs called miRNA (microRNA). These small single-stranded RNAs regulate the expression of genes. The discovery of introns led to the Nobel Prize in Physiology or Medicine in 1993 for Phillip Allen Sharp and Richard J. Roberts. The term *intron* was introduced by American biochemist Walter Gilbert:

Classification

Four classes of introns are known to exist:

- Nuclear introns
- Group I intron
- Group II intron
- Group III intron

Some introns, such as the Group I and Group II introns, after transcription possess ribozyme activity, enabling them to catalyze their own splicing out of a primary RNA transcript. These introns are thus self splicing introns and are relatively rare compared to spliceosomal introns. This self-splicing activity was discovered by Thomas Cech, who shared the 1989 Nobel Prize in Chemistry with Sidney Altman for the discovery of the catalytic properties of RNA.

Nuclear or spliceosomal introns are spliced by the spliceosome and a series of snRNAs (small nuclear RNAs). There are certain splice signals (or consensus sequences) which abet the splicing (or identification) of these introns by the spliceosome.Group II and III introns are similar and have a conserved secondary structure. A so-called lariat pathway is used in their splicing. They perform functions similar to the spliceosome and may be evolutionarily related to it. Group I introns are the only class of introns whose splicing requires a free guanine nucleoside. They possess a secondary structure different from that of group II and III introns. Many self-splicing introns code for maturases that help with the splicing process, generally only the splicing of the intron that encodes it. There are two competing theories that offer alternative scenarios for the origin and early evolution of spliceosomal introns. Other classes of introns such as self-splicing and tRNA introns are not subject to much debate, but see for the former. These are popularly referred to as the Introns-Early (IE) and the Introns-Late (IL) views.

The *IE model*, championed by Walter Gilbert, proposes that introns are extremely old and numerously present in the earliest ancestors of prokaryotes and eukaryotes (the progenote). In this model, introns were subsequently lost from prokaryotic organisms, allowing them to attain growth efficiency. A central prediction of this theory is that the early introns were mediators that facilitated the recombination of exons that represented the protein domains. This model cannot account for some observed positional variation of introns shared among related genes. The *IL model* proposes that introns were recently inserted into originally intron-less contiguous genes after the divergence of eukaryotes and prokaryotes. In this model, introns probably originated from transposable elements. This model is based on the observation that the spliceosomal introns are restricted to eukaryotes alone. However, there is considerable debate over the presence of introns in the early prokaryote-eukaryote ancestors and the subsequent intron loss-gain during eukaryotic evolution. The evolution of introns and of the intron-exon structure may be largely independent of the evolution of coding-sequences.

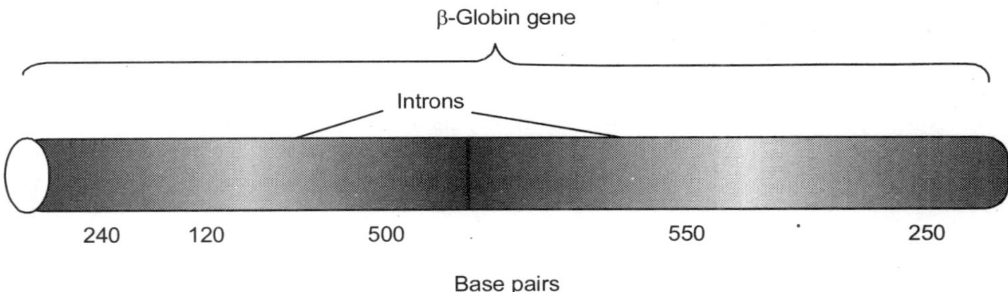

Fig. 6.32 : The Intervening Introns

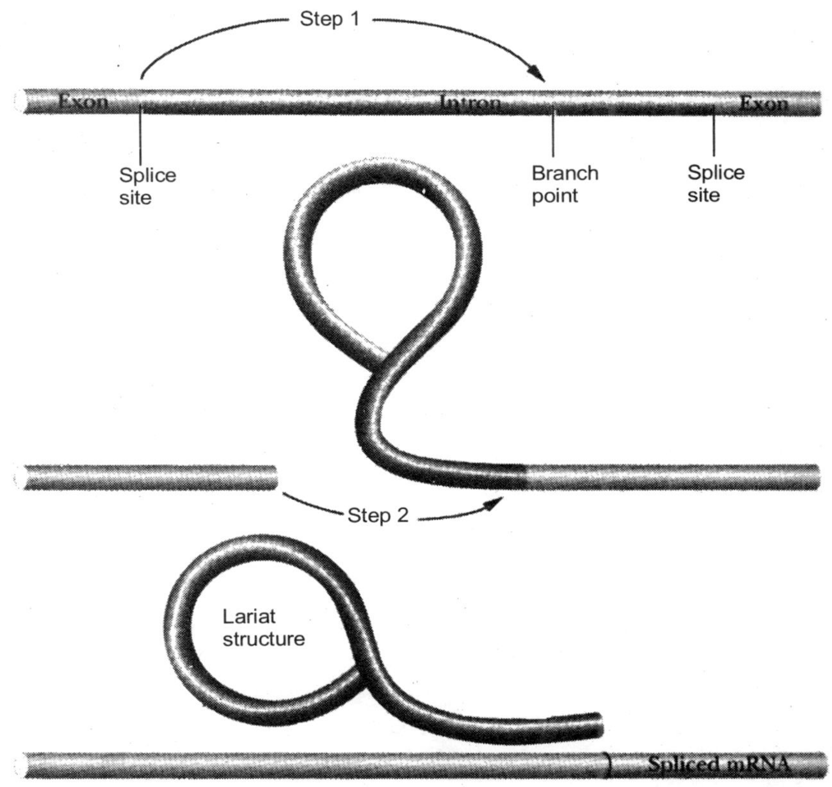

Fig. 6.33 : The formation of a lariat structure as an intermediate in the splicing of mRNA. In the first step of the splicing process, one splice site is cut and attached to a binding site near the second splice site, forming a lariat. In the course of splicing, mRNA is bound to small nuclear ribonucleoproteins (snRNPs) (not shown here), which hold the splice sites in position. In the second step, the intron lariat is excised, and the mRNA exons are spliced together. A similar lariat mechanism is involved in the self-splicing of RNA

and the coding sequences, the exons must be joined together. This process is done by the sequential action of the suitable nucleases and ligases. The number and size of the Introns vary in different genes. There is 1 intron in the gene for the muscle protein action 2 for both the α - and β - chains of hemoglobin, 3 for lysozome and so on up to as many as so introns in a single gene (the gene for one subunits for collagen).

The removal of intervening sequences takes place in the nucleus, where small nuclear robonulceoproteins, or snRNPs pronounced 'surunps' mediate the process. The snRNps as the name indicates, consist both the RNA and the proteins, such as the nucleases and ligases involved in the splicing process. The actual splicing involves a 50S to 60S particle called Spliceosome*.

It is now widely recognized that some RNAs can catalize their self splicing, the present process involving ribonucleoproteins mayhave involved from self – splicing o9f RNAs. An important similarity between the two process in that both proceed via a Lariat mechanism by which the splice sites are brought together (see fig 218). Not all the details of the process are clear, but it is established that base pairing between the mRNA to be spliced and the RNA portion of the snRNPs palyes a role in the recognition of the sequences to be spliced.

*A **spliceosome** is a complex of specialized RNA and protein subunits that removes introns from a transcribed pre-mRNA (hnRNA) segment. This process is generally referred to as splicing.

Composition

Each spliceosome is composed of five small nuclear RNA proteins, called snRNPs, (pronounced "snurnps") and a range of non-snRNP associated protein factors.

The snRNPs that make up the nuclear spliceosome are named U1, U2, U4, U5, and U6, and participate in several RNA-RNA and RNA-protein interactions. The RNA component of the snRNP is rich in uridine (the nucleoside analog of the uracil nucleotide).

The canonical assembly of the spliceosome occurs anew on each hnRNA. The hnRNA contains specific sequence elements that are recognized and utilized during spliceosome assembly. These include the 5' end splice, the branch point sequence, the polypyrimidine tract, and the 3' end splice site. The spliceosome catalyzes the removal of introns, and the ligation of the flanking exons.

Introns typically have a GU nucleotide sequence at the 5' end splice site, and an AG at the 3' end splice site. The 3' splice site can be further defined by a variable length of polypyrimidines, called the polypyrimidine tract (PPT), which serves the dual function of recruiting factors to the 3' splice site and possibly recruiting factors to the branch point sequence (BPS). The BPS contains the conserved Adenosine required for the first step of splicing.

A group of less abundant snRNPs, U11, U12, U4atac, and U6atac, together with U5, are subunits of the so-called minor spliceosome that splices a rare class of pre-mRNA introns, denoted U12-type. These snRNPs form the U12 spliceosome are located in the cytosol.

New evidence derived from the first crystal structure of a group II intron suggests that the spliceosome is actually a ribozyme, and that it uses a two–metal ion mechanism for catalysis.

Alternative Splicing

Alternative splicing (the re-combination of different exons) is a major source of genetic diversity in eukaryotes. Splice variants have been used to account for the relatively small number of genes in the human genome. For years the estimate

PROTEIN BIOSYNTHESIS IN EUCARYOTES AND PROCARYOTES

Over all protein synthesis or also called translation takes place in three stages the *initiation, elongation and termination*. During initiation the mRNA ribosome complex is formed and the first codon which is

---

widely varied with top estimates reaching 100,000 genes, but now, due to the Human Genome Project the figure is believed to be closer to 20,000 genes. One particular Drosophila gene (DSCAM) can be alternatively spliced into 38,000 different mRNA.

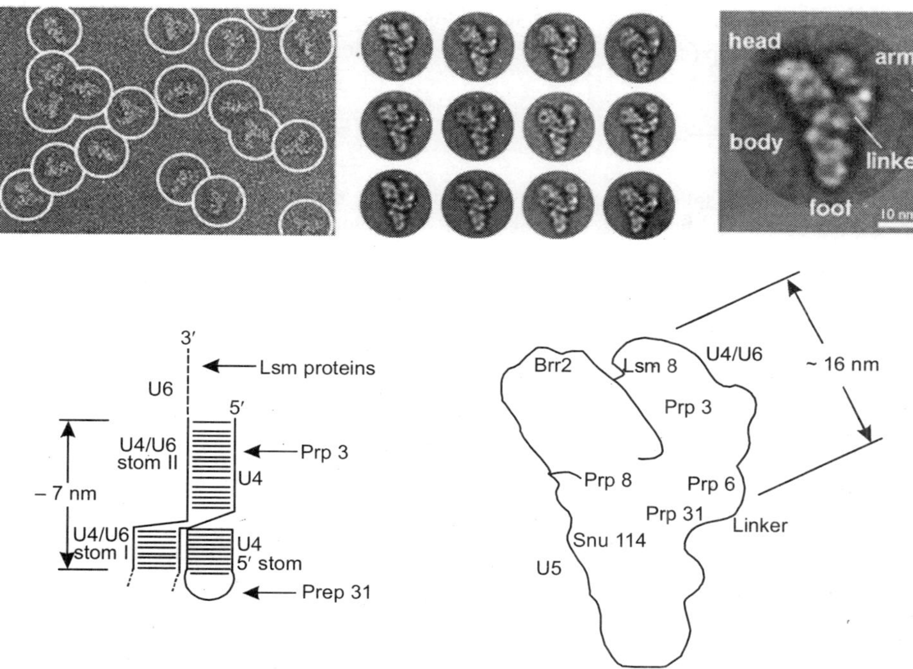

Fig. 6.34 : Above are confocal microscopy fields of negatively stained yeast (*Saccharomyces cerevisiae*) tri-snRNPs. Below left is a schematic illustration of the interaction of tri-snRNP proteins with the U4/U6 snRNA duplex. Below right is a cartoon model of the yeast tri-snRNP with shaded areas corresponding to U5 (gray), U4/U6 (orange) and the linker region (yellow)

RNA Splicing

In 1977, work by the Sharp and Roberts labs revealed that genes of higher organisms are "split" or present in several distinct segments along the DNA molecule. The coding regions of the gene are separated by non-coding DNA that is not involved in protein expression. The split gene structure was found when adenoviral mRNAs were hybridized to endonuclease cleavage fragments of single stranded viral DNA. It was observed that the mRNAs of the mRNA-DNA hybrids contained 5' and 3' tails of non-hydrogen bonded regions. When larger fragments of viral DNAs were used, forked structures of looped out DNA were observed when hybridized to the viral mRNAs. It was realized that the looped out regions, the introns, are excised from the precursor mRNAs in a process Sharp named "splicing". The split gene structure was subsequently found to be common to most eukaryotic genes. Phillip Sharp and Richard J. Roberts were awarded the 1993 Nobel Prize in Physiology or Medicine for their discovery of introns and the splicing process.

...(Contd.)

AUG binds the amino acyl tRNA or inhibitor tRNA. During this elongation phase the other codons are read sequentially and the polypeptide grows by addition of amino acids it C – terminal end. This process continues, until a termination codon which does not have a corresponding aminoacyl tRNA with which two base pair is reached, At this point protein synthesis ceases (termination phase) and the finished polypeptide is released from the ribosome, the usually at one time many ribosome molecule are translated simultaneously with mRNA, forming a structure called a polyribosome or polysome.

A tRNA molecule is about 74 – 95 nucleotides long making these some of the smallest molecule ion the cell. Each tRNA molecule has cover leaf secondary structure with the anticodon accessible at the end of the anticodon stem loop (see Fig. 6.35). Each tRNA molecule. During a synthesis of the aminoacyl tRNA.

Spliceosome Assembly

The model for formation of the spliceosome active site involves an ordered, stepwise assembly of discrete snRNP particles on the hnRNA substrate. The first recognition of hnRNAs involves U1 snRNP binding to the 5' end splice site of the hnRNA and other non-snRNP associated factors to form the commitment complex, or early (E) complex in mammals. The commitment complex is an ATP-independent complex that commits the hnRNA to the splicing pathway. U2 snRNP is recruited to the branch region through interactions with the E complex component U2AF (U2 snRNP auxiliary factor) and possibly U1 snRNP. In an ATP-dependent reaction, U2 snRNP becomes tightly associated with the branch point sequence (BPS) to form complex A. A duplex formed between U2 snRNA and the hnRNA branch region bulges out the branch adenosine specifying it as the nucleophile for the first transesterification. The presence of a pseudouridine residue in U2 snRNA, nearly opposite of the branch site, results in an altered conformation of the RNA-RNA duplex upon the U2 snRNP binding. Specifically, the altered structure of the duplex induced by the pseudouridine places the 2' OH of the bulged adenosine in a favorable position for the first step of splicing. The U4/U5/U6 tri-snRNP (see Figure 217A) is recruited to the assembling spliceosome to form complex B, and following several rearrangements, complex C (the spliceosome) is activated for catalysis. It is unclear how the triple snRNP is recruited to complex A, but this process may be mediated through protein-protein interactions and/or base pairing interactions between U2 snRNA and U6 snRNA. The U5 snRNP interacts with sequences at the 5' and 3' splice sites via the invariant loop of U5 snRNA and U5 protein components interact with the 3' splice site region. Upon recruitment of the triple snRNP, several RNA-RNA rearrangements precede the first catalytic step and further rearrangements occur in the catalytically active spliceosome. Several of the RNA-RNA interactions are mutually exclusive; however, it is not known what triggers these interactions, nor the order of these rearrangements. The first rearrangement is probably the displacement of U1 snRNP from the 5' splice site and formation of a U6 snRNA interaction. It is known that U1 snRNP is only weakly associated with fully formed spliceosomes, and U1 snRNP is inhibitory to the formation of a U6-5' splice site interaction on a model of substrate oligonucleotide containing a short 5' exon and 5' splice site. Binding of U2 snRNP to the branch point sequence (BPS) is one example of an RNA-RNA interaction displacing a protein-RNA interaction. Upon recruitment of U2 snRNP, the branch binding protein SF1 in the commitment complex is displaced since the binding site of U2 snRNA and SF1 are mutually exclusive events.Within the U2 snRNA, there are other mutually exclusive rearrangements that occur between competing conformations. For example, in the active form, stem loop IIa is favored; in the inactive form a mutually exclusive interaction between the loop and a downstream sequence predominates. It is unclear how U4 is displaced from U6 snRNAm, although RNA has been implicated in spliceosome assembly, and may function to unwind U4/U6 and promote the formation of a U2/U6 snRNA interaction. The interactions of U4/U6 stem loops I and II dissociate and the freed stem loop II region of U6 folds on itself to form an intramolecular stem loop and U4 is no longer required in further spliceosome assembly. The freed stem loop I region of U6 base pairs with U2 snRNA forming the U2/U6 helix I. However, the helix I structure is mutually exclusive with the 3' half of an internal 5' stem loop region of U2 snRNA.

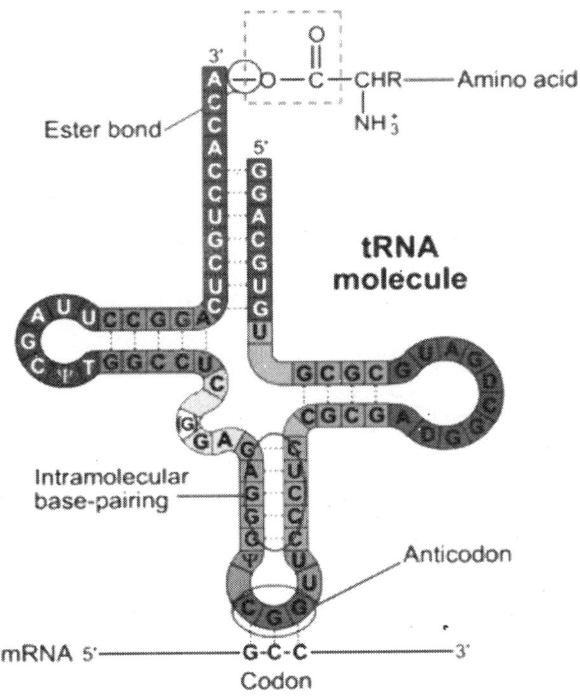

Fig. 6.35 : The Synthesis of Aminoacyl tRNA

The amino acid molecule is covalently bound to the A residue of the CCA sequence at the 3' end (see Fig. 6.35). Each tRNA molecule carries only a single amino acid, however, because of the redundancy of the genetic code, several codons may encode the same amino acids and so there will also exist in several types of tRNA with corresponding anticodons all bearing the same amino acid. The correct nomenclature is for example $tRNA^{Gly}$ for the tRNA that will accept amino acyl – tRNA is GlytRNA and is the amino acyl tRNA shown in Fig. 6.35.

Synthesis of any aminoacyl tRNA is crucially important for two reasons. First each amino acid must be covalently linked to a RNA molecule in order to take part in protein synthesis, which depends upon the 'adopter' function of tRNA to ensure that correct amino acids are incorporate. Second, the covalent bond that enables the amino acid to react with the end of the growing polypeptide chain to form a new peptide bond, for this reason the synthesis of aminoacyl tRNA is also referred to as amino acid activation. Amino acids that are not linked to tRNA cannot be added to the growing peptide.

The attachment of an amino acid to a tRNA is catalyzed by an enzyme called amino acyl tRNA synthetase. A separate amino acyl tRNA exists for every amino acid making 20 synthetase in total. The synthesis reaction occurs in two steps, the first step is the reaction of an amino acid and ATP to form an amino acid adenylate (also known as aminoacyl AMP)

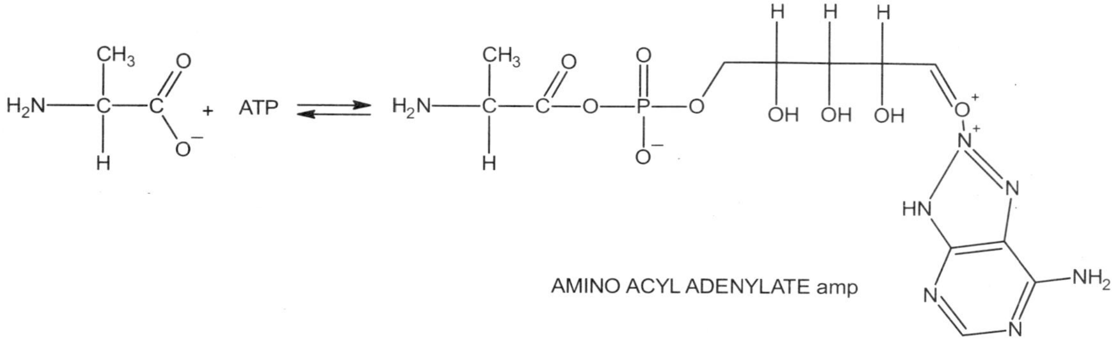

2-aminopropanoyl (5*Z*)-5-[(7-amino-3*H*-[1,2,3]triazolo[4,5-*d*]pyrimidin-2-ium-2-yl)oxonio]-2,3,4-trihydroxypentyl phosphate

In the second step without leaving the enzyme, the amino acyl group of amino acyl AMP is transferred to the 3' end of the tRNA molecule to form aminoacyl – tRNA

aminoacyl – AMP – tRNA → aminoacyl – tRNA + AMP

The overall reaction is

Amino acid + ATP + tRNA → aminoacyl – tRNA + AMP + PP_i

And is driven by the subsequent by hydrolysis of the pyrophosphate to inorganic phosphate.

The basic plan of protein synthesis in eukaryotes and archaea is similar to that in bacteria. The major structural and mechanistic themes recur in all domains of life. However, eukaryotic protein synthesis entails more protein components than does prokaryotic protein synthesis, and some steps are more intricate. Some noteworthy similarities and differences are as follows:

1. *Ribosomes.* Eukaryotic ribosomes are larger. They consist of a 60S large subunit and a 40S small subunit, which come together to form an 80S particle having a mass of 4200 kd, compared with 2700 kd for the prokaryotic 70S ribosome. The 40S subunit contains an 18S RNA that is homologous to the prokaryotic 16S RNA. The 60S subunit contains three RNAs: the 5S and 28S RNAs are the counterparts of the prokaryotic 5S and 23S molecules; its 5.8S RNA is unique to eukaryotes.
2. *Initiator tRNA.* In eukaryotes, the initiating amino acid is methionine rather than *N*-formylmethionine. However, as in prokaryotes, a special tRNA participates in initiation. This aminoacyl-tRNA is called Met-$tRNA_i$ or Met-$tRNA_f$ (the subscript "i" stands for initiation, and "f" indicates that it can be formylated in vitro).
3. *Initiation.* The initiating codon in eukaryotes is always AUG. Eukaryotes, in contrast with prokaryotes, do not use a specific purine-rich sequence on the 52 side to distinguish initiator AUGs from internal ones. Instead, the AUG nearest the 52 end of mRNA is usually selected as the start site. A 40S ribosome attaches to the cap at the 52 end of eukaryotic mRNA and searches for an AUG codon by moving step-by-step in the 32 direction (Figure 6.36).

This scanning process in eukaryotic protein synthesis is powered by helicases that hydrolyze ATP. Pairing of the anticodon of Met-tRNA$_i$ with the AUG codon of mRNA signals that the target has been found. In almost all cases, eukaryotic mRNA has only one start site and hence is the template for a single protein. In contrast, a prokaryotic mRNA can have multiple Shine-Dalgarno sequences and, hence, start sites, and it can serve as a template for the synthesis of several proteins. Eukaryotes utilize many more initiation factors than do prokaryotes, and their interplay is much more intricate. The prefix *eIF* denotes a eukaryotic initiation factor. For example, eIF-4E is a protein that binds directly to the 7-methylguanosine cap, whereas eIF-4A is a helicase. The difference in initiation mechanism between prokaryotes and eukaryotes is, in part, a consequence of the difference in RNA processing. The 52 end of mRNA is readily available to ribosomes immediately after transcription in prokaryotes. In contrast, pre-mRNA must be processed and transported to the cytoplasm in eukaryotes before translation is initiated. Thus, there is ample opportunity for the formation of complex secondary structures that must be removed to expose signals in the mature mRNA. The 52 cap provides an easily recognizable starting point..

4. *Elongation and termination.* Eukaryotic elongation factors EF1α and EF1γ are the counterparts of prokaryotic EF-Tu and EF-Ts. The GTP form of EF1á delivers aminoacyl-tRNA to the A site of the ribosome, and EF1βγ catalyzes the exchange of GTP for bound GDP. Eukaryotic EF2 mediates GTP-driven translocation in much the same way as does prokaryotic EF-G. Termination in eukaryotes is carried out by a single release factor, eRF1, compared with two in prokaryotes. Finally, eIF3, like its prokaryotic counterpart IF3, prevents the reassociation of ribosomal subunits in the absence of an initiation complex.

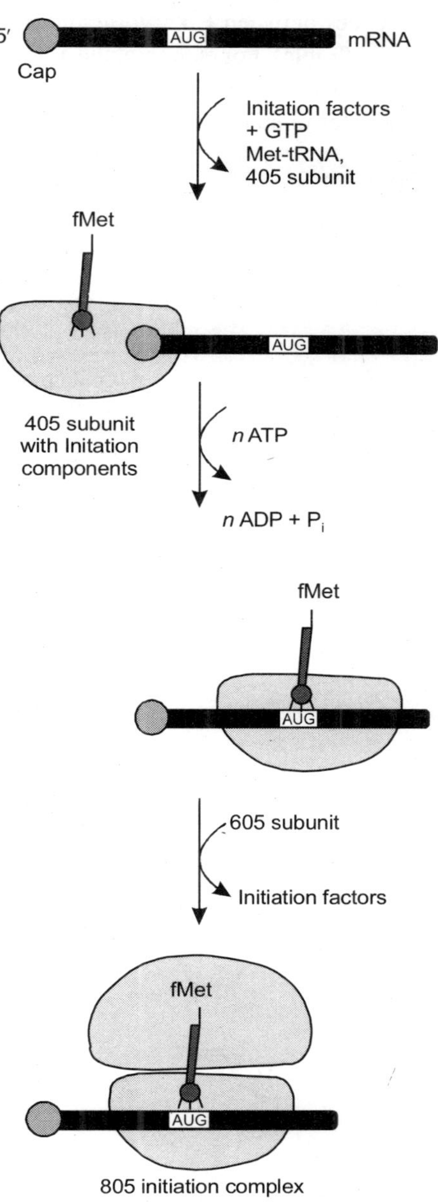

Fig. 6.36 : The Protein Biosynthesis in the Eucaryotes and Procaryotes

INHIBITORS OF PROTEIN SYNTHESIS

The differences between eukaryotic and prokaryotic ribosomes can be exploited for the development of antibiotics (Table). For example, the antibiotic *puromycin* inhibits protein synthesis by causing nascent prokaryotic polypeptide chains to be released before their synthesis is completed. Puromycin is an analog of the terminal aminoacyl-adenosine part of aminoacyl-tRNA in the formula below:

Antibiotic	Action
Streptomycin and other aminoglycosides	Inhibit initiation and cause misreading of mRNA (prokaryotes)
Tetracycline	Binds to the 30S subunit and inhibits binding of aminoacyl-tRNAs (prokaryotes)
Chloramphenicol	Inhibits the peptidyl transferase activity of the 50S ribosomal subunit (prokaryotes)
Cycloheximide	Inhibits the peptidyl transferase activity of the 60S ribosomal subunit (eukaryotes)
Erythromycin	Binds to the 50S subunit and inhibits translocation (prokaryotes)
Puromycin	Causes premature chain termination by acting as an analog of aminoacyl-tRNA (prokaryotes and eukaryotes)

It binds to the A site on the ribosome and inhibits the entry of aminoacyl-tRNA. Furthermore, puromycin contains an α-amino group. This amino group, like the one on aminoacyl-tRNA, forms a peptide bond with the carboxyl group of the growing peptide chain. The product, a peptide having a covalently attached puromycin residue at its carboxyl end, dissociates from the ribosome.

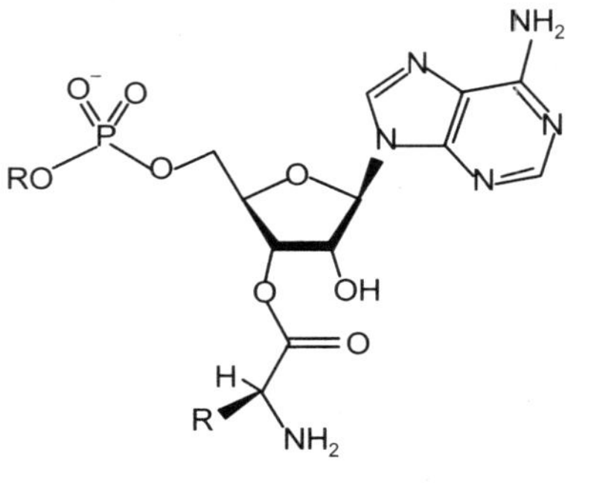

Aminoacyl-tRNA

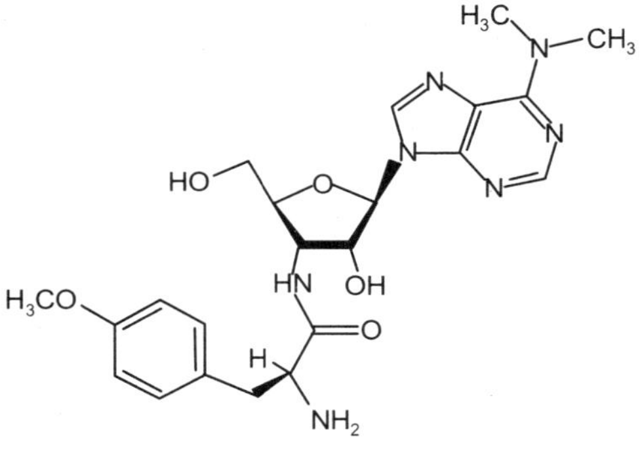

Puromycin

Streptomycin, a highly basic trisaccharide, interferes with the binding of formylmethionyl-tRNA to ribosomes and thereby prevents the correct initiation of protein synthesis. Other *aminoglycoside antibiotics* such as neomycin, kanamycin, and gentamycin interfere with the *decoding site* located near nucleotide 1492 in 16S rRNA of the 30S subunit *Chloramphenicol* acts by inhibiting peptidyl transferase activity. *Erythromycin* binds to the 50S subunit and blocks translocation. Finally, *cyclohexamide* blocks peptidyl transferase activity in eukaryotic ribosomes, making a useful laboratory tool for blocking protein synthesis in eukaryotic cells.

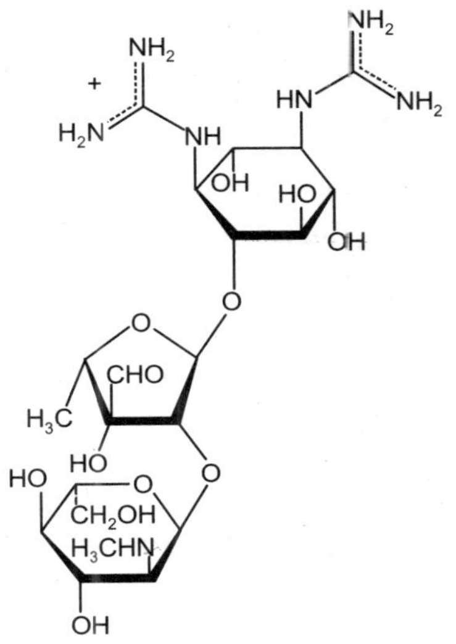

Diphtheria was a major cause of death in childhood before the advent of effective immunization. The lethal effects of this disease are due mainly to a protein toxin produced by *Corynebacterium diphtheriae*, a bacterium that grows in the upper respiratory tract of an infected person. The gene that encodes the toxin comes from a lysogenic phage that is harbored by some strains of *C. diphtheriae*. A few micrograms of diphtheria toxin is usually lethal in an unimmunized person because it inhibits protein synthesis. The toxin is cleaved shortly after entering a target cell into a 21-kd A fragment and a 40-kd B fragment. *The A fragment of the toxin catalyzes the covalent modification of an important component of the protein-synthesizing machinery, whereas the B fragment enables the A fragment to enter the cytosol of its target cell.*

Streptomycin

Fig. 6.37 : Diphtheria Toxin Blocks Protein Synthesis in Eukaryotes by Inhibiting Translocation

A single A fragment of the toxin in the cytosol can kill a cell. Why is it so lethal? The target of the A fragment is EF2, the elongation factor catalyzing translocation in eukaryotic protein synthesis. EF2 contains *diphthamide*, an unusual amino acid residue of unknown function that is formed by posttranslational modification of histidine. The A fragment catalyzes the transfer of the adenosine diphosphate ribose unit of NAD^+ to a nitrogen atom of the diphthamide ring. *This ADP-ribosylation of a single side chain of EF2 blocks its capacity to carry out translocation of the growing polypeptide chain.* Protein synthesis ceases, accounting for the remarkable toxicity of diphtheria toxin.

RETROVIRUS GENOMES

A **retrovirus** is an RNA virus that is replicated in a host cell via the enzyme reverse transcriptase to produce DNA from its RNA genome. The DNA is then incorporated into the host's genome by an integrase enzyme. The virus thereafter replicates as part of the host cell's DNA. Retroviruses are enveloped viruses that belong to the viral family *Retroviridae*.

The virus itself stores its nucleic acid in the form of a +mRNA (including the 5'cap and 3'PolyA inside the virion) genome and serves as a means of delivery of that genome into cells it targets as an obligate parasite, and constitutes the infection. Once in the host's cell, the RNA strands undergo reverse transcription in the cytosol and are integrated into the host's genome, at which point the retroviral DNA is referred to as a provirus. It is difficult to detect the virus until it has infected the host.

Virions of retroviruses consist of enveloped particles about 100 nm in diameter. The virions also contain two identical single-stranded RNA molecules 7-10 kilobases (kb) in length. Although virions of different retroviruses do not have the same morphology or biology, all the virion components are very similar. [1]

The main virion components are:

- Envelope: composed of a lipid bilayer which is obtained from the host plasma membrane during budding process.
- RNA: consisted of a dimer RNA. It has a cap at 5' end and polyadenyle at 3' end. The RNA genome also has terminal noncoding regions which is important in replication and internal regions that encode virion proteins for gene expression. The 5' end includes four regions which are R, U5, PB and L. R region is a short repeated sequence at each end of the genome during the reverse transcription in order to ensure correct end-to-end transfer in growing chain. U5, on the other hand, is a short unique sequence between R and PB. PB consists of 18 bases complementary to 3' end of tRNA primer. L region is an untranslated leader region that gives signal for packaging of genome RNA. The 3' end includes 3 regions which are PP (polypurine), U3 and R. PP is primer for plus-strand DNA synthesis during reverse transcription. U3 is a sequence between PP and R which has signal that provirus can use in transcription. Lastly, R is the terminal repeated sequence at 3' end.
- Proteins: consisted of gag proteins, protease (PR), pol proteins and env proteins. Gag proteins are major components of the viral capsid which are about 2000-4000 copies per virion. Protease, on one hand, is expressed differently in different viruses. It functions in proteolytic cleavages during virion maturation to make mature gag and pol proteins. Pol proteins are responsible for synthesis of viral DNA and integration into host DNA after infection. Finally, env proteins play role in association and entry of virion into the host cell.

When retroviruses have integrated their own genome into the germ line, their genome is passed on to a following generation. These endogenous retroviruses (ERVs), contrasted with exogenous ones, now make up 5-8% of the human genome.[3] Most insertions have no known function and are often referred to as "junk DNA". However, many endogenous retroviruses play important roles in host biology, such as control of gene transcription, cell fusion during placental development in the course of the germination of an embryo, and resistance to exogenous retroviral infection. Endogenous retroviruses have also received special attention in the research of immunology-related pathologies, such as autoimmune diseases like multiple sclerosis, although endogenous retroviruses have not yet been proven to play any causal role in this class of disease. The role of endogenous retroviruses in human gene evolution is explored in a 2005 peer-reviewed article.

While transcription was classically thought to only occur from DNA to RNA, reverse transcriptase transcribes RNA into DNA. The term "retro" in retrovirus refers to this reversal (making DNA from RNA) of the central dogma of molecular biology. Reverse transcriptase activity outside of retroviruses has been found in almost all eukaryotes, enabling the generation and insertion of new copies of retrotransposons into the host genome. These inserts are transcribed by enzymes of the host into new RNA molecules which enter the cytosol. Next, some of these RNA molecules are translated into viral proteins. For example, the *gag* gene is translated into molecules of the capsid protein, the *pol* gene is transcribed into molecules of reverse transcriptase, and the *env* gene is translated into molecules of the envelope protein. It is important to note that a retrovirus must "bring" its own reverse transcriptase in its capsid, otherwise it is unable to utilize the enzymes of the infected cell to carry out the task, due to the unusual nature of producing DNA from RNA.

Industrial drugs that are designed as protease and reverse transcriptase inhibitors can quickly be proved ineffective because the gene sequences that code for the protease and the reverse transcriptase can undergo many substitutions. These substitutions of nitrogenous bases, which make up the DNA strand, can make either the protease or the reverse transcriptase difficult to attack. The amino acid substitution enables the enzymes to evade the drug regiments because mutations in the gene sequences can cause physical or chemical change which makes them harder to detect by the drug. When the drugs that are supposed to attack enzymes, such as protease, are designed, the manufacturers target specific sites on the enzyme. One way to attack these targets can be through hydrolysis of molecular bonds, which means that the drug will add molecules of H_2O (water) to specific bonds. By adding molecules of water at a site on the virus, the drug breaks the previous bonds that were linked to each other. If several of these breaks occur, the result can lead to lysis, the death of the virus.

Because reverse transcription lacks the usual proofreading of DNA replication, a retrovirus mutates very often. This enables the virus to grow resistant to antiviral pharmaceuticals quickly, and impedes the development of effective vaccines and inhibitors for the retrovirus.

Retrovirus genomes commonly contain these three open reading frames that encode for proteins that can be found in the mature virus:

- *group-specific antigen* (gag) codes for core and structural proteins of the virus;
- *polymerase* (pol) codes for reverse transcriptase, protease and integrase; and,
- *envelope* (env) codes for the retroviral coat proteins.

This DNA can be incorporated into host genome as a provirus that can be passed on to progeny cells. In this way some retroviruses can convert normal cells into cancer cells. Some provirus remains latent in the cell for a long period of time before it is activated by the change in cell environment.

Studies of retroviruses led to the first demonstrated synthesis of DNA from RNA templates, a fundamental mode for transferring genetic material that occurs in both eukaryotes and prokaryotes. It has been speculated that the RNA to DNA transcription processes used by retroviruses may have first caused DNA to be used as genetic material. In this model, the RNA world hypothesis, cellular organisms adopted the more chemically stable DNA when retroviruses evolved to create DNA from the RNA templates.

The following genera are included here:

- Genus *Alpharetrovirus*; type species: *Avian leukosis virus*; others include *Rous sarcoma virus*
- Genus *Betaretrovirus*; type species: *Mouse mammary tumour virus*
- Genus *Gammaretrovirus*; type species: *Murine leukemia virus*; others include *Feline leukemia virus*.

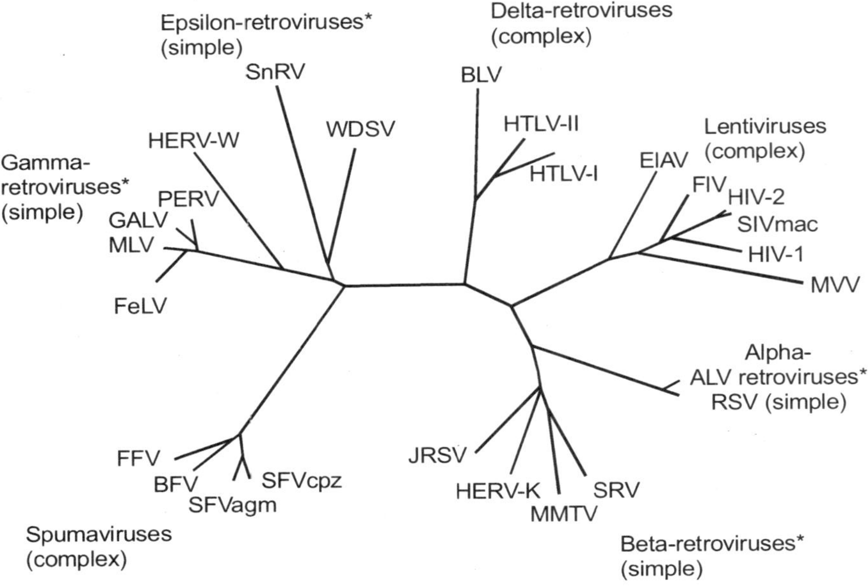

Fig. 6.38 : Showing the Phylogeny of Retroviruses.

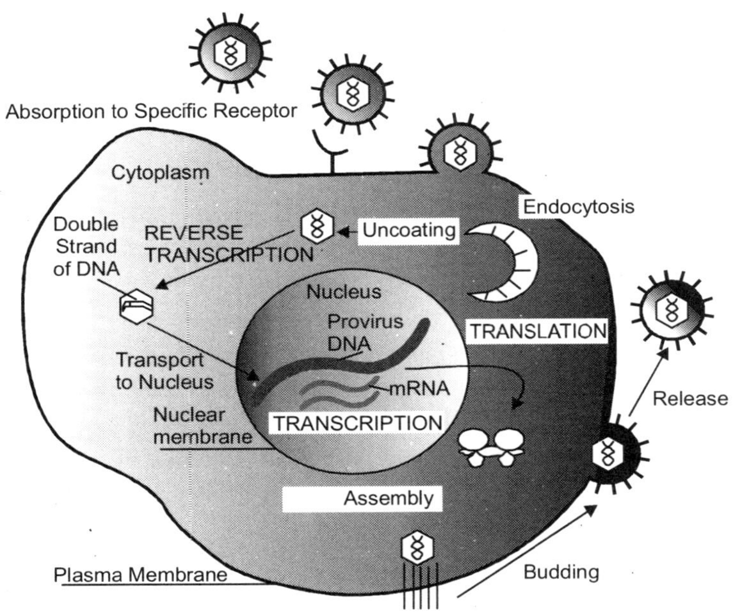

Ritrovirus relication

Fig. 6.38A: Retrivirus replication

- Genus *Deltaretrovirus*; type species: *Bovine leukemia virus*; others include the cancer-causing *Human T-lymphotropic virus*
- Genus *Epsilonretrovirus*; type species: *Walleye dermal sarcoma virus*
- Genus *Lentivirus*; type species: *Human immunodeficiency virus 1*; others include *Simian*, *Feline* immunodeficiency viruses
- Genus *Spumavirus*; type species: *Simian foamy virus*

These were previously divided into three subfamilies (*Oncovirinae*, *Lentivirinae*, and *Spumavirinae*), but with current knowledge of retroviruses, this is no longer appropriate.

Endogenous retroviruses are not formally included in this classification system, and are broadly classified into three classes, on the basis of relatedness to exogenous genera:

- Class I are most similar to the gammaretroviruses
- Class II are most similar to the betaretroviruses and alpharetroviruses
- Class III are most similar to the spumaviruses

All members of Group VI use virally encoded reverse transcriptase, an RNA-dependent DNA polymerase, to produce DNA from the initial virion RNA genome. This DNA is often integrated into the host genome, as in the case of retroviruses and pseudoviruses, where it is replicated and transcribed by the host.

Group VI includes:

- Family *Metaviridae*
- Family *Pseudoviridae*
- Family *Retroviridae* - Retroviruses, e.g. *HIV*

Both families in Group VII have DNA genomes contained within the invading virus particles. The DNA genome is transcribed into both mRNA, for use as a transcript in protein synthesis, and pre-genomic RNA, for use as the template during genome replication. Virally encoded reverse transcriptase uses the pre-genomic RNA as a template for the creation of genomic DNA.

APOPTOSIS

Apoptosis is the process of programmed cell death (PCD) that may occur in multicellular organisms. Programmed cell death involves a series of biochemical events leading to a characteristic cell morphology and death, in more specific terms, a series of biochemical events that lead to a variety of morphological changes, including blebbing, changes to the cell membrane such as loss of membrane asymmetry and attachment, cell shrinkage, nuclear fragmentation, chromatin condensation, and chromosomal DNA fragmentation. Processes of disposal of cellular debris whose results do not damage the organism differentiate apoptosis from necrosis.

In contrast to necrosis, which is a form of traumatic cell death that results from acute cellular injury, apoptosis, in general, confers advantages during an organism's life cycle. For example, the differentiation of fingers and toes in a developing human embryo occurs because cells between the fingers apoptose; the result is that the digits are separate. Between 50 billion and 70 billion cells die each day due to apoptosis in the average human adult. For an average child between the ages of 8 and 14, approximately 20 billion to 30 billion cells die a day. In a year, this amounts to the proliferation and subsequent destruction of a mass of cells equal to an individual's body weight.

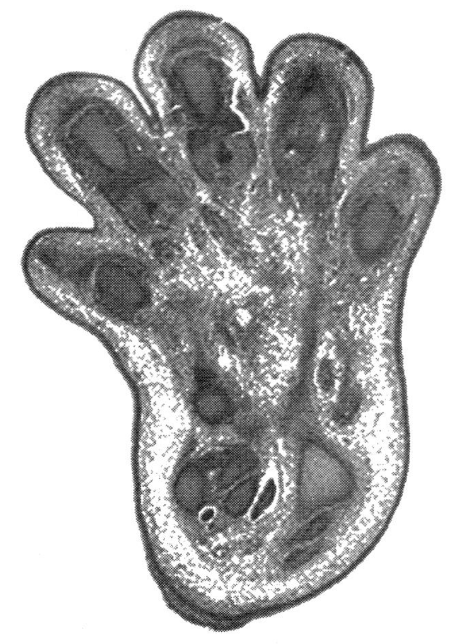

Fig. 6.39 : Showing the Histologic cross section of embryonic foot of mouse (*Mus musculus*) in 15.5 day of its development. There are still cells between fingers. (Full development of mouse lasts 27 days.) (Compare this image with image of leg of mouse.)

Research on apoptosis has increased substantially since the early 1990s. In addition to its importance as a biological phenomenon, defective apoptotic processes have been implicated in an extensive variety of diseases. Excessive apoptosis causes hypotrophy, such as in ischemic damage, whereas an insufficient amount results in uncontrolled cell proliferation, such as cancer.

Cell death is a completely normal process in living organisms and was first discovered by scientists over 100 years ago. A German scientist Carl Vogt was first to describe the principle of apoptosis in 1842. In 1885, anatomist Walther Flemming delivered a more precise description of the process of programmed cell death. However, it was not until 1965 that the topic was resurrected. While studying tissues using electron microscopy, John Foxton Ross Kerr at University of Queensland was able to distinguish apoptosis (Greek: *apo* - from, *ptosis* - falling) from traumatic cell death.[3] Following the publication of a paper describing the phenomenon, Kerr was invited to join Alastair R Currie, as well as Andrew Wyllie, who was Currie's graduate student, at University of Aberdeen. In 1972, the trio published a seminal article in the British Journal of Cancer. Kerr had initially used the term programmed cell necrosis, but in the article, the process of natural cell death was called *apoptosis*. Kerr, Wyllie and Currie credited James Cormack, a professor of Greek language at University of Aberdeen, with suggesting the term apoptosis. Kerr received the Paul Ehrlich and Ludwig Darmstaedter Prize on March 14, 2000, for his description of apoptosis. He shared the prize with Boston biologist Robert Horvitz.

In Greek, apoptosis translates to "dropping off" of petals or leaves from plants or trees. Cormack, professor of Greek language, reintroduced the term for medical use as it had a medical meaning for the Greeks over two thousand years before. Hippocrates used the term to mean "the falling off of the bones". Galen extended its meaning to "the dropping of the scabs". Cormack was no doubt aware of this usage when he suggested the name.

Apoptosis occurs when a cell is damaged beyond repair, infected with a virus, or undergoing stressful conditions such as starvation. Damage to DNA from ionizing radiation or toxic chemicals can also induce apoptosis via the actions of the tumour-suppressing gene *p53*. The "decision" for apoptosis can come from the cell itself, from the surrounding tissue, or from a cell that is part of the immune system. In these cases apoptosis functions to remove the damaged cell, preventing it from sapping further nutrients from the organism, or halting further spread of viral infection.

Apoptosis also plays a role in preventing cancer. If a cell is unable to undergo apoptosis because of mutation or biochemical inhibition, it continues to divide and develop into a tumour. For example, infection by papillomaviruses causes a viral gene to interfere with the cell's p53 protein, an important member of the apoptotic pathway. This interference in the apoptotic capability of the cell plays a role in the development of cervical cancer.

In the adult organism, the number of cells is kept relatively constant through cell death and division. Cells must be replaced when they malfunction or become diseased, but proliferation must be offset by cell death. This control mechanism is part of the homeostasis required by living organisms to maintain their internal states within certain limits. Some scientists have suggested homeodynamics as a more accurate term. The related term *allostasis* reflects a balance of a more complex nature by the body.

Homeostasis is achieved when the rate of mitosis (cell division) in the tissue is balanced by cell death. If this equilibrium is disturbed, one of two potentially fatal disorders occurs:

- the cells are dividing faster than they die, effectively developing a tumor.
- the cells are dividing slower than they die, causing cell loss.

Homeostasis involves a complex series of reactions, an ongoing process inside an organism that calls for different types of cell signaling. Any impairment can cause a disease. For example, dysregulation of signaling pathway has been implicated in several forms of cancer. The pathway, which conveys an anti-apoptotic signal, has been found to be activated in pancreatic adenocarcinoma tissues.

Programmed cell death is an integral part of both plant and animal tissue development. Development of an organ or tissue is often preceded by the extensive division and differentiation of a particular cell, the resultant mass is then "pruned" into the correct form by apoptosis. Unlike necrosis, cellular death caused by injury, apoptosis results in cell shrinkage and fragmentation. Such shrinkage and fragmentation allow the cells to be phagocytosed and their components reused without releasing potentially harmful intracellular substances such as hydrolytic enzymes into the surrounding tissue.

Research on chick embryos has suggested how selective cell proliferation, combined with selective apoptosis, sculpts developing tissues in vertebrates. During vertebrate embryo development, structures called the notochord and the floor plate secrete a gradient of the signaling molecule (Shh), and it is this gradient that directs cells to form patterns in the embryonic neural tube. Cells that receive Shh in a receptor in their membranes called patched1 (Ptc1) survive and proliferate. In the absence of Shh, one of the ends of this same Ptc1 receptor (the carboxyl-terminal, inside the membrane) is cleaved by caspase-3, an action that exposes an apoptosis-producing domain.

During development, apoptosis is tightly regulated and different tissues use different signals for inducing apoptosis. In birds, bone morphogenetic proteins (BMP) signaling is used to induce apoptosis in the interdigital tissue. In *Drosophila* flies, steroid hormones regulate cell death. Developmental cues can also induce apoptosis, such as the sex-specific cell death of hermaphrodite specific neurons in *C. elegans* males through low *TRA-1* transcription factor activity (*TRA-1* helps prevent cell death).

The development of B lymphocytes and T lymphocytes in a human body is a complex process that creates a large pool of diverse cells and subsequently eliminates those potentially damaging to the body. Apoptosis is the mechanism by which the body removes both the ineffective and the potentially damaging immature cells. In T cells, apoptosis is initiated by the withdrawal of survival signals.

Cytotoxic T cells are able to directly induce apoptosis by opening up pores in the target's membrane and releasing chemicals that bypass the normal apoptotic pathway. The pores are created by the action of secreted perforin, and the granules contain granzyme B, a serine protease that activates a variety of caspases by cleaving aspartate residues.

The process of apoptosis is controlled by a diverse range of cell signals, which may originate either extracellularly (*extrinsic inducers*) or intracellularly (*intrinsic inducers*). Extracellular signals may include toxins, hormones, growth factors, nitric oxide or cytokines, and therefore must either cross the plasma membrane or transduce to effect a response. These signals may positively or negatively induce apoptosis. The binding and subsequent initiation of apoptosis by a molecule is termed positive, whereas the active repression of apoptosis by a molecule is termed negative.

A cell initiates intracellular apoptotic signalling in response to a stress, which may bring about cell suicide. The binding of nuclear receptors by glucocorticoids, heat, radiation, nutrient deprivation, viral infection, hypoxia and increased intracellular calcium concentration, for example, by damage to the membrane

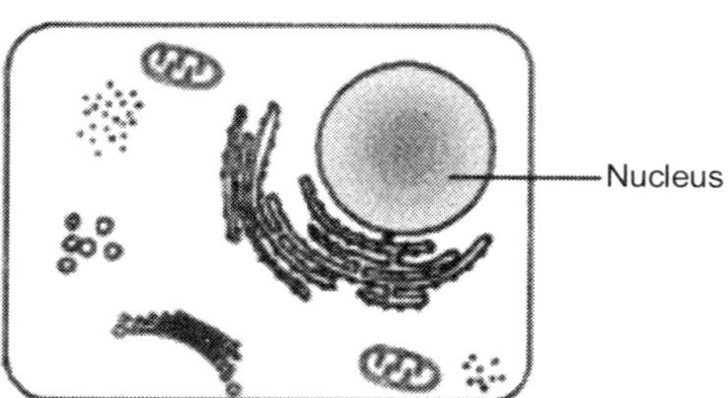

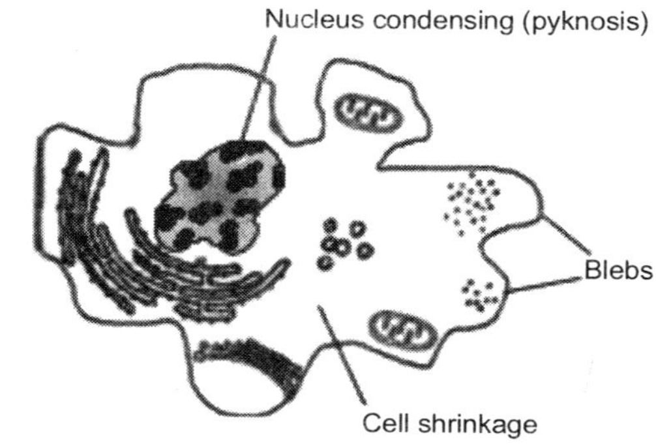

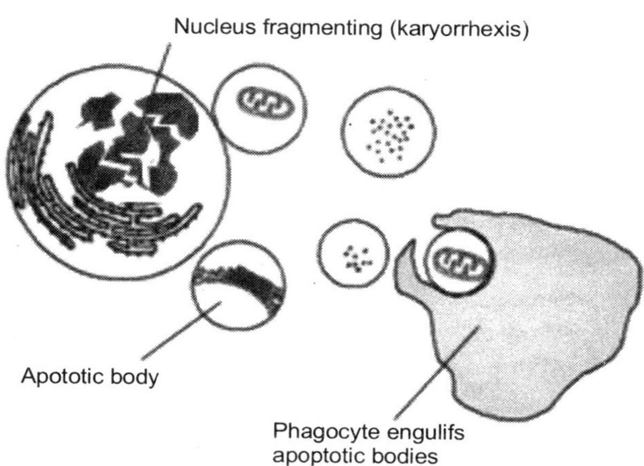

Fig. 6.40 : The Process of Apopstasis

can all trigger the release of intracellular apoptotic signals by a damaged cell. A number of cellular components, such as poly ADP ribose polymerase, may also help regulate apoptosis.

Before the actual process of cell death is carried out by enzymes, apoptotic signals must cause regulatory proteins to start off the death pathway. This step allows apoptotic signals to cause cell death, or the process to be stopped, should the cell no longer need to die. Several proteins are involved, but two main methods of regulation have been identified: targeting mitochondria functionality, or directly transducing the signal via adaptor proteins to the apoptotic mechanisms. There is another extrinsic pathway that has been noticed in several toxical studies that have shown that an increase in calcium concentration with in a cell caused by drug activity; also has the ablity to cause apoptosis via a calcium binding protease calpain.

The mitochondria are essential to multicellular life. Without them, a cell ceases to respire aerobically and quickly dies, a fact exploited by some apoptotic pathways. Apoptotic proteins that target mitochondria affect them in different ways. They may cause mitochondrial swelling through the formation of membrane pores, or they may increase the permeability of the mitochondrial membrane and cause apoptotic effectors to leak out. There is also a growing body of evidence that indicates that nitric oxide is able to induce apoptosis by helping to dissipate the membrane potential of mitochondria and therefore make it more permeable.

Mitochondrial proteins known as SMACs (second mitochondria-derived activator of caspases) are released into the cytosol following an increase in permeability. SMAC binds to *inhibitor of apoptosis proteins* (IAPs) and deactivates them, preventing the IAPs from arresting the apoptotic process and therefore allowing apoptosis to proceed. IAP also normally suppresses the activity of a group of cysteine proteases called caspases, which carry out the degradation of the cell, therefore the actual degradation enzymes can be seen to be indirectly regulated by mitochondrial permeability.

Cytochrome c is also released from mitochondria due to formation of a channel, MAC, in the outer mitochondrial membrane, and serves a regulatory function as it precedes morphological change associated with apoptosis. Once cytochrome c is released it binds with *Apaf-1* and ATP, which then bind to *pro-caspase-9* to create a protein complex known as an apoptosome. The apoptosome cleaves the pro-caspase to its active form of caspase-9, which in turn activates the effector *caspase-3*.

MAC is itself subject to regulation by various proteins, such as those encoded by the mammalian *Bcl-2* family of anti-apoptopic genes, the homologs of the *ced-9* gene found in *C. elegans*. *Bcl-2* proteins are able to promote or inhibit apoptosis either by direct action on MAC or indirectly through other proteins. It is important to note that the actions of some *Bcl-2* proteins are able to halt apoptosis even if cytochrome c has been released by the mitochondria.

Two theories of the direct initiation of apoptotic mechanisms in mammals have been suggested: the *TNF-induced* (tumour necrosis factor) model and the *Fas-Fas ligand-mediated* model, both involving receptors of the *TNF receptor* (TNFR) family coupled to extrinsic signals.

TNF is a cytokine produced mainly by activated macrophages, and is the major extrinsic mediator of apoptosis. Most cells in the human body have two receptors for TNF: *TNF-R1* and *TNF-R2*. The binding of TNF to *TNF-R1* has been shown to initiate the pathway that leads to caspase activation via the intermediate membrane proteins *TNF receptor-associated death domain* (TRADD) and *Fas-associated death domain protein* (FADD). Binding of this receptor can also indirectly lead to the activation of transcription factors involved in cell survival and inflammatory responses. The link between TNF and apoptosis shows why an abnormal production of TNF plays a fundamental role in several human diseases, especially in autoimmune diseases.

The Fas receptor (also known as *Apo-1* or *CD95*) binds the Fas ligand (FasL), a transmembrane protein part of the TNF family. The interaction between Fas and FasL results in the formation of the *death-inducing signaling complex* (DISC), which contains the FADD, caspase-8 and caspase-10. In some types of cells (type I), processed caspase-8 directly activates other members of the caspase family, and triggers the execution of apoptosis. In other types of cells (type II), the *Fas*-DISC starts a feedback loop that spirals into increasing release of pro-apoptotic factors from mitochondria and the amplified activation of caspase-8.

Following *TNF-R1* and *Fas* activation in mammalian cells a balance between pro-apoptotic (BAX,[26] BID, BAK, or BAD) and anti-apoptotic (*Bcl-Xl* and *Bcl-2*) members of the *Bcl-2* family is established. This balance is the proportion of pro-apoptotic homodimers that form in the outer-membrane of the mitochondrion. The pro-apoptotic homodimers are required to make the mitochondrial membrane permeable for the release of caspase activators such as cytochrome c and SMAC. Control of pro-apoptotic proteins under normal cell conditions of non-apoptotic cells is incompletely understood, but it has been found that a mitochondrial outer-membrane protein, VDAC2, interacts with BAK to keep this potentially-lethal apoptotic effector under control. When the death signal is received, products of the activation cascade displace VDAC2 and BAK is able to be activated.

There also exists a caspase-independent apoptotic pathway that is mediated by AIF (apoptosis-inducing factor).

Many pathways and signals lead to apoptosis, but there is only one mechanism that actually causes the death of a cell. After a cell receives stimulus, it undergoes organized degradation of cellular organelles by activated proteolytic caspases. A cell undergoing apoptosis shows a characteristic morphology:

1. Cell shrinkage and rounding are shown because of the breakdown of the proteinaceous cytoskeleton by caspases.
2. The cytoplasm appears dense, and the organelles appear tightly packed.
3. Chromatin undergoes condensation into compact patches against the nuclear envelope in a process known as pyknosis, a hallmark of apoptosis.
4. The nuclear envelope becomes discontinuous and the DNA inside it is fragmented in a process referred to as karyorrhexis. The nucleus breaks into several discrete *chromatin bodies* or *nucleosomal units* due to the degradation of DNA.
5. The cell membrane shows irregular buds known as blebs.
6. The cell breaks apart into several vesicles called *apoptotic bodies*, which are then phagocytosed.

Apoptosis progresses quickly and its products are quickly removed, making it difficult to detect or visualize. During karyorrhexis, endonuclease activation leaves short DNA fragments, regularly spaced in size. These give a characteristic "laddered" appearance on agar gel after electrophoresis. Tests for DNA laddering differentiate apoptosis from ischemic or toxic cell death.

The removal of dead cells by neighboring phagocytic cells has been termed efferocytosis. Dying cells that undergo the final stages of apoptosis display phagocytotic molecules, such as phosphatidylserine, on their cell surface. Phosphatidylserine is normally found on the cytosolic surface of the plasma membrane, but is redistributed during apoptosis to the extracellular surface by a hypothetical protein known as scramblase. These molecules mark the cell for phagocytosis by cells possessing the appropriate receptors, such as macrophages. Upon recognition, the phagocyte reorganizes its cytoskeleton for engulfment of the cell. The removal of dying cells by phagocytes occurs in an orderly manner without eliciting an inflammatory response.

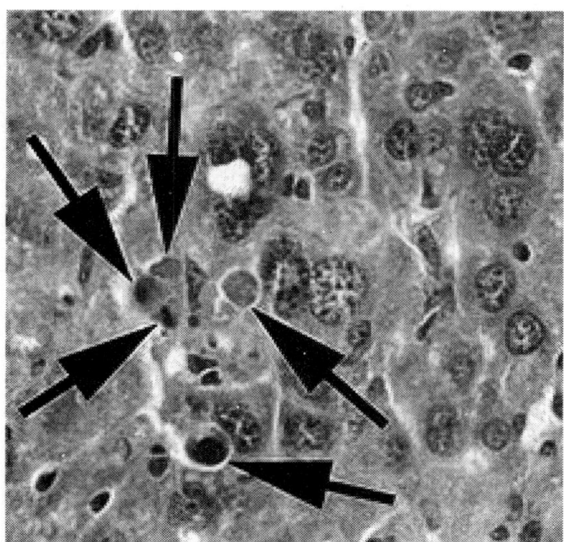

Fig. 6.41 : A section of mouse liver showing several apoptotic cells, indicated by arrows

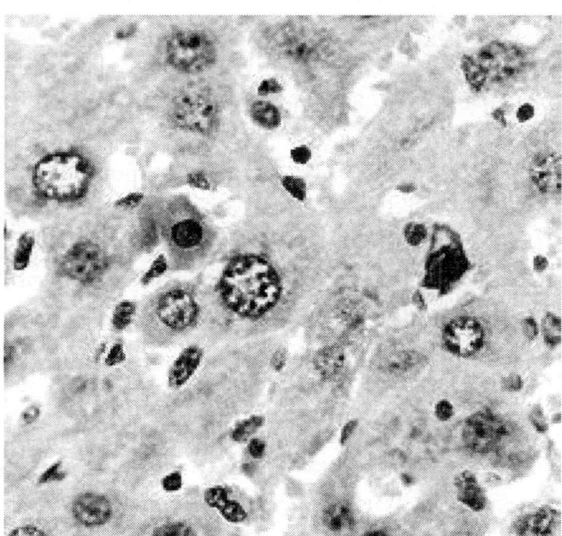

Fig. 6.42 : A section of mouse liver stained to show cells undergoing apoptosis.

The many different types of apoptotic pathways contain a multitude of different biochemical components, many of them not yet understood. As a pathway is more or less sequential in nature, it is a victim of causality; removing or modifying one component leads to an effect in another. In a living organism this can have disastrous effects, often in the form of disease or disorder. A discussion of every disease caused by modification of the various apoptotic pathways would be impractical, but the concept overlying each one is the same: the normal functioning of the pathway has been disrupted in such a way as to impair the ability of the cell to undergo normal apoptosis. This results in a cell that lives past its "use-by-date" and is able to replicate and pass on any faulty machinery to its progeny, increasing the likelihood of the cell becoming cancerous or diseased.

A recently-described example of this concept in action can be seen in the development of a lung cancer called NCI-H460. The *X-linked inhibitor of apoptosis protein* (XIAP) is overexpressed in cells of the H460 cell line. XIAPs bind to the processed form of caspase-9, and suppress the activity of apoptotic activator cytochrome c, therefore overexpression leads to a decrease in the amount of pro-apoptotic agonists. As a consequence, the balance of anti-apoptotic and pro-apoptotic effectors is upset in favour of the former, and the damaged cells continue to replicate despite being directed to die.

The tumor-suppressor protein p53 accumulates when DNA is damaged due to a chain of biochemical reactions. Part of this pathway includes alpha-interferon and beta-interferon, which induce transcription of the *p53* gene and result in the increase of p53 protein level and enhancement of cancer cell-apoptosis.[38] p53 prevents the cell from replicating by stopping the cell cycle at G1, or interphase, to give the cell time to repair, however it will induce apoptosis if damage is extensive and repair efforts fail. Any disruption to the regulation of the *p53* or interferon genes will result in impaired apoptosis and the possible formation of tumors.

The progression of the human immunodeficiency virus infection to AIDS is primarily due to the depletion of $CD4^+$ T-helper lymphocytes, which leads to a compromised immune system. One of the mechanisms by which T-helper cells are depleted is apoptosis, which results from a series of biochemical pathways:

1. HIV enzymes deactivate anti-apoptotic *Bcl-2* This does not directly cause cell death, but primes the cell for apoptosis should the appropriate signal be received.In parallel these enzymes activate pro-apoptotic *procaspase-8*, which does directly activate the mitochondrial events of apoptosis.
2. HIV may increase the level of cellular proteins which prompt Fas-mediated apoptosis.
3. HIV proteins decrease the amount of CD4 glycoprotein marker present on the cell membrane.
4. Released viral particles and proteins present in extracellular fluid are able to induce apoptosis in nearby "bystander" T helper cells.
5. HIV decreases the production of molecules involved in marking the cell for apoptosis, giving the virus time to replicate and continue releasing apoptotic agents and virions into the surrounding tissue.
6. The infected $CD4^+$ cell may also receive the death signal from a cytotoxic T cell.

Cells may also die as a direct consequence of viral infection.

Viruses can trigger apoptosis of infected cells via a range of mechanisms including:

- Receptor binding.
- Activation of protein kinase R (PKR).
- Interaction with p53.
- Expression of viral proteins coupled to MHC proteins on the surface of the infected cell, allowing recognition by cells of the immune system (such as Natural Killer and cytotoxic T cells) that then induce the infected cell to undergo apoptosis.

Most viruses encode proteins that can inhibit apoptosis.Several viruses encode viral homologs of Bcl-2. These homologs can inhibit pro-apoptotic proteins such as BAX and BAK, which are essential for the activation of apoptosis. Examples of viral Bcl-2 proteins include the Epstein-Barr virus BHRF1 protein and the adenovirus E1B 19K protein. Some viruses express caspase inhibitors that inhibit caspase activity and an example is the CrmA protein of cowpox viruses. Whilst a number of viruses can block the effects of TNF and Fas. For example the M-T2 protein of myxoma viruses can bind TNF preventing it from binding the TNF receptor and inducing a response.[43] Furthermore, many viruses express p53 inhibitors that can bind p53 and inhibit its transcriptional transactivation activity. Consequently p53 cannot induce apoptosis since it cannot induce the expression of pro-apoptotic proteins. The adenovirus E1B-55K protein and the hepatitis B virus HBx protein are examples of viral proteins that can perform such a function.

Interestingly, viruses can remain intact from apoptosis particularly in the latter stages of infection. They can be exported in the *apoptotic bodies* that pinch off from the surface of the dying cell and the fact that they are engulfed by phagocytes prevents the initiation of a host response. This favours the spread of the virus.

Programmed cell death in plants has a number of molecular similarities to animal apoptosis, but it also has differences, notably the presence of a cell wall and the lack of an immune system which removes the pieces of the dead cell. Instead of an immune response, the dying cell synthesizes substances to break itself down and places them in a vacuole which ruptures as the cell dies. Whether this whole process resembles animal apoptosis closely enough to warrant using the name *apoptosis* (as opposed to the more general *programmed cell death*) is unclear.

Caspase Independent Apoptosis

There is an extrinsic pathway that has been noticed in several toxical studies that have shown that an increase in calcium concentration with in a cell caused by drug activity; also has the ablity to cause apoptosis via a calcium binding protease calpain.

RFLP AND VNTR EXPLANATION AND THEIR APPLICATION IN MEDICINE

RFLP (often pronounced "rif lip", as if it were a word) is a method used by molecular biologists to follow a particular sequence of DNA as it is passed on to other cells. RFLPs can be used in many different settings to accomplish different objectives. RFLPs can be used in paternity cases or criminal cases to determine the source of a DNA sample. RFLPs can be used determine the disease status of an individual. RFLPs can be used to measure recombination rates which can lead to a genetic map with the distance between RFLP loci measured in centiMorgans.

On this web page, you can see how RFLPs are produced and then three examples of applying RFLP analysis: paternity, disease status, and genetic mapping.

RFLP Production

Each organism inherits its DNA from its parents. Since DNA is replicated with each generation, any given sequence can be passed on to the next generation. An RFLP is a sequence of DNA that has a restriction site on each end with a "**target**" sequence in between. A target sequence is any segment of DNA that bind to a **probe** by forming complementary base pairs. A probe is a sequence of single-stranded DNA that has been tagged with radioactivity or an enzyme so that the probe can be detected. When a probe base pairs to its target, the investigator can detect this binding and know where the target sequence is since the probe is detectable. RFLP produces a series of bands when a Southern blot is performed with a particular combination of restriction enzyme and probe sequence.

For example, let's follow a particular RFLP that is defined by the restriction enzyme EcoR I and the target sequence of 20 bases GCATGCATGCATGCATGCAT. EcoR I binds to its recognition seuqence GAATTC and cuts the double-stranded DNA as shown:

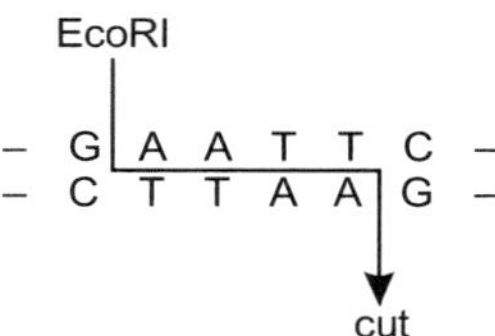

In the segement of DNA shown below, you can see the elements of an RFLP; a target sequence flanked by a pair of restriction sites. When this segment of DNA is cut by EcoR I, three restriction fragments are produced, but only one contains the target sequence which can be bound by the complementary probe sequence (**purple**).

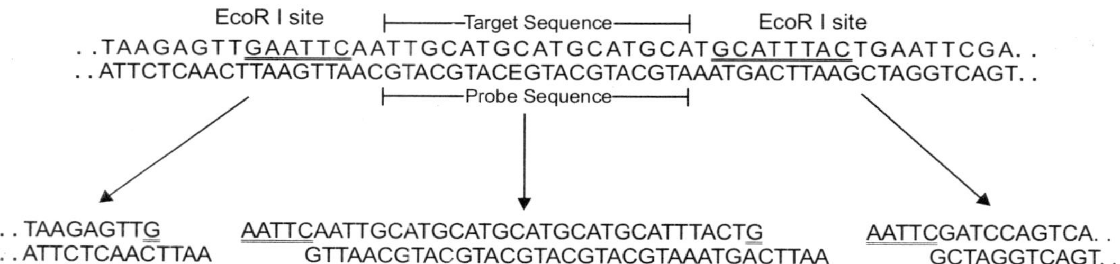

Let's look at two people and the segments of DNA they carry that contain this RFLP (for clarity, we will only see one of the two stands of DNA). Since Jack and Jill are both diploid organisms, they have two copies of this RFLP. When we examine one copy from Jack and one copy from Jill, we see that they are identical:

Jack 1: -GAATTC—(8.2 kb)—-**GCATGCATGCATGCATGCAT**—(4.2 kb)—GAATTC-

Jill 1: -GAATTC—(8.2 kb)—-**GCATGCATGCATGCATGCAT**—(4.2 kb)—GAATTC-

When we examine their second copies of this RFLP, we see that they are not identical. Jack 2 lacks an EcoR I restriction site that Jill has 1.2 kb upstream of the target sequence (difference in italics).

Jack 2: -GAATTC—(1.8 kb)-*CCCTTT*—(1.2 kb)—**GCATGCATGCATGCATGCAT**—(1.3 kb)-GAATTC-

Jill 2: -GAATTC—(1.8 kb)-*GAATTC*—(1.2 kb)—**GCATGCATGCATGCATGCAT**—(1.3 kb)-GAATTC-

Therefore, when Jack and Jill have their DNA subject to RFLP analysis, they will have one band in common and one band that does not match the other's in molecular weight:

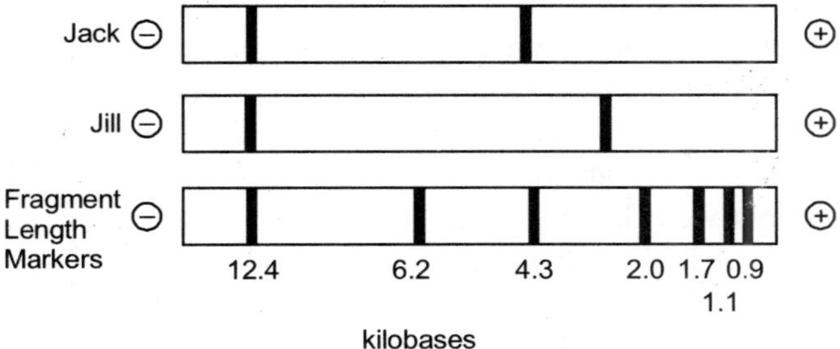

Paternity Case

Let's use RFLP technology to determine if Jack is the father of Jill's child named Payle.

In this scenario, DNA was extracted from white blood cells from all three individuals and subjected to RFLP analysis. The results are shown below:

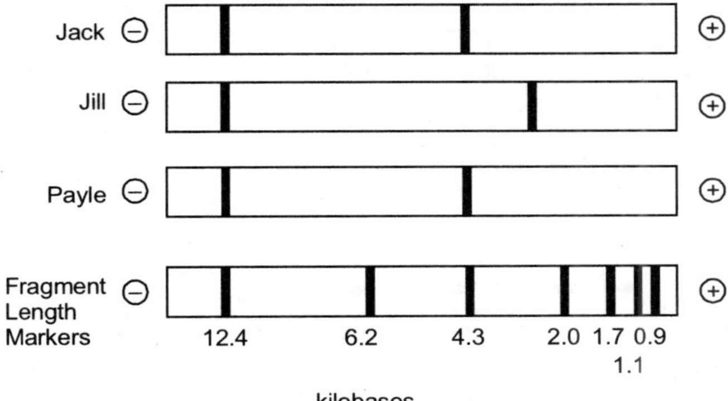

In this case, it appears that Jack could be the father, since Payle inherited the 12.4 kb fragment from Jill and the 4.3 fragment from Jack. However, it is possible that another man with similar RFLP pattern could be as well.To be certain, several more RFLP loci would be tested. It would be highly unlikely that two men (other than identical twins) would share multiple RFLP patterns and so paternity could be confirmed.

In a different scenario, DNA was extracted from white blood cells from all three individuals and subjected to RFLP analysis. The results are shown below:

This time, it can be determined that Jack is NOT the father of Payle since Payle has a band of about 6 kb and Jack does not. Therefore, it is very probable that Payle's father is not Jack, though it is possible that Payle carries a new mutation at this locus and a different sized band was produced. What could you do as an investigator to be more certain that Jack was not the father of Payle?

Disease Status

In this example, we want to know if a person carries any cystic fibrosis (CF) alleles and if so, how many. Because CF is a recessive disease, anyonne with CF must be homozygous for disease alleles.

From pedigree information, we can often determine who in this family is a carrier. However, if a couple comes to a genetic counselor, often an RFLP analysis is performed on the couple's DNA.

RFLPs are known for CF and so it would be easy to determine if a person were homozygous wild-type (*wt*), heterozygous "carrier", or homozygous disease alleles and thus have CF.

For couples expecting a child, it would be simple to test both parents and make a prediction about the eventual disease status of their fetus. For example, if both parents were homozygous *wt*, then all of their children would also be homozygous *wt*:

However, if both parents were heterozygous, they could have children with any of the three genotypes, though heterozygous children would be twice as likely as either of the homozygous genotypes.

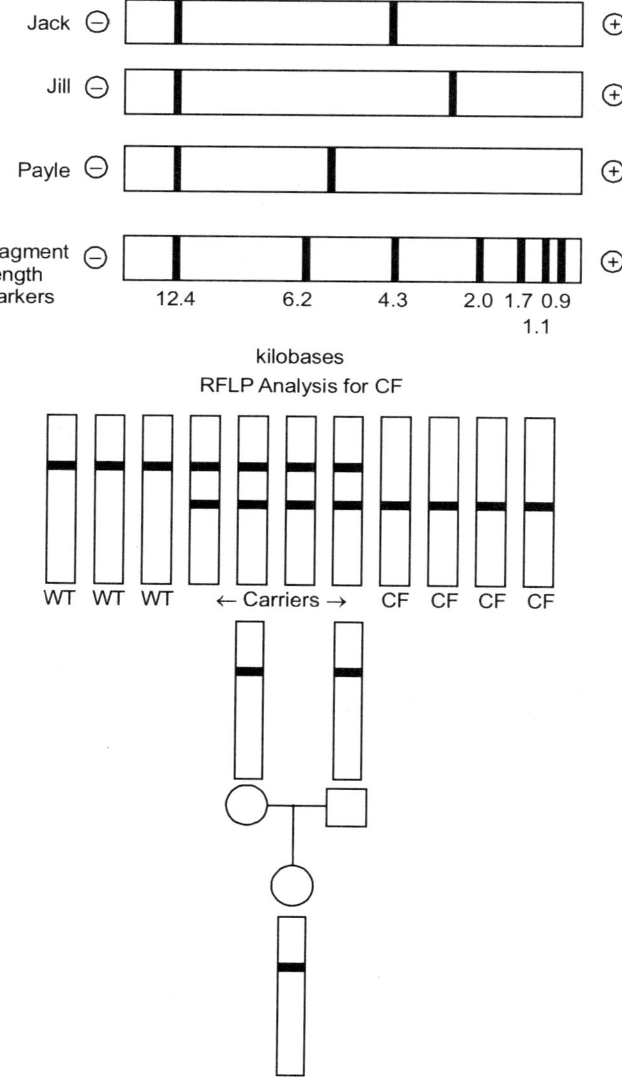

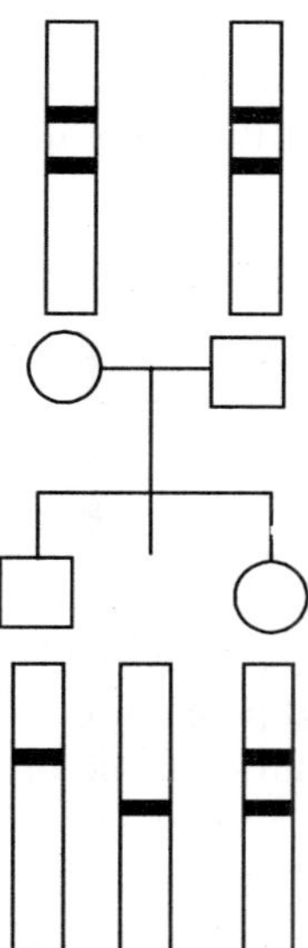

With increasing genomic sequence information, increasing numbers of genetic disease can be predicted from RFLP analyses.

Genetic Mapping

To calculate the genetic distance between to loci, you need to be able to observe recombination. Traditionally, this was performed by observing phenotypes but with RFLP analysis, it is possible to measure the genetic distance between two RFLP loci whether they are a part of genes or not.

Let's look at a simple example in fruit flies. Two RFLP loci with two RFLP bands possible at each locus:

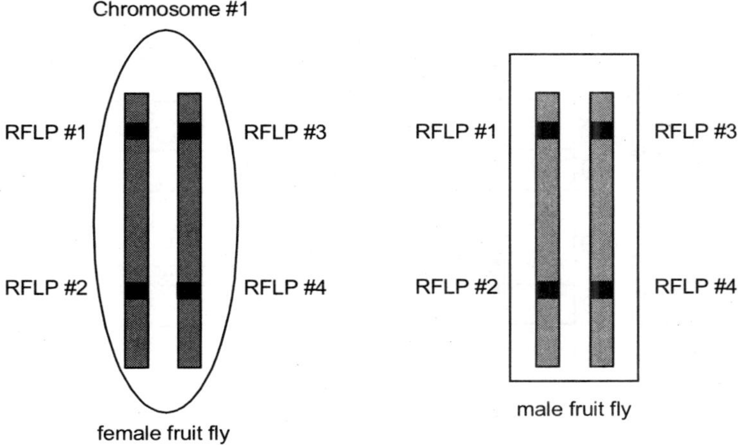

These loci are located on the same chromosome for the female (left) and the male (right). The upper locus can produce two different bands called 1 and 3. The lower locus can produce bands called 2 or 4. The male is homozygous for band 1 at the upper locus and 2 for the lower locus. The female is heterozygous at both loci. Thier RFLP banding patterns can be seen on the Southern blot below:

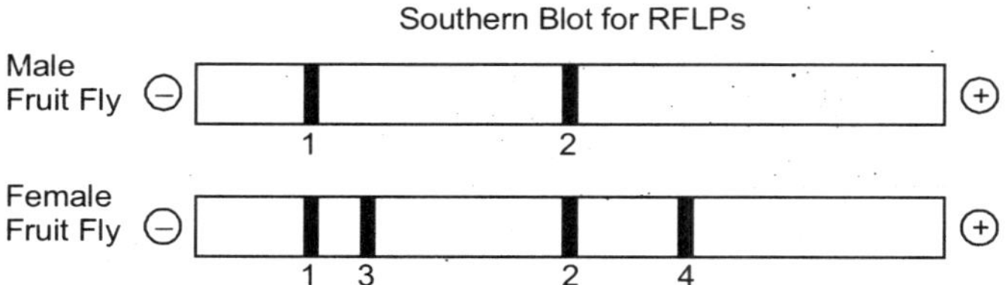

The male can only produce one type of gamete (1 and 2) but the female can produce four different gametes. Two of the possible four are called parental because they carry both RFLP bands from the same chromosome; 1 and 2 from the left chromosome or 3 and 4 from the right chromosome. The other two chromosomes are recombinant because recombination has occurred between the two loci and thus the RFLP bands are mixed so that 1 is now linked to 4 and 3 is linked to 2.

Type of Chromatid	Alleles
Parental	RFLP 1 and 2
Parental	RFLP 3 and 4
Recombinant	RFLP 1 and 4
Recombinant	RFLP 3 and 2

When these two flies mate, the frequency of the four possible progeny can be measured and from this information, the genetic distance between the two RFLP loci (upper and lower) can be determined.

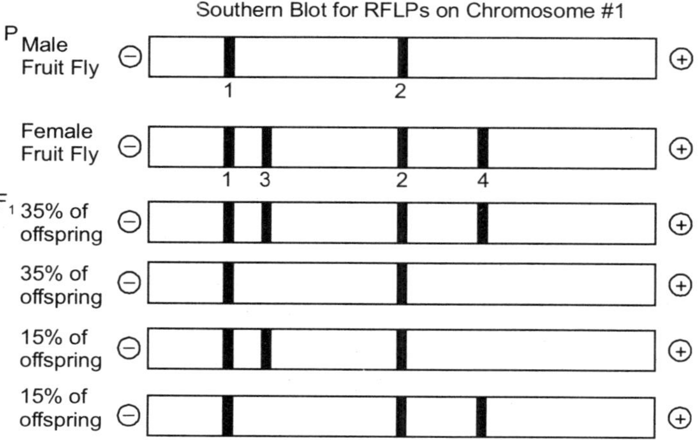

In this example, 70% of the progeny were produce from parental genotype eggs and 30% were produced by recombinant genotype eggs. Therefore, these two RFLP loci are 30 centiMorgans apart from each other.

A VARIABLE NUMBER TANDEM REPEAT (OR VNTR)

It is a location in a genome where a short nucleotide sequence is organized as a tandem repeat. These can be found on many chromosomes, and often show variations in length between individuals. Each variant acts as an inherited allele, allowing them to be used for personal or parental identification. Their analysis is useful in genetics and biology research, forensics, and DNA fingerprinting.

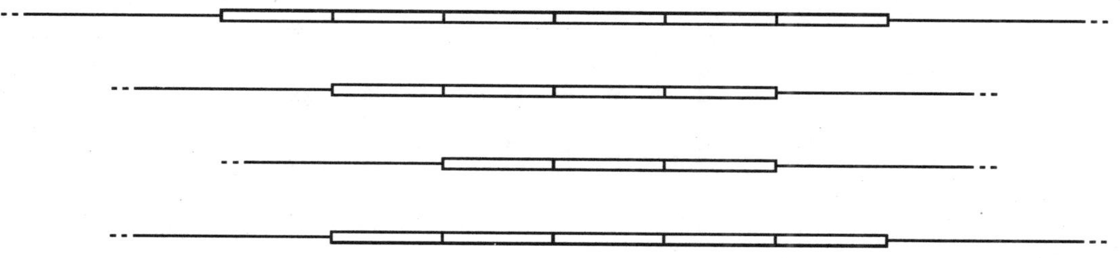

Fig. 6.43 : Schematic of a Variable Number of Tandem Repeats in 4 alleles.

In the schematic above, the rectangular blocks represent each of the repeated DNA sequences at a particular VNTR location. The repeats are tandem - they are clustered together and oriented in the same direction. Individual repeats can be removed from (or added to) the VNTR via recombination or replication errors, leading to alleles with different numbers of repeats. Flanking the repeats are segments of non-repetitive sequence (shown here as thin lines), allowing the VNTR blocks to be extracted with restriction enzymes and analyzed by RFLP, or amplified by the polymerase chain reaction (PCR) technique and their size determined by gel electrophoresis.

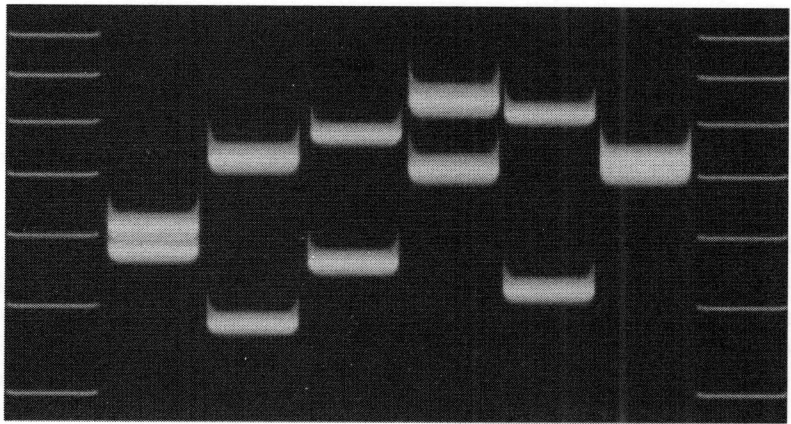

Fig. 6.44 : The Variations of VNTR (D1S80) allele lengths in 6 individuals

VNTRs were an important source of RFLP genetic markers used in linkage analysis (mapping) of genomes. Now that many genomes have been sequenced, VNTRs have become essential to forensic crime investigations, via DNA fingerprinting and the CODIS database. When removed from surrounding DNA by the PCR or RFLP methods, and their size determined by gel electrophoresis or Southern blotting, they produce a pattern of bands unique to each individual. When tested with a group of independent VNTR markers, the likelihood of two unrelated individuals having the same allelic pattern is extremely improbable. VNTR analysis is also being used to study genetic diversity and breeding patterns in populations of wild or domesticated animals.

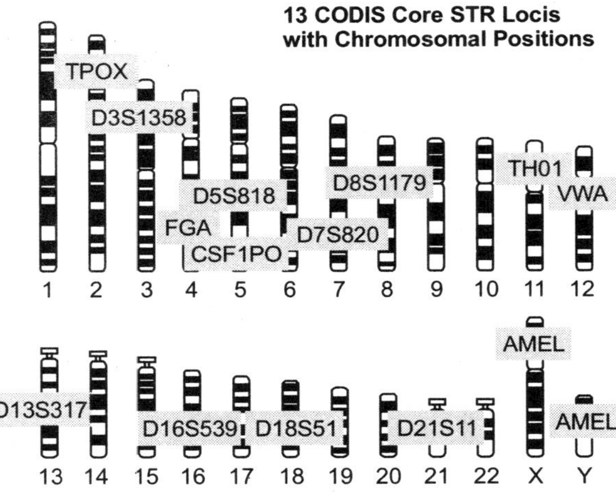

Fig. 6.45 : The Chromosomal locations of the 13 VNTR loci in the CODIS panel

VNTR Inheritance

In analyzing VNTR data, two basic genetic principles can be used:

- **Identity Matching**- both VNTR alleles from a specific location must match. If two samples are from the same individual, they must show the same allele pattern.
- **Inheritance Matching**- the VNTR alleles must follow the rules of inheritance. In matching an individual with his parents or children, a person must have an allele that matches one from each parent. If the relationship is more distant, such as a grandparent or sibling, then matches must be consistent with the degree of relatedness.

Relationship to other types of repetitive DNA

Repetitive DNA, representing over 40% of the human genome, is arranged in a bewildering array of patterns. Repeats were first identified by the extraction of Satellite DNA, which does not reveal how they are organized. The use of restriction enzymes showed that some repeat blocks were interspersed throughout the genome. DNA sequencing later showed that other repeats are clustered at specific locations, with tandem repeats being more common than inverted repeats (which may interfere with DNA replication). The VNTRs are the class of clustered tandem repeats that exhibit allelic variation in their lengths.

Classes of VNTRs

There are two principal families of VNTRs: microsatellites and minisatellites. The former are repeats of sequences less than about 5 base pairs in length (an arbitrary cutoff), while the latter involve longer blocks. Confusing this distinction is the recent use of the terms Short Tandem Repeat (STR) and Simple Sequence Repeat (SSR), which are more descriptive, but whose definitions are similar to that of microsatellites. VNTRs with very short repeat blocks may be unstable - dinucleotide repeats may vary from one tissue to another within an individual, while trinucleotide repeats have been found to vary from one generation to another (see Huntington's disease). The 13 assays used in the CODIS database are usually referred to as STRs, and most analyze VNTRs that involve repeats of 4 base pairs.

THE SOUTHERN BLOTTING

A **Southern blot** is a method routinely used in molecular biology to check for the presence of a DNA sequence in a DNA sample. Southern blotting combines agarose gel electrophoresis for size separation of DNA with methods to transfer the size-separated DNA to a filter membrane for probe hybridization. The method is named after its inventor, the British biologist Edwin Southern. Other blotting methods (i.e., western blot, northern blot, southwestern blot) that employ similar principles, but using RNA or protein, have later been named in reference to Southern's name. As the technique was eponymously named, Southern blot should be capitalized, whereas northern and western blots should not.

Method

1. Restriction endonucleases are used to cut high-molecular-weight DNA strands into smaller fragments.
2. The DNA fragments are then electrophoresed on an agarose gel to separate them by size.
3. If some of the DNA fragments are larger than 15 kb, then prior to blotting, the gel may be treated with an acid, such as dilute HCl, which depurinates the DNA fragments, breaking the DNA into smaller pieces, thus allowing more efficient transfer from the gel to membrane.
4. If alkaline transfer methods are used, the DNA gel is placed into an alkaline solution (typically containing sodium hydroxide) to denature the double-stranded DNA. The denaturation in an alkaline environment provides for improved binding of the negatively charged DNA to a positively charged membrane, separates it into single DNA strands for later hybridization to the probe and destroys any residual RNA that may still be present in the DNA.
5. A sheet of nitrocellulose (or, alternatively, nylon) membrane is placed on top of (or below, depending on the direction of the transfer) the gel. Pressure is applied evenly to the gel (either using suction, or by placing a stack of paper towels and a weight on top of the membrane and gel), to ensure good and even contact between gel and membrane. Buffer transfer by capillary action from a region of high water potential to a region of low water potential (usually filter paper and paper tissues) is then used to move the DNA from the gel on to the membrane; ion exchange interactions bind the DNA to the membrane due to the negative charge of the DNA and positive charge of the membrane.

6. The membrane is then baked, i.e., exposed to high temperature (60 to 100 °C) (in the case of nitrocellulose) or exposed to ultraviolet radiation (nylon) to permanently and covalently crosslink the DNA to the membrane.
7. The membrane is then exposed to a hybridization probe—a single DNA fragment with a specific sequence whose presence in the target DNA is to be determined. The probe DNA is labelled so that it can be detected, usually by incorporating radioactivity or tagging the molecule with a fluorescent or chromogenic dye. In some cases, the hybridization probe may be made from RNA, rather than DNA. To ensure the specificity of the binding of the probe to the sample DNA, most common hybridization methods use salmon testes (sperm) DNA for blocking of the membrane surface and target DNA, deionized formamide, and detergents such as SDS to reduce non-specific binding of the probe.
8. After hybridization, excess probe is washed from the membrane, and the pattern of hybridization is visualized on X-ray film by autoradiography in the case of a radioactive or fluorescent probe, or by development of color on the membrane if a chromogenic detection method is used.

Hybridization of the probe to a specific DNA fragment on the filter membrane indicates that this fragment contains DNA sequence that is complementary to the probe.

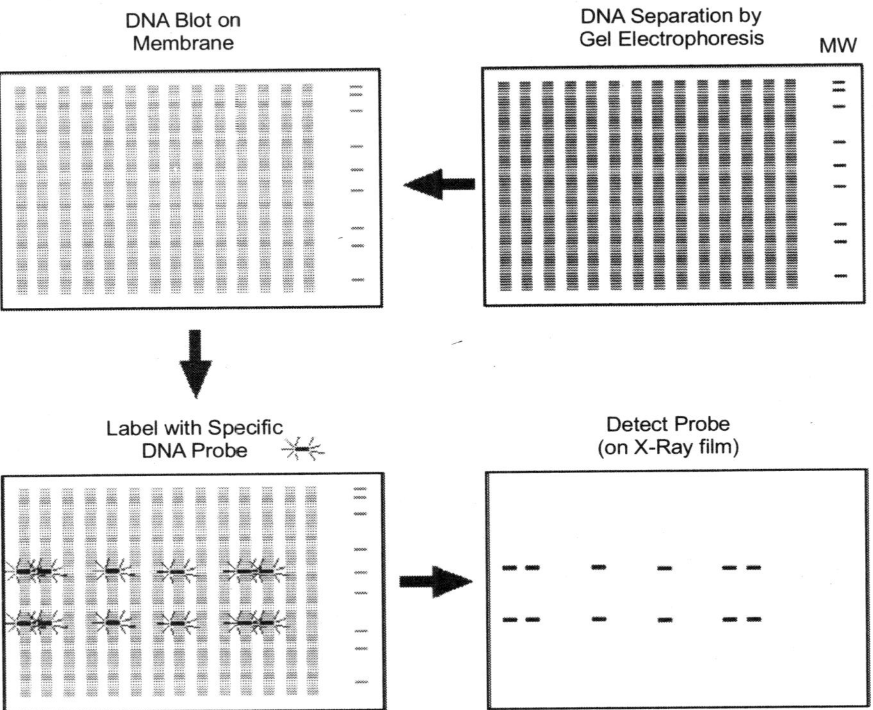

Fig. 6.46 : Southern blots allow investigators to determine the molecular weight of a restriction fragment and to measure relative amounts in different samples

The transfer step of the DNA from the electrophoresis gel to a membrane permits easy binding of the labelled hybridization probe to the size-fractionated DNA. It also allows for the fixation of the target-probe hybrids, required for analysis by autoradiography or other detection methods.

Southern blots performed with restriction enzyme-digested genomic DNA may be used to determine the number of sequences (e.g., gene copies) in a genome. A probe that hybridizes only to a single DNA segment that has not been cut by the restriction enzyme will produce a single band on a Southern blot, whereas multiple bands will likely be observed when the probe hybridizes to several highly similar sequences (e.g., those that may be the result of sequence duplication). Modification of the hybridization conditions (for example, increasing the hybridization temperature or decreasing salt concentration) may be used to increase specificity and decrease hybridization of the probe to sequences that are less than 100% similar.

This is a brief overview of how a Southern blot (more formally called an DNA blot) is performed and what type of data you can obtain form one.

Procedure

(1) DNA (genomic or other source) is digested with a restriction enzyme and separated by gel electrophoresis, usually an agarose gel. Because there are so many different restriction fragments on the gel, it usually appears as a smear rather than discrete bands. The DNA is denature into single strands by incubation with NaOH.

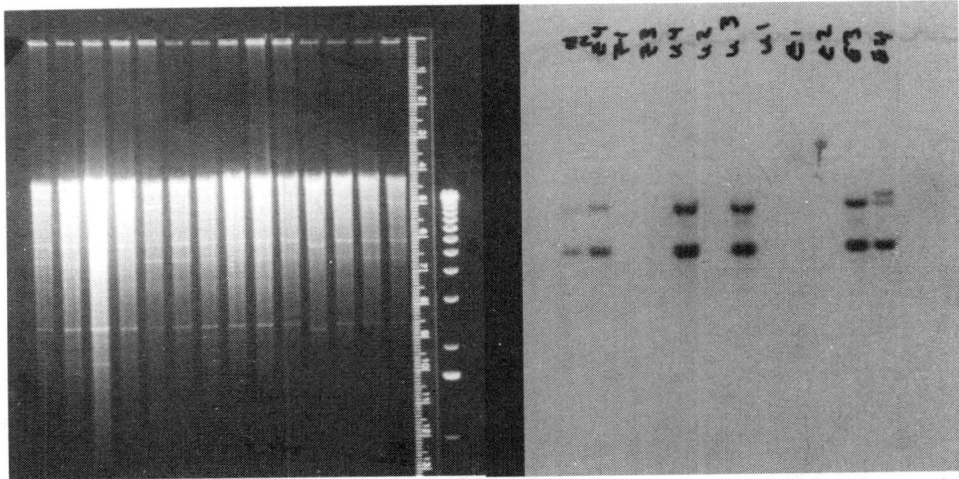

Fig. 6.47 : The figure on the left shows a photograph of a 0.7% agarose gel that has 14 different samples loaded on it (plus molecular weight marker in the far right lane and a glowing ruler used for analysis of the results). Each sample of DNA has been digested with the same restriction enzyme (EcoRI). Notice that the DNA does not appear as a series of discrete bands but rather as a smear. The DNA was transferred to nitrocellulose and then probed with a radioactive fragment of DNA that was derived from the transformed gene. The figure on the right is a copy of the X-ray film and reveals which strains contain the target DNA and which ones do not.

(2) The DNA is transfered to a membrane which is a sheet of special blotting paper. The DNA fragementsw retain the same pattern of separation they had on the gel.

(3) The blot is incubated with many copies of a probe which is single-stranded DNA. This probe will form base pairs with its complementary DNA sequence and bind to form a double-stranded DNA molecule. The probe cannot be seen but it is either radioactive or has an enzyme bound to it (e.g. alkaline phosphatase or horseradish peroxidase).

(4) The location of the probe is revealed by incubating it with a colorless substrate that the attached enzyme converts to a coloured product that can be seen or gives off light which will expose X-ray film. If the probe was labelled with radioactivity, it can expose X-ray film directly.

Below is an example of a real Southern blot used to detect the presence of a gene that was transformed into a mixed cell population. In this Southern blot, it is easy to determine which cells incorporated the gene and which ones did not.

NORTHERN BLOTTING

The **northern blot** is a technique used in molecular biology research to study gene expression by detection of RNA (or isolated mRNA) in a sample.

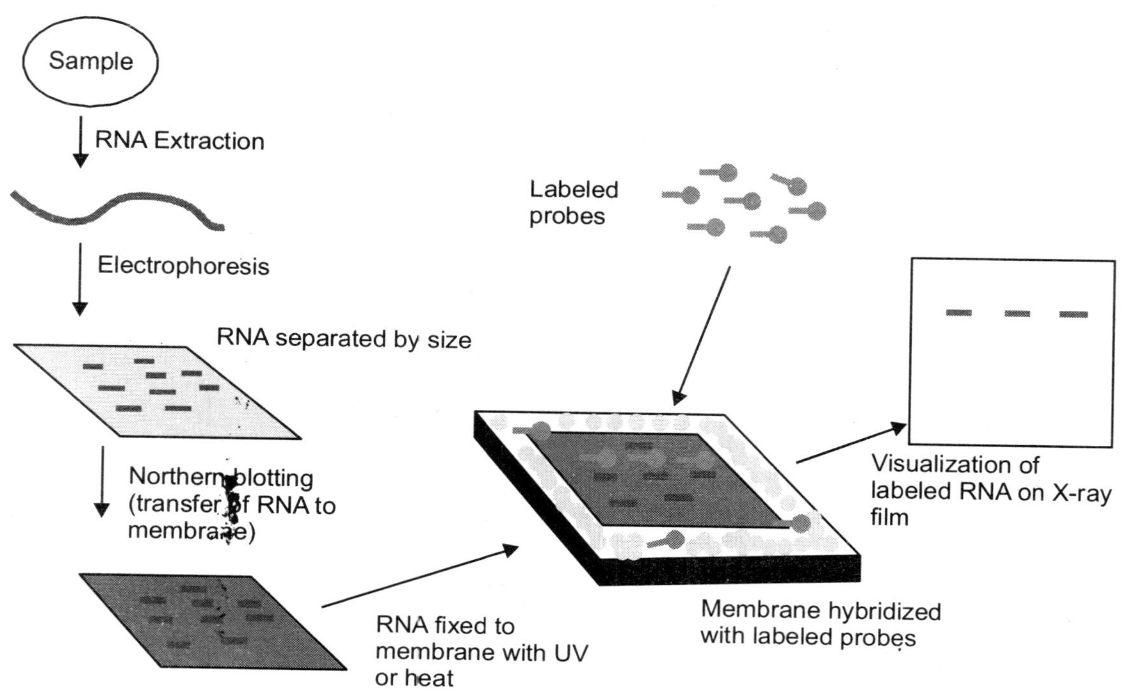

Fig. 6.48 : Flow diagram outlining the general procedure for RNA detection by northern blotting

With northern blotting it is possible to observe cellular control over structure and function by determining the particular gene expression levels during differentiation, morphogenesis, as well as abnormal or diseased conditions. Northern blotting involves the use of electrophoresis to separate RNA samples by size, and detection with a hybridization probe complementary to part of or the entire target sequence. The term 'northern blot' actually refers specifically to the capillary transfer of RNA from the electrophoresis gel to the blotting membrane, however the entire process is commonly referred to as northern blotting. The northern blot technique was developed in 1977 by James Alwine, David Kemp, and George Stark at Stanford University. Northern blotting takes its name from its similarity to the first blotting technique, the Southern blot, named for biologist Edwin Southern. The major difference is that RNA, rather than DNA, is analyzed in the northern blot.

A general blotting procedure starts with extraction of total RNA from a homogenized tissue sample. The mRNA can then be isolated through the use of oligo (dT) cellulose chromatography to maintain only those RNAs with a poly(A) tail. RNA samples are then separated by gel electrophoresis. Since the gels are fragile and the probes are unable to enter the matrix, the RNA samples, now separated by size, are transferred to a nylon membrane through a capillary or vacuum blotting system.

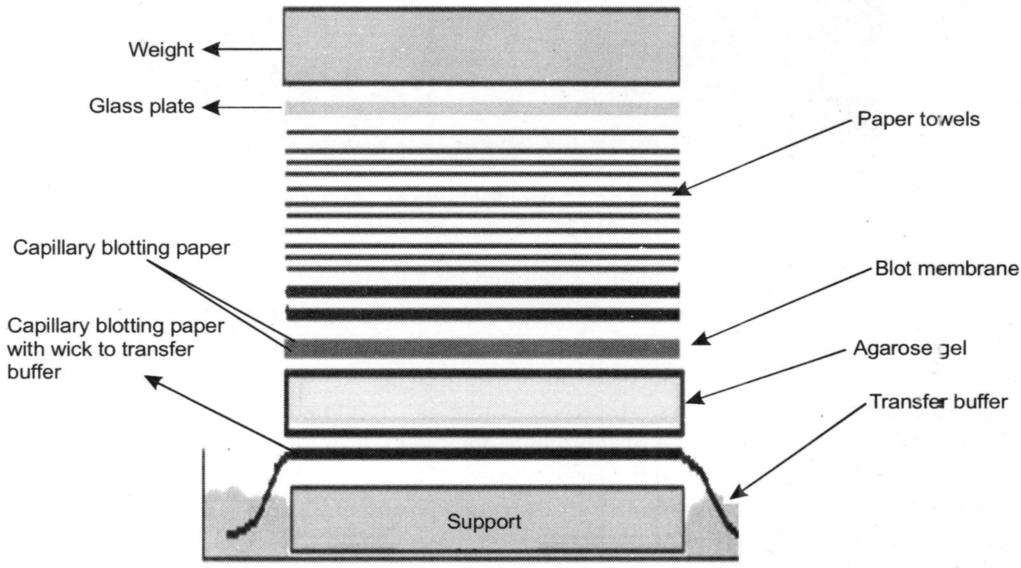

Fig. 6.49 : Capillary Blotting System set-up for the Transfer of RNA from an Electrophoresis Gel to a Blotting Membrane

A nylon membrane with a positive charge is the most effective for use in northern blotting since the negatively charged nucleic acids have a high affinity for them. The transfer buffer used for the blotting usually contains formamide because it lowers the annealing temperature of the probe-RNA interaction preventing RNA degradation by high temperatures. Once the RNA has been transferred to the membrane it is immobilized through covalent linkage to the membrane by UV light or heat. After a probe has been labelled, it is hybridized to the RNA on the membrane. Experimental conditions that can affect the efficiency and specificity of hybridization include ionic strength, viscosity, duplex length, mismatched base pairs, and base composition. The membrane is washed to ensure that the probe has bound specifically and to avoid background signals from arising. The hybrid signals are then detected by X-ray film and can be quantified by densitometry. To create controls for comparison in a northern blot, samples not displaying the gene product of interest can be used after determination by microarrays or RT-PCR.

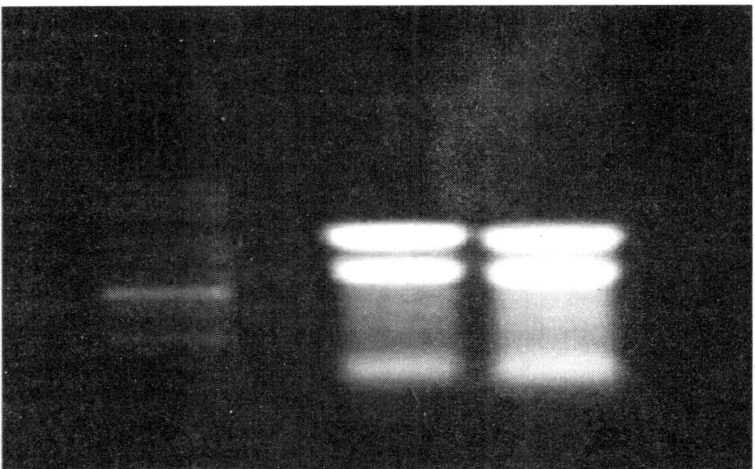

Fig. 6.50 : The RNA run on a formaldehyde agarose gel to highlight the 28S (top band) and 18S (lower band) ribosomal subunits

The RNA samples are most commonly separated on agarose gels containing formaldehyde as a denaturing agent for the RNA to limit secondary structure. The gels can be stained with ethidium bromide (EtBr) and viewed under UV light to observe the quality and quantity of RNA before blotting. Polyacrylamide gel electrophoeresis with urea can also be used in RNA separation but it is most commonly used for fragmented or microRNAs. An RNA ladder is often run alongside the samples on an electrophoresis gel to observe the size of fragments obtained but in total RNA samples the ribosomal subunits can act as size markers. Since the large ribosomal subunit is 28S (approximately 5kb) and the small ribosomal subunit is 18S (approximately 2kb) two prominent bands will appear on the gel, the larger at close to twice the intensity of the smaller.

Probes for northern blotting are composed of nucleic acids with a complementary sequence to all or part of the RNA of interest, they can be DNA, RNA, or oligonucleotides with a minimum of 25 complementary bases to the target sequence. RNA probes (riboprobes) that are transcribed in vitro are able to withstand more rigorous washing steps preventing some of the background noise. Commonly cDNA is created with labelled primers for the RNA sequence of interest to act as the probe in the northern blot. The probes need to be labelled either with radioactive isotopes (^{32}P) or with chemiluminescence in which alkaline phosphatase or horseradish peroxidase breakdown chemiluminescent substrates producing a detectable emission of light. The chemiluminescent labelling can occur in a couple ways, the probe attached to the enzyme, or the probe labelled with a ligand (e.g. biotin) for which the antibody (e.g. avidin or streptavidin) is attached to the enzyme. X-ray film can detect both the radioactive and chemiluminescent signals and many researchers prefer the chemiluminescent signals because they are faster, more sensitive, and reduce the health hazards that go along with radioactive labels. The same membrane can be probed up to five times without a significant loss of the target RNA.

APPLICATIONS IN MEDICINE

Northern blotting allows one to observe a particular gene's expression pattern between tissues, organs, developmental stages, environmental stress levels, pathogen infection, and over the course of treatment. The technique has been used to show overexpression of oncogenes and downregulation of tumor-suppressor genes in cancerous cells when compared to 'normal' tissue, as well as the gene expression in the rejection of transplanted organs. If an upregulated gene is observed by an abundance of mRNA on the northern blot the sample can then be sequenced to determine if the gene is known to researchers or if it is a novel finding. The expression patterns obtained under given conditions can provide insight into the function of that gene. Since the RNA is first separated by size, if only one probe type is used variance in the level of each band on the membrane can provide insight into the size of the product, suggesting alternative splice products of the same gene or repetitive sequence motifs. The variance in size of a gene product can also indicate deletions or errors in transcript processing, by altering the probe target used along the known sequence it is possible to determine which region of the RNA is missing.

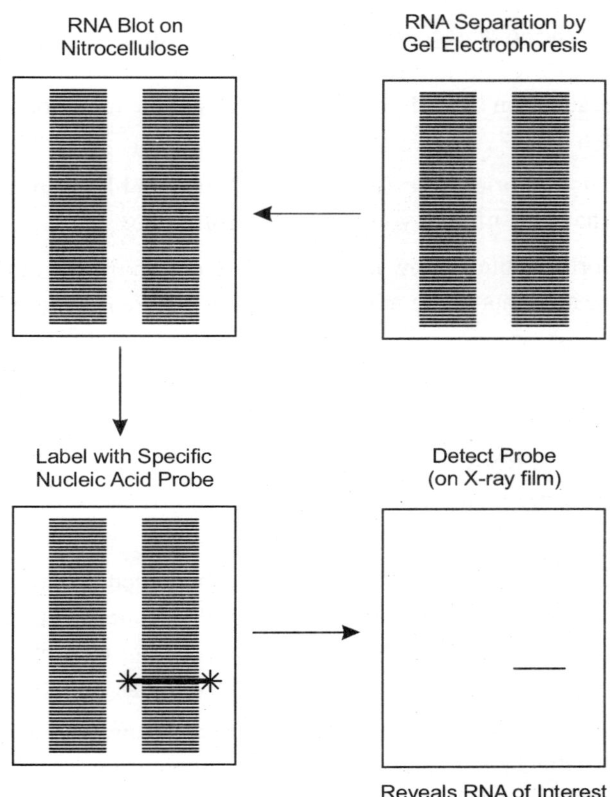

Fig. 6.51 : The Northern Blot Method

BlotBase is an online database publishing northern blots. BlotBase has over 700 published northern blots of human and mouse samples, in over 650 genes across more than 25 different tissue types. Northern blots can be searched by a blot ID, paper reference, gene identifier, or by tissue. The results of a search provide the blot ID, species, tissue, gene, expression level, blot image (if available), and links to the publication that the work originated from. This new database provides sharing of information between members of the science community that was not previously seen in northern blotting as it was in sequence analysis, genome determination, protein structure, etc.

Analysis of gene expression can be done by several different methods including RT-PCR, RNase protection assays, microarrays, serial analysis of gene expression (SAGE), as well as northern blotting. Microarrays are quite commonly used and are usually consistent with data obtained from northern blots, however at times northern blotting is able to detect small changes in gene expression that microarrays cannot. The advantage that microarrays have over northern blots is that thousands of genes can be visualized at a time while northern blotting is usually looking at one or a small number of genes. A problem in northern blotting is often sample degradation by RNases (both endogenous to the sample and through environmental contamination) which can be avoided by proper sterilization of glassware and the use of RNase inhibitors such as DEPC (diethylpyrocarbonate). The chemicals used in most northern blots can be a risk to the researcher, since formaldehyde, radioactive material, ethidium bromide, DEPC, and UV light are all harmful under certain exposures. Compared to RT-PCR northern blotting has a low sensitivity but it also has a high specificity which is important to reduce false positive results. The advantages of using northern blotting include the detection of RNA size, the observation of alternate splice products, the use of probes with partial homology, the quality and quantity of RNA can be measured on the gel prior to blotting, and the membranes can be stored and reprobed for years after blotting.

This is a brief overview of how a Northern blot (more formally called an RNA blot) is performed and what type of data you can obtain form one.

Northern blots allow investigators to determine the molecular weight of an mRNA and to measure relative amounts of the mRNA present in different samples.

(1) RNA (either total RNA or just mRNA) is separated by gel electrophoresis, usually an agarose gel. Because there are so many different RNA molecules on the gel, it usually appears as a smear rather than discrete bands.

(2) The RNA is transfered to a sheet of special blotting paper called nitrocellulose, though other types of paper, or membranes, can be used. The RNA molecules retain the same pattern of separation they had on the gel.

(3) The blot is incubated with a probe which is single-stranded DNA. This probe will form base pairs with its complementary RNA sequence and bind to form a double-stranded RNA-DNA molecule. The probe cannot be seen but it is either radioactive or has an enzyme bound to it (e.g. alkaline phosphatase or horseradish peroxidase).

(4) The location of the probe is revealed by incubating it with a colourless substrate that the attached enzyme converts to a coloured product that can be seen or gives off light which will expose X-ray film. If the probe was labelled with radioactivity, it can expose X-ray film directly.

RECOMBINANT DNA

Recombinant DNA is a form of DNA that does not exist naturally, which is created by combining DNA sequences that would not normally occur together. In terms of genetic modification, recombinant DNA (rDNA) is introduced through the addition of relevant DNA into an existing organismal DNA, such as the plasmids of bacteria, to code for or alter different traits for a specific purpose, such as antibiotic resistance. It differs from genetic recombination, in that it does not occur through processes within the cell, but is engineered. A recombinant protein is protein that is derived from recombinant DNA.

The Recombinant DNA technique was first proposed by Peter Lobban, a graduate student with A. Dale Kaiser at the Stanford University Department of Biochemistry. The technique was then realized by Lobban and Kaiser; Jackson, Symons and Berg; and Cohen, Chang, Boyer and Helling. Stanley Norman Cohen, in 1972-3. They published their findings in papers including a 1972 paper entitled "Biochemical Method for Inserting New Genetic Information into DNA of Simian Virus 40: Circular SV40 DNA Molecules Containing Lambda Phage Genes and the Galactose Operon of Escherichia coli", a 1973 paper entitled "Enzymatic end-to end joining of DNA molecules" and a 1974 paper entitled "Construction of Biologically Functional Bacterial Plasmids *in vitro*", all of which described techniques to isolate and amplify genes or DNA segments and insert them into another cell with precision, creating a transgenic bacterium. Recombinant DNA technology was made possible by the discovery of restriction endonucleases by Werner Arber, Daniel Nathans, and Hamilton Smith, for which they received the 1978 Nobel Prize in Medicine.

Cloning and Relation to Plasmids

The use of cloning is interrelated with Recombinant DNA in classical biology, as the term "clone" refers to a cell or organism derived from a parental organism, with modern biology referring to the term as a collection of cells derived from the same cell that remain identical. In the classical instance, the use of recombinant DNA provides the initial cell from which the host organism is then expected to recapitulate when it undergoes further cell division, with bacteria remaining a prime example due to the use of viral vectors in medicine that contain recombinant DNA inserted into a structure known as a plasmid. Plasmids are extrachromosomal self-replicating circular forms of DNA present in most bacteria, such as *Escherichia coli (E. Coli)*, containing genes related to catabolism and metabolic activity, and allowing the carrier bacterium to survive and reproduce in conditions present within other species and environments. These genes represent characteristics of resistance to bacteriophages and antibiotics[1] and some heavy metals, but can also be fairly easily removed or separated from the plasmid by restriction endonucleases, which regularly produce "sticky ends" and allow the attachment of a selected segment of DNA, which codes for more "reparative" substances, such as peptide hormone medications including insulin, growth hormone, and oxytocin. In the introduction of useful genes into the plasmid, the bacteria are then used as a viral vector, which are encouraged to reproduce so as to recapitulate the altered DNA within other cells it infects, and increase the amount of cells with the recombinant DNA present within them.

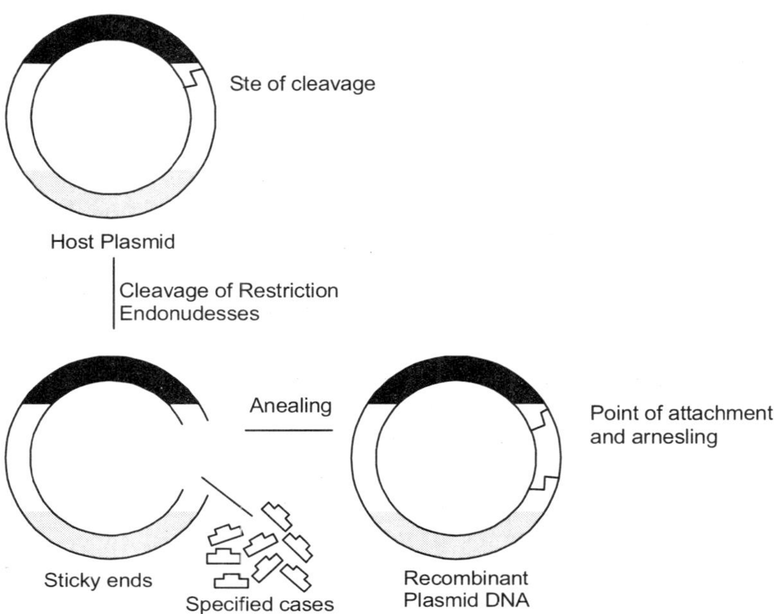

Fig. 6.52 : A Simple Example of how a Desired Gene is Inserted into a Plasmid. In this example, the gene specified in the white colour becomes useless as the new gene is added

The use of plasmids is also key within gene therapy, where their related viruses are used as **cloning vectors** or carriers, which are means of transporting and passing on genes in recombinant DNA through viral reproduction throughout an organism. Plasmids contain three common features — a **replicator**, **selectable marker** and a **cloning site**. The replicator or "ori" refers to the origin of replication with regard to location and bacteria where replication begins. The marker refers to a particular gene that usually contains resistance to an antibiotic, but may also refer to a gene that is attached alongside the desired one, such as that which confers luminescence to allow identification of successfully recombined DNA. The cloning site is a sequence of nucleotides representing one or more positions where cleavage by restriction endonucleases occurs. Most eukaryotes do not maintain canonical plasmids; yeast is a notable exception. In addition, the Ti plasmid of the bacterium *Agrobacterium tumefaciens* can be used to integrate foreign DNA into the genomes of many plants. Other methods of introducing or creating recombinant DNA in eukaryotes include homologous recombination and transfection with modified viruses.

When recombinant DNA is then further altered or changed to host additional strands of DNA, the molecule formed is referred to as "chimeric" DNA molecule, with reference to the mythological chimera, which consisted as a composite of several animals. The presence of chimeric plasmid molecules is somewhat regular in occurrence, as, throughout the lifetime of an organism, the propagation by vectors ensures the presence of hundreds of thousands of organismal and bacterial cells that all contain copies of the original chimeric DNA.

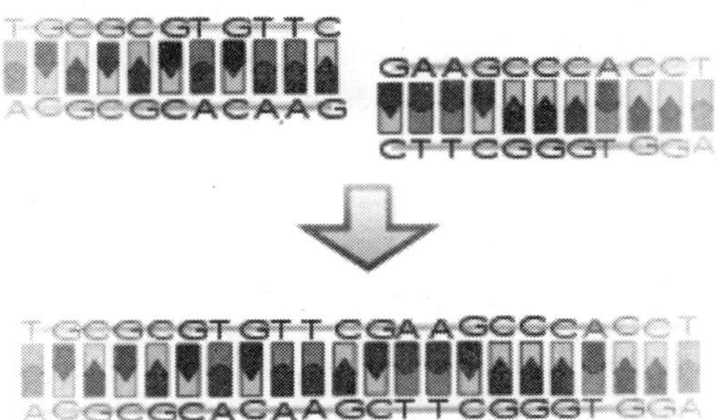

Fig. 6.53 : An Example of Chimeric Plasmid Formation from Two "Blunt ends" via the Enzyme, T4 Ligase

In the production of chimeric plasmids, the processes involved can be somewhat uncertain, as the intended outcome of the addition of foreign DNA may not always be achieved and may result in the formation of unusable plasmids. Initially, the plasmid structure is linearised to allow the addition by bonding of complementary foreign DNA strands to single-stranded "overhangs" or "sticky ends" present at the ends of the DNA molecule from staggered, or "S-shaped" cleavages produced by restriction endonucleases.

A common weiner vector used for the donation of plasmids originally was the bacterium Escherichia coli and, later, the EcoRI derivative, which was used for its versatility with addition of new DNA by "relaxed" replication when inhibited by chloramphenicol and spectinomycin, later being replaced by the pBR322 plasmid. In the case of EcoRI, the plasmid can anneal with the presence of foreign DNA via the route of sticky-end ligation, or with "blunt ends" via blunt-end ligation, in the presence of the phage T_4 ligase, which forms covalent links between 3-carbon OH and 5-carbon PO_4 groups present on blunt ends. Both sticky-end, or overhang ligation and blunt-end ligation can occur between foreign DNA segments, and cleaved ends of the original plasmid depending upon the restriction endonuclease used for cleavage.

Until the 1920s, there was no known way to produce insulin because the hormone was not officially identified until 1921. Once identified, the production problem was quickly solved when it was found that insulin from the pancreas of a cow, pig or even some species of fish could be used successfully in humans. This method was the primary solution for type 1 diabetes mellitus for decades, and manufacturing methods had steadily improved the purity of the hormone which was made from animal pancreases. However, proponents of the genetic engineering technology continued to raise what they claimed was a looming problem with traditional methods of insulin production: a supposed shortage of supply in the not-too-distant future. But in the 1987 book *Invisible Frontiers: The Race to Synthesize a Human Gene*, author Stephen S. Hall wrote that the supposed shortage is now known to be an assumption based on mistaken facts. He wrote:

To hear some tell it, there was never a supply problem with pig pancreases in the first place. "The whole thing was rubbish," insists Paul Haycook, research director at Squibb-Novo. "There was never a shortage of pig pancreases, and there never will be." Haycook blames the scare on a miscalculation by an official who had prepared projections for the Food and Drug Administration — a mistake based, ironically, on a mistake in an Eli Lilly training brochure which confused kilograms with pounds. Instead of projecting an insulin shortage by 1982, a revised FDA report predicted adequate insulin supplies through the year 2006. In any event, there is never likely to be a shortage caused by a scarcity of pancreases.

Scientists and entrepreneurs were very eager to show they could devise another way to synthesize the hormone, in part, because of competition from other researchers and also because of the promise for the fame and fortune that its so-called "discovery" could bring them. Insulin was part of a wider vision to introduce biotechnology medicines, and was chosen specifically because it is a simple hormone and was therefore relatively easy to copy.Insulin was chosen as the ideal candidate because it is a relatively simple protein, it was so widely used that if researchers could prove that biosynthetic "human" insulin was safe and effective, then the technology would be accepted as such, and it would open the flood gates for many other products to be made this way.

That was exactly what happened. One of the biggest breakthroughs in recombinant DNA technology happened in the manufacture of biosynthetic "human" insulin, which was the first medicine made via recombinant DNA technology ever to be approved by the FDA.

Henry I. Miller was an early advocate of biotechnology and drugs, and continues to be so as a fellow at the Hoover Institute even though he no longer works for the FDA. Miller began work for the FDA as head of a special department created to establish new procedures for approving drugs created through biotechnology. In his book "To America's Health: A Model for Reform of the Food and Drug Administration" (Hoover Institution Press, 2000), Miller states that he pushed for rapid approval of biosynthetic insulin from his boss, who was not comfortable approving it on such short notice, especially when it had been tested on so few people. Amazingly, Miller admits that he actually waited for his boss to go on vacation, and then took the approval to his boss' boss, who then approved the drug.

As far as technical details, the specific gene sequence, or oligonucleotide, that codes for insulin production in humans was introduced to a sample colony of *E. coli* (the bacteria found in the human intestine). Only about 1 out of 10^6 bacteria picks up the sequence. However, this is not really a problem, because the lifecycle is only about 30 minutes for *E. coli*. This means that in a 24-hour period, there may be billions of *E. coli* that are coded with the DNA sequences needed to induce insulin production.

However, a sampling of initial reaction showed that Humulin was greeted more as a technological rather than a medical breakthrough, and that this sentiment was building even before the drug reached pharmacies. As early as 1980, the British magazine *New Scientist* reported, "Other big chemical manufacturers predict that Eli Lilly's massive $40 million investment in two plants to make insulin - may be a classic example of backing a loser." *The Economist* concluded: "The first bug-built drug for human use may turn out to be a commercial flop. But the way has now been cleared-and remarkably quickly, too - for biotechnologists with interesting new products to clear the regulatory hurdles and run away with the prizes."

Ultimately, widespread consumer adoption of biosynthetic "human" insulin did not occur until the manufacturers removed highly-purified animal insulin from the market.

POLYMERIZED CHAIN REACTION (PCR)

In molecular biology, the polymerase chain reaction (PCR) is a technique to amplify a single or few copies of a piece of DNA across several orders of magnitude, generating thousands to millions of copies of a particular DNA sequence. The method relies on thermal cycling, consisting of cycles of repeated heating and cooling of the reaction for DNA melting and enzymatic replication of the DNA. Primers (short DNA fragments) containing sequences complementary to the target region along with a DNA polymerase (after which the method is named) are key components to enable selective and repeated amplification. As PCR progresses, the DNA generated is itself used as a template for replication, setting in motion a chain reaction in which the DNA template is exponentially amplified. PCR can be extensively modified to perform a wide array of genetic manipulations.

Almost all PCR applications employ a heat-stable DNA polymerase, such as Taq polymerase, an enzyme originally isolated from the bacterium *Thermus aquaticus*. This DNA polymerase enzymatically assembles a new DNA strand from DNA building blocks, the nucleotides, by using single-stranded DNA as a template and DNA oligonucleotides (also called DNA primers), which are required for initiation of DNA synthesis. The vast majority of PCR methods use thermal cycling, i.e., alternately heating and cooling the PCR sample to a defined series of temperature steps. These thermal cycling steps are necessary first to physically separate the two strands in a DNA double helix at a high temperature in a process called DNA melting. At a lower temperature, each strand is then used as the template in DNA synthesis by the DNA polymerase to selectively amplify the target DNA. The selectivity of PCR results from the use of primers that are complementary to the DNA region targeted for amplification under specific thermal cycling conditions.

Developed in 1984 by Kary Mullis, PCR is now a common and often indispensable technique used in medical and biological research labs for a variety of applications. These include DNA cloning for sequencing, DNA-based phylogeny, or functional analysis of genes; the diagnosis of hereditary diseases; the identification of genetic fingerprints (used in forensic sciences and paternity testing); and the detection and diagnosis of infectious diseases. In 1993 Mullis was awarded the Nobel Prize in Chemistry for his work on PCR.

A 1971 paper in the Journal of Molecular Biology by Kleppe and co-workers first described a method using an enzymatic assay to replicate a short DNA template with primers *in vitro*. However, this early manifestation of the basic PCR principle did not receive much attention, and the invention of the polymerase chain reaction in 1983 is generally credited to Kary Mullis.

At the core of the PCR method is the use of a suitable DNA polymerase able to withstand the high temperatures of >90 °C (>195 °F) required for separation of the two DNA strands in the DNA double helix after each replication cycle. The DNA polymerases initially employed for in vitro experiments presaging PCR were unable to withstand these high temperatures. So the early procedures for DNA replication were very inefficient, time consuming, and required large amounts of DNA polymerase and continual handling throughout the process.

The discovery in 1976 of Taq polymerase (a DNA polymerase purified from the thermophilic bacterium, *Thermus aquaticus*, which naturally occurs in hot (50 to 80 °C (120 to 175 °F)) environments) paved the way for dramatic improvements of the PCR method. The DNA polymerase isolated from *T. aquaticus* is stable at high temperatures remaining active even after DNA denaturation, thus obviating the need to add new DNA polymerase after each cycle. This allowed an automated thermocycler-based process for DNA amplification.

When Mullis developed the PCR in 1983, he was working in Emeryville, California for Cetus Corporation, one of the first biotechnology companies. There, he was responsible for synthesizing short chains of DNA. Mullis has written that he conceived of PCR while cruising along the Pacific Coast Highway one night in his car. He was playing in his mind with a new way of analyzing changes (mutations) in DNA when he realized that he had instead invented a method of amplifying any DNA region through repeated cycles of duplication driven by DNA polymerase.

In *Scientific American*, Mullis summarized the procedure: "Beginning with a single molecule of the genetic material DNA, the PCR can generate 100 billion similar molecules in an afternoon. The reaction is easy to execute. It requires no more than a test tube, a few simple reagents, and a source of heat." He was awarded the Nobel Prize in Chemistry in 1993 for his invention, seven years after he and his colleagues at Cetus first put his proposal to practice. However, some controversies have remained about the intellectual and practical contributions of other scientists to Mullis' work, and whether he had been the sole inventor of the PCR principle. The PCR technique was patented by Kary Mullis and assigned to Cetus Corporation, where Mullis worked when he invented the technique in 1983. The *Taq* polymerase enzyme was also covered by patents. There have been several high-profile lawsuits related to the technique, including an unsuccessful lawsuit brought by DuPont. The pharmaceutical company Hoffmann-La Roche purchased the rights to the patents in 1992 and currently holds those that are still protected.

A related patent battle over the Taq polymerase enzyme is still ongoing in several jurisdictions around the world between Roche and Promega. The legal arguments have extended beyond the lives of the original PCR and Taq polymerase patents, which expired on March 28, 2005

PCR is used to amplify a specific region of a DNA strand (the DNA target). Most PCR methods typically amplify DNA fragments of up to ~10 kilo base pairs (kb), although some techniques allow for amplification of fragments up to 40 kb in size.

A basic PCR set up requires several components and reagents. These components include:

- *DNA template* that contains the DNA region (target) to be amplified.
- Two *primers* that are complementary to the 3' (three prime) ends of each of the sense and anti-sense strand of the DNA target.
- *Taq polymerase* or another DNA polymerase with a temperature optimum at around 70 °C.
- *Deoxynucleoside triphosphates* (dNTPs; also very commonly and erroneously called deoxynucleotide triphosphates), the building blocks from which the DNA polymerases synthesizes a new DNA strand.

- *Buffer solution,* providing a suitable chemical environment for optimum activity and stability of the DNA polymerase.
- *Divalent cations,* magnesium or manganese ions; generally Mg^{2+} is used, but Mn^{2+} can be utilized for PCR-mediated DNA mutagenesis, as higher Mn^{2+} concentration increases the error rate during DNA synthesis
- *Monovalent cation* potassium ions.

The PCR is commonly carried out in a reaction volume of 10-200 μl in small reaction tubes (0.2-0.5 ml volumes) in a thermal cycler. The thermal cycler heats and cools the reaction tubes to achieve the temperatures required at each step of the reaction (see below). Many modern thermal cyclers make use of the Peltier effect which permits both heating and cooling of the block holding the PCR tubes simply by reversing the electric current. Thin-walled reaction tubes permit favourable thermal conductivity to allow for rapid thermal equilibration. Most thermal cyclers have heated lids to prevent condensation at the top of the reaction tube. Older thermocyclers lacking a heated lid require a layer of oil on top of the reaction mixture or a ball of wax inside the tube.

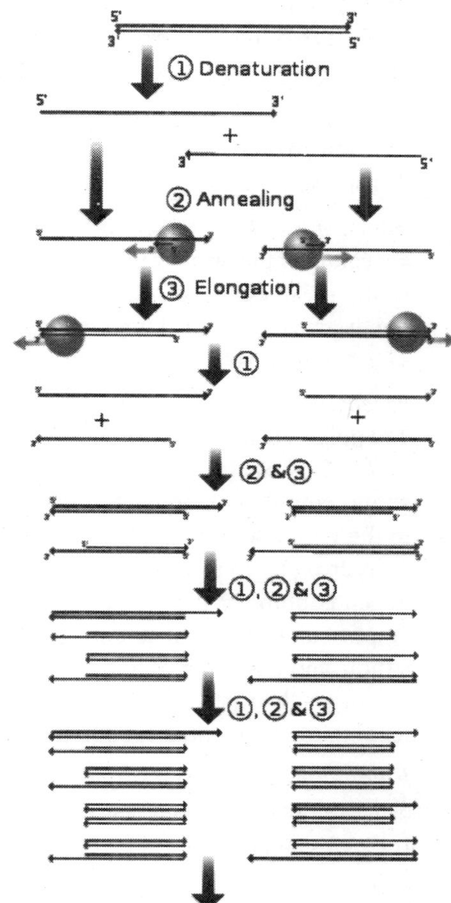

Fig. 6.54: Schematic drawing of the PCR cycle. (1) Denaturing at 94–96 °C. (2) Annealing at ~65 °C (3) Elongation at 72 °C. Four cycles are shown here. The blue lines represent the DNA template to which primers (red arrows) anneal that are extended by the DNA polymerase (light green circles), to give shorter DNA products (green lines), which themselves are used as templates as PCR progresses

The PCR usually consists of a series of 20 to 40 repeated temperature changes called cycles; each cycle typically consists of 2-3 discrete temperature steps. Most commonly PCR is carried out with cycles that have three temperature steps (Fig. 6.54). The cycling is often preceded by a single temperature step (called *hold*) at a high temperature (>90°C), and followed by one hold at the end for final product extension or brief storage. The temperatures used and the length of time they are applied in each cycle depend on a variety of parameters. These include the enzyme used for DNA synthesis, the concentration of divalent ions and dNTPs in the reaction, and the melting temperature (Tm) of the primers.

- *Initialization step*: This step consists of heating the reaction to a temperature of 94–96 °C (or 98°C if extremely thermostable polymerases are used), which is held for 1–9 minutes. It is only required for DNA polymerases that require heat activation by hot-start PCR.
- *Denaturation step*: This step is the first regular cycling event and consists of heating the reaction to 94–98°C for 20–30 seconds. It causes DNA melting of the DNA template by disrupting the hydrogen bonds between complementary bases, yielding single strands of DNA.
- *Annealing step*: The reaction temperature is lowered to 50–65 °C for 20–40 seconds allowing annealing of the primers to the single-stranded DNA template. Typically the annealing temperature is about 3-5 degrees Celsius below the Tm of the primers used. Stable DNA-DNA hydrogen bonds are only formed when the primer sequence very closely matches the template sequence. The polymerase binds to the primer-template hybrid and begins DNA synthesis.
- *Extension/elongation step*: The temperature at this step depends on the DNA polymerase used; Taq polymerase has its optimum activity temperature at 75–80 °C, and commonly a temperature of 72 °C is used with this enzyme. At this step the DNA polymerase synthesizes a new DNA strand complementary to the DNA template strand by adding dNTPs that are complementary to the template in 5' to 3' direction, condensing the 5'-phosphate group of the dNTPs with the 3'-hydroxyl group at the end of the nascent (extending) DNA strand. The extension time depends both on the DNA polymerase used and on the length of the DNA fragment to be amplified. As a rule-of-thumb, at its optimum temperature, the DNA polymerase will polymerize a thousand bases per minute. Under optimum conditions, i.e., if there are no limitations due to limiting substrates or reagents, at each extension step, the amount of DNA target is doubled, leading to exponential (geometric) amplification of the specific DNA fragment.
- *Final elongation*: This single step is occasionally performed at a temperature of 70-74 °C for 5–15 minutes after the last PCR cycle to ensure that any remaining single-stranded DNA is fully extended.
- *Final hold*: This step at 4-15 °C for an indefinite time may be employed for short-term storage of the reaction.

To check whether the PCR generated the anticipated DNA fragment (also sometimes referred to as the amplimer or amplicon), agarose gel electrophoresis is employed for size separation of the PCR products. The size(s) of PCR products is determined by comparison with a DNA ladder (a molecular weight marker), which contains DNA fragments of known size, run on the gel alongside the PCR products

The PCR process can be divided into three stages:

Exponential amplification: At every cycle, the amount of product is doubled (assuming 100% reaction efficiency). The reaction is very sensitive: only minute quantities of DNA need to be present.

Levelling off stage: The reaction slows as the DNA polymerase loses activity and as consumption of reagents such as dNTPs and primers causes them to become limiting.

Plateau: No more product accumulates due to exhaustion of reagents and enzyme.

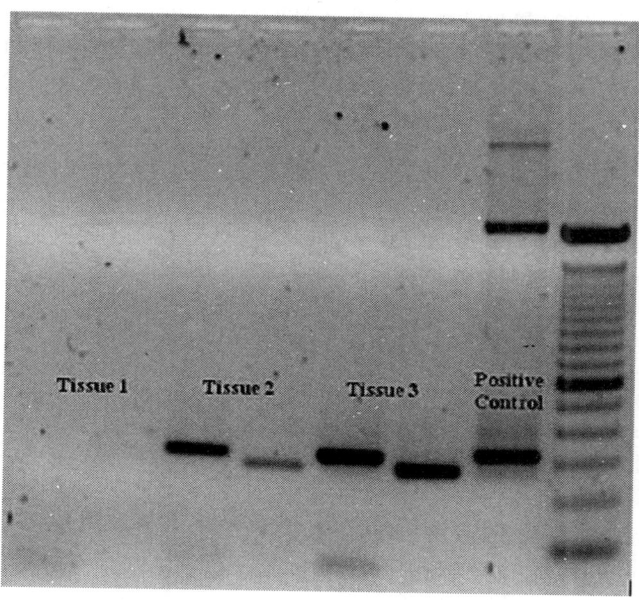

Fig. 6.55 : Ethidium Bromide-stained PCR Products after Gel Electrophoresis. Two sets of primers were used to amplify a target sequence from three different tissue samples. No amplification is present in sample #1; DNA bands in sample #2 and #3 indicate successful amplification of the target sequence. The gel also shows a positive control, and a DNA ladder containing DNA fragments of defined length for sizing the bands in the experimental PCRs.

In practice, PCR can fail for various reasons, in part due to its sensitivity to contamination causing amplification of spurious DNA products. Because of this, a number of techniques and procedures have been developed for optimizing PCR conditions. Contamination with extraneous DNA is addressed with lab protocols and procedures that separate pre-PCR mixtures from potential DNA contaminants. This usually involves spatial separation of PCR-setup areas from areas for analysis or purification of PCR products, use of disposable plasticware, and thoroughly cleaning the work surface between reaction setups. Primer-design techniques are important in improving PCR product yield and in avoiding the formation of spurious products, and the usage of alternate buffer components or polymerase enzymes can help with amplification of long or otherwise problematic regions of DNA.

PCR allows isolation of DNA fragments from genomic DNA by selective amplification of a specific region of DNA. This use of PCR augments many methods, such as generating hybridization probes for Southern or northern hybridization and DNA cloning, which require larger amounts of DNA, representing a specific DNA region. PCR supplies these techniques with high amounts of pure DNA, enabling analysis of DNA samples even from very small amounts of starting material.

Other applications of PCR include DNA sequencing to determine unknown PCR-amplified sequences in which one of the amplification primers may be used in Sanger sequencing, isolation of a DNA sequence to expedite recombinant DNA technologies involving the insertion of a DNA sequence into a plasmid or the genetic material of another organism. Bacterial colonies (*E. coli*) can be rapidly screened by PCR for correct DNA vector constructs. PCR may also be used for genetic fingerprinting; a forensic technique used to identify a person or organism by comparing experimental DNAs through different PCR-based methods.

Some PCR 'fingerprints' methods have high discriminative power and can be used to identify genetic relationships between individuals, such as parent-child or between siblings, and are used in paternity testing (Fig. 6.56). This technique may also be used to determine evolutionary relationships among organisms.

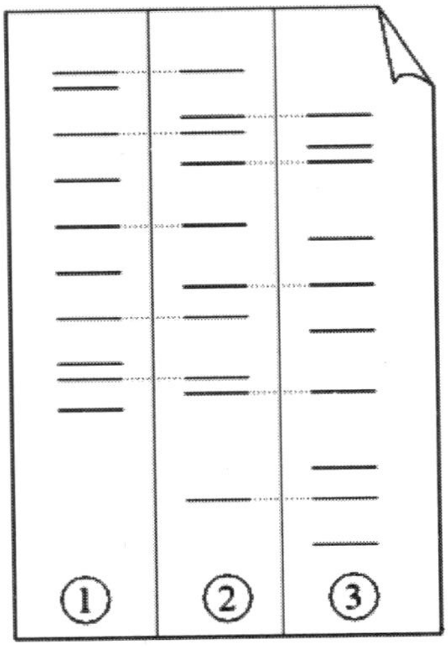

Because PCR amplifies the regions of DNA that it targets, PCR can be used to analyze extremely small amounts of sample. This is often critical for forensic analysis, when only a trace amount of DNA is available as evidence. PCR may also be used in the analysis of ancient DNA that is tens of thousands of years old. These PCR-based techniques have been successfully used on animals, such as a forty-thousand-year-old mammoth, and also on human DNA, in applications ranging from the analysis of Egyptian mummies to the identification of a Russian Tsar.

Fig. 6.56 : Electrophoresis of PCR-Amplified DNA Fragments. (1) Father. (2) Child. (3) Mother. The child has inherited some, but not all of the fingerprint of each of its parents, giving it a new, unique fingerprint

Quantitative PCR methods allow the estimation of the amount of a given sequence present in a sample – a technique often applied to quantitatively determine levels of gene expression. Real-time PCR is an established tool for DNA quantification that measures the accumulation of DNA product after each round of PCR amplification.

The PCR allows early diagnosis of malignant diseases such as leukemia and lymphomas, which is currently the highest developed in cancer research and is already being used routinely. PCR assays can be performed directly on genomic DNA samples to detect translocation-specific malignant cells at a sensitivity which is at least 10,000 fold higher than other methods.

The PCR also permits identification of non-cultivatable or slow-growing microorganisms such as mycobacteria, anaerobic bacteria, or viruses from tissue culture assays and animal models. The basis for PCR diagnostic applications in microbiology is the detection of infectious agents and the discrimination of non-pathogenic from pathogenic strains by virtue of specific genes.

Viral DNA can likewise be detected by PCR. The primers used need to be specific to the targeted sequences in the DNA of a virus, and the PCR can be used for diagnostic analyses or DNA sequencing of the viral genome. The high sensitivity of PCR permits virus detection soon after infection and even before the onset of disease. Such early detection may give physicians a significant lead in treatment. The amount of virus ("viral load") in a patient can also be quantified by PCR-based DNA quantitation techniques :

- **Allele-specific PCR**: This diagnostic or cloning technique is used to identify or utilize single-nucleotide polymorphisms (SNPs) (single base differences in DNA). It requires prior knowledge of a DNA sequence, including differences between alleles, and uses primers whose 3' ends encompass the SNP. PCR amplification under stringent conditions is much less efficient in the presence of a mismatch between template and primer, so successful amplification with an SNP-specific primer signals presence of the specific SNP in a sequence. See SNP genotyping for more information.
- **Assembly PCR** or **Polymerase Cycling Assembly (PCA)**: Assembly PCR is the artificial synthesis of long DNA sequences by performing PCR on a pool of long oligonucleotides with short overlapping segments. The oligonucleotides alternate between sense and antisense directions, and the overlapping segments determine the order of the PCR fragments thereby selectively producing the final long DNA product.
- **Asymmetric PCR**: Asymmetric PCR is used to preferentially amplify one strand of the original DNA more than the other. It finds use in some types of sequencing and hybridization probing where having only one of the two complementary strands is required. PCR is carried out as usual, but with a great excess of the primers for the chosen strand. Due to the slow (arithmetic) amplification later in the reaction after the limiting primer has been used up, extra cycles of PCR are required. A recent modification on this process, known as **L**inear-**A**fter-**T**he-**E**xponential-PCR (LATE-PCR), uses a limiting primer with a higher melting temperature (Tm) than the excess primer to maintain reaction efficiency as the limiting primer concentration decreases mid-reaction.
- **Helicase-dependent amplification**: This technique is similar to traditional PCR, but uses a constant temperature rather than cycling through denaturation and annealing/extension cycles. DNA Helicase, an enzyme that unwinds DNA, is used in place of thermal denaturation.
- **Hot-start PCR**: This is a technique that reduces non-specific amplification during the initial set up stages of the PCR. The technique may be performed manually by heating the reaction components to the melting temperature (e.g., 95°C) before adding the polymerase. Specialized enzyme systems have been developed that inhibit the polymerase's activity at ambient temperature, either by the binding of an antibody or by the presence of covalently bound inhibitors that only dissociate after a high-temperature activation step. Hot-start/cold-finish PCR is achieved with new hybrid polymerases that are inactive at ambient temperature and are instantly activated at elongation temperature.
- **Intersequence-specific PCR** (ISSR): a PCR method for DNA fingerprinting that amplifies regions between some simple sequence repeats to produce a unique fingerprint of amplified fragment lengths.
- **Inverse PCR**: a method used to allow PCR when only one internal sequence is known. This is especially useful in identifying flanking sequences to various genomic inserts. This involves

a series of DNA digestions and self ligation, resulting in known sequences at either end of the unknown sequence.

- **Ligation-mediated PCR**: This method uses small DNA linkers ligated to the DNA of interest and multiple primers annealing to the DNA linkers; it has been used for DNA sequencing, genome walking, and DNA footprinting.
- **Methylation-specific PCR** (MSP): The MSP method was developed by Stephen Baylin and Jim Herman at the Johns Hopkins School of Medicine, and is used to detect methylation of CpG islands in genomic DNA. DNA is first treated with sodium bisulfite, which converts unmethylated cytosine bases to uracil, which is recognized by PCR primers as thymine. Two PCRs are then carried out on the modified DNA, using primer sets identical except at any CpG islands within the primer sequences. At these points, one primer set recognizes DNA with cytosines to amplify methylated DNA, and one set recognizes DNA with uracil or thymine to amplify unmethylated DNA. MSP using qPCR can also be performed to obtain quantitative rather than qualitative information about methylation.
- **Miniprimer PCR**: Miniprimer PCR uses a novel thermostable polymerase (S-Tbr) that can extend from short primers ("smalligos") as short as 9 or 10 nucleotides. This method permits PCR targeting smaller primer binding regions, and is particularly useful to amplify unknown, but conserved, DNA sequences, such as the 16S (or eukaryotic 18S) rRNA gene. 16S rRNA miniprimer PCR was used to characterize a microbial mat community growing in an extreme environment, a hypersaline pond in Puerto Rico. In that study, deeply divergent sequences were discovered with high frequency and included representatives that deûned two new division-level taxa, suggesting that miniprimer PCR may reveal new dimensions of microbial diversity. By enlarging the "sequence space" that may be queried by PCR primers, this technique may enable novel PCR strategies that are not possible within the limits of primer design imposed by Taq and other commonly used enzymes.
- **Multiplex Ligation-dependent Probe Amplification (MLPA)**: permits multiple targets to be amplified with only a single primer pair, thus avoiding the resolution limitations of multiplex PCR.
- **Multiplex-PCR**: The use of multiple, unique primer sets within a single PCR mixture to produce amplicons of varying sizes specific to different DNA sequences. By targeting multiple genes at once, additional information may be gained from a single test run that otherwise would require several times the reagents and more time to perform. Annealing temperatures for each of the primer sets must be optimized to work correctly within a single reaction, and amplicon sizes, i.e., their base pair length, should be different enough to form distinct bands when visualized by gel electrophoresis.
- **Nested PCR**: increases the specificity of DNA amplification, by reducing background due to non-specific amplification of DNA. Two sets of primers are being used in two successive PCRs. In the first reaction, one pair of primers is used to generate DNA products, which besides the intended target, may still consist of non-specifically amplified DNA fragments. The product(s) are then used in a second PCR with a set of primers whose binding sites are

completely or partially different from and located 3' of each of the primers used in the first reaction. Nested PCR is often more successful in specifically amplifying long DNA fragments than conventional PCR, but it requires more detailed knowledge of the target sequences.

- **Overlap-extension PCR**: is a genetic engineering technique allowing the construction of a DNA sequence with an alteration inserted beyond the limit of the longest practical primer length.
- **Quantitative PCR (Q-PCR)**: is used to measure the quantity of a PCR product (preferably real-time). It is the method of choice to quantitatively measure starting amounts of DNA, cDNA or RNA. Q-PCR is commonly used to determine whether a DNA sequence is present in a sample and the number of its copies in the sample. The method with currently the highest level of accuracy is **Quantitative real-time PCR**. It is often confusingly known as RT-PCR (**R**eal **T**ime PCR) or RQ-PCR. QRT-PCR or RTQ-PCR are more appropriate contractions. RT-PCR commonly refers to reverse transcription PCR, which is often used in conjunction with Q-PCR. QRT-PCR methods use fluorescent dyes, such as Sybr Green, or fluorophore-containing DNA probes, such as TaqMan, to measure the amount of amplified product in real time.
- **RT-PCR**: (**R**everse **T**ranscription PCR) is a method used to amplify, isolate or identify a known sequence from a cellular or tissue RNA. The PCR is preceded by a reaction using reverse transcriptase to convert RNA to cDNA. RT-PCR is widely used in expression profiling, to determine the expression of a gene or to identify the sequence of an RNA transcript, including transcription start and termination sites and, if the genomic DNA sequence of a gene is known, to map the location of exons and introns in the gene. The 5' end of a gene (corresponding to the transcription start site) is typically identified by an RT-PCR method, named RACE-PCR, short for *Rapid Amplification of cDNA Ends*.
- **Single-cell PCR**:
- **Solid Phase PCR**: encompasses multiple meanings, including Polony Amplification (where PCR colonies are derived in a gel matrix, for example), 'Bridge PCR' (the only primers present are covalently linked to solid support surface), conventional Solid Phase PCR (where Asymmetric PCR is applied in the presence of solid support bearing primer with sequence matching one of the aqueous primers) and Enhanced Solid Phase PCR (where conventional Solid Phase PCR can be improved by employing high Tm and nested solid support primer with optional application of a thermal 'step' to favour solid support priming).
- **TAIL-PCR**: **Thermal asymmetric interlaced** PCR is used to isolate unknown sequence flanking a known sequence. Within the known sequence TAIL-PCR uses a nested pair of primers with differing annealing temperatures; a degenerate primer is used to amplify in the other direction from the unknown sequence.
- **Touchdown PCR (Step-down PCR)**: a variant of PCR that aims to reduce nonspecific background by gradually lowering the annealing temperature as PCR cycling progresses. The annealing temperature at the initial cycles is usually a few degrees (3-5°C) above the T_m of the primers used, while at the later cycles, it is a few degrees (3-5°C) below the primer T_m. The

higher temperatures give greater specificity for primer binding, and the lower temperatures permit more efficient amplification from the specific products formed during the initial cycles.

- **PAN-AC**: This method uses isothermal conditions for amplification, and may be used in living cells.
- **Universal Fast Walking**: this method allows genome walking and genetic fingerprinting using a more specific 'two-sided' PCR than conventional 'one-sided' approaches (using only one gene-specific primer and one general primer - which can lead to artefactual 'noise') by virtue of a mechanism involving lariat structure formation. Streamlined derivatives of UFW are LaNe RAGE (lariat-dependent nested PCR for rapid amplification of genomic DNA ends), 5'RACE LaNe and 3'RACE LaNe.

THE CONCEPT OF GENOMIC LIBRARY

A **genomic library** is a population of host bacteria, each of which carries a DNA molecule that was inserted into a cloning vector, such that the collection of cloned DNA molecules represents the entire genome of the source organism. This term also represents the collection of all of the vector molecules, each carrying a piece of the chromosomal DNA of the organism, prior to the insertion of these molecules into the host cells.

1. Plasmids: Have a capacity of 10 kb.
2. Phages: Have a capacity of 23 kb.
3. Cosmids: Have a capacity of 45 kb.
4. Yeast artificial chromosomes (YAC): Have a capacity of 1000 kb.

The DNA molecules of an organism of interest are isolated. The DNA molecules are then partially digested by an endonuclease restriction enzyme. Sometimes, the DNA molecules are digested for different lengths of time in order to ensure that all the DNA has been digested to manageable sizes. The digested DNA molecules are separated by size using agarose electrophoresis, and a suitable range of lengths of DNA pieces are isolated and ligated into vectors. The vectors can then be taken up by suitable hosts.The hosts are kept in liquid media and can be frozen at -80°C for a long period of time. Usually the hosts are bacteria that do not contain any plasmids, so as to be sensitive to antibiotics.

The process of subdividing genomic DNA into clonable elements and inserting them into hosts is called creating a library, a clone bank or a gene bank. A complete library of host cells will contain all of the genomic DNA of the source organism.

CHAPTER 7

Cancer Aids and Xenobiotics

THE BIOCHEMISTRY OF CANCER

In the study of the human body the terms anatomy and physiology for many years were supposed to refer only to those tissues and functions that could be observed with the eye, or the eye aided by the microscope, and the most minute particles visible with the microscope were about 1/10 micron1or 1/250,000 inch in diameter. When the study of matter was extended to include smaller particles, anatomy became physics, and physiology, chemistry. Yet the minutest fragments of matter of which we have any knowledge are physical, and chemical changes, processes in a general way known as biochemical, take place only in these minute subdivisions of matter. The extraordinary revelations of modern physics, ushered in by an understanding of those waves which are known as light, which travel approximately

Is 86,000 miles a second, have introduced us into a new realm of thought. The conception of matter today is that two particles or waves are concertized, the electron, representing an extraordinarily active negative phase of electricity, and the proton, positive and more massive, and that all matter is formed by varying degrees of fusion of electrons and protons in the form of atoms, which again are combined as molecules. Since the molecules are smaller than a ray of visible light, they do not diffract the light, but possess the characteristics of the mass. Colloids which are aggregations of molecules, and therefore larger in size, reflect rays of light. Colloidal particles include those from 1/1000 micron or 1/250000 inch in diameter to those of 1/10 micron or 1/250000 inch in diameter (visible to the eye aide by the microscope). Colloids, like molecules and atoms, are detectable only by ultramicroscopic methods.

A newer electric component is now comprehended as our understanding has enlarged, knowing as the photon, equally important with the electron and the proton. Electrons and protons are essentially indestructible. The photon has no such in destructibility. It represents the forces that are given off, under certain conditions, by atoms, consisting of masses of energy, with the same rapidity of wave motion and other characteristics of the electron and the proton, but it is destructible in the sense that the energy contained is completely dispersed as another force may be exhausted, or as it may be combined with atom. The light which makes plants grow and which gives us warmth has the double characteristics of waves and particles, and is found to consist ultimately of photons. The photon, then, has to do undoubtedly with our whole understanding of ultraviolet rays, some of the vitamins, and many of the newer conceptions which concern the chemical exchanges of the body. As far as life is concerned, the photon is as important as the two better understood and longer known particles which we speak of as the electron and the proton. The subject of light and its influence on the human body, and the potentialities of various kinds of food substances which in oily way or another are derived from light-and act as-light carriers, opens up an entirely new aspect of metabolism as related to health and disease. In all of the known combinations of electron, proton and photon in our conception of physics, the elements still maintain a place, although a precarious one. Investigations of the elements by Moseley, the brilliant young scientist, who at the age of twenty-eight was killed in the Gallipoli campaign, have brought out some interesting facts from which certain deductions can be made that apply to medicine. We know that there are ninety-two elements varying in atomic weight from hydrogen, which is the lightest, to uranium, which is the heaviest, and that between each two elements in the progression of atomic weight there is approximately the weight of an atom of the element hydrogen. It is possible, although not yet shown, that elements which are close in the scale of atomic weight can be changed, experimentally, at least, into the element above or below. Meithei and others have attempted to convert mercury (200) into gold (196) thus carrying out the (Iraqi of the alchemist.) It is scarcely necessary to call attention to the enormous value of such investigations as related to the industries. The molecule of coal contains fewer atoms of hydrogen than the molecule of petroleum oil. Bergius has been able to force additional atoms of hydrogen into certain types of coal and thus has produced petroleum oil. With the extraordinary accomplishments of physics and chemistry as related to the industries, the production of artificial rubber, artificial silk, and artificial foods, many profitable researches in human metabolism are being initiated, either in the discovery of the causative agents of disease or those factors which have to do with the development of resistance to disease. As illustrating the newer conceptions of the elements no better example can be taken than chlorophyll, the green colouring matter of plants, which is one of the most important substances in the world. Chlorophyll is able to convert the heat and energy of the sun's rays into power for transforming water and carbon dioxide in the atmosphere and certain inorganic materials from the earth into carbonaceous compounds. For this purpose. the element magnesium is essential, just as in the subsequent conversion of this material into heat and energy in the human body, the element iron is necessary. Life is maintained in the cells by combustion, an exchange of electrons which is carried out in the colloids of the cells. In the bodily economy, the element oxygen, which composes from 18.5 to 20 per cent of the blood, by volume, is associated with iron, of which there are one or more atoms in every molecule of hemoglobin. In animals below the vertebrates, such as the Amphioxus, which- in the scale of life stands between the vertebrates and the invertebrates, there is only white circulating fluid, and in this fluid the element copper takes the place of iron.

In the vertebrates copper has a powerful influence on respiration. This curious relation of copper to vital processes is seen in the treatment of pernicious anemia by the administration of liver. The extract of liver tissue appears to have the same effect as liver itself, and experimentally, in conditions of anemia, the ash has the same effect; the essential elemental constituent is copper. The liver is the great detoxication agent of the body, and to a large extent the source of the plasma of the blood. If a measured piece of the spleen or other organ is placed in the abdominal-cavity of an animal, harm may not result, but if a similar piece of liver is thus placed, the animal dies, and its liver shows the greatest effect from the resultant toxæmia. Krogh has shown that the walls of the vascular capillaries contrail contractile cells, derived from the non-striated muscle, which are to a large extent self-controlled. Under the circulatory pressure of the systole of the heart, the capillaries permit oxygen and molecular substances, such as the crystalloids and amino acids, to pass by filtration, osmosis, diffusion, and other forces, through the stomas in the vessel wall to serve vital purposes, to maintain nutrition, and produce heat, and energy. The ash of this combustion passes into the venous capillaries to be carried to the emunctories, and the blood passes to the lungs for a fresh supply of oxygen. When, however, certain toxic poisons, for example, histamine, paralyze the contractile cells, causing the stomas in the capillary walls to opens more widely, larger bodies, possibly colloids just above molecular size, and fluids of the Blood plasma, pass from the capillaries into the tissues, resulting in the condition known as shock. Carbohydrates (glucose, $C_6H_{12}O_6$ and fats (stearic acid, $C_{18}H_{36}O_2$) have no powers of building the body. The carbohydrates furnish heat, from which we recognize only the carbon dioxide as it is exhaled from the lungs, because by oxidation, hydrogen, which is a source of great value for heat and energy, is converted into water, which is used in the system. The fats are oxidized slowly because of the small amount of oxygen they contain, but they produce a much larger amount of heat and a greater quantity of water than the carbohydrates. The utilization of water produced from oxidation of fats and carbohydrates is shown in the hibernating animal, whose temperature during the winter drops to about 5O°, approximately that of the cave in which the animal takes his winter sleep. In the great retardation of all bodily functions and complete cessation of the function of the kidney, the physical needs are supplied by the slow oxidation of the store autumnal fat which gives not only the little energy necessary to carry on life, but also the essential water. Roger, a French experimenter, found evidence which led him to believe that metabolism of much of the fat is accomplished in the lungs, where oxygen is plentiful. The fat is carried from the intestinal tract through the thoracic duct to the left subclavian vein and finally deposited in the arterial capillaries of the lung, into which it finally disappears. He demonstrated that the blood from the right side of the heart contains a much larger amount of fat derivatives than the left. The experiment seems to show that oxygen is necessary for the conversion of fat, and that the conversion of fat into a form suitable for bodily use is not primarily a function of the liver.

Protein, on the contrary, contains in addition to the elements carbon, hydrogen, and oxygen, the element nitrogen and a little of the element sulphur. The nitrogen we know is necessary to the building of tissue, forming compounds in which may be deposited calcium and other elements.

Seventy-nine per cent. of the atmosphere is nitrogen. The oxidation of carbon and hydrogen within the body which gives rise to heat and energy is the outstanding feature of all animal life, but while the normal cell functions within normal limits, the carcinomatous cell oxidizes without control and without function. The analogy between the rapid growth of the fertilized ovum and growth of cancer has often been pointed out. The growth of the one, however, is orderly and with purpose, of the other disorderly and without purpose. There are two theories of biological oxidation, the oxygen activation theory of Warburg, and the activation-of-hydrogen theory of Wieland. Hopkins gives the weight of his authority to Wieland's theory as best explaining the chemistry of biologic processes. Since uncontrolled combustion of food material is the most important feature of the growth of cancer, it is worthwhile to consider how combustion normally is controlled. If a pound of glucose is burned outside the body it will produce the same amount of heat and energy as if burned in the body, but in the air it will burn only at temperatures over 4000°C. What holds the temperature of the body at 98.5° F.? The best explanation at present is that oxidation is carried on by catalysts or enzyme action. Kendall and others have shown that one of the agents which influences oxidation, glutathione, contains one atom of the element sulphur to the molecule and that if this atom of sulphur is removed; it ceases to function in this way. Sulphur itself has no such effect on metabolic processes. When a growth begins to undergo hyaline degeneration, there is a definite slowing up of the malignant process. In this connection, it is known that hyaline and keratin contain a high percentage of sulphur. This indicates two different effects of sulphur depending on its form of combination. In studying the manner of oxidation in the body, it is apparent that the size of the nucleus of the cell determines the rapidity of 7th process. Biochemically, the nucleus appears to be the agent of oxidation which enables the cell to secure nourishment. The most rapidly growing embryonic cells of the body are those derived from the chorionic villi (Langhans's cells). Langhans's cells have extremely large-nuclei and undergo the most rapid division of any normal cells in the body, but there is no peculiarity in the structure of the nucleolus and the cytoplasm of the cells. Wilson, MacCarty, and Broders have shown that the malignant cell has not only a large nucleus, but, also noteworthy, a large nucleolus. Broders's classification of the malignancy of tumours has been based on a comparison of the malignant cell with the normal cell of the same tissue. In brief, it has been found that when malignant cells are compared with the rapidly growing normal cells, the nucleolus cell as the nucleus is found to be greatly enlarged. It would appear that the size of the nucleus controls the total activity of the cell in relation to its increased blood supply, whereas the nucleolus may control the function of the cell in relation to the necessities of life. The increase, in the size of the nucleolus, as well as the nucleus, means not only rapid and uncontrolled production of cells but cells without function, which is characteristic of cancer. In other words, the greater the variation of the malignant cell over the normal cell the more malignant the growth.

The excess of nucleus in the malignant cell over and above the normal cell is evidently due to an excess of the normal activating enzyme or oxidizing agent which controls combustion. Heat and radio-active substances and some elements reduce or destroy the oxidizing power of the nucleus and the nucleolus, particularly of malignant cells which are less resistant than normal cells. Bell, of the University of Liverpool, in his work with the use of the element lead in cancer, has obtained some very interesting results, although the practical application of lead in the treatment of cancer has not proved very successful and not free from danger. His attention was called to the possible value of lead by observing its effect on workers in lead at Liverpool. Persons who had advanced cancer and yet were forced by the exigencies of life to earn, and who had taken up employment in the lead works, often experienced a decided change for the better: pain disappeared, the growth became very greatly reduced, due to the effect of lead on oxidation of the rapidly growing cell, and the malignant processes occasionally were held in abeyance for a considerable length of time. Bell is particularly interested in gynecology, and he learned that lead frequently caused sterility, usually temporary, in both men and women who worked in the lead works.

This fact led him to investigate not only the effect of lead in cancer, but also its destructive effect on the rapid growth of the syncytial cells of the chorionic villi (Langhans's cells). He was able to show marvelous resemblances between these controlled but most active embryonic chorionic cells and malignant disease. The element gold in colloidal form has been shown to have, to some extent, such an effect. Arsenic, another element, at various times has had a reputation as a cancer cure, both as an external application, an escharotics, and given internally. At times it has enjoyed a reputation as a palliative, probably through its effect on the bodily tissues encouraging resistance of the normal tissue and acting deleteriously on the lowered vitality of the cancer cell. Those who go back a quarter of a century and remember the work of Alexander, of Massachusetts, in the treatment of cancer with arsenic, cannot get away from the conviction that it had some effect, but it did not cure. While it is quite true that neither lead nor arsenic has achieved a solid reputation as a curative agent, they at least direct our attention to susceptibility of the cancer cell to certain chemical agents. Fischera, of the University of Rome, in the study of certain types of tumors associated with pregnancy, noticed the regression of tumors due to the absorption of substances derived from embryonic tissue. Acting on the known facts, first that all living animal tissue, as the result of its energy, produces metabolic substances which are poisonous to living tissue of the same kind, and, second, that the more immature the cells of the body, the more effective these poisons, Fischera manufactured serum from human embryos expelled between the second and sixth months, and injected this serum into patients suffering from cancer. The effect on the living embryonic tissues was most marked, and to some extent the experiments showed cell specificity. Extraordinary effects were sometimes produced. In certain cases the malignant growths were greatly reduced, and for a time the patients were partially relieved from some of the more serious symptoms, especially pain. In spite of this early improvement, however, the patients were not cured. 'Of late, Fischera has been experimenting to see whether, by inoculation, the tissues of the body could be rendered immune to cancer, just as the development of smallpox is prevented by vaccination. The proofs with regard to this experiment must depend on careful investigations, conducted over long periods, on an enormous number of persons, with proper control. Carrell's observation is significant, that in the growth of malignant fibroblastic tissue of fowl cancer under glass, there was acid reaction at the progressing margin of growth, whereas in the slower growth of normal tissue it was alkaline. Passing from the specific reactions connected with these chemical changes which might be multiplied by relevant observations, we come to an important question, that of the fluids of the body, especially as related to the acid-alkali content. Recent researches by McDonald, at the University of Pennsylvania, appear to show that persons whose reactions are toward the acid side do not often have cancer, as compared with those persons whose reactions are toward the alkaline state. The detection of changes in the acid-alkali balance requires most delicate methods for its determination. The alkalinity of the body tissues, while exceedingly slight, is fixed, and never in life becomes acid. Cell growth is accelerated by a preponderance of the element sodium and inhibited by an excess of the element calcium, which brings up the question of the parathyroid hormone, in relation to calcium reserve in the blood and tissues. The rather curious, relation of the acid-alkali balance of the body to growth also is deserving of study. Seventy-five per cent of the human body is composed of water, and much of this water is in a bound state. Unbound water is only slightly compressible, but when it is adsorbed with silica, a certain proportion of it becomes bound, and thereby is reduced in volume to 75 per cent. of the original volume. Gortner, the eminent agricultural chemist, of the University of Minnesota, in working out the question of bound water, found that plants which froze readily contained much unbound water which could be squeezed out in a press, and that, on the contrary, in those plants that would not freeze in continued cold weather, the water content was largely bound. This was also true of desert plants. These observations provide an easy test as to what plants would be suitable for regions of the country that had either prolonged winter or excessive dryness.

The essential alkalinity of the human body is very slight and extremely refined methods are required for differentiation. It is comparable to the difference between ordinary tap water and distilled water, and the distilled water must not be exposed to the air, or it will absorb acid from the atmospheric carbon dioxide. No gland in the human body produces an acid. We think at once of the hydrochloric acid of the gastric cells, but it has been demonstrated by Harvey and Bensley that this acid is formed not in the gland substance, but in the acini* of the glands by the union of the necessary chemical elements. Acidity of the large intestine depends upon bacteria. In the study of the cause and prevention of malignant disease biologic phenomena of a biochemical nature assume an importance which only recently has been recognized. Cancer never develops in sound tissues. Chronic irritation, by opening up an atrium for possible entrance of micro6rganisms to the body from the outside, seems to suggest an external agent. This does not explain why in certain cases in which the sources of chronic irritation are very slight, cancer develops, while in others in which the sources of chronic irritation are very extensive for great lengths of time, cancer does not develop. It is difficult to surmount the fact that when cancer has extended by metastasis to a new situation, it produces the histopathological picture of the tissues in which it originated, rather than that of the organ which became affected secondarily. If the disease were due to a foreign invader, it presumably would reproduce the type of cells of the newly invaded tissue rather than that of the primary seat of the tumor. In any event, the agents which act on the cell to produce malignancy become an integral part of the cell itself, as metastasis in the human species takes place only by transplantation of the cell. One factor of supreme importance which has not been sufficiently stressed is that individuals vary in their susceptibility to the cause or causes of cancer, whatever they may be. In no other way can we explain why go per cent of persons do not have cancer and 10 per cent. die from it. It is as logical to accept the hypothesis is that go per cent of persons have greater resistance to cancer than the 10 per cent, as to attempt to force an explanation that only 10 per cent come in contact with hypothetic causative agents. If the patient's susceptibility to the disease is the important factor in the development of cancer, the situation of the growth must be determined by the organ or tissue subjected to the insult of the precancerous lesion, and the grade of malignancy and the metastatic possibilities by the susceptibility of the body as a whole. The studies of Murray on tar painting, those of Gye and Barnard on the transplantation of the Rous fowl toxin, the work of Slye on cancer in mice, and of Bowing and Desjardins on the effect of radium and X-ray in lessening the malignant character of the growth, all point to local and general susceptibility as being

*Acini-Latin acinus meaning pin pointing by the conjugation of more than one elements.

perhaps the controlling factor in the genesis of malignant conditions, and indicate that the possibilities of increasing resistance to cancer in the more susceptible individuals is not only a possibility but a goal which every effort must be made to reach.

THE CARCINOGEN AND TYPES

A carcinogen is any substance, radionuclide or radiation that is an agent directly involved in the exacerbation of cancer or in the increase of its propagation. This may be due to the ability to damage the genome or to the disruption of cellular metabolic processes. Several radioactive substances are considered carcinogens, but their carcinogenic activity is attributed to the radiation, for example gamma rays and alpha particles, which they emit. Common examples of carcinogens are inhaled asbestos, certain dioxins, and tobacco smoke.

Cancer is a disease where damaged cells of the patient's body do not undergo programmed cell death, but their growth is no longer controlled and their metabolism is altered. Carcinogens may increase the risk of getting cancer by altering cellular metabolism or damaging DNA directly in cells, which interferes with biological processes, and induces the uncontrolled, malignant division, ultimately leading to the formation of tumors. Usually DNA damage, if too severe to repair, leads to programmed cell death, but if the programmed cell death pathway is damaged, then the cell cannot prevent itself from becoming a cancer cell.There are many natural carcinogens. Aflatoxin B_1, which is produced by the fungus *Aspergillus flavus* growing on stored grains, nuts and peanut butter, is an example of a potent, naturally-occurring microbial carcinogen. Certain viruses such as Hepatitis B and human papilloma viruses have been found to cause cancer in humans. The first one shown to cause cancer in animals is Rous sarcoma virus, discovered in 1910 by Peyton Rous.

Benzene, kepone, EDB, asbestos, and the waste rock of oil shale mining have all been classified as carcinogenic. As far back as the 1930s, industrial smoke and tobacco smoke were identified as sources of dozens of carcinogens, including benzo[*a*]pyrene, tobacco-specific nitrosamines such as nitrosonornicotine, and reactive aldehydes such as formaldehyde—which is also a hazard in embalming and making plastics. Vinyl chloride, from which PVC is manufactured, is a carcinogen and thus a hazard in PVC production.

Co-carcinogens are chemicals that do not necessarily cause cancer on their own, but promote the activity of other carcinogens in causing cancer.

After the carcinogen enters the body, the body makes an attempt to eliminate it through a process called biotransformation. The purpose of these reactions is to make the carcinogen more water-soluble so that it can be removed from the body. But these reactions can also convert a less toxic carcinogen into a more toxic one.

DNA is nucleophilic, therefore soluble carbon electrophiles are carcinogenic, because DNA attacks them. For example, some alkenes are toxicated by human enzymes to produce an electrophilic epoxide. DNA attacks the epoxide, and is bound permanently to it. This is the mechanism behind the carcinogenity of benzo[*a*]pyrene in tobacco smoke, other aromatics, aflatoxin and mustard gas.

CERCLA identifies all radionuclides as carcinogens, although the nature of the emitted radiation (alpha, beta, gamma, or neutron and the radioactive strength), its consequent capacity to cause ionization in tissues, and the magnitude of radiation exposure, determine the potential hazard. Carcinogenity of radiation depends of the type of radiation, type of exposure and penetration. For example, alpha radiation has low penetration and is not a hazard outside the body, but are carcinogenic when inhaled or ingested.

For example, Thorotrast, a (incidentally-radioactive) suspension previously used as a contrast medium in x-ray diagnostics, is a potent human carcinogen known because of its retention within various organs and persistent emission of alpha particles. Marie Curie, one of the pioneers of radioactivity, died of cancer caused by radiation exposure during her experiments.

Not all types of electromagnetic radiation are in fact carcinogenic. Low-energy waves on the electromagnetic spectrum are generally not, including radio waves, microwave radiation, infrared radiation and visible light. Higher-energy radiation, including ultraviolet radiation (present in sunlight), x-rays, and gamma radiation, generally *is* carcinogenic, if received in sufficient doses.

Several published studies suggest a link between exposure to light at night and risk of breast cancer, due to suppression of the normal nocturnal production of melatonin. The reduced production of the hormone melatonin might increase the risk of breast cancer and citing "environmental lighting" as a possible causal factor. Researchers at the National Cancer Institute (NCI) and National Institute of Environmental Health Sciences have concluded a study that suggests that artificial light during the night can be a factor for breast cancer. A good review of current knowledge of the health consequences of exposure to artificial light at night and an explanation of the causal mechanisms has been published in the Journal of Pineal Research in 2007.

Substances or foods irradiated with electrons or electromagnetic radiation (such as microwave, X-ray or gamma) are not carcinogenic. No "radiation" remains, just like no light remains in room after you turn out the light. (In contrast, non-electromagnetic neutron radiation produced inside nuclear reactors can produce secondary radiation by making bombarded substances radioactive.)

In other words, there is no possibility of getting cancer if the food is consumed.

Cooking food at high temperatures, for example grilling or barbecuing meats, can lead to the formation of minute quantities of many potent carcinogens that are comparable to those found in cigarette smoke (i.e., benzo[*a*]pyrene). Charring of food resembles coking and tobacco pyrolysis, and produces similar carcinogens. There are several carcinogenic pyrolysis products, such as polynuclear aromatic hydrocarbons, which are converted by human enzymes into epoxides, which attach permanently to DNA. Pre-cooking meats in a microwave oven for 2–3 minutes before grilling shortens the time on the hot pan, and removes heterocyclic amine (HCA) precursors, which can help minimize the formation of these carcinogens.

Reports from the Food Standards Agency have found that the known animal carcinogen acrylamide is generated in fried or overheated carbohydrate foods (such as french fries and potato chips). Studies are underway at the FDA and European regulatory agencies to assess its potential risk to humans. The charred residue on barbecued meats has been identified as a carcinogen, along with many other tars.

Dr. T. Colin Campbell argues in *The China Study* that the milk protein casein, found in milk and many prepared foods, is also a carcinogen. However, independent studies report that casein and other milk proteins protect against cancer.

Tobacco smoke contains over 4000 chemical compounds, many of which are carcinogenic or otherwise toxic.

"Shiftwork that involves circadian disruption" was listed, in 2007, as a probable carcinogen by the World Health Organization's International Agency for Research on Cancer. (IARC Press release No. 180). Multiple studies have documented a link between night shift work and the increased incidence of breast cancer. Circadian disruption by exposure to light at night suppresses the production of the hormone melatonin which leads to reduction in cellular immune defense and surveillance necessary for protection from development of cancers. Melatonin also seems to have a direct protective effect against cancer possibly in part because of its strong anti oxidant properties.

Carcinogens can be classified as genotoxic or nongenotoxic. Genotoxins cause irreversible genetic damage or mutations by binding to DNA. Genotoxins include chemical agents like N-nitroso-N-methylurea (MNU) or non-chemical agents such as ultraviolet light and ionizing radiation. Certain viruses can also act as carcinogens by interacting with DNA.

Nongenotoxins do not directly affect DNA but act in other ways to promote growth. These include hormones and some organic compounds.

Approximate Equivalences between Classification Schemes

IARC	GHS	NTP	ACGIH	EU
Group 1	Cat. 1A	Known	A1	Cat. 1
Group 2A	Cat. 1B	Reasonably suspected	A2	Cat. 2
Group 2B	Cat. 2		A3	Cat. 3
Group 3			A4	
Group 4			A5	

The International Agency for Research on Cancer (IARC) is an intergovernmental agency established in 1965, which forms part of the World Health Organization of the United Nations. It is based in Lyon, France. Since 1971 it has published a series of *Monographs on the Evaluation of Carcinogenic Risks to Humans* that have been highly influential in the classification of possible carcinogens.

- Group 1: the agent (mixture) is definitely carcinogenic to humans. The exposure circumstance entails exposures that are carcinogenic to humans.
- Group 2A: the agent (mixture) is probably carcinogenic to humans. The exposure circumstance entails exposures that are probably carcinogenic to humans.
- Group 2B: the agent (mixture) is possibly carcinogenic to humans. The exposure circumstance entails exposures that are possibly carcinogenic to humans.
- Group 3: the agent (mixture or exposure circumstance) is not classifiable as to its carcinogenicity to humans.

- Group 4: the agent (mixture) is probably not carcinogenic to humans.

The Globally Harmonized System of Classification and Labelling of Chemicals (GHS) is a United Nations initiative to attempt to harmonize the different systems of assessing chemical risk which currently exist (as of March 2009) around the world. It classifies carcinogens into two categories, of which the first may be divided again into subcategories if so desired by the competent regulatory authority:

- Category 1: known or presumed to have carcinogenic potential for humans
  - Category 1A: the assessment is based primarily on human evidence
  - Category 1B: the assessment is based primarily on animal evidence
- Category 2: suspected human carcinogens

The National Toxicology Program of the U.S. Department of Health and Human Services is mandated to produce a biennial *Report on Carcinogens*. As of March 2009, the latest edition was the 11th report (2005). It classifies carcinogens into two groups:

- Known to be a human carcinogen
- Reasonably anticipated to be a human carcinogen

The American Conference of Governmental Industrial Hygienists (ACGIH) is a private organization best known for its publication of threshold limit values (TLVs) for occupational exposure and monographs on workplace chemical hazards. It assesses carcinogenicity as part of wider assessment of the occupational hazards of chemicals.

- Group A1: Confirmed human carcinogen
- Group A2: Suspected human carcinogen
- Group A3: Confirmed animal carcinogen with unknown relevance to humans
- Group A4: Not classifiable as a human carcinogen
- Group A5: Not suspected as a human carcinogen

The European Union classification of carcinogens is contained in the Dangerous Substances Directive and the Dangerous Preparations Directive. It consists of three categories:

- Category 1: Substances known to be carcinogenic to humans.
- Category 2: Substances which should be regarded as if they are carcinogenic to humans.
- Category 3: Substances which cause concern for humans, owing to possible carcinogenic effects but in respect of which the available information is not adequate for making a satisfactory assessment.

This assessment scheme is being phased out in favour of the GHS scheme, to which it is very close in category definitions.

A procarcinogen is a precursor to a carcinogen. One example is nitrites when taken in by thdiet. They are not carcinogenic themselves, but turn into nitrosamines in the body, which are carcinogenic.

Occupational carcinogens are agents that pose a risk of cancer in several specific work-locations:

Carcinogen	Associated cancer sites or types	Occupational uses or sources
1	2	3
Arsenic and its compounds	• Lung • Skin • Hemangiosarcoma	• Smelting byproduct • Component of: – Alloys – Electrical and semiconductor devices – Medications (e.g. melarsoprol) – Herbicides – Fungicides – Animal dips
Asbestos	• Lungs • Mesothelioma • Gastrointestinal tract	Not in use, but still found in: • Constructions – Roofing papers – Floor tiles • Fire-resistant textiles • Friction linings
Benzene	• Leukemia • Hodgkin lymphoma	• Light fuel oil • Former use as solvent and fumigant • Printing • Lithography • Paint • Rubber • Dry cleaning • Adhesives • Coatings • Detergents
Beryllium and its compounds	• Lung	• Missile fuel • Lightweight alloys – Aerospace applications – Nuclear reactors
Cadmium and its compounds	• Prostate	• Yellow pigments • Phosphors • Solders • Batteries • Metal paintings and coatings
Hexavalent chromium(VI) compounds	• Lung	• Paints • Pigments • Preservatives

...(Contd.)

1	2	3
Ethylene oxide	• Leukemia	• Ripening agent for fruits and nuts • Rocket propellant • Fumigant for foodstuffs and textiles • Sterilant for hospital equipment
Nickel	• Nose • Lung	• Nickel plating • Ferrous alloys • Ceramics • Batteries • Stainless-steel welding byproduct
Radon and its decay products	• Lung	• Uranium decay – Quarries and mines – Cellars and poorly ventilated places
Vinyl chloride	• Hemangiosarcoma • Liver	• Refrigerant • Vinyl polymers • Adhesive for plastics • Former use in pressurized containers
Shiftwork that involves circadian disruption	• Breast	
Involuntary smoking (Passive smoking)	• Lung	• Tobacco smoke

ONCOGENS AND PROTO-ONCOGENS

An oncogene is a gene that, when mutated or expressed at high levels, helps turn a normal cell into a tumor cell.

Many cells normally undergo a programmed form of death (apoptosis). Activated oncogenes can cause those cells to survive and proliferate instead. Most oncogenes require an additional step, such as mutations in another gene, or environmental factors, such as viral infection, to cause cancer. Since the 1970s, dozens of oncogenes have been identified in human cancer. Many cancer drugs target those DNA sequences and their products.

PROTO-ONCOGENE

A proto-oncogene is a normal gene that can become an oncogene due to mutations or increased expression. Proto-oncogenes code for proteins that help to regulate cell growth and differentiation. Proto-oncogenes are often involved in signal transduction and execution of mitogenic signals, usually through their protein products. Upon *activation*, a proto-oncogene (or its product) becomes a tumor-inducing agent, an oncogene. Examples of proto-oncogenes include RAS, WNT, MYC, ERK, and TRK.

The first oncogene was discovered in 1970 and was termed SRC (pronounced *sarc* as in *sarcoma*). SRC was in fact first discovered as an oncogene in a chicken retrovirus. Experiments performed by Dr. G. Steve Martin of the University of California, Berkeley demonstrated that the SRC was indeed the oncogene of the virus.

In 1976 Drs. J. Michael Bishop and Harold E. Varmus of the University of California, San Francisco demonstrated that oncogenes were defective proto-oncogenes, found in many organisms including humans. For this discovery Bishop and Varmus were awarded the Nobel Prize in 1989.

The proto-oncogene can become an oncogene by a relatively small modification of its original function. There are three basic activation types:

- A mutation within a proto-oncogene can cause a change in the protein structure, causing
  - an increase in protein (enzyme) activity
  - a loss of regulation
- An increase in protein concentration, caused by
  - an increase of protein expression (through misregulation)
  - an increase of protein (mRNA) stability, prolonging its existence and thus its activity in the cell
  - a gene duplication (one type of chromosome abnormality), resulting in an increased amount of protein in the cell
- A chromosomal translocation (another type of chromosome abnormality), causing
  - an increased gene expression in the wrong cell type or at wrong times
  - the expression of a constitutively active *hybrid protein*. This type of aberration in a dividing stem cell in the bone marrow leads to adult leukemia

Mutations in microRNAs can lead to activation of oncogenes. New research indicates that small RNAs 21-25 nucleotides in length called microRNAs (miRNAs) can control expression of these genes by downregulating them. Antisense messenger RNAs could theoretically be used to block the effects of oncogenes.

There are several systems for classifying oncogenes, but there is not yet a widely accepted standard. They are sometimes grouped both spatially (moving from outside the cell inwards) and chronologically (parallelling the "normal" process of signal transduction). There are several categories that are commonly used:

Category	Examples	Description
1	2	3
Growth factors, or mitogens	c-Sis	Usually secreted by specialized cells to induce cell proliferation in themselves, nearby cells, or distant cells. An oncogene may cause a cell to secrete growth factors even though it

...(Contd.)

1	2	3
		does not normally do so. It will thereby induce its own uncontrolled proliferation (autocrine loop), and proliferation of neighboring cells. It may also cause production of growth hormones in other parts of the body.
Receptor tyrosine kinases	epidermal growth factor receptor (EGFR), platelet-derived growth factor receptor (PDGFR), and vascular endothelial growth factor receptor (VEGFR), HER2/neu	Kinases add phosphate groups to other proteins to turn them on or off. Receptor kinases add phosphate groups to receptor proteins at the surface of the cell (which receive protein signals from outside the cell and transmit them to the inside of the cell). Tyrosine kinases add phosphate groups to the amino acid tyrosine in the target protein. They can cause cancer by turning the receptor permanently on (constitutively), even without signals from outside the cell.
Cytoplasmic tyrosine kinases	Src-family, Syk-ZAP-70 family, and BTK family of tyrosine kinases, the Abl gene in CML - Philadelphia chromosome	—
Cytoplasmic Serine/ threonine kinases and their regulatory subunits	Raf kinase, and cyclin-dependent kinases (through overexpression).	—
Regulatory GTPases	Ras protein	Ras is a small GTPase which hydrolyses GTP into GDP and phosphate. Ras is activated by growth factor signaling (ie. EGF, TGFalpha) and acting like a binary switch (on/off) in growth signaling pathways. Downstream effectors of Ras include Raf, MEK, MEKK, MAPK, ERK, most of which in turn regulate genes that mediate cell proliferation.
Transcription factors	myc gene	—

There are two mechanisms by which proto-oncogenes can be converted to cellular oncogenes:

CANCER

Today, the Greek term carcinoma is the medical term for a malignant tumor derived from epithelial cells. It is Celsus who translated *carkinos* into the Latin *cancer*, also meaning crab. Galen used "*oncos*" to describe *all* tumours, the root for the modern word oncology.

Hippocrates described several kinds of cancers. He called benign tumours *ogkos*, Greek for tumour, and malignant tumours *carkinos*, Greek for crab or crayfish. This name comes from the appearance of the cut surface of a solid malignant tumour, with "the veins stretched on all sides as the animal the crab has its feet, whence it derives its name". He later added the suffix *-oma*, Greek for swelling, giving the name *carcinoma*. Since it was against Greek tradition to open the body, Hippocrates only described and made drawings of outwardly visible tumors on the skin, nose, and breasts. Treatment was based on the humor theory of four bodily fluids (black and yellow bile, blood, and phlegm). According to the patient's humor, treatment consisted of diet, blood-letting, and/or laxatives. Through the centuries it was discovered that cancer could occur anywhere in the body, but humor-theory based treatment remained popular until the 19th century with the discovery of cells.

The oldest known description and surgical treatment of cancer was discovered in Egypt and dates back to approximately 1600 B.C. The Papyrus describes 8 cases of ulcers of the breast that were treated by cauterization, with a tool called "the fire drill." The writing says about the disease, "There is no treatment."

Another very early surgical treatment for cancer was described in the 1020s by Avicenna (Ibn Sina) in *The Canon of Medicine*. He stated that the excision should be radical and that all diseased tissue should be removed, which included the use of amputation or the removal of veins running in the direction of the tumor. He also recommended the use of cauterization for the area being treated if necessary.

In the 16th and 17th centuries, it became more acceptable for doctors to dissect bodies to discover the cause of death. The German professor Wilhelm Fabry believed that breast cancer was caused by a milk clot in a mammary duct. The Dutch professor Francois de la Boe Sylvius, a follower of Descartes, believed that all disease was the outcome of chemical processes, and that acidic lymph fluid was the cause of cancer. His contemporary Nicolaes Tulp believed that cancer was a poison that slowly spreads, and concluded that it was contagious.

The first cause of cancer was identified by British surgeon Percivall Pott, who discovered in 1775 that cancer of the scrotum was a common disease among chimney sweeps. The work of other individual physicians led to various insights, but when physicians started working together they could make firmer conclusions.

With the widespread use of the microscope in the 18th century, it was discovered that the 'cancer poison' spread from the primary tumor through the lymph nodes to other sites ("metastasis"). This view of the disease was first formulated by the English surgeon Campbell De Morgan between 1871 and 1874. The use of surgery to treat cancer had poor results due to problems with hygiene. The renowned Scottish surgeon Alexander Monro saw only 2 breast tumour patients out of 60 surviving surgery for two years. In the 19th century, asepsis improved surgical hygiene and as the survival statistics went up, surgical removal of the tumor became the primary treatment for cancer. With the exception of William Coley who in the late 1800s felt that the rate of cure after surgery had been higher *before* asepsis (and who injected bacteria into tumors with mixed results), cancer treatment became dependent on the individual art of the surgeon at removing a tumor. During the same period, the idea that the body was made up of various tissues, that in turn were made up of millions of cells, laid rest the humor-theories about chemical imbalances in the body. The age of cellular pathology was born.

When Marie Curie and Pierre Curie discovered radiation at the end of the 19th century, they stumbled upon the first effective non-surgical cancer treatment. With radiation also came the first signs of multi-disciplinary approaches to cancer treatment. The surgeon was no longer operating in isolation, but worked together with hospital radiologists to help patients. The complications in communication this brought, along with the necessity of the patient's treatment in a hospital facility rather than at home, also created a parallel process of compiling patient data into hospital files, which in turn led to the first statistical patient studies.

A founding paper of cancer epidemiology was the work of Janet Lane-Claypon, who published a comparative study in 1926 of 500 breast cancer cases and 500 control patients of the same background and lifestyle for the British Ministry of Health. Her ground-breaking work on cancer epidemiology was carried on by Richard Doll and Austin Bradford Hill, who published "Lung Cancer and Other Causes of Death In Relation to Smoking. A Second Report on the Mortality of British Doctors" followed in 1956 (otherwise known as the British doctors study). Richard Doll left the London Medical Research Center (MRC), to start the Oxford unit for Cancer epidemiology in 1968. With the use of computers, the unit was the first to compile large amounts of cancer data. Modern epidemiological methods are closely linked to current concepts of disease and public health policy. Over the past 50 years, great efforts have been spent on gathering data across medical practise, hospital, provincial, state, and even country boundaries, as a way to study the interdependence of environmental and cultural factors on cancer incidence.

Cancer patient treatment and studies were restricted to individual physicians' practices until World War II, when medical research centers discovered that there were large international differences in disease incidence. This insight drove national public health bodies to make it possible to compile health data across practises and hospitals, a process that many countries do today. The Japanese medical community observed that the bone marrow of victims of the atomic bombings of Hiroshima and Nagasaki was completely destroyed. They concluded that diseased bone marrow could also be destroyed with radiation, and this led to the discovery of bone marrow transplants for leukemia. Since World War II, trends in cancer treatment are to improve on a micro-level the existing treatment methods, standardize them, and globalize them as a way to find cures through epidemiology and international partnerships.

Cancer (medical term: malignant neoplasm) is a class of diseases in which a group of cells display *uncontrolled growth* (division beyond the normal limits), *invasion* (intrusion on and destruction of adjacent tissues), and sometimes *metastasis* (spread to other locations in the body via lymph or blood). These three malignant properties of cancers differentiate them from benign tumors, which are self-limited, and do not invade or metastasize. Most cancers form a tumor but some, like leukemia, do not. The branch of medicine concerned with the study, diagnosis, treatment, and prevention of cancer is oncology.

Cancer may affect people at all ages, even fetuses, but the risk for most varieties increases with age. Cancer causes about 13% of all human deaths. According to the American Cancer Society, 7.6 million people died from cancer in the world during 2007. Cancers can affect all animals.

Nearly all cancers are caused by abnormalities in the genetic material of the transformed cells. These abnormalities may be due to the effects of carcinogens, such as tobacco smoke, radiation, chemicals, or infectious agents. Other cancer-promoting genetic abnormalities may be randomly acquired through errors in DNA replication, or are inherited, and thus present in all cells from birth. The heritability of cancers is usually affected by complex interactions between carcinogens and the host's genome. New aspects of the genetics of cancer pathogenesis, such as DNA methylation, and microRNAs are increasingly recognized as important.

Genetic abnormalities found in cancer typically affect two general classes of genes. Cancer-promoting *oncogenes* are typically activated in cancer cells, giving those cells new properties, such as hyperactive growth and division, protection against programmed cell death, loss of respect for normal tissue boundaries, and the ability to become established in diverse tissue environments. *Tumor suppressor genes* are then inactivated in cancer cells, resulting in the loss of normal functions in those cells, such as accurate DNA replication, control over the cell cycle, orientation and adhesion within tissues, and interaction with protective cells of the immune system.

Diagnosis usually requires the histologic examination of a tissue biopsy specimen by a pathologist, although the initial indication of malignancy can be symptoms or radiographic imaging abnormalities. Most cancers can be treated and some cured, depending on the specific type, location, and stage. Once diagnosed, cancer is usually treated with a combination of surgery, chemotherapy and radiotherapy. As research develops, treatments are becoming more specific for different varieties of cancer. There has been significant progress in the development of targeted therapy drugs that act specifically on detectable molecular abnormalities in certain tumors, and which minimize damage to normal cells. The prognosis of cancer patients is most influenced by the type of cancer, as well as the stage, or extent of the disease. In addition, histologic grading and the presence of specific molecular markers can also be useful in establishing prognosis, as well as in determining individual treatments.

The following closely related terms may be used to designate abnormal growths:

- **Tumor or tumour:** originally, it meant any abnormal swelling, lump or mass. In current English, however, the word tumor has become synonymous with neoplasm, specifically solid neoplasm. Note that some neoplasms, such as leukemia, do not form tumors.
- **Neoplasm:** the scientific term to describe an abnormal proliferation of genetically altered cells. Neoplasms can be benign or malignant:
  - **Malignant neoplasm** or **malignant tumor**: synonymous with **cancer**.
  - **Benign neoplasm** or **benign tumor**: a tumor (solid neoplasm) that stops growing by itself, does not invade other tissues and does not form metastases.
- **Invasive** tumor is another synonym of **cancer**. The name refers to invasion of surrounding tissues.
- **Pre-malignancy**, **pre-cancer** or **non-invasive** tumor: A neoplasm that is not invasive but has the potential to progress to cancer (become invasive) if left untreated. These lesions are, in order of increasing potential for cancer, atypia, dysplasia and carcinoma in situ.

The following terms can be used to describe a cancer:

- **Screening**: a test done on healthy people to detect tumors before they become apparent. A mammogram is a screening test.

- **Diagnosis**: the confirmation of the cancerous nature of a lump. This usually requires a biopsy or removal of the tumor by surgery, followed by examination by a pathologist.
- **Surgical excision**: the removal of a tumor by a surgeon.
  - **Surgical margins**: the evaluation by a pathologist of the edges of the tissue removed by the surgeon to determine if the tumor was removed completely ("negative margins") or if tumor was left behind ("positive margins").
- **Grade**: a number (usually on a scale of 3) established by a pathologist to describe the degree of resemblance of the tumor to the surrounding benign tissue.
- **Stage**: a number (usually on a scale of 4) established by the oncologist to describe the degree of invasion of the body by the tumor.
- **Recurrence**: new tumors that appear at the site of the original tumor after surgery.
- **Metastasis**: new tumors that appear far from the original tumor.
- **Median survival time**: a period of time, often measured in months or years, over which 50% of the cancer patients are expected to be alive.
- **Transformation:** the concept that a low-grade tumor transforms to a high-grade tumor over time. Example: Richter's transformation.
- **Chemotherapy**: treatment with drugs.
- **Radiation therapy**: treatment with radiations.
- **Adjuvant therapy**: treatment, either chemotherapy or radiation therapy, given after surgery to kill the remaining cancer cells.
- **Prognosis**: the probability of cure after the therapy. It is usually expressed as a probability of survival five years after diagnosis. Alternatively, it can be expressed as the number of years when 50% of the patients are still alive. Both numbers are derived from statistics accumulated with hundreds of similar patients to give a Kaplan-Meier curve.
- **Cure**: A cancer patient is "cured" if 95% of treated patients live a certain period of time after the date of their diagnosis of cancer. For example in the case of Hodgkin's disease this period of time is 10 years, whereas for Burkitt's lymphoma this period would be 1 year. The phrase "cure" used in oncology is based upon the statistical concept of a median survival time and disease-free median survival time.

A large invasive ductal carcinoma in a mastectomy specimen.

Cancers are classified by the type of cell that resembles the tumor and, therefore, the tissue presumed to be the origin of the tumor. These are the histology and the location, respectively. Examples of general categories include:

- **Carcinoma:** Malignant tumors derived from epithelial cells. This group represents the most common cancers, including the common forms of breast, prostate, lung and colon cancer.
- **Sarcoma:** Malignant tumors derived from connective tissue, or mesenchymal cells.

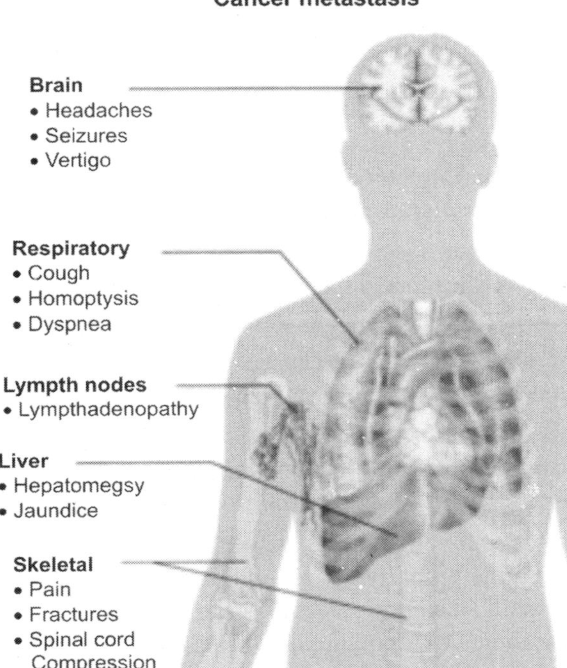

Fig. 7.1 : The Symptoms of Cancer Metastasis Depend on the Location of the Tumour.

- **Lymphoma** and **leukemia:** Malignancies derived from hematopoietic (blood-forming) cells
- **Germ cell tumor:** Tumors derived from totipotent cells. In adults most often found in the testicle and ovary; in fetuses, babies, and young children most often found on the body midline, particularly at the tip of the tailbone; in horses most often found at the poll (base of the skull).
- **Blastic tumor or blastoma:** A tumor (usually malignant) which resembles an immature or embryonic tissue. Many of these tumors are most common in children.

Malignant tumors (cancers) are usually named using **-carcinoma**, **-sarcoma** or **-blastoma** as a suffix, with the Latin or Greek word for the organ of origin as the root. For instance, a cancer of the liver is called *hepatocarcinoma*; a cancer of the fat cells is called *liposarcoma*. For common cancers, the English organ name is used. For instance, the most common type of breast cancer is called *ductal carcinoma of the breast* or *mammary ductal carcinoma*. Here, the adjective *ductal* refers to the appearance of the cancer under the microscope, resembling normal breast ducts.

Benign tumors (which are not cancers) are named using **-oma** as a suffix with the organ name as the root. For instance, a benign tumor of the smooth muscle of the uterus is called *leiomyoma* (the common name of this frequent tumor is *fibroid*). Unfortunately, some cancers also use the **-oma** suffix, examples being melanoma and seminoma.

Signs and Symptoms

Roughly, cancer symptoms can be divided into three groups:

- *Local symptoms*: unusual lumps or swelling (*tumour*), hemorrhage (bleeding), pain and/or ulceration. Compression of surrounding tissues may cause symptoms such as jaundice (yellowing the eyes and skin).
- *Symptoms of metastasis (spreading)*: enlarged lymph nodes, cough and hemoptysis, hepatomegaly (enlarged liver), bone pain, fracture of affected bones and neurological symptoms. Although advanced cancer may cause pain, it is often not the first symptom.
- *Systemic symptoms*: weight loss, poor appetite, fatigue and cachexia (wasting), excessive sweating (night sweats), anemia and specific paraneoplastic phenomena, i.e. specific conditions that are due to an active cancer, such as thrombosis or hormonal changes.

Every symptom in the above list can be caused by a variety of conditions (a list of which is referred to as the differential diagnosis). Cancer may be a common or uncommon cause of each item.

Cancer is a diverse class of diseases which differ widely in their causes and biology. Any organism, even plants, can acquire cancer. Nearly all known cancers arise gradually, as errors build up in the cancer cell and its progeny (see mechanisms section for common types of errors).

Anything which replicates (our cells) will probabilistically suffer from errors (mutations). Unless error correction and prevention is properly carried out, the errors will survive, and might be passed along to daughter cells. Normally, the body safeguards against cancer via numerous methods, such as: apoptosis, helper molecules (some DNA polymerases), possibly senescence, etc. However these error-correction methods often fail in small ways, especially in environments that make errors more likely to arise and propagate. For example, such environments can include the presence of disruptive substances called carcinogens, or periodic injury (physical, heat, etc.), or environments that cells did not evolve to withstand, such as hypoxia. Cancer is thus a *progressive* disease, and these progressive errors slowly accumulate until a cell begins to act contrary to its function in the animal.

The errors which cause cancer are often *self-amplifying*, eventually compounding at an exponential rate. For example:

- A mutation in the error-correcting machinery of a cell might cause that cell and its children to accumulate errors more rapidly
- A mutation in signaling (endocrine) machinery of the cell can send error-causing signals to nearby cells
- A mutation might cause cells to become neoplastic, causing them to migrate and disrupt more healthy cells
- A mutation may cause the cell to become immortal (see telomeres), causing them to disrupt healthy cells forever

Thus cancer often explodes in something akin to a chain reaction caused by a few errors, which compound into more severe errors. Errors which produce more errors are effectively the root cause of cancer, and also the reason that cancer is so hard to treat: even if there were 10,000,000,000 cancerous cells and one killed all but 10 of those cells, those cells (and other error-prone precancerous cells) could still self-replicate or send error-causing signals to other cells, starting the process over again. This rebellion-like scenario is an undesirable survival of the fittest, where the driving forces of evolution itself work against the body's design and enforcement of order. In fact, once cancer has begun to develop, this same force continues to drive the progression of cancer towards more invasive stages, and is called clonal evolution.

Research about cancer causes often falls into the following categories:

- Agents (e.g. viruses) and events (e.g. mutations) which cause or facilitate genetic changes in cells destined to become cancer.
- The precise nature of the genetic damage, and the genes which are affected by it.
- The consequences of those genetic changes on the biology of the cell, both in generating the defining properties of a cancer cell, and in facilitating additional genetic events which lead to further progression of the cancer.

Cancer pathogenesis is traceable back to DNA mutations that impact cell growth and metastasis. Substances that cause DNA mutations are known as mutagens, and mutagens that cause cancers are known as carcinogens. Particular substances have been linked to specific types of cancer. Tobacco smoking is associated with many forms of cancer, and causes 90% of lung cancer. Prolonged exposure to asbestos fibers is associated with mesothelioma.

Many mutagens are also carcinogens, but some carcinogens are not mutagens. Alcohol is an example of a chemical carcinogen that is not a mutagen. Such chemicals may promote cancers through stimulating the rate of cell division. Faster rates of replication leaves less time for repair enzymes to repair damaged DNA during DNA replication, increasing the likelihood of a mutation. Decades of research has demonstrated the link between tobacco use and cancer in the lung, larynx, head, neck, stomach, bladder, kidney, oesophagus and pancreas. Tobacco smoke contains over fifty known carcinogens, including nitrosamines and polycyclic aromatic hydrocarbons. Tobacco is responsible for about one in three of all cancer deaths in the developed world, and about one in five worldwide. Indeed, lung cancer death rates in the United States have mirrored smoking patterns, with increases in smoking followed by dramatic increases in lung cancer death rates and, more recently, decreases in smoking followed by decreases in lung cancer death rates in men. However, the numbers of smokers worldwide is still rising, leading to what some organizations have described as the *tobacco epidemic*.

Sources of ionizing radiation, such as radon gas, can cause cancer. Prolonged exposure to ultraviolet radiation from the sun can lead to melanoma and other skin malignancies.

Non-ionizing radio frequency radiation from mobile phones and other similar RF sources has also been proposed as a cause of cancer, but there is currently little established evidence of such a link.

Some cancers can be caused by infection with pathogens. Many cancers originate from a viral infection; this is especially true in animals such as birds, but also in humans, as viruses are responsible for 15% of human cancers worldwide. The main viruses associated with human cancers are human papillomavirus, hepatitis B and hepatitis C virus, Epstein-Barr virus, and human T-lymphotropic virus. Experimental and epidemiological data imply a causative role for viruses and they appear to be the second most important risk factor for cancer development in humans, exceeded only by tobacco usage. The mode of virally-induced tumors can be divided into two, *acutely-transforming* or *slowly-transforming*. In acutely transforming viruses, the virus carries an overactive oncogene called viral-oncogene (v-onc), and the infected cell is transformed as soon as v-onc is expressed. In contrast, in slowly-transforming viruses, the virus genome is inserts near a proto-oncogene in the host genome. The viral promoter or other transcription regulation elements then cause overexpression of that proto-oncogene. This induces uncontrolled cell division. Because the site of insertion is not specific to proto-oncogenes and the chance of insertion near any proto-oncogene is low, slowly-transforming viruses will cause tumors much longer after infection than the acutely-transforming viruses.

Hepatitis viruses, including hepatitis B and hepatitis C, can induce a chronic viral infection that leads to liver cancer in 0.47% of hepatitis B patients per year (especially in Asia, less so in North America), and in 1.4% of hepatitis C carriers per year. Liver cirrhosis, whether from chronic viral hepatitis infection or alcoholism, is associated with the development of liver cancer, and the combination of cirrhosis and viral hepatitis presents the highest risk of liver cancer development. Worldwide, liver cancer is one of the most common, and most deadly, cancers due to a huge burden of viral hepatitis transmission and disease.

Advances in cancer research have made a vaccine designed to prevent cancer available. In 2006, the U.S. Food and Drug Administration approved a human papilloma virus vaccine, called Gardasil. The vaccine protects against four HPV types, which together cause 70% of cervical cancers and 90% of genital warts. In March 2007, the US Centers for Disease Control and Prevention (CDC) Advisory Committee on Immunization Practices (ACIP) officially recommended that females aged 11–12 receive the vaccine, and indicated that females as young as age 9 and as old as age 26 are also candidates for immunization.In addition to viruses, researchers have noted a connection between bacteria and certain cancers. The most prominent example is the link between chronic infection of the wall of the stomach with *Helicobacter pylori* and gastric cancer. Although only a minority of those infected with *Helicobacter* go on to develop cancer, since this pathogen is quite common it is probably responsible for the majority of these cancers. Some hormones can act in a similar manner to non-mutagenic carcinogens in that they may stimulate excessive cell growth. A well-established example is the role of hyperestrogenic states in promoting endometrial cancer.HIV is associated with a number of malignancies, including Kaposi's sarcoma, non-Hodgkin's lymphoma, and HPV-associated malignancies such as anal cancer and cervical cancer. AIDS-defining illnesses have long included these diagnoses. The increased incidence of malignancies in HIV patients points to the breakdown of immune surveillance as a possible etiology of cancer. Certain other immune deficiency states (e.g. common variable immunodeficiency and IgA deficiency) are also associated with increased risk of malignancy. Most forms of cancer are *sporadic*, meaning that there is no inherited cause of the cancer. There are, however, a number of recognised syndromes where there is an inherited predisposition to cancer, often due to a defect in a gene that protects against tumor formation. Famous examples are:

- certain inherited mutations in the genes *BRCA1* and *BRCA2* are associated with an elevated risk of breast cancer and ovarian cancer
- tumors of various endocrine organs in multiple endocrine neoplasia (MEN types 1, 2a, 2b)

- Li-Fraumeni syndrome (various tumors such as osteosarcoma, breast cancer, soft tissue sarcoma, brain tumors) due to mutations of p53
- Turcot syndrome (brain tumors and colonic polyposis)
- Familial adenomatous polyposis an inherited mutation of the *APC* gene that leads to early onset of colon carcinoma.
- Hereditary nonpolyposis colorectal cancer (HNPCC, also known as Lynch syndrome) can include fəmilial cases of colon cancer, uterine cancer, gastric cancer, and ovarian cancer, without a preponderance of colon polyps.
- Retinoblastoma, when occurring in young children, is due to a hereditary mutation in the retinoblastoma gene.
- Down syndrome patients, who have an extra chromosome 21, are known to develop malignancies such as leukemia and testicular cancer, though the reasons for this difference are not well understood.

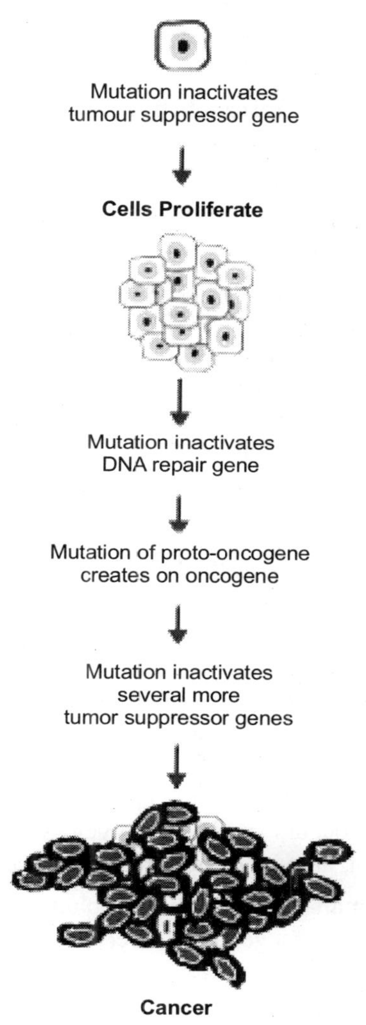

Fig. 7.2 : The Mechanism of Cancer

Excepting the rare transmissions that occur with pregnancies and only a marginal few organ donors, cancer is generally not a transmissible disease. The main reason for this is tissue graft rejection caused by MHC incompatibility. In humans and other vertebrates, the immune system uses MHC antigens to differentiate between "self" and "non-self" cells because these antigens are different from person to person. When non-self antigens are encountered, the immune system reacts against the appropriate cell. Such reactions may protect against tumour cell engraftment by eliminating implanted cells. In the United States, approximately 3,500 pregnant women have a malignancy annually, and transplacental transmission of acute leukaemia, lymphoma, melanoma and carcinoma from mother to fetus has been observed. The development of donor-derived tumors from organ transplants is exceedingly rare. The main cause of organ transplant associated tumors seems to be malignant melanoma, that was undetected at the time of organ harvest, though other cases exist. In fact, cancer from one organism will usually grow in another organism of that species, as long as they share the same histocompatibility genes, proven using mice; however this would never happen in a real-world setting except as described above.

In non-humans, a few types of cancer have been found to be caused by transmission of the tumor cells themselves. This phenomenon is seen in dogs with Sticker's sarcoma, also known as canine transmissible venereal tumor, as well as Devil facial tumour disease in Tasmanian devils.

Cancers are caused by a series of mutations. Each mutation alters the behaviour of the cell somewhat.

Cancer is fundamentally a disease of regulation of tissue growth. In order for a normal cell to transform into a cancer cell, genes which regulate cell growth and differentiation must be altered. Genetic changes can occur at many levels, from gain or loss of entire chromosomes to a mutation affecting a single DNA nucleotide. There are two broad categories of genes which are affected by these changes. Oncogenes may be normal genes which are expressed at inappropriately high levels, or altered genes which have novel properties. In either case, expression of these genes promotes the malignant phenotype of cancer cells. Tumor suppressor genes are genes which inhibit cell division, survival, or other properties of cancer cells. Tumor suppressor genes are often disabled by cancer-promoting genetic changes. Typically, changes in many genes are required to transform a normal cell into a cancer cell.

There is a diverse classification scheme for the various genomic changes which may contribute to the generation of cancer cells. Most of these changes are mutations, or changes in the nucleotide sequence of genomic DNA. Aneuploidy, the presence of an abnormal number of chromosomes, is one genomic change which is not a mutation, and may involve either gain or loss of one or more chromosomes through errors in mitosis.

Large-scale mutations involve the deletion or gain of a portion of a chromosome. Genomic amplification occurs when a cell gains many copies (often 20 or more) of a small chromosomal locus, usually containing one or more oncogenes and adjacent genetic material. Translocation occurs when two separate chromosomal regions become abnormally fused, often at a characteristic location. A well-known example of this is the Philadelphia chromosome, or translocation of chromosomes 9 and 22, which occurs in chronic myelogenous leukemia, and results in production of the BCR-abl fusion protein, an oncogenic tyrosine kinase.

Small-scale mutations include point mutations, deletions, and insertions, which may occur in the promoter of a gene and affect its expression, or may occur in the gene's coding sequence and alter the function or stability of its protein product. Disruption of a single gene may also result from integration of genomic material from a DNA virus or retrovirus, and such an event may also result in the expression of viral oncogenes in the affected cell and its descendants.

Oncogenes promote cell growth through a variety of ways. Many can produce hormones, a "chemical messenger" between cells which encourage mitosis, the effect of which depends on the signal transduction of the receiving tissue or cells. In other words, when a hormone receptor on a recipient cell is stimulated, the signal is conducted from the surface of the cell to the cell nucleus to effect some change in gene transcription regulation at the nuclear level. Some oncogenes are part of the signal transduction system itself, or the signal receptors in cells and tissues themselves, thus controlling the sensitivity to such hormones. Oncogenes often produce mitogens, or are involved in transcription of DNA in protein synthesis, which creates the proteins and enzymes responsible for producing the products and biochemicals cells use and interact with.

Mutations in proto-oncogenes, which are the normally quiescent counterparts of oncogenes, can modify their expression and function, increasing the amount or activity of the product protein. When this happens, the proto-oncogenes become oncogenes, and this transition upsets the normal balance of cell cycle regulation in the cell, making uncontrolled growth possible. The chance of cancer cannot be reduced by removing proto-oncogenes from the genome, even if this were possible, as they are critical for growth, repair and homeostasis of the organism. It is only when they become mutated that the signals for growth become excessive.

One of the first oncogenes to be defined in cancer research is the ras oncogene. Mutations in the Ras family of proto-oncogenes (comprising H-Ras, N-Ras and K-Ras) are very common, being found in 20% to 30% of all human tumours. Ras was originally identified in the Harvey sarcoma virus genome, and researchers were surprised that not only was this gene present in the human genome but that, when ligated to a stimulating control element, could induce cancers in cell line cultures.

Many tumor suppressor genes effect signal transduction pathways which regulate apoptosis, also known as "programmed cell death".

Tumor suppressor genes code for anti-proliferation signals and proteins that suppress mitosis and cell growth. Generally, tumor suppressors are transcription factors that are activated by cellular stress or DNA damage. Often DNA damage will cause the presence of free-floating genetic material as well as other signs, and will trigger enzymes and pathways which lead to the activation of tumor suppressor genes. The functions of such genes is to arrest the progression of the cell cycle in order to carry out DNA repair, preventing mutations from being passed on to daughter cells. The p53 protein, one of the most important studied tumor suppressor genes, is a transcription factor activated by many cellular stressors including hypoxia and ultraviolet radiation damage.

Despite nearly half of all cancers possibly involving alterations in p53, its tumor suppressor function is poorly understood. p53 clearly has two functions: one a nuclear role as a transcription factor, and the other a cytoplasmic role in regulating the cell cycle, cell division, and apoptosis.

The Warburg hypothesis is the preferential use of glycolysis for energy to sustain cancer growth. p53 has been shown to regulate the shift from the respiratory to the glycolytic pathway.

However, a mutation can damage the tumor suppressor gene itself, or the signal pathway which activates it, "switching it off". The invariable consequence of this is that DNA repair is hindered or inhibited: DNA damage accumulates without repair, inevitably leading to cancer.

Mutations of tumor suppressor genes that occur in germline cells are passed along to offspring, and increase the likelihood for cancer diagnoses in subsequent generations. Members of these families have increased incidence and decreased latency of multiple tumors. The tumor types are typical for each type of tumor suppressor gene mutation, with some mutations causing particular cancers, and other mutations causing others. The mode of inheritance of mutant tumor suppressors is that an affected member inherits a defective copy from one parent, and a normal copy from the other. For instance, individuals who inherit one mutant *p53* allele (and are therefore heterozygous for mutated *p53*) can develop melanomas and pancreatic cancer, known as Li-Fraumeni syndrome. Other inherited tumor suppressor gene syndromes include *Rb* mutations, linked to retinoblastoma, and *APC* gene mutations, linked to adenopolyposis colon cancer. Adenopolyposis colon cancer is associated with thousands of polyps in colon while young, leading to colon cancer at a relatively early age. Finally, inherited mutations in *BRCA1* and *BRCA2* lead to early onset of breast cancer.

Development of cancer was proposed in 1971 to depend on at least two mutational events. In what became known as the Knudson two-hit hypothesis, an inherited, germ-line mutation in a tumor suppressor gene would only cause cancer if another mutation event occurred later in the organism's life, inactivating the other allele of that tumor suppressor gene.

Usually, oncogenes are dominant, as they contain gain-of-function mutations, while mutated tumor suppressors are recessive, as they contain loss-of-function mutations. Each cell has two copies of the same gene, one from each parent, and under most cases gain of function mutations in just one copy of a particular proto-oncogene is enough to make that gene a true oncogene. On the other hand, loss of function mutations need to happen in both copies of a tumor suppressor gene to render that gene completely non-functional. However, cases exist in which one mutated copy of a tumor suppressor gene can render the other, wild-type copy non-functional. This phenomenon is called the *dominant negative effect* and is observed in many p53 mutations.

Knudson's two hit model has recently been challenged by several investigators. Inactivation of one allele of some tumor suppressor genes is sufficient to cause tumors. This phenomenon is called haploinsufficiency and has been demonstrated by a number of experimental approaches. Tumors caused by haploinsufficiency usually have a later age of onset when compared with those by a two hit process.

Cancer cell Biology

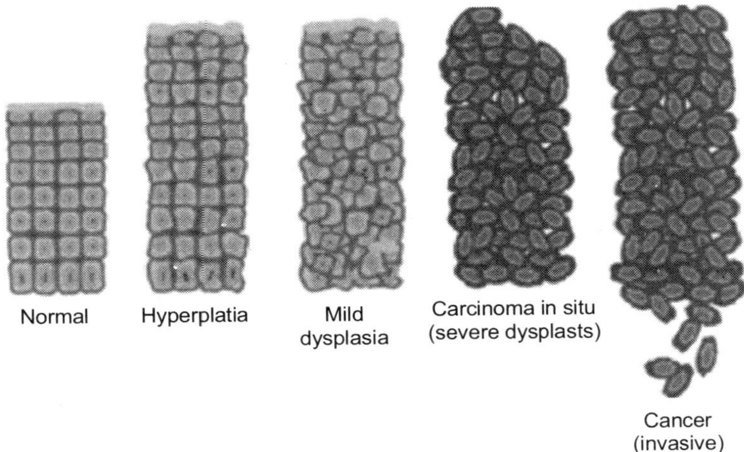

Fig. 7.3 : The Organization of tissue

Often, the multiple genetic changes which result in cancer may take many years to accumulate. During this time, the biological behaviour of the pre-malignant cells slowly change from the properties of normal cells to cancer-like properties. Pre-malignant tissue can have a distinctive appearance under the microscope. Among the distinguishing traits are an increased number of dividing cells, variation in nuclear size and shape, variation in cell size and shape, loss of specialized cell features, and loss of normal tissue organization. Dysplasia is an abnormal type of excessive cell proliferation characterized by loss of normal tissue arrangement and cell structure in pre-malignant cells. These early neoplastic changes must be distinguished from hyperplasia, a reversible increase in cell division caused by an external stimulus, such as a hormonal imbalance or chronic irritation.

The most severe cases of dysplasia are referred to as "carcinoma in situ." In Latin, the term "in situ" means "in place", so carcinoma in situ refers to an uncontrolled growth of cells that remains in the original location and has not shown invasion into other tissues. Nevertheless, carcinoma in situ may develop into an invasive malignancy and is usually removed surgically, if possible.

Just like a population of animals undergoes evolution, an unchecked population of cells also can undergo evolution. This undesirable process is called somatic evolution, and is how cancer arises and becomes more malignant.

Most changes in cellular metabolism that allow cells to grow in a disorderly fashion lead to cell death. However once cancer begins, cancer cells undergo a process of natural selection: the few cells with new genetic changes that enhance their survival or reproduction continue to multiply, and soon come to dominate the growing tumor, as cells with less favourable genetic change are out-competed. This is exactly how pathogens such as MRSA can become antibiotic-resistant (or how HIV can become drug-resistant), and the same reason why crop blights and pests can become pesticide-resistant. This evolution is why cancer recurrences will have cells which have acquired cancer-drug resistance (or in some cases, resistance to radiation from radiotherapy).

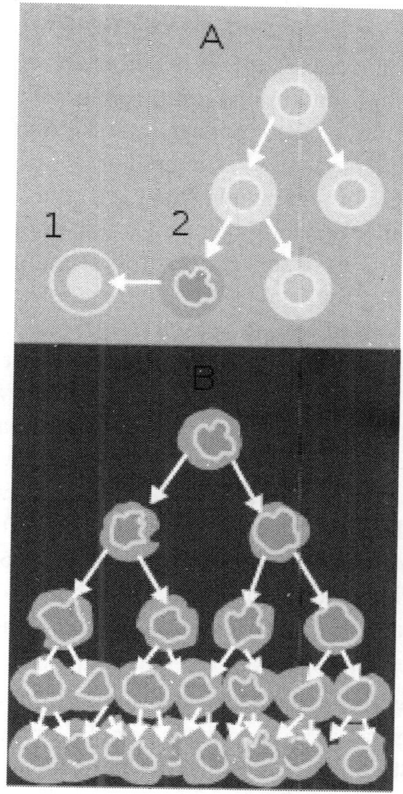

Fig. 7.4 : When normal cells are damaged beyond repair, they are eliminated by apoptosis (A). Cancer cells avoid apoptosis and continue to multiply in an unregulated manner (B).

In a 2000 article by Hanahan and Weinberg, the biological properties of malignant tumor cells were summarized as follows:

- Acquisition of self-sufficiency in growth signals, leading to unchecked growth.
- Loss of sensitivity to anti-growth signals, also leading to unchecked growth.
- Loss of capacity for apoptosis, in order to allow growth despite genetic errors and external anti-growth signals.
- Loss of capacity for senescence, leading to limitless replicative potential (immortality)
- Acquisition of sustained angiogenesis, allowing the tumor to grow beyond the limitations of passive nutrient diffusion.

- Acquisition of ability to invade neighbouring tissues, the defining property of invasive carcinoma.
- Acquisition of ability to build metastases at distant sites, the classical property of malignant tumors (carcinomas or others).

The completion of these multiple steps would be a very rare event without :

- Loss of capacity to repair genetic errors, leading to an increased mutation rate (genomic instability), thus accelerating all the other changes.

These biological changes are classical in carcinomas; other malignant tumors may not need all to achieve them all. For example, tissue invasion and displacement to distant sites are normal properties of leukocytes; these steps are not needed in the development of leukemia. The different steps do not necessarily represent individual mutations. For example, inactivation of a single gene, coding for the p53 protein, will cause genomic instability, evasion of apoptosis and increased angiogenesis. Not all the cancer cells are dividing. Rather, a subset of the cells in a tumor, called cancer stem cells, replicate themselves and generate differentiated cells.

Cancer prevention is defined as active measures to decrease the incidence of cancer. This can be accomplished by avoiding carcinogens or altering their metabolism, pursuing a lifestyle or diet that modifies cancer-causing factors and/or medical intervention (chemoprevention, treatment of pre-malignant lesions). The epidemiological concept of "prevention" is usually defined as either primary prevention, for people who have not been diagnosed with a particular disease, or secondary prevention, aimed at reducing recurrence or complications of a previously diagnosed illness. The vast majority of cancer risk factors are environmental or lifestyle-related in nature, leading to the claim that cancer is a largely preventable disease. Examples of modifiable cancer risk factors include alcohol consumption (associated with increased risk of oral, esophageal, breast, and other cancers), smoking (although 20% of women with lung cancer have never smoked, versus 10% of men), physical inactivity (associated with increased risk of colon, breast, and possibly other cancers), and being overweight / obese (associated with colon, breast, endometrial, and possibly other cancers). Based on epidemiologic evidence, it is now thought that avoiding excessive alcohol consumption may contribute to reductions in risk of certain cancers; however, compared with tobacco exposure, the magnitude of effect is modest or small and the strength of evidence is often weaker. Other lifestyle and environmental factors known to affect cancer risk (either beneficially or detrimentally) include certain sexually transmitted diseases (such as those conveyed by the human papillomavirus), the use of exogenous hormones, exposure to ionizing radiation and ultraviolet radiation, and certain occupational and chemical exposures.

Every year, at least 200,000 people die worldwide from cancer related to their workplace. Millions of workers run the risk of developing cancers such as lung cancer and mesothelioma from inhaling asbestos fibers and tobacco smoke, or leukemia from exposure to benzene at their workplaces. Currently, most cancer deaths caused by occupational risk factors occur in the developed world. It is estimated that approximately 20,000 cancer deaths and 40,000 new cases of cancer each year in the U.S. are attributable to occupation.

The consensus on diet and cancer is that obesity increases the risk of developing cancer. Particular dietary practices often explain differences in cancer incidence in different countries (e.g. gastric cancer is more common in Japan, while colon cancer is more common in the United States. In this example the preceding consideration of Haplogroups are excluded). Studies have shown that immigrants develop the risk of their new country, often within one generation, suggesting a substantial link between diet and cancer. Whether reducing obesity in a population also reduces cancer incidence is unknown.

Despite frequent reports of particular substances (including foods) having a beneficial or detrimental effect on cancer risk, few of these have an established link to cancer. These reports are often based on studies in cultured cell media or animals. Public health recommendations cannot be made on the basis of these studies until they have been validated in an observational (or occasionally a prospective interventional) trial in humans.

Proposed dietary interventions for primary cancer risk reduction generally gain support from epidemiological association studies. Examples of such studies include reports that reduced meat consumption is associated with decreased risk of colon cancer, and reports that consumption of coffee is associated with a reduced risk of liver cancer. Studies have linked consumption of grilled meat to an increased risk of stomach cancer, colon cancer, breast cancer, and pancreatic cancer, a phenomenon which could be due to the presence of carcinogens such as benzopyrene in foods cooked at high temperatures.

A 2005 secondary prevention study showed that consumption of a plant-based diet and lifestyle changes resulted in a reduction in cancer markers in a group of men with prostate cancer who were using no conventional treatments at the time. These results were amplified by a 2006 study in which over 2,400 women were studied, half randomly assigned to a normal diet, the other half assigned to a diet containing less than 20% calories from fat. The women on the low fat diet were found to have a markedly lower risk of breast cancer recurrence, in the interim report of December, 2006.

Recent studies have also demonstrated potential links between some forms of cancer and high consumption of refined sugars and other simple carbohydrates. Although the degree of correlation and the degree of causality is still debated some organizations have in fact begun to recommend reducing intake of refined sugars and starches as part of their cancer prevention regimens.

In November 2007, the American Institute for Cancer Research (AICR), in conjunction with the World Cancer Research Fund (WCRF), published *Food, Nutrition, Physical Activity and the Prevention of Cancer: a Global Perspective*, "the most current and comprehensive analysis of the literature on diet, physical activity and cancer".The WCRF/AICR Expert Report lists 10 recommendations that people can follow to help reduce their risk of developing cancer, including the following dietary guidelines: reducing intake of foods and drinks that promote weight gain, namely energy-dense foods and sugary drinks, eating mostly foods of plant origin, (3) limiting intake of red meat and avoiding processed meat, limiting consumption of alcoholic beverages, and (5) reducing intake of salt and avoiding mouldy cereals (grains) or pulses (legumes).

Some mushrooms offer an anti-cancer effect, which is thought to be linked to their ability to up-regulate the immune system. Some mushrooms known for this effect include, Reishi, *Agaricus blazei*, [1], Maitake,[2] and *Trametes versicolor*[3]. Research suggests the compounds in medicinal mushrooms most responsible for up-regulating the immune system and providing an anti-cancer effect, are a diverse collection of polysaccharide compounds, particularly beta-glucans. Beta-glucans are known as "biological response modifiers", and their ability to activate the immune system is well documented.

Specifically, beta-glucans stimulate the innate branch of the immune system. Research has shown beta-glucans have the ability to stimulate macrophage, NK cells, T cells, and immune system cytokines. The mechanisms in which beta-glucans stimulate the immune system is only partially understood. One mechanism in which beta-glucans are able to activate the immune system, is by interacting with the Macrophage-1 antigen (CD18) receptor on immune cells.

The idea that cancer can be prevented through vitamin supplementation stems from early observations correlating human disease with vitamin deficiency, such as pernicious anemia with vitamin B12 deficiency, and scurvy with Vitamin C deficiency. This has largely not been proven to be the case with cancer, and vitamin supplementation is largely not proving effective in preventing cancer. The cancer-fighting components of food are also proving to be more numerous and varied than previously understood, so patients are increasingly being advised to consume fresh, unprocessed fruits and vegetables for maximal health benefits.

Epidemiological studies have shown that low vitamin D status is correlated to increased cancer risk. However, the results of such studies need to be treated with caution, as they cannot show whether a correlation between two factors means that one causes the other (*i.e.* correlation does not imply causation). The possibility that Vitamin D might protect against cancer has been contrasted with the risk of malignancy from sun exposure. Since exposure to the sun enhances natural human production of vitamin D, some cancer researchers have argued that the potential deleterious malignant effects of sun exposure are far outweighed by the cancer-preventing effects of extra vitamin D synthesis in sun-exposed skin. In 2002, Dr. William B. Grant claimed that 23,800 premature cancer deaths occur in the US annually due to insufficient UVB exposure (apparently via vitamin D deficiency). This is higher than 8,800 deaths occurred from melanoma or squamous cell carcinoma, so the overall effect of sun exposure might be beneficial. Another research group estimates that 50,000–63,000 individuals in the United States and 19,000 - 25,000 in the UK die prematurely from cancer annually due to insufficient vitamin D.

The case of beta-carotene provides an example of the importance of randomized clinical trials. Epidemiologists studying both diet and serum levels observed that high levels of beta-carotene, a precursor to vitamin A, were associated with a protective effect, reducing the risk of cancer. This effect was particularly strong in lung cancer. This hypothesis led to a series of large randomized clinical trials conducted in both Finland and the United States (CARET study) during the 1980s and 1990s. This study provided about 80,000 smokers or former smokers with daily supplements of beta-carotene or placebos. Contrary to expectation, these tests found no benefit of beta-carotene supplementation in reducing lung cancer incidence and mortality. In fact, the risk of lung cancer was slightly, but not significantly, *increased* by beta-carotene, leading to an early termination of the study.

Results reported in the Journal of the American Medical Association (JAMA) in 2007 indicate that folic acid supplementation is not effective in preventing colon cancer, and folate consumers may be more likely to form colon polyps.

The concept that medications could be used to prevent cancer is an attractive one, and many high-quality clinical trials support the use of such chemoprevention in defined circumstances.

Daily use of tamoxifen, a selective estrogen receptor modulator (SERM), typically for 5 years, has been demonstrated to reduce the risk of developing breast cancer in high-risk women by about 50%. A recent study reported that the selective estrogen receptor modulator raloxifene has similar benefits to tamoxifen in preventing breast cancer in high-risk women, with a more favourable side effect profile.

Raloxifene is a SERM like tamoxifen; it has been shown (in the STAR trial) to reduce the risk of breast cancer in high-risk women equally as well as tamoxifen. In this trial, which studied almost 20,000 women, raloxifene had fewer side effects than tamoxifen, though it did permit more DCIS to form.

Finasteride, a 5-alpha-reductase inhibitor, has been shown to lower the risk of prostate cancer, though it seems to mostly prevent low-grade tumors. The effect of COX-2 inhibitors such as rofecoxib and celecoxib upon the risk of colon polyps have been studied in familial adenomatous polyposis patients and in the general population. In both groups, there were significant reductions in colon polyp incidence, but this came at the price of increased cardiovascular toxicity.

Genetic Testing

Genetic testing for high-risk individuals is already available for certain cancer-related genetic mutations. Carriers of genetic mutations that increase risk for cancer incidence can undergo enhanced surveillance, chemoprevention, or risk-reducing surgery. Early identification of inherited genetic risk for cancer, along with cancer-preventing interventions such as surgery or enhanced surveillance, can be lifesaving for high-risk individuals.

Gene	Cancer types	Availability
BRCA1, BRCA2	Breast, ovarian, pancreatic	Commercially available for clinical specimens
MLH1, MSH2, MSH6, PMS1, PMS2	Colon, uterine, small bowel, stomach, urinary tract	Commercially available for clinical specimens

Vaccination

Prophylactic vaccines have been developed to prevent infection by oncogenic infectious agents such as viruses, and therapeutic vaccines are in development to stimulate an immune response against cancer-specific epitopes.

As reported above, a preventive human papillomavirus vaccine exists that targets certain sexually transmitted strains of human papillomavirus that are associated with the development of cervical cancer and genital warts. The only two HPV vaccines on the market as of October 2007 are Gardasil and Cervarix. There is also a hepatitis B vaccine, which prevents infection with the hepatitis B virus, an infectious agent that can cause liver cancer. A canine melanoma vaccine has also been developed.

Cancer screening is an attempt to detect unsuspected cancers in an asymptomatic population. Screening tests suitable for large numbers of healthy people must be relatively affordable, safe, noninvasive procedures with acceptably low rates of false positive results. If signs of cancer are detected, more definitive and invasive follow up tests are performed to confirm the diagnosis.

Screening for cancer can lead to earlier diagnosis in specific cases. Early diagnosis may lead to extended life, but may also falsely prolong the lead time to death through lead time bias or length time bias.

A number of different screening tests have been developed for different malignancies. Breast cancer screening can be done by breast self-examination, though this approach was discredited by a 2005 study in over 300,000 Chinese women. Screening for breast cancer with mammograms has been shown to reduce the average stage of diagnosis of breast cancer in a population. Stage of diagnosis in a country has been shown to decrease within ten years of introduction of mammographic screening programs. Colorectal cancer can be detected through fecal occult blood testing and colonoscopy, which reduces both colon cancer incidence and mortality, presumably through the detection and removal of pre-malignant polyps. Similarly, cervical cytology testing (using the Pap smear) leads to the identification and excision of precancerous lesions. Over time, such testing has been followed by a dramatic reduction of cervical cancer incidence and mortality. Testicular self-examination is recommended for men beginning at the age of 15 years to detect testicular cancer. Prostate cancer can be screened using a digital rectal exam along with prostate specific antigen (PSA) blood testing, though some authorities (such as the US Preventive Services Task Force) recommend against routinely screening all men.

Screening for cancer is controversial in cases when it is not yet known if the test actually saves lives. The controversy arises when it is not clear if the benefits of screening outweigh the risks of follow-up diagnostic tests and cancer treatments. For example: when screening for prostate cancer, the PSA test may detect small cancers that would never become life threatening, but once detected will lead to treatment. This situation, called overdiagnosis, puts men at risk for complications from unnecessary treatment such as surgery or radiation. Follow up procedures used to diagnose prostate cancer (prostate biopsy) may cause side effects, including bleeding and infection. Prostate cancer treatment may cause incontinence (inability to control urine flow) and erectile dysfunction (erections inadequate for intercourse). Similarly, for breast cancer, there have recently been criticisms that breast screening programs in some countries cause more problems than they solve. This is because screening of women in the general population will result in a large number of women with false positive results which require extensive follow-up investigations to exclude cancer, leading to having a high number-to-treat (or number-to-screen) to prevent or catch a single case of breast cancer early.

Cervical cancer screening via the Pap smear has the best cost-benefit profile of all the forms of cancer screening from a public health perspective as, being largely caused by a virus, it has clear risk factors (sexual contact), and the natural progression of cervical cancer is that it normally spreads slowly over a number of years therefore giving more time for the screening program to catch it early. Moreover, the test itself is easy to perform and relatively cheap.

For these reasons, it is important that the benefits and risks of diagnostic procedures and treatment be taken into account when considering whether to undertake cancer screening.

Use of medical imaging to search for cancer in people without clear symptoms is similarly marred with problems. There is a significant risk of detection of what has been recently called an *incidentaloma* - a benign lesion that may be interpreted as a malignancy and be subjected to potentially dangerous investigations. Recent studies of CT scan-based screening for lung cancer in smokers have had equivocal results, and systematic screening is not recommended as of July 2007. Randomized clinical trials of plain-film chest X-rays to screen for lung cancer in smokers have shown no benefit for this approach.

Canine cancer detection has shown promise, but is still in the early stages of research.

Most cancers are initially recognized either because signs or symptoms appear or through screening. Neither of these lead to a definitive diagnosis, which usually requires the opinion of a pathologist, a type of physician (medical doctor) who specializes in the diagnosis of cancer and other diseases. People with suspected cancer are investigated with medical tests. These commonly include blood tests, X-rays, CT scans and endoscopy.

A cancer may be suspected for a variety of reasons, but the definitive diagnosis of most malignancies must be confirmed by histological examination of the cancerous cells by a pathologist. Tissue can be obtained from a biopsy or surgery. Many biopsies (such as those of the skin, breast or liver) can be done in a doctor's office. Biopsies of other organs are performed under anesthesia and require surgery in an operating room.

The tissue diagnosis given by the pathologist indicates the type of cell that is proliferating, its histological grade, genetic abnormalities, and other features of the tumor. Together, this information is useful to evaluate the prognosis of the patient and to choose the best treatment. Cytogenetics and immunohistochemistry are other types of testing that the pathologist may perform on the tissue specimen. These tests may provide information about the molecular changes (such as mutations, fusion genes, and numerical chromosome changes) that has happened in the cancer cells, and may thus also indicate the future behaviour of the cancer (prognosis) and best treatment.

Cancer can be treated by surgery, chemotherapy, radiation therapy, immunotherapy, monoclonal antibody therapy or other methods. The choice of therapy depends upon the location and grade of the tumor and the stage of the disease, as well as the general state of the patient (performance status). A number of experimental cancer treatments are also under development.

Complete removal of the cancer without damage to the rest of the body is the goal of treatment. Sometimes this can be accomplished by surgery, but the propensity of cancers to invade adjacent tissue or to spread to distant sites by microscopic metastasis often limits its effectiveness. The effectiveness of chemotherapy is often limited by toxicity to other tissues in the body. Radiation can also cause damage to normal tissue.

Because "cancer" refers to a class of diseases, it is unlikely that there will ever be a single "cure for cancer" any more than there will be a single treatment for all infectious diseases. Angiogenesis inhibitors were once thought to have potential as a "silver bullet" treatment applicable to many types of cancer, but this has not been the case in practice.

In theory, non-hematological cancers can be cured if entirely removed by surgery, but this is not always possible. When the cancer has metastasized to other sites in the body prior to surgery, complete surgical excision is usually impossible. In the Halstedian model of cancer progression, tumors grow locally, then spread to the lymph nodes, then to the rest of the body. This has given rise to the popularity of local-only treatments such as surgery for small cancers. Even small localized tumors are increasingly recognized as possessing metastatic potential.

Examples of surgical procedures for cancer include mastectomy for breast cancer and prostatectomy for prostate cancer. The goal of the surgery can be either the removal of only the tumor, or the entire organ. A single cancer cell is invisible to the naked eye but can regrow into a new tumor, a process called recurrence. For this reason, the pathologist will examine the surgical specimen to determine if a margin of healthy tissue is present, thus decreasing the chance that microscopic cancer cells are left in the patient.

In addition to removal of the primary tumor, surgery is often necessary for staging, e.g. determining the extent of the disease and whether it has metastasized to regional lymph nodes. Staging is a major determinant of prognosis and of the need for adjuvant therapy.

Occasionally, surgery is necessary to control symptoms, such as spinal cord compression or bowel obstruction. This is referred to as palliative treatment.

Radiation therapy (also called radiotherapy, X-ray therapy, or irradiation) is the use of ionizing radiation to kill cancer cells and shrink tumors. Radiation therapy can be administered externally via external beam radiotherapy (EBRT) or internally via brachytherapy. The effects of radiation therapy are localised and confined to the region being treated. Radiation therapy injures or destroys cells in the area being treated (the "target tissue") by damaging their genetic material, making it impossible for these cells to continue to grow and divide. Although radiation damages both cancer cells and normal cells, most normal cells can recover from the effects of radiation and function properly. The goal of radiation therapy is to damage as many cancer cells as possible, while limiting harm to nearby healthy tissue. Hence, it is given in many fractions, allowing healthy tissue to recover between fractions.

Radiation therapy may be used to treat almost every type of solid tumor, including cancers of the brain, breast, cervix, larynx, lung, pancreas, prostate, skin, stomach, uterus, or soft tissue sarcomas. Radiation is also used to treat leukemia and lymphoma. Radiation dose to each site depends on a number of factors, including the radiosensitivity of each cancer type and whether there are tissues and organs nearby that may be damaged by radiation. Thus, as with every form of treatment, radiation therapy is not without its side effects.

CHEMOTHERAPY

Chemotherapy is the treatment of cancer with drugs ("anticancer drugs") that can destroy cancer cells. In current usage, the term "chemotherapy" usually refers to *cytotoxic* drugs which affect rapidly dividing cells in general, in contrast with *targeted therapy*. Chemotherapy drugs interfere with cell division in various possible ways, e.g. with the duplication of DNA or the separation of newly formed chromosomes. Most forms of chemotherapy target all rapidly dividing cells and are not specific to cancer cells, although some degree of specificity may come from the inability of many cancer cells to repair DNA damage, while normal cells generally can. Hence, chemotherapy has the potential to harm healthy tissue, especially those tissues that have a high replacement rate (e.g. intestinal lining). These cells usually repair themselves after chemotherapy.

Because some drugs work better together than alone, two or more drugs are often given at the same time. This is called "combination chemotherapy"; most chemotherapy regimens are given in a combination.

The treatment of some leukaemias and lymphomas requires the use of high-dose chemotherapy, and total body irradiation (TBI). This treatment ablates the bone marrow, and hence the body's ability to recover and repopulate the blood. For this reason, bone marrow, or peripheral blood stem cell harvesting is carried out before the ablative part of the therapy, to enable "rescue" after the treatment has been given. This is known as autologous stem cell transplantation. Alternatively, hematopoietic stem cells may be transplanted from a matched unrelated donor (MUD).

Targeted therapy, which first became available in the late 1990s, has had a significant impact in the treatment of some types of cancer, and is currently a very active research area. This constitutes the use of agents specific for the deregulated proteins of cancer cells. Small molecule targeted therapy drugs are generally inhibitors of enzymatic domains on mutated, overexpressed, or otherwise critical proteins within the cancer cell. Prominent examples are the tyrosine kinase inhibitors imatinib (Gleevec/Glivec) and gefitinib (Iressa).

Monoclonal antibody therapy is another strategy in which the therapeutic agent is an antibody which specifically binds to a protein on the surface of the cancer cells. Examples include the anti-HER2/neu antibody trastuzumab (Herceptin) used in breast cancer, and the anti-CD20 antibody rituximab, used in a variety of B-cell malignancies.

Targeted therapy can also involve small peptides as "homing devices" which can bind to cell surface receptors or affected extracellular matrix surrounding the tumor. Radionuclides which are attached to these peptides (e.g. RGDs) eventually kill the cancer cell if the nuclide decays in the vicinity of the cell. Especially oligo- or multimers of these binding motifs are of great interest, since this can lead to enhanced tumor specificity and avidity.

Photodynamic therapy (PDT) is a ternary treatment for cancer involving a photosensitizer, tissue oxygen, and light (often using lasers). PDT can be used as treatment for basal cell carcinoma (BCC) or lung cancer; PDT can also be useful in removing traces of malignant tissue after surgical removal of large tumors.

Cancer immunotherapy refers to a diverse set of therapeutic strategies designed to induce the patient's own immune system to fight the tumor. Contemporary methods for generating an immune response against tumours include intravesical BCG immunotherapy for superficial bladder cancer, and use of interferons and other cytokines to induce an immune response in renal cell carcinoma and melanoma patients. Vaccines to generate specific immune responses are the subject of intensive research for a number of tumours, notably malignant melanoma and renal cell carcinoma. Sipuleucel-T is a vaccine-like strategy in late clinical trials for prostate cancer in which dendritic cells from the patient are loaded with prostatic acid phosphatase peptides to induce a specific immune response against prostate-derived cells.

Allogeneic hematopoietic stem cell transplantation ("bone marrow transplantation" from a genetically non-identical donor) can be considered a form of immunotherapy, since the donor's immune cells will often attack the tumor in a phenomenon known as graft-versus-tumor effect. For this reason, allogeneic HSCT leads to a higher cure rate than autologous transplantation for several cancer types, although the side effects are also more severe.

The growth of some cancers can be inhibited by providing or blocking certain hormones. Common examples of hormone-sensitive tumors include certain types of breast and prostate cancers. Removing or blocking estrogen or testosterone is often an important additional treatment. In certain cancers, administration of hormone agonists, such as progestogens may be therapeutically beneficial.

Angiogenesis inhibitors prevent the extensive growth of blood vessels (angiogenesis) that tumors require to survive. Some, such as bevacizumab, have been approved and are in clinical use. One of the main problems with anti-angiogenesis drugs is that many factors stimulate blood vessel growth in cells normal or cancerous. Anti-angiogenesis drugs only target one factor, so the other factors continue to stimulate blood vessel growth. Other problems include route of administration, maintenance of stability and activity and targeting at the tumor vasculature.

Although the control of the symptoms of cancer is not typically thought of as a treatment directed at the cancer, it is an important determinant of the quality of life of cancer patients, and plays an important role in the decision whether the patient is able to undergo other treatments. Although doctors generally have the therapeutic skills to reduce pain, nausea, vomiting, diarrhoea, hemorrhage and other common problems in cancer patients, the multidisciplinary specialty of palliative care has arisen specifically in response to the symptom control needs of this group of patients. This is an especially important aspect of care for those patients whose disease is not a good candidate for other forms of treatment. As most treatments for cancer involve significantly unpleasant side effects, a patient with little realistic hope of a cure may choose to seek palliative care only, eschewing more radical therapies in exchange for a prolonged period of normal living.

Pain medication, such as morphine and oxycodone, and antiemetics, drugs to suppress nausea and vomiting, are very commonly used in patients with cancer-related symptoms. Improved antiemetics such as ondansetron and analogues, as well as aprepitant have made aggressive treatments much more feasible in cancer patients.

Chronic pain due to cancer is almost always associated with continuing tissue damage due to the disease process or the treatment (i.e. surgery, radiation, chemotherapy). Although there is always a role for environmental factors and affective disturbances in the genesis of pain behaviours, these are not usually the predominant etiologic factors in patients with cancer pain. Furthermore, many patients with severe pain associated with cancer are nearing the end of their lives and palliative therapies are required. Issues such as social stigma of using opioids, work and functional status, and health care consumption are not likely to be important in the overall case management. Hence, the typical strategy for cancer pain management is to get the patient as comfortable as possible using opioids and other medications, surgery, and physical measures. Doctors have been reluctant to prescribe narcotics for pain in terminal cancer patients, for fear of contributing to addiction or suppressing respiratory function. The palliative care movement, a more recent offshoot of the hospice movement, has engendered more widespread support for preemptive pain treatment for cancer patients.

Fatigue is a very common problem for cancer patients, and has only recently become important enough for oncologists to suggest treatment, even though it plays a significant role in many patients' quality of life.Clinical trials, also called research studies, test new treatments in people with cancer. The goal of this research is to find better ways to treat cancer and help cancer patients. Clinical trials test many types of treatment such as new drugs, new approaches to surgery or radiation therapy, new combinations of treatments, or new methods such as gene therapy.

A clinical trial is one of the final stages of a long and careful cancer research process. The search for new treatments begins in the laboratory, where scientists first develop and test new ideas. If an approach seems promising, the next step may be testing a treatment in animals to see how it affects cancer in a living being and whether it has harmful effects. Of course, treatments that work well in the lab or in animals do not always work well in people. Studies are done with cancer patients to find out whether promising treatments are safe and effective.Patients who take part may be helped personally by the treatment they receive. They get up-to-date care from cancer experts, and they receive either a new treatment being tested or the best available standard treatment for their cancer. At the same time, new treatments also may have unknown risks, but if a new treatment proves effective or more effective than standard treatment, study patients who receive it may be among the first to benefit. There is no guarantee that a new treatment being tested or a standard treatment will produce good results. In children with cancer, a survey of trials found that those enrolled in trials were on average not more likely to do better or worse than those on standard treatment; this confirms that success or failure of an experimental treatment cannot be predicted.

Complementary and alternative medicine (CAM) treatments are the diverse group of medical and health care systems, practices, and products that are not part of conventional medicine. "Complementary medicine" refers to methods and substances used along with conventional medicine, while "alternative medicine" refers to compounds used instead of conventional medicine. CAM use is common among people with cancer; a 2000 study found that 69% of cancer patients had used at least one CAM therapy as part of their cancer treatment. Most complementary and alternative medicines for cancer have not been rigorously studied or tested. Some alternative treatments which have been investigated and shown to be ineffective continue to be marketed and promoted.

The incidence of concurrent cancer during pregnancy has risen due to the increasing age of pregnant mothers and due to the incidental discovery of maternal tumors during prenatal ultrasound examinations.

Cancer treatment needs to be selected to do least harm to both the woman and her embryo/fetus. In some cases a therapeutic abortion may be recommended.

Radiation therapy is generally out of the question, and chemotherapy always poses the risk of miscarriage and congenital malformations Little is known about the effects of medications on the child.

Even if a drug has been tested as not crossing the placenta to reach the child, some cancer forms can harm the placenta and make the drug pass over it anyway. Some forms of skin cancer may even metastasize to the child's body.

Diagnosis is also made more difficult, since computed tomography is infeasible because of its high radiation dose. Still, magnetic resonance imaging works normally. However, contrast media cannot be used, since they cross the placenta.

As a consequence of the difficulties to properly diagnose and treat cancer during pregnancy, the alternative methods are either to perform a Cesarean section when the child is viable in order to begin a more aggressive cancer treatment, or, if the cancer is malignant enough that the mother is unlikely to be able to wait that long, to perform an abortion in order to treat the cancer.

Fetal tumors are sometimes diagnosed while still in utero. Teratoma is the most common type of fetal tumor, and usually is benign.

Ultrasound energy is being studied as a form of therapy.

Cancer has a reputation for being a deadly disease. While this certainly applies to certain particular types, the truths behind the historical connotations of cancer are increasingly being overturned by advances in medical care. Some types of cancer have a prognosis that is substantially better than nonmalignant diseases such as heart failure and stroke.

Progressive and disseminated malignant disease has a substantial impact on a cancer patient's quality of life, and many cancer treatments (such as chemotherapy) may have severe side-effects. In the advanced stages of cancer, many patients need extensive care, affecting family members and friends. Palliative care solutions may include permanent or "respite" hospice nursing.

Many local organizations offer a variety of practical and support services to people with cancer. Support can take the form of support groups, counseling, advice, financial assistance, transportation to and from treatment, films or information about cancer. Neighbourhood organizations, local health care providers, or area hospitals may have resources or services available.

Counselling can provide emotional support to cancer patients and help them better understand their illness. Different types of counselling include individual, group, family, peer counselling, bereavement, patient-to-patient, and sexuality.

Many governmental and charitable organizations have been established to help patients cope with cancer. These organizations are often involved in cancer prevention, cancer treatment, and cancer research.

Cancer is responsible for about 25% of all deaths in the U.S., and is a major public health problem in many parts of the world. In the U.S., lung cancer causes about 30% of cancer deaths but only about 15% of new cancer cases; the most commonly occurring cancer in men is prostate cancer (about 25% of new cases) and in women is breast cancer (also about 25%). Cancer can also occur in young children and adolescents, but it is rare (about 150 cases per million in the U.S.), with leukemia being the most common. In the first year of life the incidence is about 230 cases per million in the U.S., with the most common being neuroblastoma.

Over a third of cancer deaths worldwide are due to potentially modifiable risk factors, which are headed by tobacco smoking, alcohol use, and diets low in fruit and vegetables. In developed countries overweight and obesity is also a leading cause of cancer, and in low-and-middle-income countries sexual transmission of human papillomavirus is a leading risk factor for cervical ca

XENOBIOTICS

A xenobiotic is a chemical which is found in an organism but which is not normally produced or expected to be present in it. It can also cover substances which are present in much higher concentrations than are usual. Specifically, drugs such as antibiotics are xenobiotics in humans because the human body does not produce them itself, nor are they part of a normal diet.

Natural compounds can also become xenobiotics if they are taken up by another organism, such as the uptake of natural human hormones by fish found downstream of sewage treatment plant outfalls, or the chemical defenses produced by some organisms as protection against predators.

However, the term **xenobiotics** is very often used in the context of pollutants such as dioxins and polychlorinated biphenyls and their effect on the biota, because xenobiotics are understood as substances foreign to an entire biological system, i.e. artificial substances, which did not exist in nature before their synthesis by humans.

The body removes xenobiotics by xenobiotic metabolism. This consists of the deactivation and the secretion of xenobiotics, and happens mostly in the liver. Secretion routes are urine, feces, breath, and sweat. Hepatic enzymes are responsible for the metabolism of xenobiotics by first activating them (oxidation, reduction, hydrolysis and/or hydration of the xenobiotic), and then conjugating the active secondary metabolite with glucuronic or sulphuric acid, or glutathione, followed by excretion in bile or urine. An example of a group of enzymes involved in xenobiotic metabolism is hepatic microsomal cytochrome P450. These enzymes that metabolize xenobiotics are very important for the pharmaceutical industry, because they are responsible for the breakdown of medications.

Organisms can also evolve to tolerate xenobiotics. An example is the co-evolution of the production of tetrodotoxin in the rough-skinned newt and the evolution of tetrodotoxin resistance in its predator, the common garter snake. In this predator-prey pair, an evolutionary arms race has produced high levels of toxin in the newt and correspondingly high levels of resistance in the snake. This evolutionary response is based on the snake evolving modified forms of the ion channels that the toxin acts upon, so becoming resistant to its effects. Xenobiotic substances are becoming an increasingly large problem in Sewage Treatment systems, since they are relatively new substances and are very difficult to categorize. Antibiotics, for example, were derived from plants originally, and so mimic naturally occurring substances. This, along with the natural monopoly nature of municipal Waste Water Treatment Plants makes it nearly impossible to remove this new pollutant load.

Some xenobiotics are resistant to degradation. For example, they may be synthetic organochlorides such as plastics and pesticides, or naturally occurring organic chemicals such as polyaromatic hydrocarbons (PAHs) and some fractions of crude oil and coal. However, it is believed that microorganisms are capable of degrading all the different complex and resistant xenobiotics found on the earth.

The term **xenobiotic** is also used to refer to organs transplanted from one species to another. For example, some researchers hope that hearts and other organs could be transplanted from pigs to humans. Many people die every year whose lives could have been saved if a critical organ had been available for transplant. Kidneys are currently the most commonly transplanted organ. Xenobiotic organs would need to be developed in such a way that they would not be rejected by the immune system.

DETOXIFICATION

Body cleansing or detoxification is a treatment in alternative medicine which proponents claim rid the body of accumulated harmful substances that exert a negative effect on individual health. Critics argue that such cleansings are often unnecessary, and are based on questionable or disproved scientific claims.

The premise of body cleansing is based on the Ancient Egyptian and Greek idea of autointoxication, in which foods consumed or in the humoral theory of health that the four humours themselves can putrefy and produce toxins that harm the body. Biochemistry and microbiology appeared to support the theory in the 19th century, but by the early twentieth century, detoxification based approaches quickly fell out of favour. Despite abandonment by mainstream medicine, the idea has persisted in the popular imagination and amongst alternative medicine practitioners. In recent years, notions of body cleansing have undergone something of a resurgence, along with many other alternative medical approaches. Nonetheless, mainstream medicine continues to view the field as unscientific and anachronistic.

Various modalities of body cleansing are currently employed, ranging from physical treatments (e.g. colon cleansing), to dietary restrictions (e.g. avoiding foods) or dietary supplements. Some variants involve the use of herbs and supplements that purportedly speed or increase the effectiveness of the process of cleansing. Several naturopathic and homeopathic preparations are also promoted for cleansing; such products are often marketed as targeting specific organs, such as fiber for the colon or juices for the kidneys.

Detox diets are dietary plans that claim to have detoxifying effects. Scientists, dietitians, and doctors, while generally judging 'detox diets' harmless (unless nutritional deficiency results), often dispute the value and need of 'detox diets' due to lack of supporting factual evidence.

"Detox" diets usually suggest that water, or fruits and vegetables compose a majority of one's food intake. Limiting this to unprocessed (and sometimes also non-GM foods is often advocated. Limiting or eliminating alcohol is also a major factor, and drinking more water is similarly recommended.

Methods to modify the diet for the purpose of detoxification include:

- Eliminating foods such as caffeine, alcohol, processed food (including any bread), pre-made or canned food, salt, sugar, wheat, red meat, pork, fried and deep fried food, yellow cheese, cream, butter and margarine, shortening, etc., while focusing on pure foods such as raw fruits and vegetables, whole grains, legumes, raw nuts and seeds, fish, vegetable oils, herbs and herbal teas, water, etc.
- Master Cleanse, also known as the *lemonade diet*, terms that refer to the fasting diet advocated by Stanley Burroughs

Some proponents of detox diets emphasize it as a lifestyle rather than a diet, Others have touted spiritual and psychological benefits of regular detox dieting.

Body cleansing and detoxification have been referred to as an elaborate hoax used by con artists to cure nonexistent illnesses. Most doctors contend that the 'toxins' in question do not even exist. In response, alternative medicine proponents frequently cite heavy metals or pesticides as the source of toxification, however no evidence exists that detoxification approaches have a measurable effect on these or any other chemical levels. Medical experts state that body cleansing is unnecessary as the human body is naturally capable of maintaining itself, with several organs dedicated to cleansing the blood and gut.

Professor Alan Boobis OBE, Toxicologist, Division of Medicine, Imperial College London states that "The body's own detoxification systems are remarkably sophisticated and versatile. They have to be, as the natural environment that we evolved in is hostile. It is remarkable that people are prepared to risk seriously disrupting these systems with unproven 'detox' diets, which could well do more harm than good."

Highly restrictive detox diets such as Water fasting or the Master Cleanse are not the safest form of weight loss. These diets, if done improperly or for too long, may result in nutrient deficiencies. Of particular concern is lack of protein, which may result in wasting of muscle tissue, due to insufficient amino acids for repair. With less lean muscle tissue, the body's metabolic needs decrease, which hampers weight loss efforts unless calories are lessened further in the diet.

Finally, while many testimonial and anecdotal accounts exist of health improvements following a "detox", these are more likely attributable to the placebo effect; where people actually believe that they are doing something good and healthy. Yet, there is a severe lack of quantitative data. Some changes recommended in certain "detox" lifestyles are also found in mainstream medical advice (such as consuming a diet high in fruits and vegetables). These changes can often produce beneficial effects in and of themselves, and it is accordingly difficult to separate these effects from those caused by the more controversial detoxification recommendations.

IMMUNOCHEMISTRY AND CHEMOTHERAPY

The diseases caused by the microorganisms can be divided into three groups animal parasites, such as protozoa, vegetable parasites, such as bacteria and fungi, and filter passing viruses.

Amebic dysentery, malaria and sleeping – sickness all three common in tropical countries are caused by protozoa. Since the spirochetes are usually classed as protozoa, syphilis would be included here.

In temperate climate the chief diseases are caused by bacteria. Some of these bacteria and the disease they give rise to are the following.

Streptococcus Pyogens or β-hemolytic streptococcus includes some thirteen different strains of bacteria which give rise to septic wounds, blood poisoning erysipelas scarlet fever etc.

Diplococcus Pneumoniæ or pneumococcus, includes some thirty four strains which are the common cause of pneumonia.

Staphylococcus Aureus, commonly found in the skin, usually causes boils.

Neisseria Gonorrhœæ or Gonococcus gives rise to gonorrhea.

Neisseria Intracellularis or menigococcus causes the cerebrospinal meningitis

Various bacteria cause diphtheria, cholera, plague, lock jaws anthrax etc.

Diseases due to viruses – particles which resemble bacteria but unlike them pass through unglazed porcelain – are small-pox infantile paralysis typos fever and probably measles mumps influenza etc.

So far, greater success has attended the treatment of protozoan and bacterial diseases than the virus diseases.

BODY'S DEFENSE MECHANISM

The skin and the acid of the stomach are among these defenses – skin prevents the entrance of the bacteria and acid kills it, unfortunately still many bacteria enter the body. The large intestine is full of bacteria but fortunately these cannot penetrate through the walls of the intestine and enter into the body.

However there is some other way by which the bacteria can penetrate they are through mouth (that is throat) and through the cut on the skin or as result of 'low body resistance' bacteria do penetrate the blood stream and tissue but to counter this menace certain cells they are called macrophages counter these bacteria and are usually strong enough to kill them.

These macrophages are found in the lining of the blood vessels of the liver, bone marrow etc (reticulo-endithelial system) and some wander and may even circulate in the blood. The white cells (leukocytes) of the blood have similar functions to the macrophages.

We have seen how the body handles certain toxic substances by detoxicating them and then eliminating them, the detoxified products, when however the toxic substances enter the blood stream it is chemically complex in nature approaching, say that of a complex protein – body tries to resist the effects of such toxicity by building up for its defense:" antibodies" which tend to offset the effects of the toxic substances.

The substances which give rise to these antibodies are called "antigens" these antigens include bacteria, protozoa molds and "foreign" to the body that is to say, proteins which are not normally found in the blood-serum of the animal, gives rise to antibodies, Certain polysaccharides may also stimulate antibody production.

So far all attempts to prepare antibodies free from protein have been unsuccessful. When a suspension of living or dead bacteria (pneumococci), for example is injected into an animal characteristically when mixed with an extract of bacteria. The characteristic reaction resulting from the mixing of the serum with the bacteria originally used for the injection may give s rise to a clumping or "agglutination"[1] to a

[1] **Agglutination** is the clumping of particles. The word *agglutination* comes from the Latin *agglutinare*, meaning "to glue to."

This occurs in biology in three main examples:

1. The clumping of cells such as bacteria or red blood cells in the presence of an antibody. The antibody or other molecule binds multiple particles and joins them, creating a large complex.
2. The coalescing of small particles that are suspended in solution; these larger masses are then (usually) precipitated.
3. An allergic reaction type occurrence where cells become more compacted together to prevent foreign materials entering them. This is usually the result of an antigen in the vicinity of the cells.

dissolving or "lysis"[2] or to precipitate "precipitation" depending upon conditions. The agglutinins lysins[3] and precipitinins[4] are all antibodies.

The antigenic properties of the bacteria are usually ascribed to the proteins which they contain, Proteins in general, which the notable exception of gelatin, gives to rise to antibodies and these substances are removed from the serum with the fraction containing the globulins a fact which is used as evidence that the antibodies may resemble the serum globulins.

These antibodies, then are in all probability globulins, but differ from the ordinary serum globulins in one important respect; their reactivity with the antigen; which means that the antibody molecule differs in some structural way from the serum globulins.

The specifically of these immunological test for human blood will be discussed in the chapter 8 the hematochemistry.

With the help of a good confocal microscope we can see the actual antigen – antibody reaction.

SUB-SECTION 7.5B—THE ANTIGEN – ANTIBODY REACTION

Antibodies are proteins

- synthesized and secreted by B cells that
- bind to antigens. Most antigens are macromolecules: proteins, polysaccharides, even DNA and RNA.

The interaction occurs:

- by noncovalent forces (like that between enzymes and their substrate)
- between
  - the antigen-combining site on the antibody and
  - a portion of the antigen called the **antigenic determinant** or **epitope**.
- (a) The tube contains antibodies to the Type III pneumococcal polysaccharide isolated from the capsule surrounding the bacteria.
- (b) A solution of the polysaccharide is added, and
- (c) the formation of insoluble antigen-antibody complexes is revealed by the almost instantaneous appearance of turbidity.
- (d) After an hour, the complexes settle out as a precipitate. If the proportion of antigen to antibody in the mixture is selected properly, the fluid above the precipitate will be devoid of both.

---

[2] **Lysis** refers to the death of a cell by breaking of the cellular membrane, causing the contents to spill out, often by viral, enzymic or osmotic mechanisms that compromise its integrity. A fluid containing the contents of lysed cells is called a "lysate".Many species of bacteria are subject to lysis by the enzyme lysozyme, found in animal saliva, egg white and other secretions. Phage lytic enzymes (lysins) produced during bacteriophage infection are responsible for the ability of these viruses to lyse bacterial cells. Penicillin and related â-lactam antibiotics cause the death of bacteria through enzyme-mediated lysis that occurs after the drug causes the bacterium to form a defective cell wall.

[3] **Agglutinin** is a protein found in cow's milk. Because agglutinin clumps, cow's milk requires homogenization in order to remain smooth. The process by which agglutinin clumps is called agglutination.

[4] A **precipitin** is an antibody which can precipitate out of a solution upon antigen binding

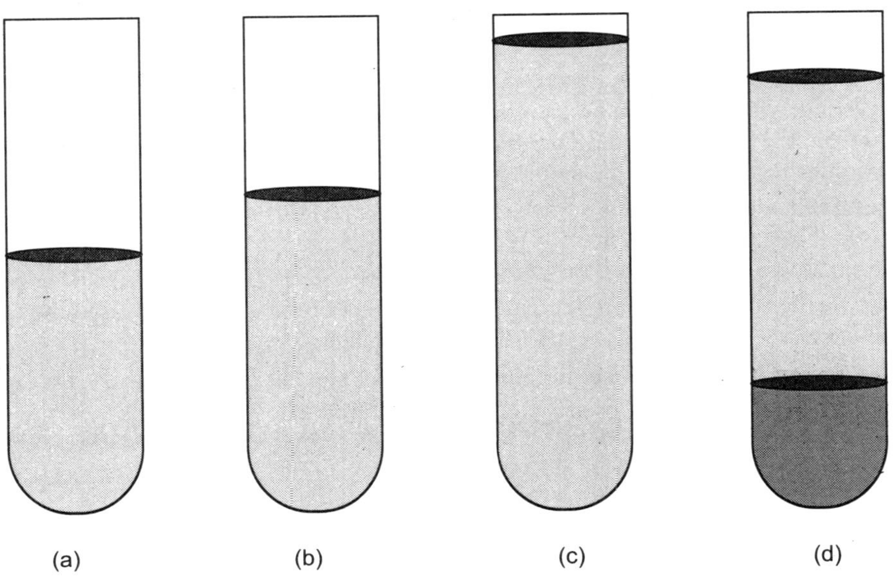

Fig. 7.5 : One type of Interaction precipitation between antibodies and antigen.

In the human body, this binding can literally be life-saving.

The capsule that surrounds pneumococci protects them from phagocytosis. (Pneumococci that fail to make a capsule "R" forms do not cause disease.

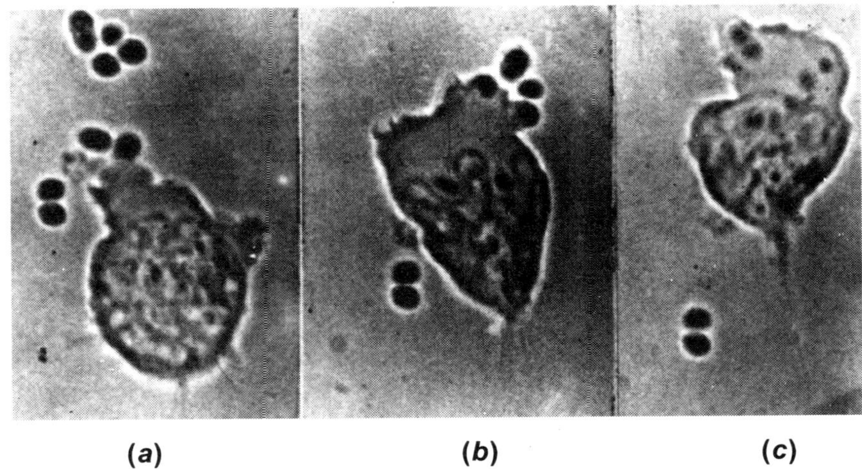

Fig. 7.6: (*a*) A neutrophil Extends a Pseudopod Toward Two Pneumococci. (*b*) these bacteria have been engulfed (arrows), and the neutrophil is beginning to engulf four more pneumococci at the upper right. (*c*) Two pneumococci have escaped.

If the appropriate antibodies are present in the body, they combine with the capsule. Coated with protein instead of polysaccharide, the pneumococci are now easy to ingest.

These photomicrographs show phagocytosis of antibody-coated pneumococci.

In the days before antibiotics, the start of antibody production by the immune system of the patient marked the turning point in the progression of the disease.

The Antigen-Combining Site

The diagram on the Fig. 7.7 shows the primary structure of an IgG antibody. Different IgG antibodies differ most markedly at the so-called hypervariable regions (shown in red):

- three in the heavy chain and
- three in the light chain

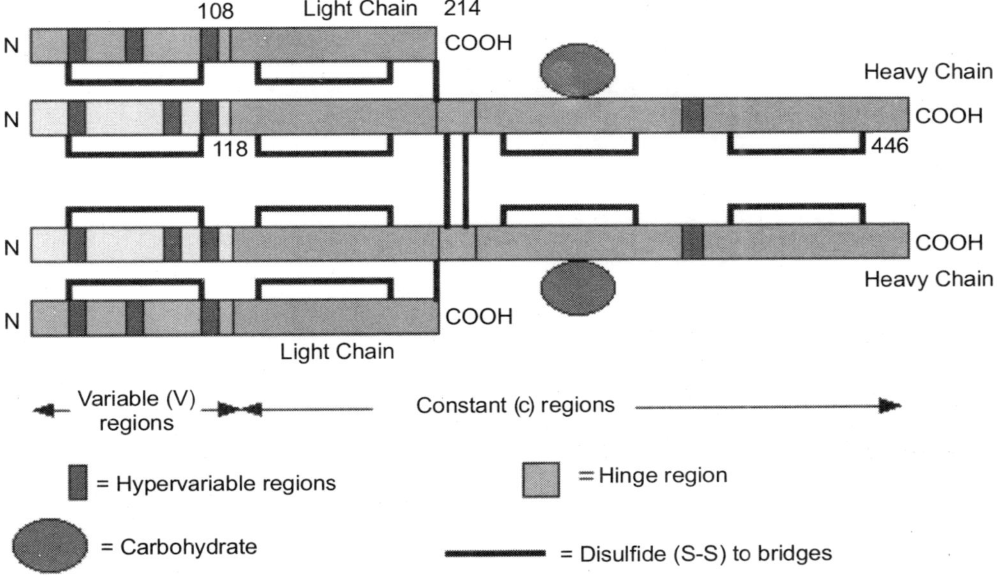

Fig. 7.7 : The shows the primary structure of an IgG Antibody

In the three-dimensional (tertiary) structure of the molecule, the 6 hypervariable regions are brought close together and make up the antigen-combining site. For this reason, the hypervariable regions are also called **complementarity determining regions** (**CDR**s).

Marked are the

- constant region of the light chain (C_L) and the first constant region of the heavy chain(C_H);
- variable regions of the heavy (V_H) and light chains (V_L);

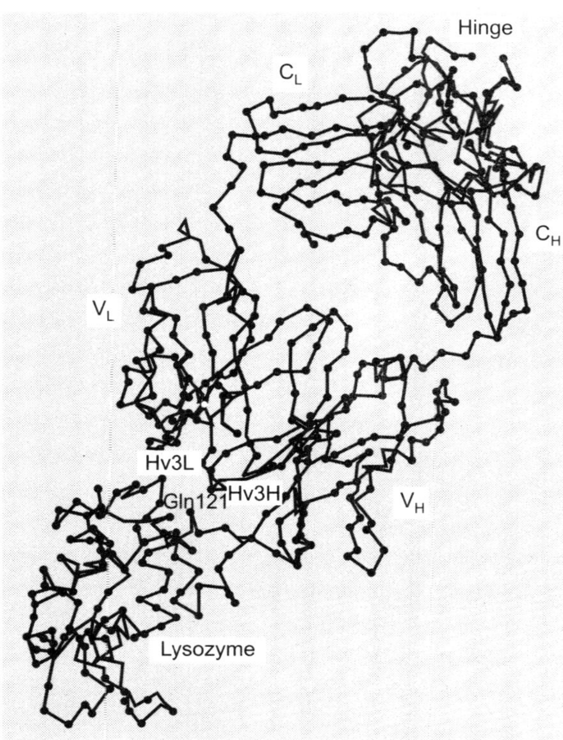

Fig. 7.8 : The Tertiary Structure of One Arm of an Antibody Specific for the lysozyme Found in Hen's Egg white (upper right) Bound to lysozyme (lower left)

- the third hypervariable region of the heavy (Hv3H) and light (Hv3L) chains. The importance of the pair of third hypervariable or complementarity determining regions in binding the epitope is typical of both antibodies and of T cell receptors for antigen. This makes good sense as the opportunities for genetic diversity at those sites is far greater than for the first and second CDRs.

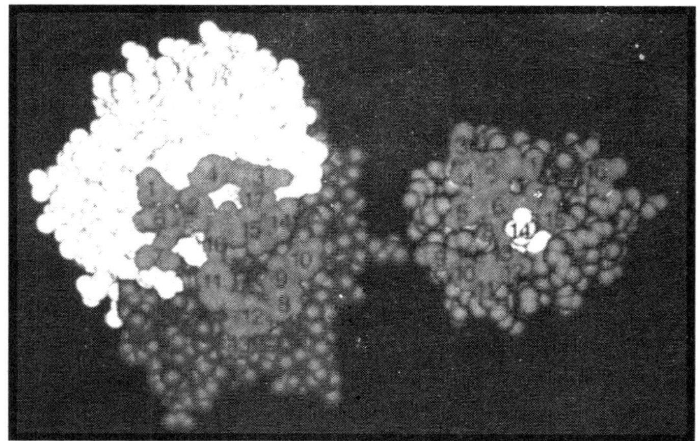

Fig. 7.9 : The Space Filling Model of the Antibody Structure

- Gln121 is a glutamine that is a dominant feature of the epitope on lysozyme.

Fig. 7.9 is a space-filling model of the same antibody with the light chain in yellow, the heavy chain in blue. The view is looking down on the epitope-binding surface (left) and the epitope on lysozyme (right). In the antigen-antibody complex, the two surfaces fit snugly together. The 17 amino acid residues that contact lysozyme in the antigen-antibody complex are numbered (7 on the L chain, 17 on the H chain. The 16 amino acid residues of lysozyme that contact the antibody; that is, that make up its epitope are also numbered. Number 14 is Gln-121.

The complementarity of the antigen-binding site and the epitope, i.e.

- their respective shapes and
- the opportunities for multiple non- covalent interactions

determine how strongly the two bind together.

The **strength** of the binding of an antibody to its antigen is called its **affinity.**

THE ANTIBODY CLASSES

There are different classes of antibody which deals with the antigens in the different parts of the body.

IgG Immunoglobulins of which there are four subclasses (IgG1, IgG2 IgG3, IgG4) provide the built of immunity to most blood born infection agents, and are the only antibody class to cross the placenta to provide the humeral immunity top the infant.

IgA This is the first line of defense against the microbes entering through the mucosal surfaces (the respiratory)

the gastrointestinal and genitourinary tracts) Secratory (dimeric). IgA is synthesized locally by plasma cells, which binds to the poly Ig receptor on epithelial cells and is transported through these cells to the luminal surface where it is released with a portion of the poly-Ig receptor (secratory componet) Sc. This antibody prevents colonization of mucosal surfaces by pathogens and medicates their phagocytosis.

IgM is an antigen receptor on B cells and the first antibody produced in an immune response. In circulation IgM is composed of five four chain units with 10 combining sites, it thus has high eagerness for antigens.

It is very efficient in the molecular level dealing with the pathogens especially early in the immune response before sufficient quantities of IgG have been produced.

B-Cells

The abbreviation "B", in **B cell**, comes from the bursa of Fabricius in birds, where they mature. In mammals, immature B cells are formed in the bone marrow

B cells are lymphocytes that play a large role in the humoral immune response (as opposed to the cell-mediated immune response, which is governed by T cells). The principal functions of B cells are to make antibodies against antigens, perform the role of Antigen Presenting Cells (APCs) and eventually develop into memory B cells after activation by antigen interaction. B cells are an essential component of the adaptive immune system.

Development of B Cells

Immature B cells are produced in the bone marrow of most mammals. Rabbits are an exception; their B cells develop in the appendix-sacculus rotundus. After reaching the IgM+ immature stage in the bone marrow, these immature B cells migrate to the spleen, where they are called transitional B cells, and some of these cells differentiate into mature B lymphocytes.

B cell development occurs through several stages, each stage representing a change in the genome content at the antibody loci. An antibody is composed of two identical light (L) and two identical heavy (H) chains, and the genes specifying them are found in the 'V' (Variable) region and the 'C' (Constant) region. In the heavy-chain 'V' region there are three segments; V, D and J, which recombine randomly, in a process called VDJ recombination, to produce a unique variable domain in the immunoglobulin of each individual B cell. Similar rearrangements occur for light-chain 'V' region except there are only two segments involved; V and J. The list below describes the process of immunoglobulin formation at the different stages of B cell development.

Stage	Heavy chain	Light chain	Ig	IL-7 receptor?	CD19?
Progenitor B cells	germline	germline	-	No	No
Early Pro-B cells	undergoes D-J rearrangement	germline	-	No	No
Late Pro-B cells	undergoes V-DJ rearrangement	germline	-	No	Yes[2]
Large Pre-B cells	is VDJ rearranged	germline	IgM in cytoplasm	Yes[3]	Yes
Small Pre-B cells	is VDJ rearranged	undergoes V-J rearrangement	IgM in cytoplasm	Yes	Yes
Immature B cells	is VDJ rearranged	VJ rearranged	IgM on surface	Yes	Yes
Mature B cells	is VDJ rearranged	VJ rearranged	IgM and IgD on surface	Yes	Yes

When the B cell fails in any step of the maturation process, it will die by a mechanism called apoptosis, here called clonal deletion. If it recognizes self-antigen during the maturation process, the B cell will become suppressed (known as anergy) or undergo apoptosis (also termed negative selection). B cells are continuously produced in the bone marrow. When B cell receptors on the surface of the cell matches the detected antigens present in the body, the B cell proliferates and secretes a free form of those receptors (antibodies) with identical binding sites as the ones on the original cell surface. After activation, the cell proliferates and B memory cells would form to recognise the same antigen. This information would then be used as a part of the adaptive immune system for a more efficient and more powerful immune response for future encounters with that antigen.

B cell membrane receptors evolve and change throughout the B cell life span. TACI, BCMA and BAFF-R are present on both immature B cells and mature B cells. All of these 3 receptors may be inhibited by Belimumab. CD20 is expressed on all stages of B cell development except the first and last; it is present from pre-pre B cells through memory cells, but not on either pro-B cells or plasma cells.

Functions

The human body makes millions of different types of B cells each day that circulate in the blood and lymphatic system performing the role of immune surveillance. They do not produce antibodies until they become fully activated. Each B cell has a unique receptor protein [referred to as the B cell receptor (BCR)] on its surface that will bind to one particular antigen. The BCR is a membrane-bound immunoglobulin, and it is this molecule that allows the distinction of B cells from other types of lymphocyte, as well as being the main protein involved in B cell activation. Once a B cell encounters its cognate antigen and receives an additional signal from a T helper cell, it can further differentiate into one of the two types of B cells listed below (plasma B cells and memory B cells). The B cell may either become one of these cell types directly or it may undergo an intermediate differentiation step, the germinal center reaction, where the B cell will hypermutate the variable region of its immunoglobulin gene ("somatic hypermutation") and possibly undergo class switching.

Clonality

B cells exist as clones. All B cells derive from a particular cell, and thus, the antibodies their differentiated progenies (see below) produce can recognize and/or bind the same components (epitope) of a given antigen. Such clonality has important consequences, as immunogenic memory relies on it. The great diversity in immune response comes about because there are up to 10^9 clones with specificities for recognizing different antigens. A single B cell or a clone of cells with shared specificity upon encountering its specific antigen divides to produce many B cells. Most of such B cells differentiate into plasma cells that secrete antibodies into blood that bind the same epitope that elicited proliferation in the first place. A small minority survives as *memory cells* that can recognize only the same epitope. However, with each cycle, the number of surviving memory cells **increases**. The increase is accompanied by *affinity maturation* which induces the survival of B cells that bind to the particular antigen with high affinity. This subsequent amplification with improved specificity of immune response is known as *secondary immune response*. B cells that encounter antigen for the first time are known as *naive* B cells.

B Cell Types

A suspected plasma cell. Plasma cells are normally detected in tissue rather than circulation and are hard to distinguish from other lymphocytes

- **Plasma B cells** (also known as *plasma cells*) are large B cells that have been exposed to antigen and are producing and secreting large amounts of antibodies, which assist in the destruction of microbes by binding to them and making them easier targets for phagocytes and activation of the complement system. They are sometimes referred to as *antibody factories*. An electron micrograph of these cells reveals large amounts of rough endoplasmic reticulum, responsible for synthesizing the antibody, in the cell's cytoplasm. These are short lived cells and undergo apoptosis when the inciting agent that induced immune response is eliminated. This occurs because of cessation of continuous exposure to various colony stimulating factors required for survival.

- **Memory B cells** are formed from activated B cells that are specific to the antigen encountered during the primary immune response. These cells are able to live for a long time, and can respond quickly following a second exposure to the same antigen.
- **B-1 cells** express IgM in greater quantities than IgG and its receptors show polyspecificity, meaning that they have low affinities for many different antigens, but have a preference for other immunoglobulins, self antigens and common bacterial polysaccharides. B-1 cells are present in low numbers in the lymph nodes and spleen and are instead found predominantly in the peritoneal and pleural cavities.
- **B-2 cells** are the conventional B cells most texts refer to.
- **Marginal-zone B cells**
- **Follicular B Cells**

Recognition of Antigen by B Cells

T cell-dependent B cell activation, showing a TH2-cell (left), B cell (right), and several interaction molecules

A critical difference between B cells and T cells is how each lymphocyte recognizes its antigen. B cells recognize their cognate antigen in its native form. They recognize free (soluble) antigen in the blood or lymph using their BCR or membrane bound-immunoglobulin. In contrast, T cells recognize their cognate antigen in a processed form, as a peptide fragment presented by an antigen presenting cell's MHC molecule to the T cell receptor.

Activation of B Cells (see Fig. 7.10)

B cell recognition of antigen is not the only element necessary for B cell activation (a combination of clonal proliferation and terminal differentiation into plasma cells). B cells that have not been exposed to antigen, also known as *naïve B cells*, can be activated in a T cell-dependent or -independent manner.

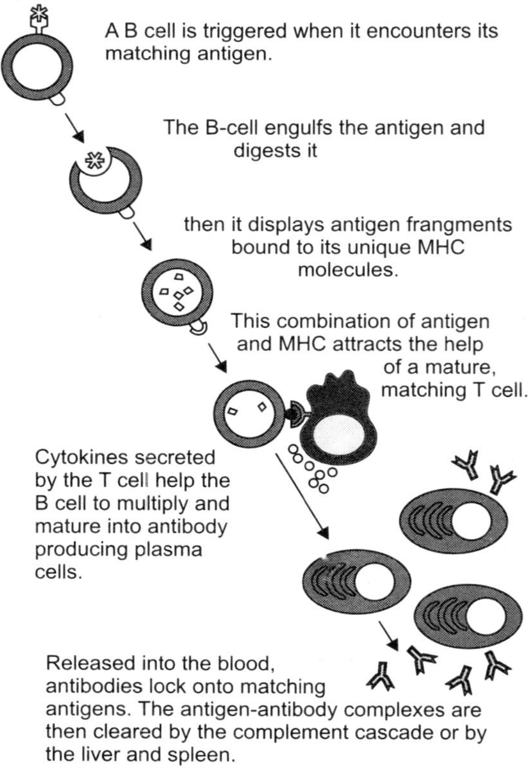

Fig. 7.10 : Showing the Activation of B Cells

T Cell-dependent Activation

Once a pathogen is ingested by an antigen-presenting cell such as a macrophage or dendritic cell, the pathogen's proteins are then digested to peptides and attached to a class II MHC protein. This complex is then moved to the outside of the cell membrane. The macrophage is now activated to deliver multiple signals to a specific T cell that recognizes the peptide presented. The T cell is then stimulated to produce autocrines (Refer to Autocrine signalling), resulting in the proliferation and differentiation to effector and memory T cells. Helper T cells (i.e CD4+ T cells) then activate specific B cells through a phenomenon known as an Immunological synapse. Activated B cells subsequently produce antibodies which assist in inhibiting pathogens until phagocytes (i.e macrophages, neutrophils) or the complement system for example clears the host of the pathogen(s).

Most antigens are T-dependent, meaning T cell help is required for maximal antibody production. With a T-dependent antigen, the first signal comes from antigen cross linking the B cell receptor (BCR) and the second signal comes from co-stimulation provided by a T cell. T dependent antigens contain proteins that are presented on B cell Class II MHC to a special subtype of T cell called a Th2 cell. When a B cell processes and presents the *same* antigen to the *primed* T_h cell, the T cell secretes cytokines that activate the B cell. These cytokines trigger B cell proliferation and differentiation into plasma cells. Isotype switching to IgG, IgA, and IgE and memory cell generation occur in response to T-dependent antigens. This isotype switching is known as Class Switch Recombination (CSR). Once this switch has occurred, that particular B cell will usually no longer make the earlier isotypes, IgM or IgD.

T cell-independent activation

Many antigens are T cell-independent in that they can deliver both of the signals to the B cell. Mice without a thymus (nude or athymic mice that do not produce any T cells) can respond to T independent antigens. Many bacteria have repeating carbohydrate epitopes that stimulate B cells, by cross-linking the IgM antigen receptors in the B cell, responding with IgM synthesis in the absence of T cell help. There are two types of T cell independent activation; *Type 1 T cell-independent* (polyclonal) activation, and *type 2 T cell-independent* activation (in which macrophages present several of the same antigen in a way that causes cross-linking of antibodies on the surface of B cells).

The Ancestral roots of B Cells

In an October 2006 issue of Nature Immunology, certain B cells of primitive vertebrates (like fish and amphibians) were shown to be capable of phagocytosis, a function usually associated with cells of the innate immune system. The authors postulate that these phagocytic B cells represent the ancestral history shared between macrophages and lymphocytes. B cells may have evolved from macrophage-like cells during the formation of the adaptive immune system.

B cells in humans (and other vertebrates) are nevertheless able to endocytose antibody-fixed pathogens, and it is through this route that MHC Class II presentation by B cells is possible, allowing Th2 help and stimulation of B cell proliferation. This is purely for the benefit of MHC Class II presentation, not as a significant method of reducing the pathogen load.

IgD This immunoglobulin functions primarily as an antigen receptor in B Cells.

IgE Allergic reactions are predominantly associated with IgE. Antigen reintroduced into a previously sensitized individual binds to antigen specific IGE on 'armed' mast cells and triggers release of the pharmacologically active agents (e.g. histamine) involved in immediate hypersensitivity syndrome such as hay fever and asthma.

The Properties of Immunoglobulin

Properties	Immunoglobulin Classes				
	IgG	IgA	IgM	IgD	IgE
Serum concentration(g/100ml)	1.2	0.4	0.12	0.003	<0.0005
Molecular Weight	140,000	160,000	900,000	180,000	200,000
Electrophoretic Mobility	γ	Slow β	Between γ and	Between γ and	Slow
H – chains	γ	α	μ	β	ε
L – Chains	λ or κ	λ or κ	λ or κ	σ	λ or κ
Complement fixation	Yes	No	Yes	λ or κ	No
Placement Transfer	Yes	No	No	No	No
Percent intravascular	40	40	70	No	—
Antibody Activity	23	6	5	—	2.5
	Most Ab to infections major part of secondary response Rh isoagglutinins LE factor isoagglutinins	Protein External secretions	First AB formed ABO agglutinins the rematoid factor	3 Antibody activity is visible found only on lympho-site surface	Region sensitizes for amophylax is

Mast Cells

A **mast cell** (or *mastocyte*) is a resident cell of several types of tissues and contains many granules rich in histamine and heparin. Although best known for their role in allergy and anaphylaxis, mast cells play an important protective role as well, being intimately involved in wound healing and defense against pathogens.

Mast cells were first described by Paul Ehrlich in his 1878 doctoral thesis on the basis of their unique staining characteristics and large granules. These granules also led him to the mistaken belief that they existed to nourish the surrounding tissue, and he named them "Mastzellen" (from the Greek *maso*, "chew"). They are now considered to be part of the immune system.

Mast cells are very close to basophil granulocytes (a class of white blood cells) in blood; the similarities between mast cells and basophils has led many to speculate that mast cells are basophils that have "homed in" on tissues. However, current evidence suggests that they are generated by different precursor cells in the bone marrow. Nevertheless, both mast cells and basophils are thought to originate from bone marrow precursors expressing the CD34 molecule. The basophil leaves the bone marrow already mature, whereas the mast cell circulates in an immature form, only maturing once in a tissue site. The tissue site an immature mast cell chooses to settle in probably determines its precise characteristics. Two types of mast cells are recognized, those from connective tissue and a distinct set of mucosal mast cells. The activities of the latter are dependent on T-cells.

Mast cells are present in most tissues in the vicinity of blood vessels, and are especially prominent near the boundaries between the outside world and the internal milieu, such as the skin, mucosa of the lungs and digestive tract, as well as in the mouth, conjunctiva and nose.

Physiology

Mast cells play a key role in the inflammatory process. When activated, a mast cell rapidly releases its characteristic *granules* and various hormonal mediators into the interstitium. Mast cells can be stimulated to degranulate by direct injury (e.g. physical or chemical), cross-linking of Immunoglobulin E (IgE) receptors, or by activated complement proteins.

Mast cells express a high-affinity receptor (FcεRI) for the Fc region of IgE, the least-abundant member of the antibodies. This receptor is of such high affinity that binding of IgE molecules is essentially irreversible. As a result, mast cells are coated with IgE. IgE is produced by B-cells (the antibody-producing cells of the immune system). IgE molecules, like all antibodies, are specific to one particular antigen.

In allergic reactions, mast cells remain inactive until an allergen binds to IgE already in association with the cell. Other membrane activation events can either prime mast cells for subsequent degranulation or can act in synergy with FceRI signal transduction. Allergens are generally proteins or polysaccharides. The allergen binds to the antigen-binding sites, which are situated on the variable regions of the IgE molecules bound to the mast cell surface. It appears that binding of two or more IgE molecules (cross-linking) is required to activate the mast cell. The clustering of the intracellular domains of the cell-bound Fc receptors, which are associated with the cross-linked IgE molecules, causes a complex sequence of reactions inside the mast cell that lead to its activation. Although this reaction is most well understood in terms of allergy, it appears to have evolved as a defense system against intestinal worm infestations (tapeworms, etc).

The molecules thus released into the extracellular environment include:

- preformed mediators (from the granules):
  - histamine (2-5 pg/cell)
  - serotonin
  - proteoglycans, mainly heparin (active as anticoagulant)
  - serine proteases, such as tryptase
- newly formed lipid mediators (eicosanoids):

  - prostaglandin D2
  - leukotriene C4
  - platelet-activating factor
- cytokines
  - Eosinophil chemotactic factor

Histamine dilates post capillary venules, activates the endothelium, and increases blood vessel permeability. This leads to local edema (swelling), warmth, redness, and the attraction of other inflammatory cells to the site of release. It also irritates nerve endings (leading to itching or pain). Cutaneous signs of histamine release are the "flare and wheal"-reaction. The bump and redness immediately following a mosquito bite are a good example of this reaction, which occurs seconds after challenge of the mast cell by an allergen.

The other physiologic activities of mast cells are much less well-understood. Several lines of evidence suggest that mast cells may have a fairly fundamental role in innate immunity – they are capable of elaborating a vast array of important cytokines and other inflammatory mediators, they express multiple "pattern recognition receptors" thought to be involved in recognizing broad classes of pathogens, and mice without mast cells seem to be much more susceptible to a variety of infections.

Mast cell granules carry a variety of bioactive chemicals. These granules have been found to be transferred to adjacent cells of the immune system and neurons via transgranulation via their pseudopodia.

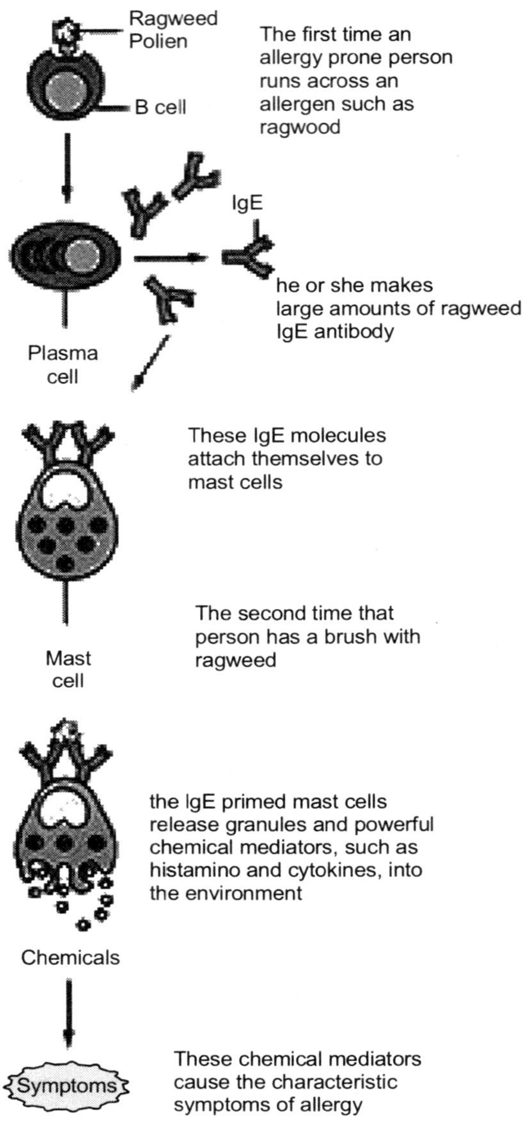

Ragweed Polien
B cell
The first time an allergy prone person runs across an allergen such as ragwood
IgE
Plasma cell
he or she makes large amounts of ragweed IgE antibody
These IgE molecules attach themselves to mast cells
Mast cell
The second time that person has a brush with ragweed
the IgE primed mast cells release granules and powerful chemical mediators, such as histamino and cytokines, into the environment
Chemicals
Symptoms
These chemical mediators cause the characteristic symptoms of allergy

Fig. 7.11 : The Role of Mast Cells in the Development of Allergy

Role in Disease

Many forms of cutaneous and mucosal allergy are mediated for a large part by *mast cells*; they play a central role in asthma, eczema, itch (from various causes) and allergic rhinitis and allergic conjunctivitis. Antihistamine drugs act by blocking the action of histamine on nerve endings. Cromoglicate-based drugs (sodium cromoglicate, nedocromil) block a calcium channel essential for mast cell degranulation, stabilizing the cell and preventing release of histamine and related mediators. Leukotriene antagonists (such as montelukast and zafirlukast) block the action of leukotriene mediators, and are being used increasingly in allergic diseases.

Anaphylaxis

In anaphylaxis (a severe systemic reaction to allergens, such as nuts, bee stings or drugs), body-wide degranulation of mast cells leads to vasodilation and, if severe, symptoms of life-threatening shock.

Autoimmunity

Mast cells are implicated in the pathology associated with the autoimmune disorders rheumatoid arthritis, bullous pemphigoid, and multiple sclerosis. They have been shown to be involved in the recruitment of inflammatory cells to the joints (e.g. rheumatoid arthritis) and skin (e.g. bullous pemphigoid) and this activity is dependent on antibodies and complement components.

Mast Cell Disorders

Mastocytosis is a rare condition featuring proliferation of mast cells. It exists in a *cutaneous* and *systemic* form, with the former being limited to the skin and the latter involving multiple organs. Mast cell tumors are often seen in dogs and cats.

CHEMOTHERAPY

Aside from the defense mechanism of the body for protecting itself against the effects of bacterial invasion other possibilities suggest themselves. One method was to apply heat, sufficient heat destroys bacteria. It is obvious, however, that it is particularly increase the temperature to a degree necessary to destroy the organisms without harming the tissues. The same applies to the action of many chemicals such as Phenol, Formaldehyde, Iodine, Chlorine, Benzene, Calcium Chloride and Mercuric Chloride are all excellent disinfectants. Drugs which would destroy the organism without materially harming the tissues and this is the problem of chemotherapy.

A partial answer to this problem was obtained when it was shown that malaria could be cured to some extent at least with quinine and syphilis with mercury compounds. This work also suggested that a specific chemical substance "was definite in its attack against that particular organism.

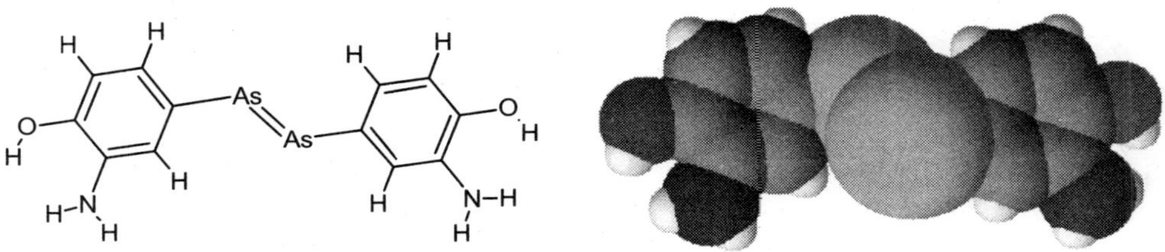

Fig. 7.11A : The Formula and 3D model of Salversan

The triumph of chemotherapy may be said to date back to 1910 when Ehrlich discovered Salversan as the cure of syphilis.

From 1910 on rapid strides were made in the field of chemotherapy, but until 1935, particularly all of the successful chemical substances in use were remedies for tropical diseases caused by protozoa.

For example one of the most striking of these diseases for which chemotherapy found an answer was malaria, a disease caused by the protozoan, plasmodium transmitted by the female anopheles mosquito to man. It had been known for a long time that the bark of the cinchona tree was the remedy for malaria, but it took years to isolate the important alkaloid responsible for the cure, the quinine (one of the 20 alkaloids present in the bark)

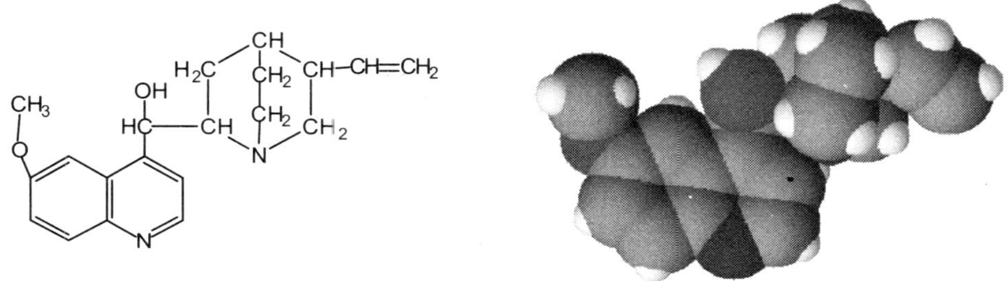

Fig. 7.12: The Formula of Quinine and 3d Model of Quinine

This valuable drug is obtained largely from the Dutch East Indies during the Second world War, but the natural quinine was scarce so the alternative synthetic substances, such as plasmochin (a quinoline derivative) and atebrin (an acidine derivative) came into use.

The synthesis of itself, for years a stumbling block to organic chemists was accomplished by Woodward and Doering in 1944. It was not until 1935, with the use of sulphanilamide that the first that the first of a series of substance was used which was effective against bacteria the bacteria which gives rise to the commonest diseases.

Sulphonamide Compounds

Domagk in an extension of the applied dyes to chemotherapy, discovered that Prontosil injected into mice infected with streptococci has a definite curative effect.

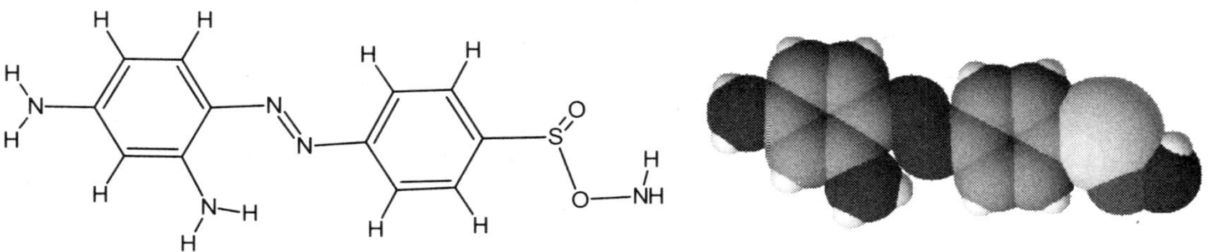

Fig. 7.13 : Formula and Structure of Prontosil

Further work made clear that Prontosil was an effective therapeutic agent against the β - hemolytic streptococcus an agent which destroys red blood cells or which gives rise to the red rashes on the skin (erysipelas and scarlet fever). It was also apparently judged by the use of hundreds of different substances, that the most effective section attached to positions (1) and (7)

When the N is in the *p* position is in the position (4) and the other N therefore be in the position (1), we may say that the best results so far obtained are form substitution of hydrogen on the N(1) group

It was next shown that Prontosil was converted in the body into sulphanilamide, a very well known organic substance and that sulphanilamide itself was most potent.

4-aminobenzenesulphonamide
Sulphanilamide

Extremely effective compounds were obtained by replacing the hydrogen in position (11) with pyridine thiozole and other groups.

The steps in the synthesis of some of these sulphonamides graphically in the set of equations shown the Fig. 7.15 below:

The sulphonamide drugs are administered orally and in fairly in large quantities. A gram may be given every three to four hours and the dosage continued for five to ten days during the critical period of the illness.

One or more of the drugs have been used sometimes with amazing success in specific wounds hematological poisoning puerperal fever (disease of child birth) erysipelas meningitis, malignant endocarditis (infection of the heart) pneumonia, gonorrhea, certain kidney infections etc.

But quite a number of bacterial diseases exist which, apparently are not influences by the "sulpha drugs" Among them are tuberculosis, typhoid paratyphoid, cholera bacillary dysentery and whopping cough.

Also the main body of viruses diseases exist which, apparently are not influences by the treatment with the sulphonamides. These include small pox, measles, infantile paralysis, cold and influenza.

An antisulphaniamide factor obtained from extract of yeast, was traced to p – amino benzoic acid, it was seen that a concentration of 1:10,000 was active to inhibit the bacteriostatic effect of sulphanilamide at a concentration of 1: 20,000

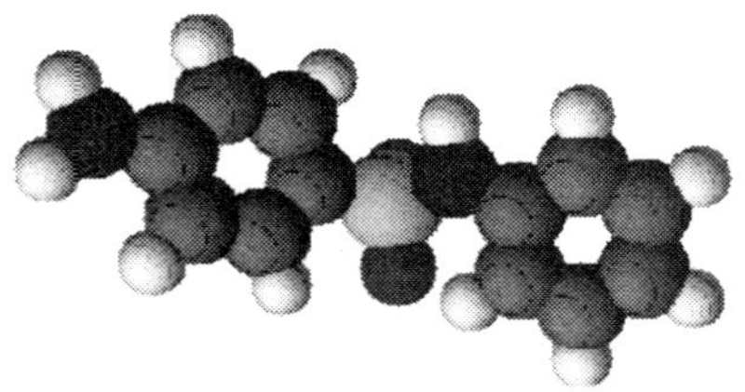

4-amino-*N*-phenylbenzene sulphonamide
Sulphapyridine

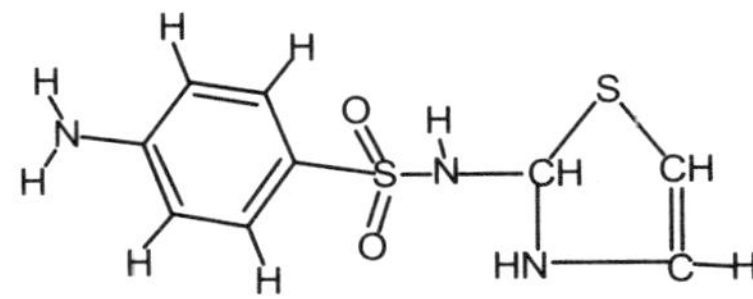

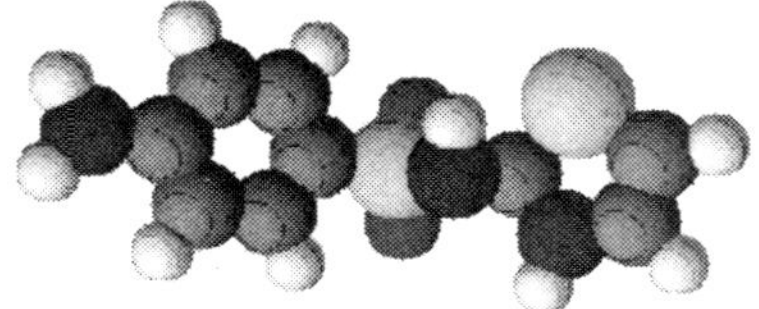

4-amino-N-(2, 3-dihydro-1, 3-thiazol-2-yl) benzene sulphonamide
Sulphathiazole

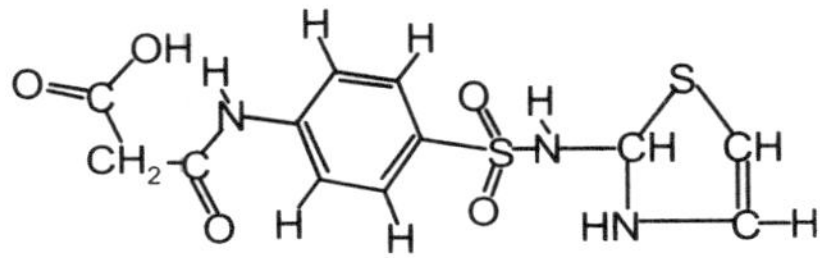

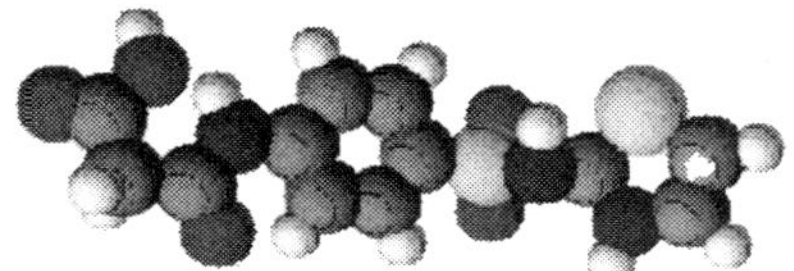

3-{[4-(2, 3-dihydro-1,3-thiazol-2-ylsulphamoyl) phenyl] amino-3-oxopropanoic acid
Succinylsulphathiazole

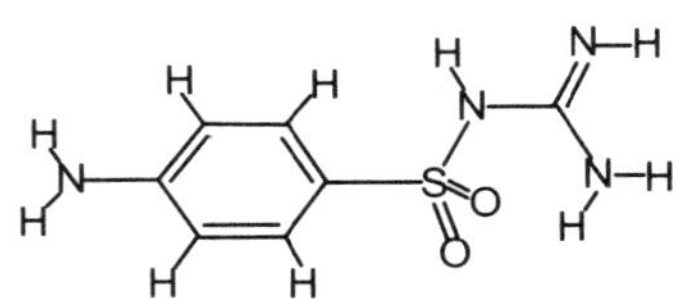

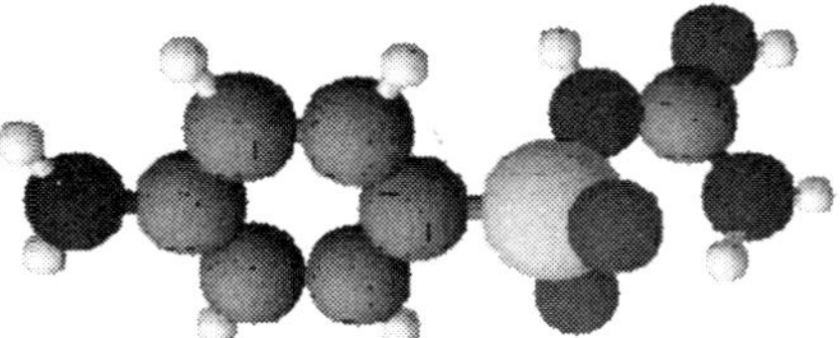

4-amino-N-carbamimidoylbenzenesulphonamide
phaguanidine

Fig. 7.14 : The different Derivatives of Sulphanilamide drug

This has suggested to Woods and to Fildes a possible mechanism of the action of sulphonamide compounds, p-Aminobenzoic acid is regarded as an essential compound synthesized by the bacterial cells from amino acids, or obtained by the cells from their environment. To be properly utilized by the cells, the p-aminobenzoic acid is acted upon by the enzyme, and there by prevents the normal metabolism of the bacterial cell. The bacteria fail to get their proper nutrition and are ultimately destroyed.

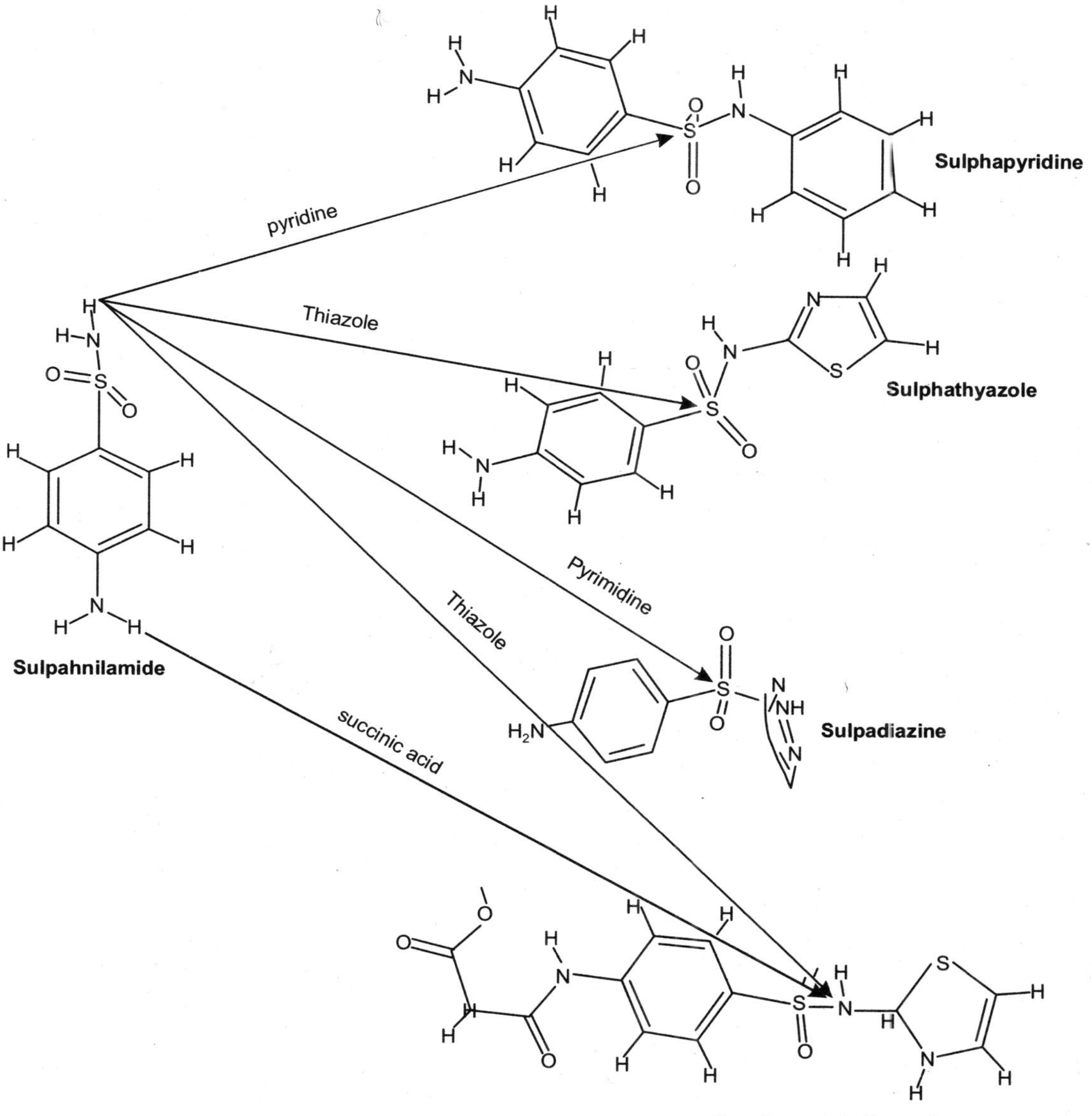

Fig. 7.15 : The Synthesis of the Derivatives of Sulphanilamides

These findings of Woods and others have told to some very suggestive work with regard to the relationship of chemical structure to physiological action, for example, pyridine-3-sulphinic acid which is related to nicotinic acid in a manner similar to the relationship of sulphanilamide to p-aminobenzoic acid inhibits bacterial growth. The same is true with thiopanic acid (pantoyltaurine) the sulphur analogue of panthothenic acid.

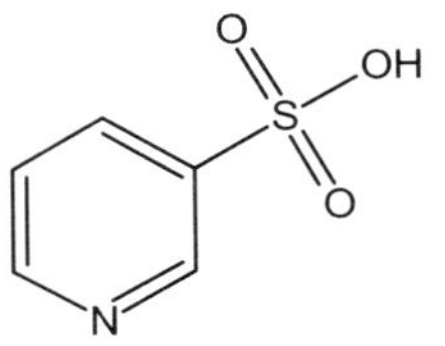

pyridine-3-sulfonic acid

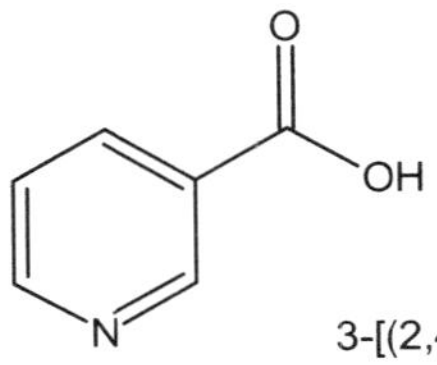

pyridine-3-carboxylic acid
nicotinic acid

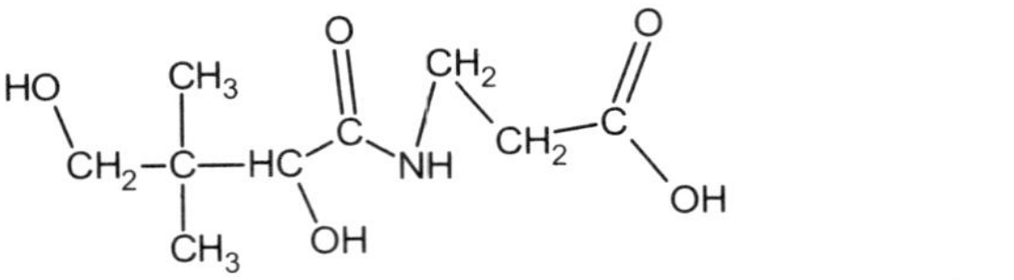

3-[(2,4-dihydroxy-3,3-dimethylbutanoyl)amino]propanoic acid
pantothenic acid

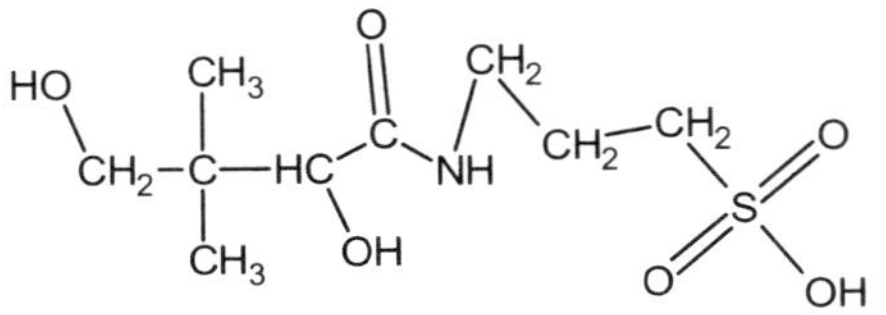

3-[(2,4-dihydroxy-3,3-dimethylbutanoyl)amino]propane-1-sulfonic acid
thiopanic acid

The possibilities spread still further Woolley and others have shown that in some cases by using compounds, modified in structure in the place of the normal (vitamin) compounds specific vitamin deficiencies may be produced. For example isoriboflavin (the 5, 6 dimethyl modification) inhibits the growth promoting action of riboflavin (the 6, 7 dimethyl) from.

It is evident that to the conclusion that the sulphonamides act by preventing the growth of bacteria (bacteriostatic) rather than by destroying them.

The sulphonamides are detoxicated in various ways in the body, presumably in the liver. They are excreted through the kidneys and appear in the urine in one or more of the following modifications as an N_4 acetyl derivative as a monohydroxy derivative as a glucuronides; as an ethereal sulphate.

Sulphaguanidine and succinylsulphathiazole are very poorly absorbed; they have therefore been used intentional antiseptics.

Side effects of chemotherapy

Experiments on animals have shown that these sulpha compounds, by destroying organisms in the intestine, often prevent to a large extent the synthesis of a number of vitamins which are normally brought about with the help of these organisms. Some of these vitamins are vitamin K and a few belonging to vitamin B complex (biotin folic acid p-aminobenzoic acid)

THE CANCER CHEMOTHERAPY

The effective use of cancer chemotherapy requires an understanding of the principles of tumor biology, cellular kinetics, pharmacology, and drug resis- tance. Thanks to the development of new, effective chemotherapeutic agents, coupled with our expanding knowledge about the administration and combination of these agents, we now are able to cure almost 20% of all new cases of cancer through chemotherapy alone.

This subhead focuses on the principles responsible for the development of modern combination chemotherapy regimens. This discussion is followed by descriptions of the major classes of chemotherapeutic drugs and their mechanisms of action.

Cellular Kinetics

Cytokinetic studies have shown how the kinetics of cellular growth defines the characteristics of tumor growth and, in part, explains the biological behaviour and heterogeneity of tumors.

Normal cell Cycle

Inherent to cytokinetic principles is the concept of the cell cycle. Daughter cells formed as a result of mitosis consist of three subpopulations: (1) cells that are nondividing and terminally differentiated, (2) cells that are continually proliferating, and (3) cells that are resting but may be recruited into the cell cycle (i.e., stem cells). All three populations exist simultaneously in tumors.

The cell cycle is composed of four phases during which the cell prepares for and effects mitosis. Cells that are committed to divide again enter the G_1 phase. Preliminary synthetic cellular processes occur that prepare the cell to enter the DNA synthetic (S) phase. Specific protein signals regulate the cell cycle and allow replication of the genome where the DNA content becomes tetraploid (4N). After completion of the S phase, the cell enters a second resting phase, G_2, prior to undergoing mitosis. The cell progresses to the mitotic (M) phase, in which the chromosomes condense and separate and the cell divides, producing two daughter cells.

Chemotherapeutic agents can be classified according to the phase of the cell cycle in which they are active. Agents that are cell-cycle-phase-nonspecific (e.g., alkylating agents) have a linear

S phase–dependent	**M phase–dependent**
Antimetabolites	Vinca alkaloids[a] Capecitabine
	Vinblastine Cytarabine
	Vincristine Doxorubicin
	Vinorelbine Fludarabine Podophyllotoxins
Floxuridine	Etoposide Fluorouracil
	Teniposide Gemcitabine Taxanes
Hydroxyurea	Docetaxel Mercaptopurine
	Paclitaxel
Methotrexate	**G2 phase–dependent**
Prednisone	Bleomycin Procarbazine Irinotecan
Thioguanine	Mitoxantrone
	Topotecan
	G1 phase–dependent Asparaginase Corticosteroids

[a] Have greatest effects in S phase and possibly late G2 phase; cell blockade or death, however, occurs in early mitosis.
Adapted, with permission, from Dorr RT, Von Hoff DD (eds):The Cancer Chemotherapy Handbook, 2nd ed, p 5. East Norwalk, Connecticut, Appleton & Lange, 1993.

dose-response curve; that is, the greater the dose of drug, the greater is the fraction of cell kill. However, cell-cycle-phase–specific drugs have a plateau with respect to cell killing ability, and cell kill will not increase with further increases in drug dosage.

Tumour Kinetics

The rate of growth of a tumor is a reflection of the proportion of actively dividing cells (the growth fraction), the length of the cell cycle (doubling time), and the rate of cell loss. Variations in these three factors are responsible for the variable rates of tumor growth observed among tumors of differing histologies, as well as among metastatic and primary tumors of the same histology.Tumors characteristically exhibit a sigmoid-shaped Gompertzian growth curve, in which tumor doubling time varies with tumor size. Tumors grow most rapidly at small tumor volumes. As tumors become larger, growth slows based on a complex process dependent on cell loss and tumor blood and oxygen supply. In order to have the best chance for cure, chemotherapy must achieve a fractional cell kill in a logarithmic fashion i.e., 1-log-kill is 90% of cells, 2-log-kill is 99% of cells). From these concepts, chemotherapy models have been developed utilizing alternating non-cross-resistant therapies, induction-intensification approaches, and adjuvant chemotherapy regimens.

Principles of Combination Chemotherapy

Using kinetic principles, a set of guidelines for designing modern combination chemotherapy regimens have been derived. Combination chemotherapy accomplishes three important objectives not possible with single-agent therapy: (1) It provides maximum cell kill within the range of toxicity tolerated by the host for each drug; (2) it offers a broader range of coverage of resistant cell lines in a heterogeneous tumor population; and (3) it prevents or slows the development of new drug-resistant cell lines.

Selection of Drugs for Combination Regimens

The following principles have been established to guide drug selection in combination regimens:

- Drugs known to be active as single agents should be selected for combinations. Preferentially, drugs that induce complete remissions should be included.
- Drugs with different mechanisms of action should be combined in order to allow for additive or synergistic effects on the tumor.
- Drugs with differing dose-limiting toxicities should be combined to allow each drug to be given at full or nearly full therapeutic doses.
- Drugs should be used in their optimal dose and schedule.
- Drugs should be given at consistent intervals. The treatment-free interval between cycles should be the shortest possible time for recovery of the most sensitive normal tissue.
- Drugs with different patterns of resistance should be combined to minimize cross-resistance.

Terminology used in Describing Chemotherapy

Chemotherapy is administered with a variety of treatment schedules designed according to the intent and responsiveness of therapy. Definitions of chemotherapy are generally based on the purpose of achieving certain therapeutic goals as described in Table 7.2.

Definitions of Response

Tumors can be classified according to their general sensitivity to chemotherapy. Response to chemotherapy is defined precisely as complete response, partial response, minimal response (stable disease), and progression. Complete response is defined as the disappearance of all evidence of disease and no appearance of new disease for a specified interval (usually 4 weeks).

Induction: High-dose, usually combination, chemotherapy given with the intent of inducing complete remission when initiating a curative regimen.The term is usuallly applied to hematologic malignancies but is equally applicable to solid tumors.

Consolidation: Repetition of the induction regimen in a patient who has achieved a complete remission after induction, with the intent of increasing cure rate or prolonging remission.

Intensification: Chemotherapy after complete remission with higher doses of the same agents used for induction or with different agents at high doses with the intent of increasing cure rate or remission duration.

Maintenance: Long-term, low-dose, single or combination chemotherapy in a patient who has achieved a complete remission, with the intent of delaying the regrowth of residual tumor cells.

Adjuvant: A short course of high-dose, usually combination chemotherapy in a patient with no evidence of residual cancer after surgery or radiotherapy, given with the intent of destroying a low number of residual tumor cells.

Neoadjuvant: Adjuvant chemotherapy given in the preoperative or perioperative period.

Palliative: Chemotherapy given to control symptoms or prolong life in a patient in whom cure is unlikely.

Salvage: A potentially curative, high-dose, usually combination, regimen given in a patient who has failed or recurred following a different curative regimen.

Partial response is defined as a reduction by at least 50% in the sum of the products of the two longest diameters of all lesions, maintained for at least one course of therapy, with no appearance of new disease. Minimal response is any response less than a partial response and is usually not reported in clinical trials. Progression is defined as growth of existing disease or appearance of new disease during chemotherapy.

The NCI (National Cancer Institute) has adopted standardized response criteria and is requiring their use by all cooperative groups. These criteria, called RECIST (Response Evaluation Criteria in

Solid Tumors), were developed and recently revised by the World Health Organization (WHO). The goals are consistency of evaluation and comparison of regimens within a single trial and regimens of different trials. A comparison of RECIST and WHO guidelines is listed in Table.

Dose Intensity

Kinetic principles predict that, for drug-sensitive cancers, the factor limiting the capacity to cure is proper dosing. Reduction in dose is associated with a decrease in cure rate before a significant reduction in the complete remission rate occurs. A dose reduction of approximately 20% can lead to a loss of up to 50% of the cure rate. Conversely, a twofold increase in dose can be associated with a 10- fold (1-log) increase in tumor cell kill in animal models.

> Tariquidar (XR9576) is an investigational intravenous drug that is a potent p-glycoprotein inhibitor that reverses MDR associated with common chemotherapy drugs. It can be safely and conveniently administered with full doses of paclitaxel, doxorubicin, and vinorelbine with no compromise of pharmacokinetics. Phase III studies in lung cancer are ongoing to evaluate the efficacy and response of tariquidar in combination therapy (Boniface G, Ferry D, Atsmon J, et al: Proc Am Soc Clin Oncol [abstract] 21:90b, 2002).

Overcoming Chemotherapy Resistance

There are multiple reasons for chemotherapy failure in cancer patients, involving a variety of anatomic, pharmacologic, and biochemical mechanisms. Tumor sanctuary sites (brain, testes) and blood flow to the tumor represent anatomic barriers; pharmacologic and biochemical explanations include altered drug ac tivation/inactivation in normal tissues, decreased drug accumulation, increased repair of drug-induced damage to the cell, altered drug targets, and altered gene expression.

Overexpression of the MDR_1 (multidrug resistance) gene is the most notable mediator of drug resistance and encodes a 170-kd transmembrane *p*-glycoprotein. *p*-Glycoprotein is an energy-dependent pump that serves to remove toxins or endogenous metabolites from the cell. A high level of MDR_1 expression is reliably correlated with resistance to cytotoxic agents. Tumours that intrinsically express the MDR1 gene prior to chemotherapy characteristically display poor durable responses.

Comparison of RECIST and WHO guidelines

Characteristic	Recist	Who
1	2	3
Objective response (LD is the longest diameter)	Target lesions [change in sum of LDs, maximum 5 per organ up to 10 total (more than one organ)]	Measurable disease (change in the sum of the prodcuts of LDs and greatest perpendicular diameters, no maximum number of lesions specified)

...(Contd.)

1	2	3
Complete response (CR)	Disappearance of all target lesions, confirmed at ≥ 4 weeks	Disappearance of all known disease, confirmed at ≥ 4 weeks
Partial response (PR)	≥ 30% decrease from baseline, confirmed at ≥ 4 weeks	≥ 50% decrease from baseline, confirmed at ≥ 4 weeks
Progressive disease (PD)	≥ 20% increase over smallest sum observed or appearance of new lessions	≥ 25% increase in one or more lesions or appearance of new lessions
Stable disease (SD)	Neither PR nor PD criteria met	Neither PR nor PD criteria met (no change)

RECIST = Response Evaluation Criteria in Solid Tumors: WHO = World Health Organization

Advantages of Liposomal Drug Delivery

- Provides selective passive targeting to tumor sites
- Increases efficacy and therapeutic index
- Improves delivery of hydrophobic molecules
- Reduces the toxicities of the encapsulated agent
- Avoids accumulation in vital organs and tissues
- Improves pharmacokinetics (reduced elimination, increased drug exposure time)
- Increases stability via encapsulation
- Enhances intracellular drug delivery to overcome drug resistance

Cancer Management: A Multi-disciplinary Approach

Chemotherapy agents subject to MDR1- mediated resistance include the anthracyclines, vinca alkaloids, taxanes, and topoisomerase inhibitors. Targeted therapies that inhibit *p*-glycoprotein are under evaluation in combination with cytotoxic drugs subject to MDR (see sidebar on previous page).

Liposomal formulations of chemotherapeutic drugs are a promising new approach to overcoming these resistance mechanisms. Liposomes are well defined lipid and lipoprotein vesicles that offer immense potential for targeting drugs to tumors. The advantages of these drug carriers over the conventional administration of chemotherapy agents are described in Table 7.4.

FDA-approved liposomal preparations of doxorubicin (Doxil), daunorubicin (DaunoXome), cytarabine (DepoCyt), and amphotericin B (Abelcet) have proven to be attractive, less toxic alternatives to the conventional drug formulations. Liposomal daunorubicin and amphotericin B have clearly shown less cardiac and renal damage, respectively. Daunorubicin liposomal is efficacious in induction regimens for acute leukemia; cytarabine liposomal appears to be superior to intrathecal cytarabine for treatment of CNS leukemia and lymphoma.

An additional advantage of the liposomal delivery system is the ability to encapsulate and stabilize very hydrophobic molecules such as paclitaxel (Taxol). Nanoparticle, albumin-stabilized paclitaxel has allowed much higher doses of drug to be given with far fewer side effects than paclitaxel which contains the toxic carrier material cremophor. Currently, a phase III randomized study comparing the efficacy of nanoparticle paclitaxel with conventional paclitaxel in metastatic breast cancer is under way.

Several liposomal formulations of conventional anticancer drugs are currently in phase I/II evaluation, including liposomal vincristine, platinum, mitoxantrone, all-*trans* retinoic acid (ATRA), and lurtotecan. There is a strong probability that these drug carriers will allow better administration of poorly soluble cancer drugs, enhance drug delivery and uptake in the tumor, and boost dose intensity, subsequently improving antitumor response, overcoming drug resistance, and decreasing chemotherapy toxicities.

Alkylating Agents and their Uses, Dosages, and Toxicities

Drug and its uses'[1]	Dosages	Toxicities[b]
1	2	3
Nitrogen mustards		
Chlorambucil CLL, HD, NHL, ovarian cancer, choriocarcinoma, lymphosarcoma	0.1 -0.2 mg/kg PO daily for 3-6 wk as required (usually 4-10 rng/d) or intermittent 0.4 mg/kg every 3-4 wk; Increase by 0.1 mg/kg until control of disease or toxicity	Bone morrow depression, gonadal dysfunction, leukemia, hyperuri-cemia, pulmonary fibrosis
Cyclophosphamide AML.ALL, CLL, HD, and NHL, multiple myeloma, mycosis fungoides, neurobfostoma,ovarian and breast cancers, retinobtetoma. lung, cesticular.and bladder cancers, sarcoma	40-50 mg/kg IV in divided doses over 2-5 d to start, followed by 10-15 mg/kg IV every 7-10d;or 3-5 mg/kg IV twice weekly; or I -5 mg/kg/d PO	Bone marrow depression, hernorrhagic cystitis, immuno suppression, alopecia, stomatitis, SIADH
Estramustine Prostate, renal cell carcinoma	14 mg/kg/d PO in 3-4 equally divided doses; 300 mg/d IV for 3-4 wk, followed by 300-450 mg/wk IV over 3-8 wk	Bone morrow depression, Ischemic heart disease, thromboembolism. thrombophlebitis gynecomastia, nausea and vomiting, hepatotoxicity
Ifosfamide Genn-ceM testiculor cancer, sarcoma. NHL, lung cancer	1.2g/m^2/dviaslow IV infusion for 5 consecutive days; repeat every 3 wk; give with mesna	Bone morrow depression, hemorrhaglc cystitis, confusion, somnolence
Mechlorethamine HD, NHL, CML, CLL, mycosis fungoi des, brondiogenic carcinoma, lymphosarcoma, polycythemia vera. malignant effusions (intracavitary)	0.4 mg/kg ideal body weight given as single dose or in divided doses of O.I-0.2 mg/kg/d	Bone marrow depression, nausea and vomiting, local phlebitis, severe skin necrosis if extravasated, gonadal dysfunction

...(Contd.)

1	2	3
Melphalan Multiple myeloma, breast and ovarian cancers, sarcoma, testicular and lung cancers	Continuous therapy: 6 mg PO daily for 2-3 wk, no therapy for 2-4 wk, then maintenance with 2-4 mg PO daily Pulse: 10 nig/m^2 PO daily for 4 d every 4-6 wk	Bone marrow depression, anorexia, nausea and vomiting.gonadal testicuiar dysfunction, leukemia
Aziridine		
Thiotepa Ovarian, breast, and superficial bladder ctfncers. HD.CMLCLL bronchogenic carcinoma, malignant effusions (Intra-cavitary). BMT for refrac-tory leukemia, lymphomas	IV; 0.3-0.4 mg/kg by rapid IV infusion Intraveslcal: 60 mg/60 ml sterile water instilled and retained in bladder for 2 h; repeat weekly for 4 wl< I ntracavitary: 0.6-0.8 mg/kg	Bone marrow depression, nausea and vomiting, mucositis, skin rashes
Alkyl sulfonate		
Busulfan CML BMT for refractory leukemia, lymphomas	2-8 mg PO daily for remission induction; adjust dosage to WBC count; 1-3 mg PO daily for maintenance: withhold induction if WBC count < 15,000/ML; resume therapy when WBC count > 50,000/uL	Bone marrow depression, pulmonary fbrasis, aplastic anemia, amenorrhea.gynecornastia, skin hyperpigmentation
Nitrosoureas		
Carmustine Brain tumor, multiple myeloma. HD, NHL melanoma, BMT for refractory solid tumors and lymphomas	150-200 mg/m^2 IV every 6-8 wk	Delayed bone marrow depression, nausea and vomiting, reversible hepato-toxicity. local phlebitis, pul-monary and renal damage (high dose)
Gliadel wafers G/iob/flstomo muttiforme	Up to 8 wafers placed in the brain cavity created by tumor removal	Fever, pain, and abnormal healing
Lomustine Brain tumors, HD, GI carcinomas.NSCLC	130 mg/m^2 PO every 6 wk; adjust dose in combination chemotherapy	Delayed bone marrow depression, nausea and vomiting, reversible hepatotoxicity, pulmonary and renal damage, neurologic reactions, leukemia
Streptozocin Pancreatic isfet-cel/, carcinoid, colon, hepntoma, NSCLC. HD	Daily 500 mg/m^2 IV for 5 d every 6 wk until maximum benefit or toxfdty Weekly: 1.000 mg/m^2 IV weekly for first 2 wk, then escalate dose to response or toxicity, not to exceed a single dose of 1.500 mg/m^2	Renal damage, nausea and vomiting, diarrhea, altered glucose metabolism, liver dysfunction
Platinum complexes		
Carboplatin Ovarian cancer, endo-metrial. head and neck, lung, cesticular. and breast cancers, relapsed acute leukemia, NHL	Single agent: 360 mg/m^2 IV every 4 wk Combination; 300 mg/m^2 IV every 4 wk calvert formula: Total dose (mg) = Target AUC x (GFR + 25)	Bone marrow depression, nausea and vomiting, peripheral neuropathy, ototoxlcfty

...(Contd.)

1	2	3
Cisplatin Test/alter, ovarian, bladder, uterine, cervical, and lung cancers, squamous cell cancer of the head and neck, sarcoma. NHL	50 mg/m² IV or more every 3 wk; or 20 mg/m² IV daily for 4-5 d every 3-4 wk;glve vigorous hydration before and after chemotherapy	Renal damage, nausea and vomiting, electrolyte disturbance, peripheral neuropathy, bone marrow depression, ototoxicity, radiosensltizer
Oxaliplatin Colorectal (second-line)	85 mg/m² IV over 120 rnln on d I followed by infusional 5-FU and leucovorin on d I -2, every 2 wk	Bone marrow depression, diarrhea, nausea and vomiting, neuropathies exacerbated by cold exposure, pharyn-golaryngeal dysesthesia
Nonclassic alkylators		
Altretamine Ovarian, lung, breast, and cervical cancers. NHL	4-12 mg/kg/d or 260 mg/m², PO divided in 3-4 doses for 14-21 d of a 28-d regimen	Nausea and vomiting, bone marrow depression, paresthesias, CNS toxicity
Dacarbazine Malignant melanoma, HD, soft-tissue sarcomas, neuroblnstoma	Melanoma: 2.0-4.5 mg/kg/d IV for 10 d every 4 wk; or 250 mg/m²/d IV for 5 d every 3 wk HD: 375 mg/m² IVon d I, repeated every 15 d (single agent); 150 mg/m²/d IV for 5 d every 4 wk (combination therapy)	Bone morrow depression, nausea and vomiting, flulike syndrome, transient hepato-toxlclty.local irritation, facial flushing, alopecia
Procarbazine HD, NHL, brain tumors, king cancer	Single agent: 4-6 mg/kg/d PO until maximum response	Bone marrow depression, nausea and vomiting, lethargy, depression, paresthesias. headache, flulike symptoms
Temozolomide Anapiastic astrocytoma fre/apsed). renal cell cancer, melanoma	HD (MOPP): l00mg/m²/d PO for 14 d l50rng/m²/dPO for 5 d every 28 d	Bone morrow depression, nausea and vomiting

[a]PDA-approved uses in italics; neoplasms are carcinomas unless otherwise indicated

[b]Dose-limiting effects in italics

ALL = acute lymphoblasdc leukeniia;AML = acute myelogenous leukemia:AUC = area under the curve: BMT = bone marrow transplantation; CLL = chronic lymphocytic leukemia; CML = chronic myelogenous leukemia; CMML - chronic nlyelomacrocytic leukemia; S-FU = fluorouracll; GFR = glomerular filtration rate; HD = Hodgkin's disease; MDS = myelodysplastic syndromes; MOPP = mechlorethamine. Oncovin. procarbazine.and prednisone; NHL = nort-Hodgkin's lymphoma;NSCLC = non-small-cell lung cancer; SIADH = syndrome of inappropriate antidiuretic hormone secretion; WBC = white blood ceil

Antimetabolites and their uses, Dosages and Toxicities

Drug and its uses'[1]	Dosages	Toxicities[b]
1	2	3
Folate analog		
Methotrexate Breost, head and neck. Gl, and king cancers, AU.CNSleukemia (intrat/h-	Numerous dosing schedules with combination therapy: Low dose: 2.5-5.0 mg PO daily, or 5-25 mg/rn² PO, IM.	Mucositis, Gulceniion (may produce hemorrhage or perforation), bone marrow depression, pulmonary

...(Contd.)

1	2	3
ecalj, ges-tofionol/trophoblastic tumors, NHL (advanced stage), Burkitt's lymphoma, osteosarcoma, mycosis fungoides	IV twice weekly; or 50 mg/m^2 IV every 2-3 wk High dose; 1-12g/m^2 IV with leucovorin rescue every I -3 wk Intratheeal: 5-10 mg/rn^2 (up to 15 rng) every 3-7 d	fibrosis (pre-viously irradiated area), nerve root irritation and convulsion (intrathecal). liver cirrhosis and osteo-porosis (chronic therapy), renal damage (high dose), diarrhea, skin erythema
Purine analogs		
Fludarabine CLL. AML, NHL (low-grade)	25 mg/m^2/d IV over 30 min for 5 d: repeat every 28 d	Bone marrow dep-ession, nausea and vomiting, fever, malaise, pulmonary infiltrates, tumor lysis syndrome, CMS effects (high dose)
Mercaptopurine ALL, CML.AML	I.5-2.5mg/kg/dPO (100-200 mg in average adult) until response or toxic effects are seen; may increase dose to 5 mg/kg/d; adjust for maintenance dose; reduce dose by 50%-75% if given with allopurinol or if renal or hepatic insufficiency ensues	Bone marrow depression, nausea and vomiting, anorexia, diarrhea, cholestasis
Thioguanine AML, ALL. CML, advanced cotorectal cancer,multiple myeloma	2 mg/kg/d PO until response or toxic effects are seen; may cautiously increase to 3 mg/kg/d	Bone marrow depression, liver damage, stomatitis
Adenosine analogs		
Cladribine Hairy-cell leukemia, NHL, mycosis fungoides, AML, CML, CLL	0.09 mg/kg/d (4 mg/m^2/d) by continuous IV infusion for 7 consecutive days	Bone morrow depression, febrile episodes, rash, infections, septicemia
Pentostatin Hairy<eli leukemia, ALLCLLlymphoblastlc lymphoma, mycosis fungoides	4 mg/m^2 IV over 30 min every other week or for 3 consecutive weeks; give vigorous hydration before and after chemotherapy	Nephrotoxidty.CNS depression, bone marrow depression, nausea and vomiting, conjunctivitis
Pyrimidine analogs		
Capecitabine Breast cancer (relopsed), coloretal cancer, and other GI malignancies	1.250 mg/m^2 bid PO with food (2 weeks on drug, I week of rest)	Diarrhea, stomatitis, nausea and vomiting, fatigue, hand-foot syndrome, bone marrow depression (minimal)
Cytarabine AML,ALL,CML,NHL, CMS leukemia (intrathecal)	AML induction: 100 mg/m^2/d by continuous IV Infusion on days 1 -7; or 100 mg/m^2 IV every 12 h on days I -7 Relapsed ALL 3 g/m^2 IV over I -3 h every 12 h for 4 doses	Bone marrow depression, nausea and vomiting, diarrhea, arachnoiditis (intrathecal), stomatitis, hepatic dysfunction, fever, conjunctivitis, confusion, somnolence, cerebellar toxicity

...(Contd.)

1	2	3
DepoCyt (liposomal cytarabine) CNS leukemia/lymphoma	*Intrathecal*: DepoCyt, 50 mg over I -5 min every 14 d. with dexarnethasone, 4 mg PO bid x 5 d	
Floxuridine GI adenocardnomas meta-static to liver, including oral, pancreatic, biliary, colon, and hepatic cancers, and metastatic breast cancer	0.1-0.6 mg/kg/d over several days via continuous arterial infusion supplying well-defined tumor; treat-ments given over I -6 wk	Stomatitis and CI ulcers, bone morrow depression, abdominal pain, nausea and vomiting, diarrhea, liver dysfunction (transient)
Fluorouracil Colon, rectal, stomach, pancreas, breast, head and neck, renal cell, pros-tate,and ovarian cancers, squamous cell carcinoma of esophagus, basal and squamous cell carcinoma of skin (topical), hepatic cancer (intra-arterial)	Numerous dosing schedules with combination therapy: *Loading dose*: 300-500 mg/m^2; or 12 mg/kg IV daily for 3-5 d.followed by weekly maintenance *Maintenance*: 10-15 mg/kg IV weekly, as toxicity permits *Infusion*: 20-25 mg/kg by continuous IV infusion over 24 h daily for 4-5 d, every 4 wk	Stornatft/s and GI ulcers (Infusion) bone marrow depression (bolus), diarrhea, nausea and vomiting, esophagitis, angina, cerebellar ataxia, radiosensitizer
Gemcitabine Pancreatic cancer, lung, ovarian, breast, and bladder cancers	1,000 mg/m^2 IV over 30 min, once weekly for up to 7 weeks (or until toxicity necessitates reducing or withholding a dose), followed by I week of rest Subsequent cycles: Infusions once weekly for 3 consecutive weeks out of every 4 weeks	Bone marrow depression, transient fever,flulike syndrome, skin rash, mild nausea and vomiting
Substituted urea Hydroxyurea CML, acute leukemia (emergent treatment), head and neck cancer, ovarian cancer, melanoma, essential thrombocytosis. polycythemia vera	*Intermittent'* 80 mg/kg PO every third day *Continuous*: 20-30 mg/kg PO daily	Bone marrow depression, mild nausea and vomiting, skin rashes, radiosensitizer (See Table 5 for abbreviations)

[a] FDA-approved uses in italics; neoplasms are carcinomas unless otherwise indicated
*> Dose-limiting effects In italics

Natural products and their uses, dosages, and toxicities

Drug and its uses'[1]	Dosages	Toxicities[b]
1	2	3
Antiturnor antibiotics Bleomycin Testicular cancer. HD, reticulum cell sarcoma, lympliosor-coma. squa-maus ceil cancer of the head and neck, skin, cervix, wlva. and penis	1 0-20 U/m^2 given IV, IM, or SC weekly or twice weekly; maximum total dose. 400 U:a 2-U test dose should be given because of a possible anaphylactoid reaction	Pneumonitis and pulmonary fibrosis, fever and allergic reactions, anaphylaxis. hyper-pigmentation, Raynaud's phenomenon, alopecia

...(Contd.)

1	2	3
Dactinomycin Testicularconcer-gestational trophoblastic tumors. Wilms' tumor, rhabdomyosarcoma, Ewing's sarcoma	0.010-0.015 mg/kg IV daily for 5 d every 3 wk (usual adult dose, 0.5 mg). or 2 mg/m^2 IV as a single dose every 3-4 wk	Stomatitis, bone morrow depression, anorexia, nausea and vomiting. diarrhea, alopecia, skin changes.anaphylactold reaction
Daunorubicin AMLALL	Remission Induction: 30-45 mg/rn^2/d IV for 3 d in combination therapy; total cumulative dose, 550 mg/m^2	Bone marrow depression, cardiotoxiclty, alopecia, nausea and vomiting, diarrhea, stomatitis, fever, dermatitis at previously irradiated sites, red urine, anaphylactoid reaction
DaunoXome (liposomal daunorubicin) Kaposi's sarcoma	Liposomal preparation: 40 mg/m^2 IV every 2 wk	
Doxorubicin ALL,AML, breast, ovarian, bladder cancers. HD, NHL.SCLC.gastric cancer, sarcoina.Wilms' tumor, neuroblastoma. thyroid cancer	60-90 mg/m^2 single IV injection every 21 d. 20-30 mg/m^2/d IV for 3 d every 3-4 wk.or 20 mg/m^2 IV weekly; total cumulative dose of 550 mg/m^2; reduce dose for liver dysfunction	Bone marrow depression, cardiotoxiclty, stomatitis (continuous Infusion), alopecia, nausea and vomiting, diarrhea, fever, dermatitis at previously irradiated sites, red urine, anaphylactoid reaction
Doxil (liposomal doxorubicin) Ovarian cancer (refractory to pacitaxel-and platinum-based regimens). Kaposi's sarcoma	50 mg/m^2 IV every 4 wk 20 mg/m^2 IV every 3 wk	Bone marrow depression, hand-foot syndrome Bone marrow depression, hand-foot syndrome
Epirubicin Breast cancer	100 mg/m^2 IV on day I, or 60 mg/m^2 IV on days 1 and 8 In combination therapy	Bone morow depression, card/otoxicitv, stomatitis, alopecia
Idarubicin AML, CML (blast phase). ALL	I2mg/m^2/d IV for 3d every 3 wk in combination therapy	Bone marrow depression, nausea and vomiting. stomatitis, alopecia. cardiotoxiclty
Mitoxantrone AML,prosrote, ALL.CML. breast and ovarian cancers	Remission induction: I2mg/m^2/dl Vfor3days,in combination with Ara-C	Bone morrow depression, cardiotoxicity, alopecia, stomatitis, nausea and vomiting, blue urine and sclera
Mitomycin Gastric, colorectol, 'pancreatic oderio-cardflomas. NSCLC, breast, uterine, cervical, and head and neck cancers	20 mg/m^2 IV every 6-8 wk as a single agent, or 5-10 mg/m^2 IV every 6 wk In combination therapy	Bone morrow depression (cumulative), nausea and vomiting.anorexia, alopecia, stomatitis,fever, pulmonary fibrosis
Valrubicin Bladder	800 mg IV once a week for 6 wk	Local bladder symptoms

...(Contd.)

1	2	3
Epipodophyllotoxins		
Etoposide Testicular cancer (refractory), SCLC, HD.NHLAML gestational tropho-biastic tumors	Testicular: 50-100 mg/m^2/d IV for 5 d, or l00mg/m^2/d IV on days 1,3,and 5 Lung:. 35-50 mg/m^2/d IV for 5 d, or l00mg/m^2/d PO for 5 d For both indications, given with combination therapy and repeated evety 3-4 wk	Bone marrow depression, nausea and vomiting, diarrhea, fever, hypotension with rapid infusion, alopecia, rash
Teniposide Relapsed ALL in children, SCLC	ALL: 100 mg/m^2 once or twice weekly, or 20-60 mg/m^2/d for 5 days in combination with Ara-C Lung: 80-90 mg/m^2/d for 5 days as a single agent	Bone morrow depression, nausea and vomiting, alopecia, hypotension with rapid infusion, increased liver enzymes
Microtubule agents		
Docetaxel Breost cancer (relapsed), lung, ovarian, pancreatic cancer, head and neck, esophagus, stomach, cervical. Kaposi's sarcoma, uterine, prostate, and bladder	60-100 mg/m^2 IV over I hour every 21 days; or up to 42 mg/m^2 IV every week	Bone marrow depression, fluid retention, hypersensitivity reaction, paresthesias, rash,alopecia, myalgias
Paclitaxel Ovarian cancer (relapsed), NSCLC (in combination with (ispatln). Kaposi's sarcoma, breast cancer (relapsed), head and neck, gastric, colon, esophagus, uterine, prostate, bladder cancers and melanomas	13 5-175 mg/m^2 by IV infusion (ranging from 3-96 h) every 3 wk: or 80 mg/m^2 IV every week	Bone marrow depression peripheral neuropathy, alopecia,mucositis, anaphylaxis, dyspnea, myalgias

[3] PDA-approved uses in italics; neoplasms are carcinomas unless otherwise indicated Dose-limiting effects in italics

Chemotherapeutic Agents Classified by Mechanism of Action

Alkylating Agents

The alkylating agents impair cell function by forming covalent bonds with the amino, carboxyl, sulfhydryl, and phosphate groups in biologically important molecules. The most important sites of alkylation are DNA, RNA, and proteins. The electron-rich nitrogen at the 7 position of guanine in DNA is particularly susceptible to alkylation.

Alkylating agents depend on cell proliferation for activity but are not cell-cycle- phase–specific. A fixed percentage of cells are killed at a given dose. Tumor resistance probably occurs through efficient glutathione conjugation or by enhanced DNA repair mechanisms. Alkylating agents are classified according to their chemical structures and mechanisms of covalent bonding; this drug class includes the nitrogen mustards, nitrosoureas, and platinum complexes, among other agents (see Table 7.5).

Nitrogen mustards: The nitrogen mustards, which include such drugs as mechlorethamine (Mustargen), cyclophosphamide, ifosfamide (Ifex), and chlorambucil (Leukeran), are powerful local vesicants; as such, they can cause problems ranging from local tissue necrosis, to pulmonary fibrosis, to hemor- rhagic cystitis. The metabolites of these compounds are highly reactive in aqueous solution, in which an active alkylating moiety, the ethylene imonium ion, binds to DNA. The hematopoietic system is especially susceptible to these compounds.

Nitrosoureas: The nitrosoureas are distinguished by their high lipid solubility and chemical instability. These agents rapidly and spontaneously decompose into two highly reactive intermediates: chloroethyl diazohydroxide and isocyanate. The lipophilic nature of the nitrosoureas enables free passage across membranes; therefore, they rapidly penetrate the blood-brain barrier, achieving effective CNS concentrations. As a consequence, these agents are used for a variety of brain tumors.

Platinum agents: Cisplatin (Platinol) is an inorganic heavy metal complex that has activity typical of a cell-cycle-phase–nonspecific alkylating agent. The compound produces intrastrand and interstrand DNA cross-links and forms DNA adducts, thereby inhibiting the synthesis of DNA, RNA, and proteins. Carboplatin (Paraplatin) has the same active diamine platinum moiety as cisplatin, but it is bonded to an organic carboxylate group that allows increased water solubility and slower hydrolysis to the alkylating aqueous platinum complex, thus altering toxicity profiles. Oxaliplatin (Eloxatin) is distinguished from the other platinum compounds by a di-amino-cyclohexane ring bound to the platinum molecule, which interferes with resistance mechanisms to the drug.

Antimetabolites

Antimetabolites are structural analogs of the naturally occurring metabolites involved in DNA and RNA synthesis. As the constituents of these metabolic pathways have been elucidated, a large number of structurally similar drugs that alter the critical pathways of nucleotide synthesis have been developed.

Antimetabolites exert their cytotoxic activity either by competing with normal metabolites for the catalytic or regulatory site of a key enzyme or by substituting for a metabolite that is normally incorporated into DNA and RNA. Because of this mechanism of action, antimetabolites are most active when cells are in the S phase and have little effect on cells in the G_0 phase. Consequently,these drugs are most effective against tumors that have a high growth fraction.

Antimetabolites have a nonlinear dose-response curve, such that after a certain dose, no more cells are killed despite increasing doses (fluorouracil [5-FU] is an exception). The antimetabolites can be divided into folate analogs, purine analogs, adenosine analogs, pyrimidine analogs, and substituted ureas (see Table 7.6).

Natural Products

A wide variety of compounds possessing antitumor activity have been isolated from natural substances, such as plants, fungi, and bacteria. Likewise, selected compounds have semisynthetic and synthetic designs based on the active chemical structure of the parent compounds, and they, too, have cytotoxic effects (see Table 7.6).

Antitumor antibiotics: Bleomycin (Blenoxane) preferentially intercalates DNA at guanine-cytosine and guanine-thymine sequences, resulting in spontaneous oxidation and formation of free oxygen radicals that cause strand breakage.

Anthracyclines: The anthracycline antibiotics are products of the fungus *Streptomyces percetus* var *caesius*. They are chemically similar, with a basic anthracycline structure containing a glycoside bound to an amino sugar, daunosamine. The anthracyclines have several modes of action. Most notable are intercalation between D NA base pairs and inhibition of D NA topoisomerases I and II. Oxygen free radical formation from reduced doxorubicin intermediates is thought to be a mechanism associated with cardiotoxicity.

Epipodophyllotoxins: Etoposide is a semisynthetic epipodophyllotoxin extracted from the root of *Podophyllum peltatum* (mandrake). It inhibits topoisomerase II activity by stabilizing the DNA-topoisomerase II complex; this process ultimately results in the inability to synthesize DNA, and the cell cycle is stopped in the G_1 phase.

Epipodophyllotoxins: Etoposide is a semisynthetic epipodophyllotoxin extracted from the root of *Podophyllum peltatum* (mandrake). It inhibits topoisomerase II activity by stabilizing the DNA-topoisomerase II complex; this process ultimately results in the inability to synthesize DNA, and the cell cycle is stopped in the G_1 phase.

Vinca alkaloids: The vinca alkaloids are derived from the periwinkle plant *Vinca rosea*. Upon entering the cell, vinca alkaloids bind rapidly to the tubulin. The binding occurs in the S phase at a site different from that associated with paclitaxel and colchicine. Thus, polymerization of microtubules is blocked, resulting in impaired mitotic spindle formation in the M phase.

Taxanes: Paclitaxel and docetaxel (Taxotere) are semisynthetic derivatives of extracted precursors from the needles of yew plants. These drugs have a novel 14-member ring, the taxane. Unlike the vinca alkaloids, which cause microtubular disassembly, the taxanes promote microtubular assembly and stability, therefore blocking the cell cycle in mitosis. Docetaxel is more potent than paclitaxel in enhancing microtubular assembly and also induces apoptosis.

Camptothecin Analogs

Camptothecin analogs include irinotecan (CPT-11 [Camptosar]) and topotecan (Hycamtin). These semisynthetic analogs of the alkaloid camptothecin, derived from the Chinese ornamental tree *Camptotheca acuminata*, inhibit topoisomerase I and interrupt the elongation phase of DNA replication.

AIDS AND ITS DETECTION

History of known Cases and Spread

The oldest documented possible case of the then-unknown syndrome was thought to have been detected in 1959, when a 25-year-old British printer who had travelled in the navy between 1955 and 1957 (but apparently not to Africa) sought help at the Royal Infirmary of Manchester, England. He reported to have been suffering from puzzling symptoms, among them purplish skin lesions, for nearly two years. His condition had taken a turn for worse during Christmas 1958, when he started suffering from shortness of breath, extreme fatigue, rapid weight loss, night sweats and high fever. The doctors thought he might be suffering from tuberculosis and, even though they found no evidence of bacterial infection, they treated him for tuberculosis just to be safe, to no avail. The printer continued to weaken and he died shortly after in August 1959. His autopsy revealed evidence of two unusual infections, cytomegalovirus and *Pneumocystis carinii* pneumonia (PCP, later, when redetermined as *P. jirovecii*, renamed *Pneumocystis* pneumonia), very rare at the time but now commonly associated with AIDS patients. His case had puzzled his doctors, who preserved tissue samples from him and for years retained some interest in solving the mystery.Sir Robert Platt, then president of the Royal College of Physicians, wrote in the printer's hospital chart that he wondered *"If we are in for a new wave of virus disease now that the bacterial illnesses are so nearly conquered"*. It was only 31 years later, after the AIDS pandemic had become well-known and widespread, that they decided to perform HIV-tests on the preserved tissues of the printer, which initially turned out a positive result. While the case was reported in the July 7, 1990 issue of the British medical journal *The Lancet*, their claim was retracted in a letter in the January 20, 1996 issue, in which they reported that the tissue sample had become contaminated in the laboratory.

1981-2: From GRID to AIDS

The AIDS epidemic officially began on June 5, 1981, when the U.S. Centers for Disease Control and Prevention in its *Morbidity and Mortality Weekly Report* newsletter reported unusual clusters of Pneumocystis pneumonia (PCP) caused by a form of *Pneumocystis carinii* (now recognized as a distinct species *Pneumocystis jirovecii*) in five homosexual men in Los Angeles. Over the next 18 months, more PCP clusters were discovered among otherwise healthy men in cities throughout the country, along with other opportunistic diseases (such as Kaposi's sarcoma and persistent, generalized lymphadenopathy]), common in immunosuppressed patients.

In June 1982, a report of a group of cases amongst gay men in Southern California suggested that a sexually transmitted infectious agent might be the etiological agent, and the syndrome was initially termed "GRID", or gay-related immune deficiency. Health authorities soon realized that nearly half of the people identified with the syndrome were not homosexual men. The same opportunistic infections were also reported among hemophiliacs, heterosexual intravenous drug users, and Haitian immigrants. By August 1982, the disease was being referred to by its new CDC-coined name: Acquired Immune Deficiency Syndrome (AIDS).

Identification of the virus May 1983: LAV

In May 1983, doctors from Dr. Luc Montagnier's team at the Pasteur Institute in France, reported that they had isolated a new retrovirus from lymphoid ganglions that they believed was the cause of AIDS. The virus was later named lymphadenopathy-associated virus (LAV) and a sample was sent to the U.S. Centers for Disease Control, which was later passed to the National Cancer Institute (NCI). HIV, the virus that causes AIDS, originated in non-human primates in sub-Saharan Africa and was transferred to humans during the late 19th or early 20th century.Two types of HIV infect humans: HIV-1 and HIV-2. HIV-1 is more virulent, is easily transmitted and is the cause of the majority of HIV infections globally. HIV-1 is closely related to a virus found in chimpanzees, and molecular phylogenetics indicates that the HIV-1 virus appeared sometime between 1884 and 1924 in equatorial Africa. HIV-2 is less transmittable and is largely confined to West Africa, along with its closest relative, a virus of the Sooty Mangabey (*Cercocebus atys*), an Old World monkey of Guinea-Bissau, Gabon, and Cameroon.

Most HIV researchers agree that HIV evolved from the closely related Simian Immunodeficiency Virus (SIV), and that HIV was transferred from non-human primates to humans in the recent past (as a type of zoonosis). Research in this area is conducted using molecular phylogenetics, comparing viral genomic sequences to determine relatedness.

HIV-1 spread from Chimpanzees

Because most groups of HIV-1 are most closely related to a strain of the simian immunodeficiency virus that infects the chimpanzee subspecies *Pan troglodytes troglodytes* (SIVcpz), scientists generally accept that the virus originated in populations of wild chimpanzees in West-Central Africa. Exactly where this occurred—in the southeastern rain forests of Cameroon (modern East Province) near the Sanaga River, or further south near Kinshasa in the Democratic Republic of the Congo—has been a matter of scientific discussion.

Using HIV-1 sequences preserved in human biological samples along with estimates of viral mutation rates, scientists calculate that the jump from chimpanzee to human probably happened during the late 19th or early 20th century, a time of rapid urbanisation and colonisation in equatorial Africa. Exactly when the zoonosis occurred is not known. Some estimates suggest that HIV-1 (group M) entered the human population in the early 20th century, probably between 1915 and 1941. A study published in 2008, analyzing viral sequences recovered from a recently-discovered 1960 biopsy along with previously-known sequences, suggested a common ancestor between 1884 and 1924.

Genetic recombination had earlier been thought to "seriously confound" such phylogenetic analysis, but later "work has suggested that recombination is not likely to systematically bias [results]", although recombination is "expected to increase variance". The results of phlyogenetics study supported the later work and indicated that HIV evolves "fairly reliably".

According to the 'Hunter Theory', the "simplest and most plausible explanation for the cross-species transmission" the virus was transmitted from a chimpanzee to a human when a bushmeat hunter was bitten or cut while hunting or butchering an animal. The resulting exposure of the hunter to blood or other bodily fluids of the chimpanzee could have resulted in infection.

Method of Spread

Zoonosis (transfer of a pathogen from non-human animals to humans) and subsequent spread of the pathogen between humans, requires the following conditions:

1. a human population;
2. a nearby population of a host animal;
3. an infectious pathogen in the host animal that can spread from animal to human;
4. interaction between the species to transmit enough of the pathogen to humans to establish a human foothold, which could have taken millions of individual exposures;

5. ability of the pathogen to spread from human to human (perhaps acquired by mutation);
6. some method allowing the pathogen to disperse widely, preventing the infection from "burning out" by either killing off its human hosts or provoking immunity in a local population of humans.

Conditions that facilitated the establishment and spread of infection in human populations may have included the reuse of needles for injectable antibiotics, antimalarial drugs and vaccines; an increase in bushmeat hunting and immune suppression as a result of the harsh conditions, forced labor and displacement associated with colonialism, particularly in French Equatorial Africa; rapid urbanisation that brought infected people into close contact with others; and colonial commerce that provided opportunities for further geographical spread.

SIV in non-human primates tends to cause a non-fatal disease. Comparison of the gene sequence of SIV with HIV should therefore give us information about the factors necessary to cause disease in humans. The factors that determine the virulence of HIV as compared to most SIVs are only now being elucidated. Non-human SIVs contain a *nef* gene that down-regulates CD3, CD4, and MHC class I expression; most non-human SIVs therefore do not induce immunodeficiency; the HIV *nef* gene however has lost its ability to down-regulate CD3, which results in the immune activation and apoptosis that is characteristic of chronic HIV infection.

Genetic Studies

According to a 2008 Proceedings of the National Academy of Sciences study, a team lead by Robert Shafer at Stanford University School of Medicine has discovered that the Gray Mouse Lemur has an endogenous lentivirus (the genus to which HIV belongs) in its genetic makeup. This suggests that lentiviruses have existed for at least 14 million years, much longer than the currently known existence of HIV. In addition, the time frame falls into place when Madagascar was still yet connected to what is now the African continent; the said Lemurs later developed immunity to the virus strain and survived an era when the lentivirus was widespread among other mammalia. The study is being hailed as crucial, because it fills the blanks in the origin of the virus, as well as in its evolution, and may be important in the development of new antiviral drugs. Several alternative hypotheses for the origin of AIDS have been proposed. AIDS denialism argues that HIV or AIDS does not exist or that AIDS is not caused by HIV; some of its proponents believe that AIDS is caused by lifestyle, including sexuality or drug use. Some conspiracy theories allege that HIV was created in a bioweapons laboratory, perhaps as an agent of genocide or an accident. Scientific studies contradict these hypotheses, which have been rejected by scientific consensus.

Symptoms

The symptoms of AIDS are primarily the result of conditions that do not normally develop in individuals with healthy immune systems. Most of these conditions are infections caused by bacteria, viruses, fungi and parasites that are normally controlled by the elements of the immune system that HIV damages.

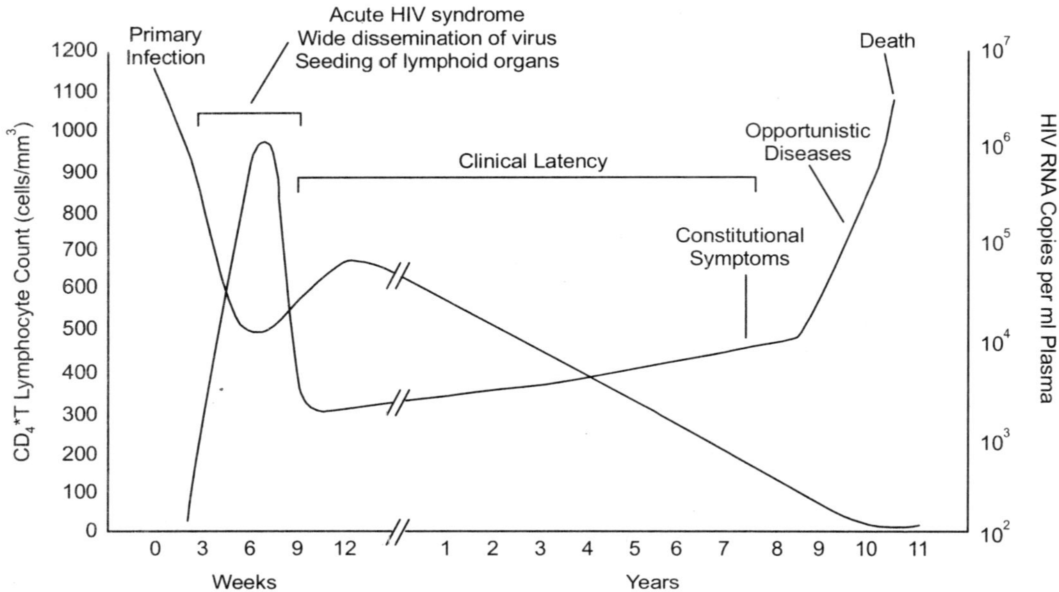

Fig. 7.16 : A generalized graph of the relationship between HIV copies (viral load) and CD4 counts over the average course of untreated HIV infection; any particular individual's disease course may vary considerably. CD4⁺ T Lymphocyte count (cells/mm³) HIV RNA copies per mL of plasma

Opportunistic infections are common in people with AIDS. HIV affects nearly every organ system. People with AIDS also have an increased risk of developing various cancers such as Kaposi's sarcoma, cervical cancer and cancers of the immune system known as lymphomas. Additionally, people with AIDS often have systemic symptoms of infection like fevers, sweats (particularly at night), swollen glands, chills, weakness, and weight loss. The specific opportunistic infections that AIDS patients develop depend in part on the prevalence of these infections in the geographic area in which the patient lives.

Pulmonary Infections

Pneumocystis pneumonia (originally known as *Pneumocystis carinii* pneumonia, and still abbreviated as PCP, which now stands for Pneumocystis pneumonia) is relatively rare in healthy, immunocompetent people, but common among HIV-infected individuals. It is caused by *Pneumocystis jirovecii*. Before the advent of effective diagnosis, treatment and routine prophylaxis in Western countries, it was a common immediate cause of death. In developing countries, it is still one of the first indications of AIDS in untested individuals, although it does not generally occur unless the CD4 count is less than 200 cells per μL of blood. Tuberculosis (TB) is unique among infections associated with HIV because it is transmissible to immunocompetent people via the respiratory route, is easily treatable once identified, may occur in early-stage HIV disease, and is preventable with drug therapy. However, multidrug resistance is a potentially serious problem. Even though its incidence has declined because of the use of directly observed therapy and other improved practices in Western countries, this is not the case in developing countries where HIV is most prevalent. In early-stage HIV infection (CD4 count >300 cells per μL), TB typically presents as a pulmonary disease. In advanced HIV infection, TB often presents atypically with extrapulmonary (systemic) disease a common feature. Symptoms are usually constitutional and are not localized to one particular site, often affecting bone marrow, bone, urinary and gastrointestinal tracts, liver, regional lymph nodes, and the central nervous system.

Gastrointestinal Infections

Esophagitis is an inflammation of the lining of the lower end of the esophagus (gullet or swallowing tube leading to the stomach). In HIV infected individuals, this is normally due to fungal (candidiasis) or viral (herpes simplex-1 or cytomegalovirus) infections. In rare cases, it could be due to mycobacteria. Unexplained chronic diarrhea in HIV infection is due to many possible causes, including common bacterial (*Salmonella*, *Shigella*, *Listeria* or *Campylobacter*) and parasitic infections; and uncommon opportunistic infections such as cryptosporidiosis, microsporidiosis, *Mycobacterium avium* complex (MAC) and viruses, astrovirus, adenovirus, rotavirus and cytomegalovirus, (the latter as a course of colitis). In some cases, diarrhea may be a side effect of several drugs used to treat HIV, or it may simply accompany HIV infection, particularly during primary HIV infection. It may also be a side effect of antibiotics used to treat bacterial causes of diarrhea (common for *Clostridium difficile*). In the later stages of HIV infection, diarrhea is thought to be a reflection of changes in the way the intestinal tract absorbs nutrients, and may be an important component of HIV-related wasting.

Neurological and Psychiatric Involvement

HIV infection may lead to a variety of neuropsychiatric sequelae, either by infection of the now susceptible nervous system by organisms, or as a direct consequence of the illness itself.

Toxoplasmosis is a disease caused by the single-celled parasite called *Toxoplasma gondii*; it usually infects the brain, causing toxoplasma encephalitis, but it can also infect and cause disease in the eyes and lungs. Cryptococcal meningitis is an infection of the meninx (the membrane covering the brain and spinal cord) by the fungus *Cryptococcus neoformans*. It can cause fevers, headache, fatigue, nausea, and vomiting. Patients may also develop seizures and confusion; left untreated, it can be lethal. Progressive multifocal leukoencephalopathy (PML) is a demyelinating disease, in which the gradual destruction of the myelin sheath covering the axons of nerve cells impairs the transmission of nerve impulses. It is caused by a virus called JC virus which occurs in 70% of the population in latent form, causing disease only when the immune system has been severely weakened, as is the case for AIDS patients. It progresses rapidly, usually causing death within months of diagnosis.

AIDS dementia complex (ADC) is a metabolic encephalopathy induced by HIV infection and fueled by immune activation of HIV infected brain macrophages and microglia. These cells are productively infected by HIV and secrete neurotoxins of both host and viral origin. Specific neurological impairments are manifested by cognitive, behavioural, and motor abnormalities that occur after years of HIV infection and are associated with low $CD4^+$ T cell levels and high plasma viral loads. Prevalence is 10–20% in Western countries but only 1–2% of HIV infections in India. This difference is possibly due to the HIV subtype in India. AIDS related mania is sometimes seen in patients with advanced HIV illness; it presents with more irritability and cognitive impairment and less euphoria than a manic episode associated with true bipolar disorder. Unlike the latter condition, it may have a more chronic course. This syndrome is less often seen with the advent of multi-drug therapy.

Tumors and Malignancies

Patients with HIV infection have substantially increased incidence of several cancers. This is primarily due to co-infection with an oncogenic DNA virus, especially Epstein-Barr virus (EBV), Kaposi's sarcoma-associated herpesvirus (KSHV) (also known as human herpesvirus-8 [HHV-8]), and human papillomavirus (HPV).

Kaposi's sarcoma (KS) is the most common tumor in HIV-infected patients. The appearance of this tumor in young homosexual men in 1981 was one of the first signals of the AIDS epidemic. Caused by a gammaherpes virus called Kaposi's sarcoma-associated herpes virus (KSHV), it often appears as purplish nodules on the skin, but can affect other organs, especially the mouth, gastrointestinal tract, and lungs. High-grade B cell lymphomas such as Burkitt's lymphoma, Burkitt's-like lymphoma, diffuse large B-cell lymphoma (DLBCL), and primary central nervous system lymphoma present more often in HIV-infected patients. These particular cancers often foreshadow a poor prognosis. Epstein-Barr virus (EBV) or KSHV cause many of these lymphomas. In HIV-infected patients, lymphoma often arises in extranodal sites such as the gastrointestinal tract. When they occur in an HIV-infected patient, KS and aggressive B cell lymphomas confer a diagnosis of AIDS. Invasive cervical cancer in HIV-infected women is also considered AIDS-defining. It is caused by human papillomavirus (HPV).

In addition to the AIDS-defining tumors listed above, HIV-infected patients are at increased risk of certain other tumors, notably Hodgkin's disease, anal and rectal carcinomas, hepatocellular carcinomas, head and neck cancers, and lung cancer. Some of these are causes by viruses, such as Hodgkin's disease (EBV), anal/rectal cancers (HPV), head and neck cancers (HPV), and hepatocellular carcinoma (hepatitis B or C). Other contributing factors include exposure to carcinogens (cigarette smoke for lung cancer), or living for years with subtle immune defects.

Interestingly, the incidence of many common tumors, such as breast cancer or colon cancer, does not increase in HIV-infected patients. In areas where HAART is extensively used to treat AIDS, the incidence of many AIDS-related malignancies has decreased, but at the same time malignant cancers overall have become the most common cause of death of HIV-infected patients. In recent years, an increasing proportion of these deaths have been from non-AIDS-defining cancers.

Other Infections

AIDS patients often develop opportunistic infections that present with non-specific symptoms, especially low-grade fevers and weight loss. These include opportunistic infection with *Mycobacterium avium-intracellulare* and cytomegalovirus (CMV). CMV can cause colitis, as described above, and CMV retinitis can cause blindness. Penicilliosis due to *Penicillium marneffei* is now the third most common opportunistic infection (after extrapulmonary tuberculosis and cryptococcosis) in HIV-positive individuals within the endemic area of Southeast Asia. An infection that often goes unrecognized in AIDS patients is Parvovirus B19. Its main consequence is anemia, which is difficult to distinguish from the effects of antiretroviral drugs used to treat AIDS itself.

Cause

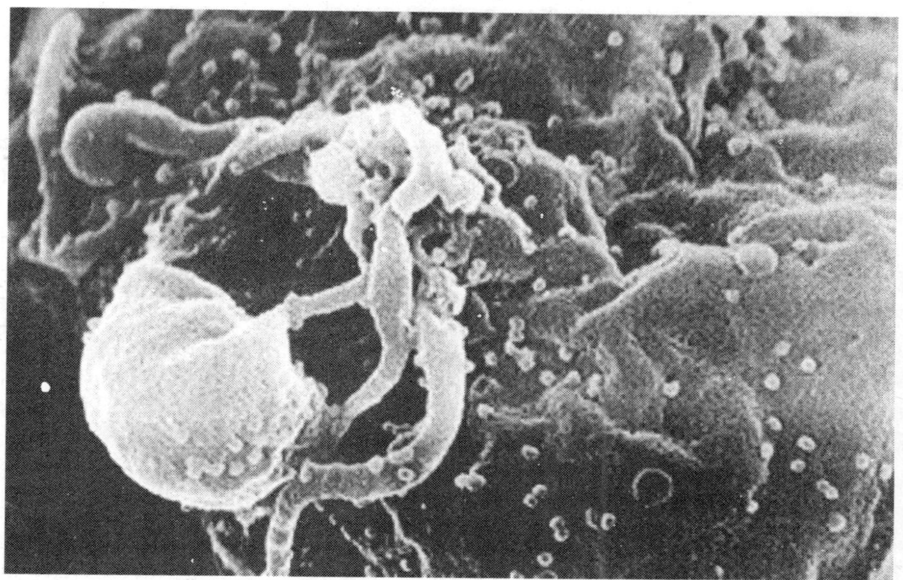

Fig. 7.17 : The Scanning Electron Micrograph of HIV-1, Coloured Green, Budding from a Cultured Lymphocyte

AIDS is the most severe acceleration of infection with HIV. HIV is a retrovirus that primarily infects vital organs of the human immune system such as $CD4^+$ T cells (a subset of T cells), macrophages and dendritic cells. It directly and indirectly destroys $CD4^+$ T cells. Once HIV has killed so many $CD4^+$ T cells that there are fewer than 200 of these cells per microliter (μL) of blood, cellular immunity is lost. Acute HIV infection progresses over time to clinical latent HIV infection and then to early symptomatic HIV infection and later to AIDS, which is identified either on the basis of the amount of $CD4^+$ T cells remaining in the blood, and/or the presence of certain infections, as noted above. In the absence of antiretroviral therapy, the median time of progression from HIV infection to AIDS is nine to ten years, and the median survival time after developing AIDS is only 9.2 months. However, the rate of clinical disease progression varies widely between individuals, from two weeks up to 20 years.Many factors affect the rate of progression. These include factors that influence the body's ability to defend against HIV such as the infected person's general immune function. Older people have weaker immune systems, and therefore have a greater risk of rapid disease progression than younger people.Poor access to health care and the existence of coexisting infections such as tuberculosis also may predispose people to faster disease progression. The infected person's genetic inheritance plays an important role and some people are resistant to certain strains of HIV. An example of this is people with the homozygous

CCR5-Δ32 variation are resistant to infection with certain strains of HIV. HIV is genetically variable and exists as different strains, which cause different rates of clinical disease progression.

Sexual Transmission

Sexual transmission occurs with the contact between sexual secretions of one person with the rectal, genital or oral mucous membranes of another. Unprotected receptive sexual acts are riskier than unprotected insertive sexual acts, and the risk for transmitting HIV through unprotected anal intercourse is greater than the risk from vaginal intercourse or oral sex.However, oral sex is not entirely safe, as HIV can be transmitted through both insertive and receptive oral sex. Sexual assault greatly increases the risk of HIV transmission as condoms are rarely employed and physical trauma to the vagina or rectum occurs frequently, facilitating the transmission of HIV. Other sexually transmitted infections (STI) increase the risk of HIV transmission and infection, because they cause the disruption of the normal epithelial barrier by genital ulceration and/or microulceration; and by accumulation of pools of HIV-susceptible or HIV-infected cells (lymphocytes and macrophages) in semen and vaginal secretions. Epidemiological studies from sub-Saharan Africa, Europe and North America suggest that genital ulcers, such as those caused by syphilis and/or chancroid, increase the risk of becoming infected with HIV by about fourfold. There is also a significant although lesser increase in risk from STIs such as gonorrhea, chlamydia and trichomoniasis, which all cause local accumulations of lymphocytes and macrophages. Transmission of HIV depends on the infectiousness of the index case and the susceptibility of the uninfected partner. Infectivity seems to vary during the course of illness and is not constant between individuals. An undetectable plasma viral load does not necessarily indicate a low viral load in the seminal liquid or genital secretions.However, each 10-fold increase in the level of HIV in the blood is associated with an 81% increased rate of HIV transmission. Women are more susceptible to HIV-1 infection due to hormonal changes, vaginal microbial ecology and physiology, and a higher prevalence of sexually transmitted diseases. People who have been infected with one strain of HIV can still be infected later on in their lives by other, more virulent strains.Infection is unlikely in a single encounter. High rates of infection have been linked to a pattern of overlapping long-term sexual relationships. This allows the virus to quickly spread to multiple partners who in turn infect their partners. A pattern of serial monogamy or occasional casual encounters is associated with lower rates of infection. HIV spreads readily through heterosexual sex in Africa, but less so elsewhere. One possibility being researched is that schistosomiasis, which affects up to 50 per cent of women in parts of Africa, damages the lining of the vagina.

Exposure to blood-borne Pathogens

This transmission route is particularly relevant to intravenous drug users, hemophiliacs and recipients of blood transfusions and blood products. Sharing and reusing syringes contaminated with HIV-infected blood represents a major risk for infection with HIV. Needle sharing is the cause of one third of all new HIV-infections in North America, China, and Eastern Europe. The risk of being infected with HIV from a single prick with a needle that has been used on an HIV-infected person is thought to be about 1 in 150. Post-exposure prophylaxis with anti-HIV drugs can further reduce this risk. This route can also affect people who give and receive tattoos and piercings. Universal precautions are frequently not followed in both sub-Saharan Africa and much of Asia because of both a shortage of supplies and inadequate training. The WHO estimates that approximately 2.5% of all HIV infections in sub-Saharan Africa are transmitted through unsafe healthcare injections. Because of this, the United Nations General Assembly has urged the nations of the world to implement precautions to prevent HIV transmission by health workers. The risk of transmitting HIV to blood transfusion recipients is extremely low in developed countries where improved donor selection and HIV screening is performed. However, according to the WHO, the overwhelming majority of the world's population does not have access to safe blood and between 5% and 10% of the world's HIV infections come from transfusion of infected blood and blood products.

Perinatal Transmission

The transmission of the virus from the mother to the child can occur *in utero* during the last weeks of pregnancy and at childbirth. In the absence of treatment, the transmission rate between a mother and her child during pregnancy, labour and delivery is 25%. However, when the mother takes antiretroviral therapy and gives birth by caesarean section, the rate of transmission is just 1%. The risk of infection is influenced by the viral load of the mother at birth, with the higher the viral load, the higher the risk. Breastfeeding also increases the risk of transmission by about 4 %.

Misconceptions

A number of misconceptions have arisen surrounding HIV/AIDS. Three of the most common are that AIDS can spread through casual contact, that sexual intercourse with a virgin will cure AIDS, and that HIV can infect only homosexual men and drug users. Other misconceptions are that any act of anal intercourse between gay men can lead to AIDS infection, and that open discussion of homosexuality and HIV in schools will lead to increased rates of homosexuality and AIDS.

Pathophysiology

The pathophysiology of AIDS is complex, as is the case with all syndromes. Ultimately, HIV causes AIDS by depleting $CD4^+$ T helper lymphocytes. This weakens the immune system and allows opportunistic infections. T lymphocytes are essential to the immune response and without them, the body cannot fight infections or kill cancerous cells. The mechanism of $CD4^+$ T cell depletion differs in the acute and chronic phases. During the acute phase, HIV-induced cell lysis and killing of infected cells by cytotoxic T cells accounts for $CD4^+$ T cell depletion, although apoptosis may also be a factor. During the chronic phase, the consequences of generalized immune activation coupled with the gradual loss of the ability of the immune system to generate new T cells appear to account for the slow decline in $CD4^+$ T cell numbers. Although the symptoms of immune deficiency characteristic of AIDS do not appear for years after a person is infected, the bulk of $CD4^+$ T cell loss occurs during the first weeks of infection, especially in the intestinal mucosa, which harbors the majority of the lymphocytes found in the body. The reason for the preferential loss of mucosal $CD4^+$ T cells is that a majority of mucosal $CD4^+$ T cells express the CCR5 coreceptor, whereas a small fraction of $CD4^+$ T cells in the bloodstream do so. HIV seeks out and destroys CCR5 expressing $CD4^+$ cells during acute infection. A vigorous immune response eventually controls the infection and initiates the clinically latent phase. However,

$CD4^+$ T cells in mucosal tissues remain depleted throughout the infection, although enough remain to initially ward off life-threatening infections. Continuous HIV replication results in a state of generalized immune activation persisting throughout the chronic phase. Immune activation, which is reflected by the increased activation state of immune cells and release of proinflammatory cytokines, results from the activity of several HIV gene products and the immune response to ongoing HIV replication. Another cause is the breakdown of the immune surveillance system of the mucosal barrier caused by the depletion of mucosal $CD4^+$ T cells during the acute phase of disease. This results in the systemic exposure of the immune system to microbial components of the gut's normal flora, which in a healthy person is kept in check by the mucosal immune system. The activation and proliferation of T cells that results from immune activation provides fresh targets for HIV infection. However, direct killing by HIV alone cannot account for the observed depletion of $CD4^+$ T cells since only 0.01–0.10% of $CD4^+$ T cells in the blood are infected.A major cause of $CD4^+$ T cell loss appears to result from their heightened susceptibility to apoptosis when the immune system remains activated. Although new T cells are continuously produced by the thymus to replace the ones lost, the regenerative capacity of the thymus is slowly destroyed by direct infection of its thymocytes by HIV. Eventually, the minimal number of $CD4^+$ T cells necessary to maintain a sufficient immune response is lost, leading to AIDS

Cells Affected

The virus, entering through which ever route, acts primarily on the following cells:

- Lymphoreticular system:
  - CD_4+ T-Helper cells
  - Macrophages
  - Monocytes
  - B-lymphocytes
- Certain endothelial cells
- Central nervous system:
  - Microglia of the nervous system
  - Astrocytes
  - Oligodendrocytes
  - Neurones – indirectly by the action of cytokines and the gp-120

The virus has cytopathic effects but how it does it is still not quite clear. It can remain inactive in these cells for long periods, though. This effect is hypothesized to be due to the CD_4-gp120 interaction.

- The most prominent effect of HIV is its T-helper cell suppression and lysis. The cell is simply killed off or deranged to the point of being function-less (they do not respond to foreign antigens). The infected B-cells can not produce enough antibodies either. Thus the immune system collapses leading to the familiar AIDS complications, like infections and neoplasms.

- Infection of the cells of the CNS cause acute aseptic meningitis, subacute encephalitis, vacuolar myelopathy and peripheral neuropathy. Later it leads to even AIDS dementia complex.
- The CD_4-gp120 interaction is also permissive to other viruses like Cytomegalovirus, Hepatitis virus, Herpes simplex virus, etc. These viruses lead to further cell damage i.e. cytopathy.

Diagnosis

The diagnosis of AIDS in a person infected with HIV is based on the presence of certain signs or symptoms. Since June 5, 1981, many definitions have been developed for epidemiological surveillance such as the Bangui definition and the 1994 expanded World Health Organization AIDS case definition. However, clinical staging of patients was not an intended use for these systems as they are neither sensitive, nor specific. In developing countries, the World Health Organization staging system for HIV infection and disease, using clinical and laboratory data, is used and in developed countries, the Centers for Disease Control (CDC) Classification System is used.

WHO Disease Staging System

In 1990, the World Health Organization (WHO) grouped these infections and conditions together by introducing a staging system for patients infected with HIV-1. An update took place in September 2005. Most of these conditions are opportunistic infections that are easily treatable in healthy people.

- Stage I: HIV infection is asymptomatic and not categorized as AIDS
- Stage II: includes minor mucocutaneous manifestations and recurrent upper respiratory tract infections
- Stage III: includes unexplained chronic diarrhea for longer than a month, severe bacterial infections and pulmonary tuberculosis
- Stage IV: includes toxoplasmosis of the brain, candidiasis of the esophagus, trachea, bronchi or lungs and Kaposi's sarcoma; these diseases are indicators of AIDS.

CDC Classification System

There are two main definitions for AIDS, both produced by the Centers for Disease Control and Prevention (CDC). The older definition is to referring to AIDS using the diseases that were associated with it, for example, lymphadenopathy, the disease after which the discoverers of HIV originally named the virus. In 1993, the CDC expanded their definition of AIDS to include all HIV positive people with a $CD4^+$ T cell count below 200 per μL of blood or 14% of all lymphocytes. The majority of new AIDS cases in developed countries use either this definition or the pre-1993 CDC definition. The AIDS diagnosis still stands even if, after treatment, the $CD4^+$ T cell count rises to above 200 per μL of blood or other AIDS-defining illnesses are cured.

HIV Test

Many people are unaware that they are infected with HIV. Less than 1% of the sexually active urban population in Africa has been tested, and this proportion is even lower in rural populations. Furthermore, only 0.5% of pregnant women attending urban health facilities are counseled, tested or receive their test results. Again, this proportion is even lower in rural health facilities. Therefore, donor blood and blood products used in medicine and medical research are screened for HIV. HIV tests are usually performed on venous blood. Many laboratories use *fourth generation* screening tests which detect anti-HIV antibody (IgG and IgM) and the HIV p24 antigen. The detection of HIV antibody or antigen in a patient previously known to be negative is evidence of HIV infection. Individuals whose first specimen indicates evidence of HIV infection will have a repeat test on a second blood sample to confirm the results.The window period (the time between initial infection and the development of detectable antibodies against the infection) can vary since it can take 3–6 months to seroconvert and to test positive. Detection of the virus using polymerase chain reaction (PCR) during the window period is possible, and evidence suggests that an infection may often be detected earlier than when using a fourth generation EIA screening test.Positive results obtained by PCR are confirmed by antibody tests. Routinely used HIV tests for infection in neonates and infants (i.e., patients younger than 2 years), born to HIV-positive mothers, have no value because of the presence of maternal antibody to HIV in the child's blood. HIV infection can only be diagnosed by PCR, testing for HIV pro-viral DNA in the children's lymphocytes.

Prevention

The three main transmission routes of HIV are sexual contact, exposure to infected body fluids or tissues, and from mother to fetus or child during perinatal period. It is possible to find HIV in the saliva, tears, and urine of infected individuals, but there are no recorded cases of infection by these secretions, and the risk of infection is negligible. The majority of HIV infections are acquired through unprotected sexual relations between partners, one of whom has HIV. The primary mode of HIV infection worldwide is through sexual contact between members of the opposite sex. During a sexual act, only male or female condoms can reduce the chances of infection with HIV and other STDs and the chances of becoming pregnant. The best evidence to date indicates that typical condom use reduces the risk of heterosexual HIV transmission by approximately 80% over the long-term, though the benefit is likely to be higher if condoms are used correctly on every occasion. The male latex condom, if used correctly without oil-based lubricants, is the single most effective available technology to reduce the sexual transmission of HIV and other sexually transmitted infections. Manufacturers recommend that oil-based lubricants such as petroleum jelly, butter, and lard not be used with latex condoms, because they dissolve the latex, making the condoms porous. If necessary, manufacturers recommend using water-based lubricants. Oil-based lubricants can however be used with polyurethane condoms. The female condom is an alternative to the male condom and is made from polyurethane, which allows it to be used in the presence of oil-based lubricants. They are larger than male condoms and have a stiffened ring-shaped opening, and are designed to be inserted into the vagina.The female condom contains an inner ring, which keeps the condom in place inside the vagina-inserting the female condom requires squeezing this ring. However, at present availability of female condoms is very low and the price remains prohibitive for many women.Preliminary studies suggest that, where female condoms are available, overall protected sexual acts increase relative to unprotected sexual acts, making them an important HIV prevention strategy. Studies on couples where one partner is infected show that with consistent condom use, HIV infection rates for the uninfected partner are below 1% per year. Prevention strategies are well-known in developed countries, but epidemiological and behavioural studies in Europe and North America suggest that a substantial minority of young people continue to engage in high-risk practices despite HIV/AIDS knowledge, underestimating their own risk of becoming infected with HIV. Randomized controlled trials have shown that male circumcision lowers the risk of HIV infection among heterosexual men by up to 60%. It is expected that this procedure will be actively promoted in many of the countries affected by HIV, although doing so will involve confronting a number of practical, cultural and attitudinal issues. However, programs to encourage condom use, including providing them free to those in poverty, are estimated to be 95 times more cost effective than circumcision at reducing the rate of HIV in sub-Saharan Africa. Some experts fear that a lower perception of vulnerability among circumcised men may result in more sexual risk-taking behaviour, thus negating its preventive effects. However, one randomized controlled trial indicated that adult male circumcision was not associated with increased HIV risk behaviour.

Exposure to Infected Body Fluids

Health care workers can reduce exposure to HIV by employing precautions to reduce the risk of exposure to contaminated blood. These precautions include barriers such as gloves, masks, protective eyeware or shields, and gowns or aprons which prevent exposure of the skin or mucous membranes to blood borne pathogens. Frequent and thorough washing of the skin immediately after being contaminated with blood or other bodily fluids can reduce the chance of infection. Finally, sharp objects like needles, scalpels and glass, are carefully disposed off to prevent needlestick injuries with contaminated items. Since intravenous drug use is an important factor in HIV transmission in developed countries, harm reduction strategies such as needle-exchange programmes are used in attempts to reduce the infections caused by drug abuse.

Mother-to-Child Transmission (MTCT)

Current recommendations state that when replacement feeding is acceptable, feasible, affordable, sustainable and safe, HIV-infected mothers should avoid breast-feeding their infant. However, if this is not the case, exclusive breast-feeding is recommended during the first months of life and discontinued as soon as possible. It should be noted that women may breastfeed other children who are not their own;

Treatment

There is currently no publicly available vaccine for HIV or cure for HIV or AIDS. The only known methods of prevention are based on avoiding exposure to the virus or, failing that, an antiretroviral treatment directly after a highly significant exposure, called post-exposure prophylaxis (PEP). PEP has a very demanding four week schedule of dosage. It also has very unpleasant side effects including diarrhea, malaise, nausea and fatigue.

Antiviral Therapy

Current treatment for HIV infection consists of highly active antiretroviral therapy, or HAART. This has been highly beneficial to many HIV-infected individuals since its introduction in 1996 when the protease inhibitor-based HAART initially became available. Current optimal HAART options consist of combinations (or "cocktails") consisting of at least three drugs belonging to at least two types, or "classes," of antiretroviral agents. Typical regimens consist of two nucleoside analogue reverse transcriptase inhibitors (NARTIs or NRTIs) plus either a protease inhibitor or a non-nucleoside reverse transcriptase inhibitor (NNRTI). Because HIV disease progression in children is more rapid than in adults, and laboratory parameters are less predictive of risk for disease progression, particularly for young infants, treatment recommendations are more aggressive for children than for adults. In developed countries where HAART is available, doctors assess the viral load, rapidity in CD4 decline, and patient readiness while deciding when to recommend initiating treatment. Standard goals of HAART include improvement in the patient's quality of life, reduction in complications, and reduction of HIV viremia below the limit of detection, but it does not cure the patient of HIV nor does it prevent the return, once treatment is stopped, of high blood levels of HIV, often HAART resistant. Moreover, it would take more than the lifetime of an individual to be cleared of HIV infection using HAART. Despite this, many HIV-infected individuals have experienced remarkable improvements in their general health and quality of life, which has led to the plummeting of HIV-associated morbidity and mortality. In the absence of HAART, progression from HIV infection to AIDS occurs at a median of between nine to ten years and the median survival time after developing AIDS is only 9.2 months. HAART is thought to increase survival time by between 4 and 12 years. For some patients, which can be more than fifty percent of patients, HAART achieves far less than optimal results, due to medication intolerance/side effects, prior ineffective antiretroviral therapy and infection with a drug-resistant strain of HIV. Non-adherence and non-persistence with therapy are the major reasons why some people do not benefit from HAART. The reasons for non-adherence and non-persistence are varied. Major psychosocial issues include poor access to medical care, inadequate social supports, psychiatric disease and drug abuse. HAART regimens can also be complex and thus hard to follow, with large numbers of pills taken frequently. Side effects can also deter people from persisting with HAART, these include lipodystrophy, dyslipidaemia, diarrhoea, insulin resistance, an increase in cardiovascular risks and birth defects. Anti-retroviral drugs are expensive, and the majority of the world's infected individuals do not have access to medications and treatments for HIV and AIDS.

Experimental and Proposed Treatments

It has been postulated that only a vaccine can halt the pandemic because a vaccine would possibly cost less, thus being affordable for developing countries, and would not require daily treatments. However, even after almost 30 years of research, HIV-1 remains a difficult target for a vaccine. Research to improve current treatments includes decreasing side effects of current drugs, further simplifying drug regimens to improve adherence, and determining the best sequence of regimens to manage drug resistance. A number of studies have shown that measures to prevent opportunistic infections can be beneficial when treating patients with HIV infection or AIDS. Vaccination against hepatitis A and B is advised for patients who are not infected with these viruses and are at risk of becoming infected. Patients with substantial immunosuppression are also advised to receive prophylactic therapy for Pneumocystis jiroveci pneumonia (PCP), and many patients may benefit from prophylactic therapy for toxoplasmosis and Cryptococcus meningitis as well. Researchers have discovered an abzyme that can destroy the protein gp120 CD4 binding site. This protein is common to all HIV variants as it is the attachment point for B lymphocytes and subsequent compromising of the immune system. In Berlin, Germany, a 42-year-old leukemia patient infected with HIV for more than a decade was given an experimental transplant of bone marrow with cells that contained an unusual natural variant of the CCR5 cell-surface receptor. This CCR5-Δ32 variant has been shown to make some cells from people who are born with it resistant to infection with some strains of HIV. Almost two years after the transplant, and even after the patient reportedly stopped taking antiretroviral medications, HIV has not been detected in the patient's blood.

Alternative Medicine

Various forms of alternative medicine have been used to treat symptoms or alter the course of the disease. Current studies indicate that alternative medicine therapies have little effect on the mortality or morbidity of the disease, but may improve the quality of life of individuals with AIDS. The psychological benefits of these therapies are the most important use. Acupuncture has been used to alleviate some symptoms with no success and cannot cure the HIV infection. Several randomized clinical trials testing the effect of herbal medicines have shown that there is no evidence that these herbs have any effect on the progression of the disease, but may instead produce serious side-effects. Morbidity and mortality among HIV-infected adults with adequate dietary nutritional intake is unaffected by multivitamin supplementation. A large Tanzanian trial in immunologically and nutritionally compromised pregnant and lactating women showed a number of benefits to daily multivitamin supplementation for both mothers and children. Dietary intake of micronutrients at RDA levels by HIV-infected adults is recommended by the World Health Organization. There is some evidence that vitamin A supplementation in children reduces mortality and improves growth. Daily doses of selenium can suppress HIV viral burden with an associated improvement of the CD4 count. Selenium can be used as an adjunct therapy to standard antiviral treatments, but cannot itself reduce mortality and morbidity.

Prognosis

Without treatment, the net median survival time after infection with HIV is estimated to be 9 to 11 years, depending on the HIV subtype, and the median survival rate after diagnosis of AIDS in resource-limited settings where treatment is not available ranges between 6 and 19 months, depending on the study. In areas where it is widely available, the development of HAART as effective therapy for HIV infection and AIDS reduced the death rate from this disease by 80%, and raised the life expectancy for a newly diagnosed HIV-infected person to about 20 years. As new treatments continue to be developed and because HIV continues to evolve resistance to treatments, estimates of survival time are likely to continue to change. Without antiretroviral therapy, death normally occurs within a year. Most patients die from opportunistic infections or malignancies associated with the progressive failure of the immune system. The rate of clinical disease progression varies widely between individuals and has been shown to be affected by many factors such as host susceptibility and immune function health care and co-infections, as well as which particular strain of the virus is involved. Even with anti-retroviral treatment, over the long-term HIV-infected patients may experience neurocognitive disorders, osteoporosis, neuropathy, cancers, nephropathy, and cardiovascular disease. It is not always clear whether these conditions result from the infection, related cc mplications, or are side effects of treatment.

HIV

Human immunodeficiency virus (HIV) is a lentivirus (a member of the retrovirus family) that can lead to *acquired immunodeficiency syndrome* (AIDS), a condition in humans in which the immune system begins to fail, leading to life-threatening opportunistic infections. Previous names for the virus include **human T-lymphotropic virus-III (HTLV-III)**, **lymphadenopathy-associated virus** (LAV), and **AIDS-associated retrovirus** (ARV).

Infection with HIV occurs by the transfer of blood, semen, vaginal fluid, pre-ejaculate, or breast milk. Within these bodily fluids, HIV is present as both free virus particles and virus within infected immune cells. The four major routes of transmission are unprotected sexual intercourse, contaminated needles, breast milk, and transmission from an infected mother to her baby at birth (Vertical transmission). Screening of blood products for HIV has largely eliminated transmission through blood transfusions or infected blood products in the developed world. HIV infection in humans is now pandemic. As of January 2006, the Joint United Nations Programme on HIV/AIDS (UNAIDS) and the World Health Organization (WHO) estimate that AIDS has killed more than 25 million people since it was first recognized on December 1, 1981. It is estimated that about 0.6 percent of the world's population is infected with HIV. In 2005 alone, AIDS claimed an estimated 2.4–3.3 million lives, of which more than 570,000 were children. A third of these deaths are occurring in sub-Saharan Africa, retarding economic growth and increasing poverty. According to current estimates, HIV is set to infect 90 million people in Africa, resulting in a minimum estimate of 18 million orphans. Antiretroviral treatment reduces both the mortality and the morbidity of HIV infection, but routine access to antiretroviral medication is not available in all countries.

HIV primarily infects vital cells in the human immune system such as helper T cells (specifically $CD4^+$ T cells), macrophages, and dendritic cells. HIV infection leads to low levels of $CD4^+$ T cells through three main mechanisms: firstly, direct viral killing of infected cells; secondly, increased rates of apoptosis in infected cells; and thirdly, killing of infected $CD4^+$ T cells by CD8 cytotoxic lymphocytes that recognize infected cells. When $CD4^+$ T cell numbers decline below a critical level, cell-mediated immunity is lost, and the body becomes progressively more susceptible to opportunistic infections.

Eventually most HIV-infected individuals develop AIDS. These individuals mostly die from opportunistic infections or malignancies associated with the progressive failure of the immune system. Without treatment, about 9 out of every 10 persons with HIV will progress to AIDS after 10-15 years. Many progress much sooner. Treatment with anti-retrovirals increases the life expectancy of people infected with HIV. Even after HIV has progressed to diagnosable AIDS, the average survival time with antiretroviral therapy (as of 2005) is estimated to be more than 5 years. Without antiretroviral therapy, death normally occurs within a year. It is hoped that current and future treatments may allow HIV-infected individuals to achieve a life expectancy approaching that of the general public.

Classification

HIV is a member of the genus Lentivirus, part of the family of Retroviridae. Lentiviruses have many common morphologies and biological properties. Many species are infected by lentiviruses, which are characteristically responsible for long-duration illnesses with a long incubation period. Lentiviruses are transmitted as single-stranded, positive-sense, enveloped RNA viruses. Upon entry of the target cell, the viral RNA genome is converted to double-stranded DNA by a virally encoded reverse transcriptase that is present in the virus particle. This viral DNA is then integrated into the cellular DNA by a virally encoded integrase, along with host cellular co-factors, so that the genome can be transcribed. Once the virus has infected the cell, two pathways are possible: either the virus becomes latent and the infected cell continues to function, or the virus becomes active and replicates, and a large number of virus particles are liberated that can then infect other cells.

There are two strains of HIV known to exist: HIV-1 and HIV-2. HIV-1 is the virus that was initially discovered and termed LAV. It is more virulent, relatively easily transmitted, and is the cause of the majority of HIV infections globally. HIV-2 is less transmittable and is largely confined to West Africa.

Comparison of HIV species

Species	Virulence	Transmittability	Prevalence	Purported origin
HIV-1	High	High	Global	Common Chimpanzee
HIV-2 History Origin	Lower	Low	West Africa	Sooty Mangabey

HIV is thought to have originated in non-human primates in sub-Saharan Africa and transferred to humans early in the 20th century. The first paper recognizing a pattern of opportunistic infections was published on 4 June 1981.

Two species of HIV infect humans: HIV-1 and HIV-2. Both species of the virus are believed to have originated in West-Central Africa and jumped species (zoonosis) from a non-human primate to humans. HIV-1 is thought to have originated in southern Cameroon after jumping from wild chimpanzees (Pan troglodytes troglodytes) to humans during the twentieth century. It evolved from a Simian Immunodeficiency Virus (SIVcpz) HIV-2, on the other hand, may have originated from the Sooty Mangabey (Cercocebus atys), an Old World monkey of Guinea-Bissau, Gabon, and Cameroon.

New World Monkeys are an interesting exception to the transmission of HIV. Their immunity is believed to be caused by retrotransposition of the Cyclophilin gene into an intron of TRIM5. The result is fusion gene that provides the owl monkey with resistance to HIV-1 infection.

Transmission

Estimated per-act risk for acquisition of HIV by exposure route.

Exposure Route	Estimated infectionsper 10,000 exposuresto an infected source
Blood Transfusion	9,000
Childbirth	2,500
Needle-sharing injection drug use	67
Percutaneous needle stick	30
Receptive anal intercourse*	50
Insertive anal intercourse*	6.5
Receptive penile-vaginal intercourse*	10
Insertive penile-vaginal intercourse*	5
Receptive oral intercourse*§	1
Insertive oral intercourse*§	0.5

* assuming no condom use

§ source refers to oral intercourse performed on a man

Three main transmission routes for HIV have been identified. HIV-2 is transmitted much less frequently by the mother-to-child and sexual route than HIV-1.

Structure and Genome

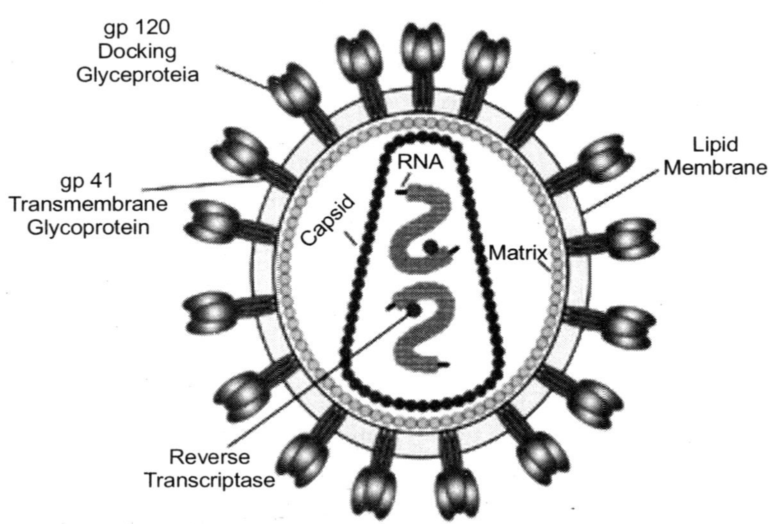

Fig. 7.18 : The Diagram of HIV

HIV is different in structure from other retroviruses. It is roughly spherical with a diameter of about 120 nm, around 60 times smaller than a red blood cell, yet large for a virus. It is composed of two copies of positive single-stranded RNA that codes for the virus's nine genes enclosed by a conical capsid composed of 2,000 copies of the viral protein p24. The single-stranded RNA is tightly bound to nucleocapsid proteins, p7 and enzymes needed for the development of the virion such as reverse transcriptase, proteases, ribonuclease and integrase. A matrix composed of the viral protein p17 surrounds the capsid ensuring the integrity of the virion particle. This is, in turn, surrounded by the viral envelope which is composed of two layers of fatty molecules called phospholipids taken from the membrane of a human cell when a newly formed virus particle buds from the cell. Embedded in the viral envelope are proteins from the host cell and about 70 copies of a complex HIV protein that protrudes through the surface of the virus particle. This protein, known as Env, consists of a cap made of three molecules called glycoprotein (gp) 120, and a stem consisting of three gp41 molecules that anchor the structure into the viral envelope. This glycoprotein complex enables the virus to attach to and fuse with target cells to initiate the infectious cycle. Both these surface proteins, especially gp120, have been considered as targets of future treatments or vaccines against HIV.

The RNA genome consists of at least 7 structural landmarks (LTR, TAR, RRE, PE, SLIP, CRS, INS) and nine genes (*gag*, *pol*, and *env*, *tat*, *rev*, *nef*, *vif*, *vpr*, *vpu*, and *tev*) encoding 19 proteins. Three of these genes, *gag*, *pol*, and *env*, contain information needed to make the structural proteins for new virus particles. For example, *env* codes for a protein called gp160 that is broken down by a viral enzyme to form gp120 and gp41. The six remaining genes, *tat*, *rev*, *nef*, *vif*, *vpr*, and *vpu* (or *vpx* in the case of HIV-2), are regulatory genes for proteins that control the ability of HIV to infect cells, produce new copies of virus (replicate), or cause disease. The two Tat proteins (p16 and p14) are transcriptional transactivators for the LTR promoter acting by binding the TAR RNA element. The Rev protein (p19) is involved in shuttling RNAs from the nucleus and the cytoplasm by binding to the RRE RNA element. The Vif protein (p23) prevents the action of APOBEC3G (a cell protein which deaminates DNA:RNA hybrids and/or interferes with the Pol protein). The Vpr protein (p14) arrests cell division at G2/M. The Nef protein (p27) downregulates CD4 (the major viral receptor), as well as the MHC class I and class II molecules. Nef also interacts with SH3 domains. The Vpu protein (p16) influences the release of new virus particles from infected cells. The ends of each strand of HIV RNA contain an RNA sequence called the long terminal repeat (LTR). Regions in the LTR act as switches to control production of new viruses and can be triggered by proteins from either HIV or the host cell. The Psi element is involved in viral genome packaging and recognized by Gag and Rev proteins. The SLIP element (TTTTTT) is involved in the frameshift in the Gag-Pol reading frame required to make functional Pol.

Tropism

The term viral tropism refers to which cell types HIV infects. HIV can infect a variety of immune cells such as $CD4^+$ T cells, macrophages, and microglial cells. HIV-1 entry to macrophages and $CD4^+$ T cells is mediated through interaction of the virion envelope glycoproteins (gp120) with the CD4 molecule on the target cells and also with chemokine coreceptors.

Macrophage (M-tropic) strains of HIV-1, or non-syncitia-inducing strains (NSI) use the β-chemokine receptor CCR5 for entry and are thus able to replicate in macrophages and $CD4^+$ T cells.

This CCR5 coreceptor is used by almost all primary HIV-1 isolates regardless of viral genetic subtype. Indeed, macrophages play a key role in several critical aspects of HIV infection. They appear to be the first cells infected by HIV and perhaps the source of HIV production when $CD4^+$ cells become depleted in the patient. Macrophages and microglial cells are the cells infected by HIV in the central nervous system. In tonsils and adenoids of HIV-infected patients, macrophages fuse into multinucleated giant cells that produce huge amounts of virus.

T-tropic isolates, or syncitia-inducing (SI) strains replicate in primary $CD4^+$ T cells as well as in macrophages and use the α-chemokine receptor, CXCR4, for entry. Dual-tropic HIV-1 strains are thought to be transitional strains of the HIV-1 virus and thus are able to use both CCR5 and CXCR4 as co-receptors for viral entry.

The α-chemokine SDF-1, a ligand for CXCR4, suppresses replication of T-tropic HIV-1 isolates. It does this by down-regulating the expression of CXCR4 on the surface of these cells. HIV that use only the CCR5 receptor are termed R5, those that only use CXCR4 are termed X4, and those that use both, X4R5. However, the use of coreceptor alone does not explain viral tropism, as not all R5 viruses are able to use CCR5 on macrophages for a productive infection and HIV can also infect a subtype of myeloid dendritic cells, which probably constitute a reservoiur that maintains infection when $CD4^+$ T cell numbers have declined to extremely low levels.

Some people are resistant to certain strains of HIV. One example of how this occurs is people with the CCR5-Δ32 mutation; these people are resistant to infection with R5 virus as the mutation stops HIV from binding to this coreceptor, reducing its ability to infect target cells.

Sexual intercourse is the major mode of HIV transmission. Both X4 and R5 HIV are present in the seminal fluid which is passed from a male to his sexual partner. The virions can then infect numerous cellular targets and disseminate into the whole organism. However, a selection process leads to a predominant transmission of the R5 virus through this pathway. How this selective process works is still under investigation, but one model is that spermatozoa may selectively carry R5 HIV as they possess both CCR3 and CCR5 but not CXCR4 on their surface and that genital epithelial cells preferentially sequester X4 virus. In patients infected with subtype B HIV-1, there is often a co-receptor switch in late-stage disease and T-tropic variants appear that can infect a variety of T cells through CXCR4. These variants then replicate more aggressively with heightened virulence that causes rapid T cell depletion, immune system collapse, and opportunistic infections that mark the advent of AIDS. Thus, during the course of infection, viral adaptation to the use of CXCR4 instead of CCR5 may be a key step in the progression to AIDS. A number of studies with subtype B-infected individuals have determined that between 40 and 50% of AIDS patients can harbour viruses of the SI, and presumably the X4, phenotype.

Replication Cycle

Entry to the Cell

HIV enters macrophages and $CD4^+$ T cells by the adsorption of glycoproteins on its surface to receptors on the target cell followed by fusion of the viral envelope with the cell membrane and the release of the HIV capsid into the cell.

Entry to the cell begins through interaction of the trimeric envelope complex (gp160 spike) and both CD4 and a chemokine receptor (generally either CCR5 or CXCR4, but others are known to interact) on the cell surface. gp120 binds to integrin $\alpha_4\beta_7$ activating LFA-1 the central integrin involved in the establishment of virological synapses, which facilitate efficient cell-to-cell spreading of HIV-1. The gp160 spike contains binding domains for both CD4 and chemokine receptors. The first step in fusion involves the high-affinity attachment of the CD4 binding domains of gp120 to CD4. Once gp120 is bound with the CD4 protein, the envelope complex undergoes a structural change, exposing the chemokine binding domains of gp120 and allowing them to interact with the target chemokine receptor. This allows for a more stable two-pronged attachment, which allows the N-terminal fusion peptide gp41 to penetrate the cell membrane. Repeat sequences in gp41, HR1 and HR2 then interact, causing the collapse of the extracellular portion of gp41 into a hairpin. This loop structure brings the virus and cell membranes close together, allowing fusion of the membranes and subsequent entry of the viral capsid.

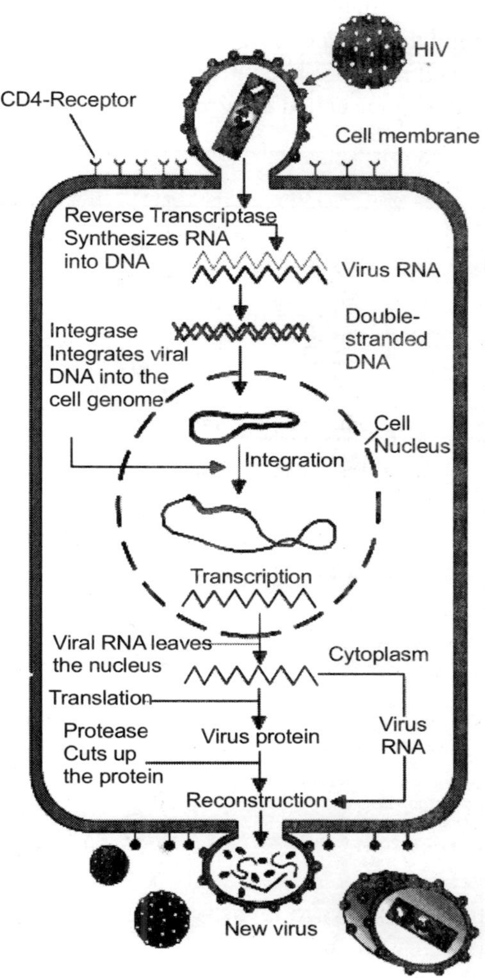

Fig. 7.19 : The HIV replication cycle

Once HIV has bound to the target cell, the HIV RNA and various enzymes, including reverse transcriptase, integrase, ribonuclease and protease, are injected into the cell. During the microtubule based transport to the nucleus, the viral single strand RNA genome is transcribed into double strand DNA, which is then integrated into a host chromosome.

HIV can infect dendritic cells (DCs) by this CD4-CCR5 route, but another route using mannose-specific C-type lectin receptors such as DC-SIGN can also be used. DCs are one of the first cells encountered by the virus during sexual transmission. They are currently thought to play an important role by transmitting HIV to T cells once the virus has been captured in the mucosa by DCs.

Replication and Transcription

Once the viral capsid enters the cell, an enzyme called *reverse transcriptase* liberates the single-stranded (+)RNA from the attached viral proteins and copies it into a complementary DNA. This process of reverse transcription is extremely error-prone and it is during this step that mutations may occur. Such mutations may cause drug resistance. The reverse transcriptase then makes a complementary DNA strand to form a double-stranded viral DNA intermediate (vDNA). This vDNA is then transported into the cell nucleus. The integration of the viral DNA into the host cell's genome is carried out by another viral enzyme called *integrase*.

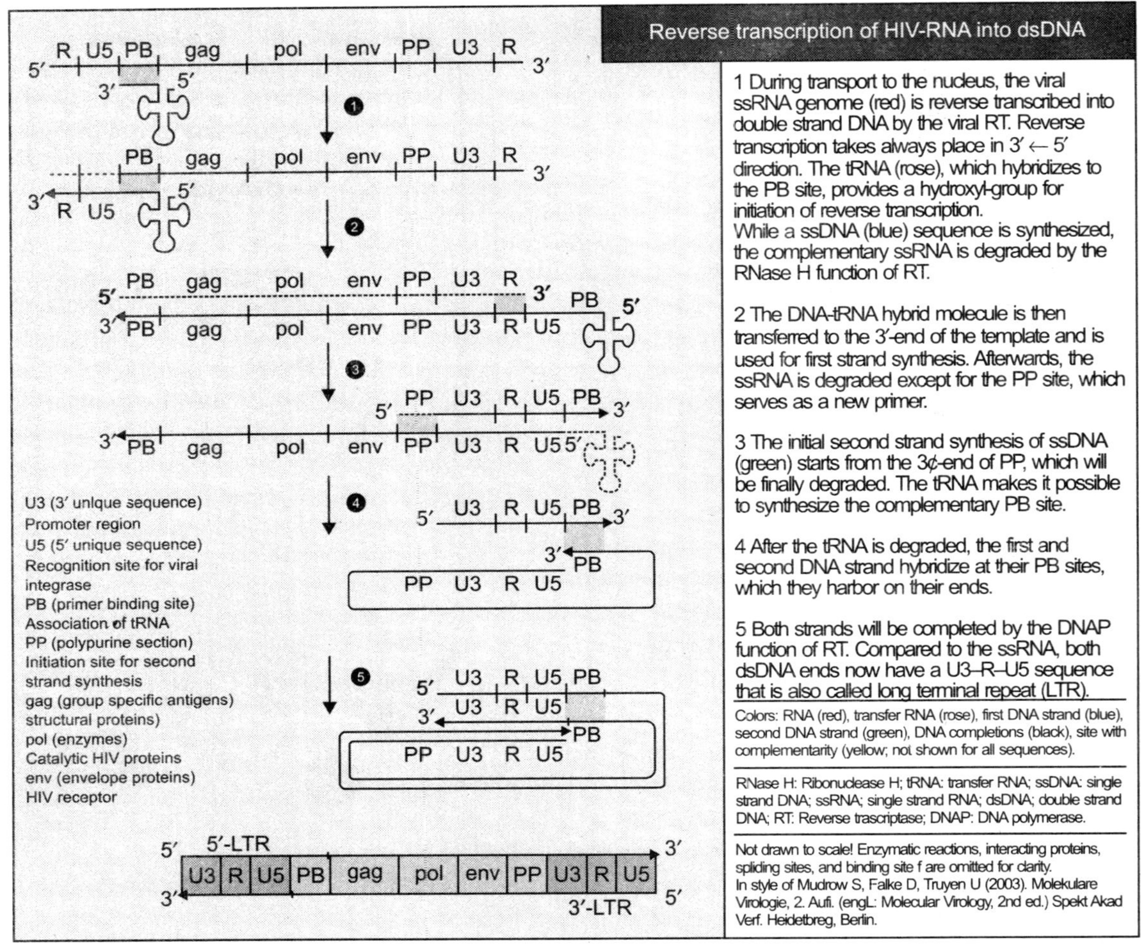

Fig. 7.20 : The Reverse transcription of the HIV genome into double strand DNA

This integrated viral DNA may then lie dormant, in the latent stage of HIV infection. To actively produce the virus, certain cellular transcription factors need to be present, the most important of which is NF-κB (NF kappa B), which is upregulated when T cells become activated. This means that those cells most likely to be killed by HIV are those currently fighting infection.

In this replication process, the integrated provirus is copied to mRNA which is then spliced into smaller pieces. These small pieces produce the regulatory proteins Tat (which encourages new virus production) and Rev. As Rev accumulates it gradually starts to inhibit mRNA splicing. At this stage, the structural proteins Gag and Env are produced from the full-length mRNA. The full-length RNA is actually the virus genome; it binds to the Gag protein and is packaged into new virus particles. HIV-1 and HIV-2 appear to package their RNA differently; HIV-1 will bind to any appropriate RNA whereas HIV-2 will preferentially bind to the mRNA which was used to create the Gag protein itself. This may mean that HIV-1 is better able to mutate (HIV-1 infection progresses to AIDS faster than HIV-2 infection and is responsible for the majority of global infections).

Assembly and Release

The final step of the viral cycle, assembly of new HIV-1 virons, begins at the plasma membrane of the host cell. The Env polyprotein (gp160) goes through the endoplasmic reticulum and is transported to the Golgi complex where it is cleaved by protease and processed into the two HIV envelope glycoproteins gp41 and gp120. These are transported to the plasma membrane of the host cell where gp41 anchors the gp120 to the membrane of the infected cell. The Gag (p55) and Gag-Pol (p160) polyproteins also associate with the inner surface of the plasma membrane along with the HIV genomic RNA as the forming virion begins to bud from the host cell. Maturation either occurs in the forming bud or in the immature virion after it buds from the host cell. During maturation, HIV proteases cleave the polyproteins into individual functional HIV proteins and enzymes. The various structural components then assemble to produce a mature HIV virion. This cleavage step can be inhibited by protease inhibitors. The mature virus is then able to infect another cell.

Genetic Variability

HIV differs from many viruses in that it has very high genetic variability. This diversity is a result of its fast replication cycle, with the generation of 10^9 to 10^{10} virions every day, coupled with a high mutation rate of approximately 3×10^{-5} per nucleotide base per cycle of replication and recombinogenic properties of reverse transcriptase. This complex scenario leads to the generation of many variants of HIV in a single infected patient in the course of one day. This variability is compounded when a single cell is simultaneously infected by two or more different strains of HIV. When simultaneous infection occurs, the genome of progeny virions may be composed of RNA strands from two different strains. This hybrid virion then infects a new cell where it undergoes replication. As this happens, the reverse transcriptase, by jumping back and forth between the two different RNA templates, will generate a newly synthesized retroviral DNA sequence that is a recombinant between the two parental genomes. This recombination is most obvious when it occurs between subtypes. The closely related simian immunodeficiency virus (SIV) exhibits a somewhat different behavior: in its natural hosts, African green monkeys and sooty mangabeys, the retrovirus is present in high levels in the blood, but evokes only a mild immune response, does not cause the development of simian AIDS, and does not undergo the extensive mutation and recombination typical of HIV. By contrast, infection of heterologous hosts (rhesus or cynomologus macaques) with SIV results in the generation of genetic diversity that is on the same order as HIV in infected humans; these heterologous hosts also develop simian AIDS. The relationship, if any, between genetic diversification, immune response, and disease progression is

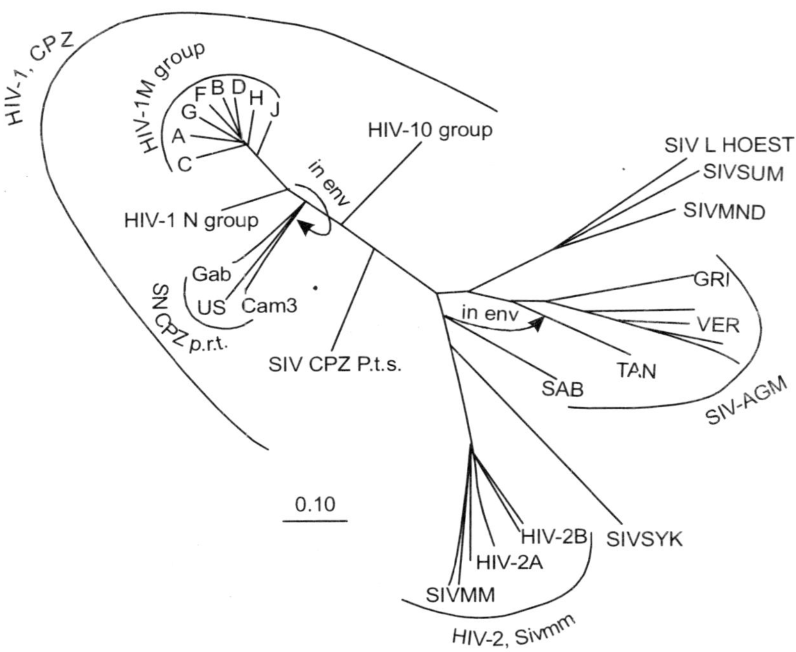

Fig. 7.21 : The Phylogenetic Tree of the SIV and HIV.

unknown.Three groups of HIV-1 have been identified on the basis of differences in *env*: M, N, and O. Group M is the most prevalent and is subdivided into eight subtypes (or clades), based on the whole genome, which are geographically distinct. The most prevalent are subtypes B (found mainly in North America and Europe), A and D (found mainly in Africa), and C (found mainly in Africa and Asia); these subtypes form branches in the phylogenetic tree representing the lineage of the M group of HIV-1. Coinfection with distinct subtypes gives rise to circulating recombinant forms (CRFs). In 2000, the last year in which an analysis of global subtype prevalence was made, 47.2 percent of infections worldwide were of subtype C, 26.7 percent were of subtype A/CRF02_AG, 12.3 percent were of subtype B, 5.3 percent were of subtype D, 3.2 percent were of CRF_AE, and the remaining 5.3 percent were composed of other subtypes and CRFs. Most HIV-1 research is focused on subtype B; few laboratories focus on the other subtypes.

The genetic sequence of HIV-2 is only partially homologous to HIV-1 and more closely resembles that of SIV than HIV-1.

The Clinical Course of Infection

Infection with HIV-1 is associated with a progressive decrease of the $CD4^+$ T cell count and an increase in viral load. The stage of infection can be determined by measuring the patient's $CD4^+$ T cell count, and the level of HIV in the blood. HIV infection has basically four stages: incubation period, acute infection, latency stage and AIDS. The initial incubation period upon infection is asymptomatic and usually lasts between two and four weeks. The second stage, acute infection, which lasts an average of 28 days and can include symptoms such as fever, lymphadenopathy (swollen lymph nodes), pharyngitis (sore throat), rash, myalgia (muscle pain), malaise, and mouth and esophageal sores. The latency stage, which occurs third, shows few or no symptoms and can last anywhere from two weeks to twenty years and beyond. AIDS, the fourth and final stage of HIV infection shows as symptoms of various opportunistic infections.

Acute HIV Infection

The initial infection with HIV generally occurs after transfer of body fluids from an infected person to an uninfected one. The first stage of infection, the primary, or acute infection, is a period of rapid viral replication that immediately follows the individual's exposure to HIV leading to an abundance of virus in the peripheral blood with levels of HIV commonly approaching several million viruses per mL. This response is accompanied by a marked drop in the numbers of circulating $CD4^+$ T cells. This acute viremia is associated in virtually all patients with the activation of $CD8^+$ T cells, which kill HIV-infected cells, and subsequently with antibody production, or seroconversion. The $CD8^+$ T cell response is thought to be important in controlling virus levels, which peak and then decline, as the $CD4^+$ T cell counts rebound to around 800 cells per μL (the normal blood value is 1200 cells per μL). A good $CD8^+$

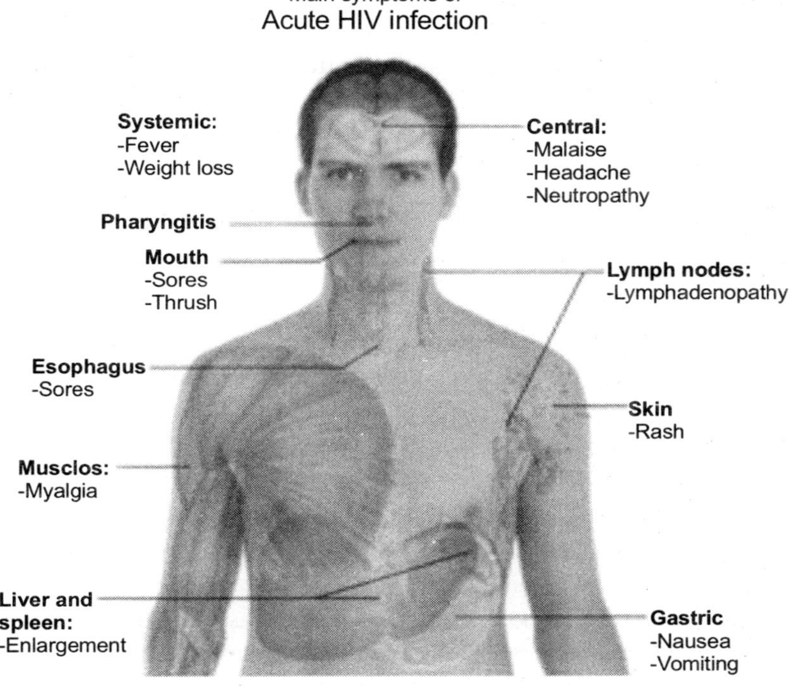

Fig. 7.22 : The Main Symptoms of Acute HIV Infection

T cell response has been linked to slower disease progression and a better prognosis, though it does not eliminate the virus. During this period (usually 2-4 weeks post-exposure) most individuals (80 to 90%) develop an influenza or mononucleosis-like illness called acute HIV infection, the most common symptoms of which may include fever, lymphadenopathy, pharyngitis, rash, myalgia, malaise, mouth and esophagal sores, and may also include, but less commonly, headache, nausea and vomiting, enlarged liver/spleen, weight loss, thrush, and neurological symptoms. Infected individuals may experience all, some, or none of these symptoms. The duration of symptoms varies, averaging 28 days and usually lasting at least a week. Because of the nonspecific nature of these symptoms, they are often not recognized as signs of HIV infection. Even if patients go to their doctors or a hospital, they will often be misdiagnosed as having one of the more common infectious diseases with the same symptoms. Consequently, these primary symptoms are not used to diagnose HIV infection as they do not develop in all cases and because many are caused by other more common diseases. However, recognizing the syndrome can be important because the patient is much more infectious during this period.

Latency Stage

A strong immune defense reduces the number of viral particles in the blood stream, marking the start of the infection's *clinical latency* stage. Clinical latency can vary between two weeks and 20 years. During this early phase of infection, HIV is active within lymphoid organs, where large amounts of virus become trapped in the follicular dendritic cells (FDC) network. The surrounding tissues that are rich in $CD4^+$ T cells may also become infected, and viral particles accumulate both in infected cells and as free virus. Individuals who are in this phase are still infectious. During this time, $CD4^+$ $CD45RO^+$ T cells carry most of the proviral load.

ELISA

An Enzyme-Linked Immunosorbent Assay (ELISA) for IgG antibodies to Thyroid Peroxidase (TPO)

Intended Use

The Diagnotic Automation, Inc. TPO IgG ELISA test system is intended for the qualitative and semi-quantitative detection of IgG-class antibody to thyroid peroxidase (TPO) in human serum. The test system is intended to be used as an aid in the diagnosis of thyroid diseases. This test is for *in vitro* diagnostic use only.

Significance and Background

Thyroid antibodies are a characteristic finding in patients with Hashimoto's and Graves' diseases (1). The presence of thyroid antibodies in the sera of 80% of patients with these two diseases led to the recommendation that some type of thyroid antibody testing be a feature of the work-up of any patient with a goiter (1). Although thyroid antibodies are predominantly associated with Hashimoto's or Graves' diseases, they may be found in the sera of patients with other diseases such as myxedma, granulomatous thyroiditis, nontoxic nodular goiter, and thyroid carcinoma (1). Thyroid antibodies are also found in most cases of lymphocytic thyroiditis in children (2), and rarely in patients with pernicious anemia and Sjögren's Syndrome (3-4).

Principle of the Elisa Assay

The Diagnotic Automation, Inc. TPO ELISA test system is designed to detect IgG class antibodies to TPO (thyroid peroxidase) antibodies in human sera. Wells of plastic microwell strips are sensitized by passive absorption with TPO antigen. The test procedure involves three incubation steps:

1. Test sera (properly diluted) are incubated in antigen coated microwells. Any antigen specific antibody in the sample will bind to the immobilized antigen. The plate is washed to remove unbound antibody and other serum components.
2. Peroxidase conjugated goat anti-human IgG (γ chain specific) is added to the wells and the plate is incubated. The conjugate will react with IgG antibody immobilized on the solid phase in step 1. The wells are washed to remove unreacted conjugate.
3. The microwells containing immobilized peroxidase conjugate are incubated with peroxidase substrate solution. Hydrolysis of the substrate by peroxidase produces a colour change. After a period of time the reaction is stopped and the colour intensity of the solution is measured photometrically. The colour intensity of the solution depends upon the antibody concentration in the original test sample.

Kit Components

Reactive Reagents

1. Twelve, 1 × 8-well strips coated with purified human thyroid peroxidase (>98% pure). The strips are packaged in a strip holder and sealed in an envelope with desiccant.
2. Horseradish peroxidase conjugated goat anti-human IgG. Ready to use. One 15 mL vial. Preservative added.
3. Human positive serum control. One 0.35mL vial. Preservative added.
4. Human negative serum control. One 0.35mL vial. Preservative added.
5. Calibrator A. One 0.35 mL vial. Preservative added.
6. Calibrator B. One 0.35 mL vial. Preservative added.
7. Calibrator C. One 0.35 mL vial. Preservative added.
8. Calibrator D. One 0.35 mL vial. Preservative added.
9. Sample diluent. One 30mL bottle containing Tween-20 bovine serum albumin, and phosphate-buffered-saline, pH 7.2 ± 0.2. Ready to use. (Green solution with a green cap; Product No. 005N).
10. TMB: One 15mL amber bottle (amber cap) containing 3,3',5,5'-tetramethylbenzidine (TMB). Ready to use. Contains DMSO < 15% (w).
11. Stop solution: One 15mL bottle (red cap) containing 1M H_2SO_4, 0.7M HCl. Ready to use.
12. Wash buffer (10X): One 100mL bottle containing phosphate-buffered-saline and Tween-20. 10X concentrate. Dilute 1 part concentrate and 9 parts deionized or distilled water. NOTE: 1X solution will have a pH of 7.2 ± 0.2. Preservative added.

Note: Kit also contains

1. Data labels: One label is adhered to the inside lid of the kit box and a second label is inside the kit box.
2. Package insert providing instructions for use.

Materials required but not provided

- ELISA microwell reader capable of reading at a wavelength of 450nm
- Pipettes capable of accurately delivering 10 to 200mL.
- Multichannel pipette capable of accurately delivering (50-200mL).
- Reagent reservoirs for multichannel pipettes.
- Wash bottle or microwell washing system
- Distilled or deionized water.
- One liter graduated cylinder.
- Serological pipettes.
- Disposable pipette tips.
- Paper towels.
- Laboratory timer to monitor incubation steps
- Disposal basin and disinfectant. (example: 10% household bleach, 0.5% sodium hypochlorite.)

Precautions

Safety Precautions

1. The Diagnostic Automation, Inc. Anti-TPO IgG ELISA kit is for *in vitro* diagnostic use. Normal precautions exercised in handling laboratory reagents should be followed. In case of contact with eyes, rinse immediately with plenty of water and seek medical advice. Wear suitable protective clothing, gloves, and eye/face protection. Do not breathe vapor. Dispose of waste observing all local, state, and federal laws.
2. The TPO plate wells do not contain viable organisms. However, the strips should be considered **POTENTIALLY BIOHAZARDOUS MATERIALS** and handled accordingly.
3. The controls are **POTENTIALLY BIOHAZARDOUS MATERIALS**. Source materials from which these products were derived were found negative for HIV-1 antigen, HbsAg. and for antibodies against HCV and HIV by approved test methods. However, since no test method can offer complete assurance that infectious agents are absent, these products should be handled at the Biosafety Level 2 as recommended for any potentially infectious human serum or blood specimen in the Centers for Disease Control/National Institutes of Health manual "Biosafety in Microbiological and Biomedical Laboratories":1993, pp 12-16, 3rd edition; and OSHA's Standard for Bloodborne Pathogens (5).

4. Controls, Sample Diluent, Conjugate, and Wash Buffer contain preservative (thimerosal, 0.04% (w/v)) which may be toxic if ingested.
5. The sample diluent contains sodium azide at a concentration of 0.1% (w/v). Sodium azide has been reported to form lead or copper azides in laboratory plumbing which may cause explosions on hammering. To prevent, rinse sink thoroughly with water after disposing of solution containing sodium azide.
6. Stop solution is TOXIC. Causes burns. Toxic by inhalation, in contact with skin and if swallowed. In case of accident or if you feel unwell, seek medical advice immediately.
7. TMB Substrate is HARMFUL. Irritating to eyes, respiratory system and skin.
8. Wash Buffer (10X) is an IRRITANT. Irritating to eyes, respiratory system and skin.

Procedure Precautions

1. For *in vitro* diagnostic use.
2. Wipe bottom of plate free of residual liquid and/or fingerprints which can alter optical density (OD) readings.
3. Dilution or adulteration of these reagents may generate erroneous results.
4. Reagents from other sources or manufacturers should not be used.
5. TMB solution should be colourless, very pale yellow, very pale green, or very pale blue when used. Contamination of the TMB substrate solution with conjugate or other oxidants will cause the solution to change colour prematurely. Do not use substrate solution if it is noticeably blue in colour. To help reduce the possibility of contamination, refer to Test Procedure, Substrate Incubation section to determine the amount of substrate solution to be used.
6. Never pipette by mouth. Avoid contact of reagents and patient specimens with skin and mucous membranes.
7. Avoid microbial contamination of reagents. Incorrect results may occur.
8. Cross contamination of reagents and/or samples could cause erroneous results
9. Reusable glassware must be washed and thoroughly rinsed free of all detergents.
10. Strict adherence to the specified time and temperature of incubations is essential for accurate results. **All reagents must be allowed to reach room temperature (20-25°C) before starting the assay.** Return unused reagents to refrigerated temperature immediately after use.
11. Improper washing could cause false positive or false negative results. Be sure to minimize the amount of any residual wash solution; (e.g., by blotting or aspiration), before adding conjugate or substrate. Do not allow the wells to dry out between incubations.
12. Avoid splashing or generation of aerosols.
13. Do not expose reagents to strong light during storage or incubation.
14. Allowing the multiwell strips and holder to equilibrate to room temperature prior to opening the protective envelope will protect the wells from condensation.

15. Wash solution should be collected in a disposal basin. Treat the waste solution with 10% household bleach (0.5% sodium hypochlorite). Avoid exposure of reagents to bleach fumes.
16. Caution: Liquid waste at acid pH should be neutralized before adding to bleach solution.
17. Do not use ELISA plate if the indicator strip on the desiccant pouch has turned from blue to pink.
18. Do not allow the conjugate to come in contact with containers or instruments which may have previously contained a solution utilizing sodium azide as a preservative. Residual amounts of sodium azide may destroy the conjugate's enzymatic activity.
19. Do not expose any of the reactive reagents to bleach-containing solutions or to any strong odors from bleach-containing solutions. Trace amounts of bleach (sodium hypochlorite) may destroy the biological activity of many of the reactive reagents within this kit.

Storage Conditions

1. Store the unopened kit at 2-8°C,
2. Coated microwell strips: Store at 2-8°C. Extra strips should be immediately resealed with desiccant and returned to storage at 2-8°C. Strips are stable for 60 days after the envelope has been opened and properly resealed and the indicator strip on the desiccant pouch remains blue.
3. Peroxidase conjugated goat anti-human antibody: Store at 2-8°C. DO NOT FREEZE.
4. Human controls: Store at 2-8°C.
5. TMB substrate solution: Store at 2-8°C.
6. Wash buffer: Store at 2-25°C. Diluted wash buffer is stable for 30 days at 2-8°C, or 7 days at 2-25°C.
7. Sample diluent: Store at 2-8°C.
8. Stop Solution: Store at 2-25°C

Specimen Collection

It is recommended that specimen collection be carried out in accordance with NCCLS document M29-A. No known test method can offer complete assurance that human blood samples will not transmit infection. Therefore, all blood derivatives should be considered potentially infectious.

Only freshly drawn and properly refrigerated sera obtained by approved aseptic venipuncture procedures should be used in this assay (6, 7). No anticoagulants or preservatives should be added. Avoid using hemolyzed, lipemic, or bacterially contaminated sera.

Store sample at room temperature for no longer than 8 hours. If testing is not performed within 8 hours, sera may be stored at 2-8°C for no longer than 48 hours. If delay in testing is anticipated, store test sera at -20°C or lower. Avoid multiple freeze/thaw cycles which may cause loss of antibody activity and give erroneous results.

Procedure

Preparation of Reagents

1. Wash Buffer: Dilute the 100mL of 10X concentrate with 900mL of distilled or deionized water. Mix thoroughly to dissolve any crystals that may be present.
2. Sample diluent, substrate solution, conjugate, and stop solution are ready to use.

Test Procedure

Set-up of the Assay

Step 1 **Reagents preparation:** Allow the reagents and samples to reach room temperature **(18-30°C)** for at least 15-30minutes. Check the Wash buffer concentrate for the presence of salt crystals. If crystals have formed in the solution, resolubilize by warming at 37°C until crystals dissolve. Dilute the stock Wash Buffer with distilled or deionized water. Use only clean vessels to dilute the buffer.

Step 2 **Numbering Wells:** Set the strips needed in strip-holder and number sufficient number of wells including three Negative controls **(e.g. B1, C1, D1)** two Positive controls **(e.g. E1, F1)** and one Blank **(e.g. A1,** neither samples nor HRP-Conjugate should be added into the Blank well). If the results will be determined by using dual wavelength plate reader, the requirement for use of Blank well could be omitted. Use only number of strips required for the test.

Step 3 Prepare a 1:21 dilution (e.g.: 10µL of serum + 200µL of Sample Diluent. NOTE: Shake well Before Use) of the Negative Control, Calibrator, Positive Control, and each patient Serum. The sample diluent will undergo a colour change confirming that the specimen has been combined with the diluent.

Example of controls/samples dispensing scheme

	1	2	3	4	5	6	7	...	...	12
A	Blank	3S								
B	Neg.	...								
C	Neg	...								
D	Neg.									
E	Pos.									
F	Pos.									
G	S1									
H	S2									

Step 4 To individual wells, add 100µL of each diluted control, calibrator and sample. Ensure that the samples are properly mixed. Use a different pipette tip for each sample.

Step 5 **Add 100µL of Sample Diluent to well A1 as a reagent blank. Check software and reader requirements for the correct reagent blank well configuration.**

Step 6 **Incubate the plate at room temperature (20-25°C) for 25±5 minutes.**

Step 7 **Washing:** At the end of the incubation, remove and discard the plate cover. Wash each well **5times** with Wash buffer. Each time, allow the microwells to soak for 30-60seconds. After the final washing cycle, turn down the plate onto blotting paper or clean towel, and tap it to remove any remaining liquids. Visually inspect the plate to ensure that no residual wash solution remains. Collect wash solution in a disposable basin and treat with 0.5% sodium hypochlorite (bleach) at the end of the days run.

Step 8 Add 100µL of the Conjugate to each well, including reagent blank well, at the same rate and in the same order as the specimens were added.

Step 9 Incubate the plate at room temperature (20-25°C) for 25 ± 5 min.

Step 10 Wash the microwells by following the procedure as described in step 7.

Step 11 Add 100µL of TMB to each well, including the reagent blank well, at the same rate and in the same order as the specimens were added.

Step 12 **Incubate** the plate at room temperature for 10 to 15 minutes.

Step 13 Stop the reaction by adding 50µL to Stop Solution to each well, including reagent blank well, at the same rate and in the same order as the TMB was added. Positive samples will turn from blue to yellow. After adding the Stop solution, tap the plate several times to ensure that the samples are thoroughly mixed.

Step 14 Set the microwell reader to read at a wavelength of 450nm and measure the optical density (OD) of each well against the reagent blank. The plate should be read within 30 minutes after the addition of the Stop Solution.

Quality Control

1. Each time the assay is run, the positive control, negative control, calibrator A, B, C, and D must be included. Additionally, a reagent blank must be included.
2. The specifications for the positive and negative controls are as follows:

Positive Control	Must be > 30 IU/mL
Negative Control	Must be < 15 IU/mL

   a. The OD of the negative control divided by the OD of the positive control should be < 0.9.

   b. If the control values are not within the above ranges, the test should be considered invalid and the test should be repeated.

4. The positive control is intended to monitor for substantial reagent failure and will not ensure precision at the assay cut-off.
5. Additional controls may be tested according to guidelines or requirements of local, state, and/or federal regulations or accrediting organizations.

Interpretation of Results

1. Calibrator

Based upon testing of normal sera, disease state sera, and the World Health Organization (WHO) international standard (66/387), a maximum normal IU/mL value has been determined by Diagnostic Automation Inc., and correlated to the calibrators. The calibrators will allow you to determine the unit value for each of the test samples evaluated. The unit values are determined for each lot of kit produced, and are printed on the data label included with each kit.

2. Conversion of Optical Density to IU/mL

Optical densities of the specimens are determined from the standard curve generated from the calibrators. A standard curve should be generated using the paired data points for each of the four calibrators (OD on the Y axis and corresponding IU/mL value on the X axis). Using the best fit, point to point curve, determine the IU/mL value for each of the specimens tested by extrapolation.

3. Quality Control

Refer to the specification sheet included with each kit. This sheet describes the lot specific specifications for each of the calibrators. If any of the calibrators are out of range, the results are considered invalid, and the patient results may not be reported.

Interpretation of Results

Using normal healthy individuals, disease-state specimens, and the WHO standard, Diagnostic Automation, Inc., has established the following guidelines for interpretation of patient results:

< 25 IU/mL	Negative
25-30 IU/mL	Equivocal*
> 30 IU/mL	Positive

* Specimens which are repeatedly equivocal should be tested using an alternate serological method.

Limitation of the Assay

1. A diagnosis should not be made solely on the basis of the ELISA result. Test results for anti-thyroglobulin should be interpreted in conjunction with the clinical evaluation and the results of other diagnostic procedures.
2. Reproducible results with an ELISA system require careful pipetting, strict adherence to incubation periods and temperatures requirements, as well as thorough washing of the test wells and thorough mixing of all solutions.

3. Hemolytic, icteric, or lipemic samples may interfere with this ELISA. Use of these types of specimens should be avoided.

Expected Results

The clinical investigation included 80 random normal donor specimens. With respect to this group, four (5%) were positive and 76 (95%) were negative.

Performance Characteristics

Comparative Study

An in-house comparative study was performed to determine the equivalence of the Diagnostic Automation, Inc. TPO IgG ELISA test system to another commercially available TPO IgG ELISA test system. Performance was evaluated using 248 specimens and the results have been summarized in Table below.

Summary of the comparative investigation, DAI ELISA test system versus a commercially available ELISA test system

		DAI TPO IgG ELISA Test System			
		Positive	**Equivocal***	**Negative**	**Totals**
Commercial	Positive	104	12	7	123
TPO IgG ELISA	Negative	7	1	117	125
Test System	Totals	111	13	124	248

*Equivocal specimens were excluded from calculations below.

Relative* Sensitivity = 104/111 = 93.7% - 95% Confidence Interval of 89.1 to 98.2%
Relative Specificity = 117/124 = 94.4% - 95% Confidence Interval of 90.3 to 98.4%
Relative Agreement = 221/235 = 94.0% - 95% Confidence Interval of 91.0 to 97.1%

* Please be advised that the term 'relative' refers to the comparison of this assay's results to that of a similar assay. There was not an attempt to correlate the assay's results with disease presence or absence. No judgement can be made on the comparison assay's accuracy to predict disease.

Reproducibility

A study was conducted in-house to determine reproducibility. Briefly, six specimens were tested; two negative, two strong positive, and two positive specimens that were near the assay cut off. Each specimen was tested in eight replicate wells on each day, for a total of three days. The resulting data was used to determine both intra and inter assay reproducibility. A summary of the study appears in Table.

Results of three day reproducibility study

	Intra-Assay				Inter-Assay			
	Day One		Day Two		Day Three		Three Days Combined	
	Mean (IU/mL)	% CV	Mean (IU/mL)	% CV	Mean (IU/mL)	% CV	Mean (IU/mL)	% CV
Sample 1	77	3.1	55	3.9	79	5.9	69	15
Sample 2	112	5.8	88	16.4	109	3.0	103	12.6
Sample 3	32	9.2	37	3.8	35	11.8	34	10.7
Sample 4	34	4.9	40	1.2	35	12.2	36	12.1
Sample 5	13	2.4	12	1.9	9	10.4	11	18.4
Sample 6	9	6.0	6	1.3	5	17.6	6	30.6

Cross Reactivity

To evaluate the system for potential cross reactivity to other autoantibodies, seventeen specimens which were positive for antibodies to nuclear antigens (ANA) on HEp-2 cells were tested. Of the specimens tested, none were positive on the TPO IgG ELISA. This study indicates that the potential for interference due to cross reactive autoantibodies is unlikely.

Correlation to the World Health Standard (NIBSC 66/387)

The World Health Standard (NIBSC 66/387) was tested on the DAI assay to determine the correlation of the result obtained to the expected result. The data from this study in presented in Table below.

Correlation to the World Health Standard; (NIBSC 66/387)

Dilution of the Standard	IU/mL asTested	OD(450 nm)	Result:(IU/mL)
neat	1000	> 3.000	278
1:2	500	> 3.000	275
1:4	250	2.710	243
1:8	125	1.740	144
1:16	62	0.890	61
1:32	32	0.457	30
1:64	16	0.202	14
1:128	8	0.112	8

CHROMOSOME 3

Chromosome 3 is one of the 23 pairs of chromosomes in humans. People normally have two copies of this chromosome. Chromosome 3 spans almost 200 million base pairs (the building material of DNA) and represents about 6.5 percent of the total DNA in cells.

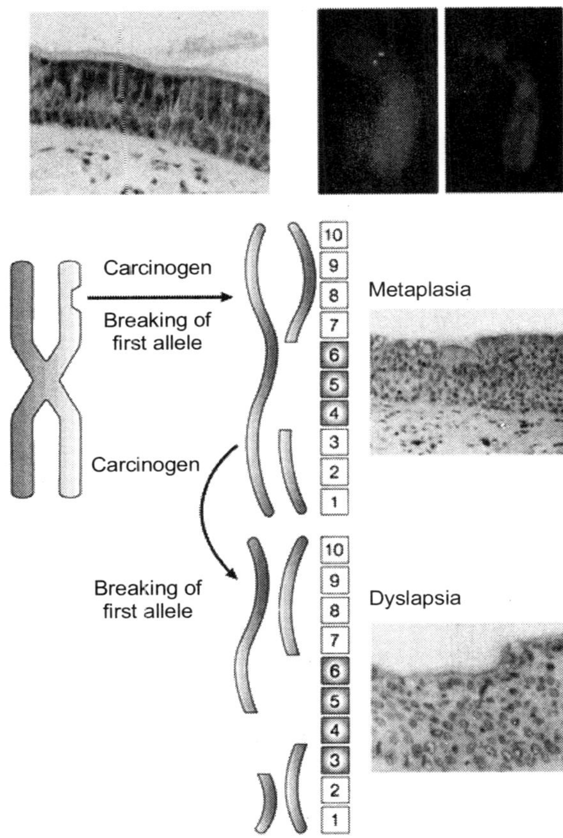

Fig. 7.23 : The Chromosome 3

Identifying genes on each chromosome is an active area of genetic research. Because researchers use different approaches to predict the number of genes on each chromosome, the estimated number of genes varies. Chromosome 3 likely contains between 1,100 and 1,500 genes.

Genes

The following are some of the genes located on chromosome 3:

- ALAS1: aminolevulinate, delta-, synthase 1
- BTD: biotinidase

- CCR5: chemokine (C-C motif) receptor 5
- COL7A1: Collagen, type VII, alpha 1 (epidermolysis bullosa, dystrophic, dominant and recessive)
- CPOX: coproporphyrinogen oxidase (coproporphyria, harderoporphyria)
- HGD: homogentisate 1,2-dioxygenase (homogentisate oxidase)
- MCCC1: methylcrotonoyl-Coenzyme A carboxylase 1 (alpha)
- MITF: microphthalmia-associated transcription factor
- MLH1: mutL homolog 1, colon cancer, nonpolyposis type 2 (E. coli)
- PCCB: propionyl Coenzyme A carboxylase, beta polypeptide
- PDCD10: programmed cell death 10
- PIK3CA: phosphoinositide-3-kinase, catalytic, alpha polypeptide
- PTHR1: parathyroid hormone receptor 1
- RAB7: RAB7, member RAS oncogene family
- SCN5A: sodium channel, voltage-gated, type V, alpha (long QT syndrome 3)
- SLC25A20: solute carrier family 25 (carnitine/acylcarnitine translocase), member 20
- SOX2: transcription factor
- TMIE: transmembrane inner ear
- USH3A: Usher syndrome 3A
- VHL: von Hippel-Lindau tumor suppressor
- ZNF9: zinc finger protein 9 (a cellular retroviral nucleic acid binding protein)

Aimatochemistry

Introduction

Aima* the Greek word for blood, considered as the river of life. This chapter is therefore all about the blood and all about its constitution.

The unicellular organisms such as amœba obtain food and dispose of waste products through processes of diffusion which the aqueous media in which the organisms exist, If the surrounding fluid is of the proper composition so that the process of diffusion can maintain a rather definite composition of the fluids within the (homeostasis) then the organism thrives. Changes in the composition of the surrounding fluid due to the accumulation of waste products the depletion of food substances or other causes the prevention and the organism ceases to function.

With the evolution of larger multi-cellular and multi-organ forms of animal life the composition of cellular fluids could not be maintained through the processes of simple diffusion alone. To meet this situation the circulatory tissue, blood was evolved.

In the higher animal form the tissues cells are bathed in interstitial fluid form which essential substances are withdrawn and to which waste products are added by the process of diffusion. The composition of the interstitial fluid in turn is maintained within proper limits by diffusion between it and the circulating blood through the capillary walls. The composition of blood is maintained within physiological limits by the provision of special mechanism to add foods to the blood (gastrointestinal

* Aima of Greek, has become hæme in English due to the wrong phonetics so this subject is wrongly pronounced as Hæmatology or Hematology, but fortunately author has followed the original Greek phonetics and so it is not Hæmatology but Aimatology (before consulting with author please see the Greek dictionary of Oxford by J.T. Pring page number 4.)

system, lungs) to remove waste products from it (kidneys, lungs, and intestines, skin) and to produce the characteristic blood cells, proteins, hormones, and other constituents not directly obtainable from the food supply.

Thus the higher animal forms are produced with the means for maintaining their cellular environment relatively independent of external influences.

Characteristics of Blood

The total blood mass of the healthy adults makes up about 6 to 8% of the body weight, is higher for men than for women and is subject to considerable individual variation. The normal blood volume for woman is 65ml and that of man is 79 ml per kg of the body weight.

The blood volume generally varies from 2000 to 2200ml per sqm of the body surface and is more nearly proportional to surface area that to body weight. The composition of blood as of any other tissue is tremendously complex. The composition of blood is shown below:

- The Average adult has a blood volume of approximately 5 litres, which composes about 8% of the body's weight.
- Osmolality of blood is 275-295 milliosmoles per kg.
- Plasma constitutes approximately 55% of blood's volume.
- Plasma Composition:
  - 91.5% Water
  - 7% Protein
  - 1.5% Inorganic Salts
  - Sodium 135-146 mM
  - Potassium 3.5-5.2 mM
  - Calcium 2.1-2.7 mM
  - Carbonate 23-31 mM
  - Phosphate 0.7-1.4 mM
- 1.1% organic substances
- It is estimated that plasma may contain as many as 40,000 different proteins from about 500 gene products. Approximately 1,000 proteins have been detected.
- Plasma contains 50-70 mg of protein per ml.
- Approx. 70% Albumin (35-50 mg/ml)
- Approx. 10% IgG (5-7 mg/ml)

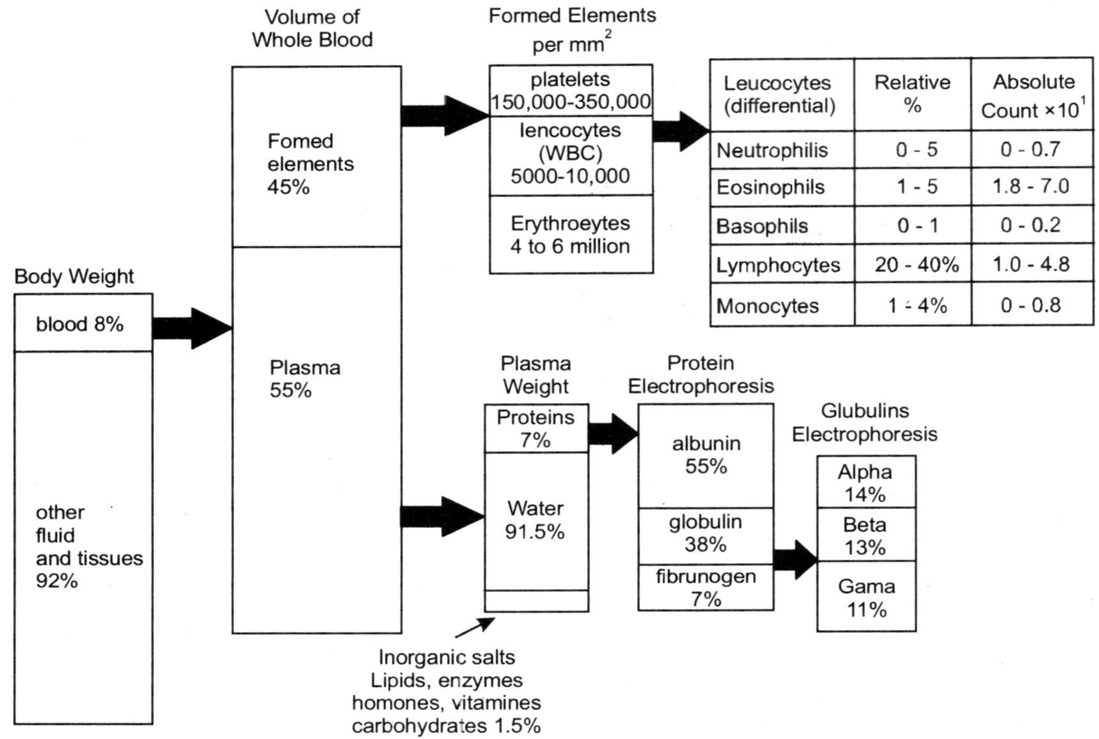

Fig. 8.1 : The Composition of Blood

List of Human Blood Components and Their Concentrations

Compound	Comments	In whole blood (g/cm³)	In plasma or serum (g/cm³)
1	2	3	4
Water	Solvent	0.81-0.86	0.93-0.95
Acetoacetate	Produced in liver	8-40 × 10-7	$4\text{-}43 \times 10^{-7}$
Acetone	product of bodyfat breakdown	-	$3\text{-}20 \times 10^{-6}$
Acetylcholine	Neurotransmitter	-	$6.6\text{-}8.2 \times 10^{-8}$
Adenosine triphosphate	Energy storage	-	-
"	total	$3.1\text{-}5.7 \times 10^{-4}$	-
"	phosphorus	$5\text{-}10 \times 10^{-5}$	-
Adrenocorticotrophic hormone			
Stimulates the adrenal cortex	-	$2.5\text{-}12 \times 10^{-11}$	
"	@ 6AM, mean	-	5.5×10^{-11}

...*(Contd.)*

1	2	3	4
Adrenocorticotrophic hormone	@ 6AM, maximum	-	12×10^{-11}
"	@ 6PM, mean	-	3.5×10^{-11}
"	@ 6PM, maximum	-	$<7.5 \times 10^{-11}$
Alanine	Amino acid	$2.7\text{-}5.5 \times 10^{-5}$	$2.4\text{-}7.6 \times 10^{-5}$
Albumin	Blood plasma protein	-	$3.5\text{-}5.2 \times 10^{-2}$
Aluminum	-	$1\text{-}40 \times 10^{-8}$	$1\text{-}88 \times 10^{-8}$
Aldosterone	Regulates electrolyte balance	-	-
"	supine	-	$3\text{-}10 \times 10^{-11}$
"	standing, male	-	$6\text{-}22 \times 10^{-11}$
"	standing, female	-	$5\text{-}30 \times 10^{-11}$
Amino acids	Protein building blocks	-	-
"	total	$3.8\text{-}5.3 \times 10^{-4}$	-
"	nitrogen	$4.6\text{-}6.8 \times 10^{-5}$	$3.0\text{-}5.5 \times 10^{-5}$
alpha-Aminobutyric acid	-	$1\text{-}2 \times 10^{-6}$	$1\text{-}2 \times 10^{-6}$
d-Aminolevulinic acid	-	-	$1.5\text{-}2.3 \times 10^{-7}$
Ammonia nitrogen	-	$1\text{-}2 \times 10^{-6}$	$1.0\text{-}4.9 \times 10^{-7}$
cAMP	Intracellular signal transduction molecule	-	-
"	male	-	$5.6\text{-}10.9 \times 10^{-9}$
"	female	-	$3.6\text{-}8.9 \times 10^{-9}$
Androstenedione	Steroid hormone	-	
"	male >18 yrs	-	$2\text{-}30 \times 10^{-10}$
"	female >18 yrs	-	$8\text{-}30 \times 10^{-10}$
Androsterone	Steroid hormone	-	1.5×10^{-7}
Angiotensin I	Angiotensin II precursor	-	$1.1\text{-}8.8 \times 10^{-11}$
Angiotensin II	Vasoconstrictor	-	$1.2\text{-}3.6 \times 10^{-11}$
Alpha 1-antitrypsin	Serine protease inhibitor	-	$7.8\text{-}20 \times 10^{-4}$
Arginine	Amino acid	$6\text{-}17 \times 10^{-6}$	$1.3\text{-}3.6 \times 10^{-5}$
Arsenic	-	-	-
"	normal range	$2\text{-}62 \times 10^{-9}$	-
"	chronic poisoning	$100\text{-}500 \times 10^{-9}$	-
"	acute poisoning	$600\text{-}9300 \times 10^{-9}$	-
Ascorbic acid (Vitamin C)	Important vitamin	$1\text{-}15 \times 10^{-6}$	$6\text{-}20 \times 10^{-6}$
Aspartic acid	Amino acid	-	$0\text{-}3 \times 10^{-6}$
Aspartic acid (in WBCs)	-	$2.5\text{-}4.0 \times 10^{-4}$	$9\text{-}12 \times 10^{-6}$

...(Contd.)

1	2	3	4
Bicarbonate	Buffer in blood	-	$5\text{-}5.7 \times 10^{-4}$
Bile acids	Digestive function, bilirubin excretion	$2\text{-}30 \times 10^{-6}$	$3\text{-}30 \times 10^{-6}$
Bilirubin	Hæmoglobin metabolite	$2\text{-}14 \times 10^{-6}$	$1\text{-}10 \times 10^{-6}$
Biotin (Vitamin H)	Gluconeogenesis,metabolize leucine,fatty acid synthesis	$7\text{-}17 \times 10^{-9}$	$9\text{-}16 \times 10^{-9}$
Blood Urea Nitrogen (BUN)	-	-	$8\text{-}23 \times 10^{-5}$
Bradykinin	-	-	7×10^{-11}
Bromide	-	-	$7\text{-}10 \times 10^{-9}$
Cadmium	-	-	-
"	normal	$1\text{-}5 \times 10^{-9}$	-
"	toxic	$0.1\text{-}3 \times 10^{-6}$	-
Calciferol (Vitamin D2)	Maintain calcium and phosphorus levels	-	$1.7\text{-}4.1 \times 10^{-8}$
Calcitonin (CT)	Hormone	-	$<1.0 \times 10^{-10}$
Calcium	Bones, Ca^{2+}	-	-
"	ionized	$4.48\text{-}4.92 \times 10^{-5}$	$4.25\text{-}5.25 \times 10^{-5}$
"	total	-	$8.4\text{-}11.5 \times 10^{-5}$
Carbon dioxide	Respiratory gas	-	-
"	arterial	$8.8\text{-}10.8 \times 10^{-4}$	$3.0\text{-}7.9 \times 10^{-5}$
"	venous	$9.8\text{-}11.8 \times 10^{-4}$	$3.3\text{-}8.3 \times 10^{-5}$
Carboxyhæmoglobin(as HbCO)	-	-	-
"	nonsmokers	0.5-1.5% total Hb	-
"	smokers, 1-2 packs/day	4-5% total Hb	-
"	smokers, 2 packs/day	8-9% total Hb	-
"	toxic	>20% total Hb	-
"	lethal	>50% total Hb	-
Carcinoembryonic antigen	-	-	$<2.5 \times 10^{-9}$
beta-Carotene	Vitamin A dimer	-	$3\text{-}25 \times 10^{-7}$
Carotenoids	Antioxidant	$2.4\text{-}23.1 \times 10^{-7}$	-
Cephalin	-	$3\text{-}11.5 \times 10^{-4}$	$0\text{-}1 \times 10^{-4}$
Ceruloplasmin	-	-	$1.5\text{-}6 \times 10^{-4}$
Chloride, as NaCl	-	$4.5\text{-}5 \times 10^{-3}$	$3.5\text{-}3.8 \times 10^{-3}$
Cholecalciferol (Vitamin D3)	-	-	-
"	1, 25-dihydroxy	-	$2.5\text{-}4.5 \times 10^{-11}$
"	24,25-dihydroxy	-	1.5×10^{-9}
"	25-hydroxy	-	$1.4\text{-}8 \times 10^{-8}$

...(Contd.)

1	2	3	4
Cholecystokinin (pancreozymin)	Stimulates fat and protein digestion	-	6.04×10^{-11}
Cholesterol	Steroid lipid	-	-
"	LDLC	-	$0.5\text{-}2.0 \times 10^{-3}$
"	HDLC	-	$2.9\text{-}9.0 \times 10^{-4}$
"	total	$1.15\text{-}2.25 \times 10^{-3}$	$1.2\text{-}2 \times 10^{-3}$
Choline, total	-	$1.1\text{-}3.1 \times 10^{-4}$	$3.6\text{-}3.5 \times 10^{-4}$
Chorionic gonadotropin	Progesterone secretion during pregnancy	-	-
"	Menstrual	-	$0\text{-}3 \times 10^{-11}$
"	Pregnancy, 1st trimester	-	$5\text{-}3300 \times 10^{-10}$
"	Pregnancy, 2nd trimester	-	$20\text{-}1000 \times 10^{-10}$
"	Pregnancy, 3rd trimester	-	$20\text{-}50 \times 10^{-10}$
"	Menopausal	-	$3\text{-}30 \times 10^{-11}$
Citric acid	-	$1.3\text{-}2.5 \times 10^{-5}$	$1.6\text{-}3.2 \times 10^{-5}$
Citrulline	-	-	$2\text{-}10 \times 10^{-6}$
Coagulation Factors	-	-	-
"	Fibrinogen	$1.2\text{-}1.6 \times 10^{-3}$	$2\text{-}4 \times 10^{-3}$
"	Prothrombin	-	1×10^{-4}
"	Tissue thromboplastin	-	1×10^{-6}
"	Proaccelerin	-	$5\text{-}12 \times 10^{-6}$
"	Proconvertin	-	1×10^{-6}
"	Antihemophilic factor	-	1×10^{-7}
"	Christmas factor	-	4×10^{-6}
"	Stuart factor	-	5×10^{-6}
"	Plasma thrmb. anteced.	-	4×10^{-6}
"	Hageman factor	-	2.9×10^{-5}
"	Fibrin stabilizing factor	-	1×10^{-5}
"	Fibrin split products	-	$<1 \times 10^{-5}$
"	Fletcher factor	-	5×10^{-5}
"	Fitzgerald factor	-	7×10^{-5}
"	von Willebrand factor	-	7×10^{-6}
Cobalamin (Vitamin B12)	Needed for nerve cells, red blood cells, and to make DNA	$6\text{-}14 \times 10^{-10}$	$1\text{-}10 \times 10^{-10}$
Cocarboxylase	-	-	$7\text{-}9 \times 10^{-8}$

...(Contd.)

1	2	3	4
Complement system	-	-	-
"	C1q	-	5.8-7.2 × 10^{-5}
"	C1r	-	2.5-3.8 × 10^{-5}
"	C1s (C1 esterase)	-	2.5-3.8 × 10^{-5}
"	C2	-	2.2-3.4 × 10^{-5}
"	C3(b1C-globulin)	-	8-15.5 × 10^{-4}
"	factor B (C3 proactivator)		
"	2-4.5 × 10^{-4}		
"	C4 (b1E-globulin)	-	1.3-3.7 × 10^{-4}
"	C4 binding protein	-	1.8-3.2 × 10^{-4}
"	C5 (b1F-globulin)	-	5.1-7.7 × 10^{-5}
"	C6	-	4.8-6.4 × 10^{-5}
"	C7	-	4.9-7 × 10^{-5}
"	C8	-	4.3-6.3 × 10^{-5}
"	C9	-	4.7-6.9 × 10^{-5}
"	Properdin		
-	2.4-3.2 × 10^{-5}		
Compound S	-	-	1-3 × 10^{-9}
Copper	-	9-15 × 10^{-7}	-
"	male	-	7-14 × 10^{-7}
"	female	-	8-15.5 × 10^{-7}
Corticosteroids	Steroid hormones	-	1-4 × 10^{-6}
Corticosterone	-	-	4-20 × 10^{-9}
Cortisol	Inhibits CRH secretion	-	3-23 × 10^{-8}
"	8 AM	-	6-23 × 10^{-8}
"	4 PM	-	3-15 × 10^{-8}
"	10 PM	-	~50% of 8 AM value
C-peptide	-	-	-
"	fasting	-	0.5-2.0 × 10^{-9}
"	maximum	-	4 × 10^{-9}
C-reactive protein	Plasma protein	-	6.8-820 × 10^{-8}
Creatine	Assists muscle cell	-	-
	energy supply		
"	male	-	1.7-5.0 × 10^{-6}
"	female	-	3.5-9.3 × 10^{-6}

...(Contd.)

1	2	3	4
Creatinine	-	-	-
"	male	-	0.8-1.5 × 10^{-5}
"	female	-	0.7-1.2 × 10^{-5}
Cyanide	-	-	-
"	nonsmokers	-	4 × 10^{-9}
"	smokers	-	6 × 10^{-9}
"	nitroprusside therapy	-	10-60 × 10^{-9}
"	toxic	-	>100 × 10^{-9}
"	lethal	-	>1000 × 10^{-9}
Cysteine	Amino acid	6-12 × 10^{-6}	1.8-5 × 10^{-5}
Dehydroepiandrosterone(DHEA)	Steroid hormone	-	-
"	aged 1-4 yrs	-	0.2-0.4 × 10^{-9}
"	aged 4-8 yrs	-	0.1-1.9 × 10^{-9}
"	aged 8-10 yrs	-	0.2-2.9 × 10^{-9}
"	aged 10-12 yrs	-	0.5-9.2 × 10^{-9}
"	aged 12-14 yrs	-	0.9-20 × 10^{-9}
"	aged 14-16 yrs	-	2.5-20 × 10^{-9}
"	male	-	0.8-10 × 10^{-9}
"	female, premenopausal	-	2.0-15 × 10^{-9}
DHEA sulfate	male	-	1.99-3.34 × 10^{-6}
DHEA sulfate	female	-	-
"	newborn	-	1.67-3.64 × 10^{-6}
"	pre-pubertal children	-	1.0-6.0 × 10^{-7}
"	premenopausal	-	8.2-33.8 × 10^{-7}
"	pregnancy	-	2.3-11.7 × 10^{-7}
"	postmenopausal	-	1.1-6.1 × 10^{-7}
11-Deoxycortisol	-	-	1-7 × 10^{-8}
Dihydrotestosterone (DHT)	-	-	-
"	male	-	3-8 × 10^{-9}
"	female	-	1-10 × 10^{-10}
Diphosphoglycerate (phosphate)	-	8-16 × 10^{-5}	-
DNA	The molecule of heredity	-	0-1.6 × 10^{-5}
Dopamine	Neurotransmitter	-	<1.36 × 10^{-10}
Enzymes, total	-	-	<6 × 10^{-5}
Epidermal growth factor (EGF)	-	-	<1 × 10^{-11}

...(Contd.)

1	2	3	4
Epinephrine	Neurotransmitter	-	-
"	after 15 min rest	-	3.1-9.5 × 10^{-11}
"	when emitted	3.8 × 10^{-9}	2-2.5 × 10^{-9}
Ergothioneine	-	1-20 × 10^{-5}	-
Erythrocytes (#/cm³)	-	-	-
"	adult male, avg. (range)	5.2 (4.6-6.2) × 109	
"	adult female, avg. (range)	4.6 (4.2-5.4) × 109	
"	children, varies with age	4.5-5.1 × 109	
"	reticulocytes	25-75 × 106	
Erythropoietin	-	-	-
"	adult, normal	-	0.5-2.5 × 10^{-10}
"	pregnant	-	2.7-6.2 × 10^{-10}
"	hypoxia or anemia	-	0.8-8.0 × 10^{-8}
Estradiol (E2)	-	-	-
"	male	-	8-36 × 10^{-12}
"	female, follicular(days 1-10)	-	1-9 × 10^{-11}
"	female, mean	-	5 × 10^{-11}
"	female, pre-fertile (days 10-12)	-	10-15 × 10^{-11}
"	female, fertile (days 12-14)	-	35-60 × 10^{-11}
"	female, luteal (days 15-28)	-	20-40 × 10^{-11}
"	female, pregnancy	-	3-70 × 10^{-7}
"	female, postmenopausal	-	1-3 × 10^{-11}
Estriol (E3)	-	-	-
"	nonpregnant	-	<2 × 10^{-9}
"	pregnancy, weeks 22-30	-	3-5 × 10^{-9}
"	pregnancy, weeks 32-37	-	6-11 × 10^{-9}
"	pregnancy, weeks 38-41	-	25-170 × 10^{-9}
Estrogen	-	-	-
"	male	-	4-11.5 × 10^{-11}
"	female, prepubertal	-	<4 × 10^{-11}
"	female, 1-10 days	-	6.1-39.4 × 10^{-11}
"	female, 11-20 days	-	12.2-43.7 × 10^{-11}
"	female, 21-30 days	-	15.6-35 × 10^{-11}
"	female, postmenopausal	-	<4 × 10^{-11}

...(Contd.)

1	2	3	4
Estrone (E1)	-	-	-
"	male	-	$2.9\text{-}17 \times 10^{-11}$
"	female, follicular	-	$2\text{-}15 \times 10^{-11}$
"	female, 1-10 days of cycle	-	$4.3\text{-}18 \times 10^{-11}$
"	female, 11-20 days of cycle	-	$7.5\text{-}19.6 \times 10^{-11}$
"	female, 20-29 days of cycle	-	$13.1\text{-}20.1 \times 10^{-11}$
"	pregnancy, weeks 22-30	-	$3\text{-}5 \times 10^{-9}$
"	pregnancy, weeks 32-37	-	$5\text{-}6 \times 10^{-9}$
"	pregnancy, weeks 38-41	-	$7\text{-}10 \times 10^{-9}$
Ethanol	-	-	-
"	social high	-	0.5×10^{-3}
"	reduced coordination	-	0.8×10^{-3}
"	depression of CNS	-	$>1 \times 10^{-3}$
"	confusion, falling down	-	2.0×10^{-3}
"	loss of consciousness	-	3.0×10^{-3}
"	coma, death	-	$>4 \times 10^{-3}$
Fat, neutral	see Triglycerides	-	-
Fatty acids, nonesterified (free)		-	$8\text{-}25 \times 10^{-5}$
Fatty acids, esterified	-	$2.5\text{-}3.9 \times 10^{-3}$	$7\text{-}20 \times 10^{-5}$
Fatty acids, total	-	-	$1.9\text{-}4.5 \times 10^{-3}$
Ferritin		-	
"	male	-	$1.5\text{-}30 \times 10^{-8}$
"	female	-	$0.9\text{-}18 \times 10^{-8}$
alpha-1-Fetoprotein	-	-	$0\text{-}2 \times 10^{-8}$
Flavine adenine dinucleotide	-	-	$8\text{-}12 \times 10^{-8}$
Fluoride	-	$1\text{-}4.5 \times 10^{-7}$	$1\text{-}4.5 \times 10^{-7}$
Folate	-	-	$2.2\text{-}17.3 \times 10^{-9}$
"	in erythrocyte	-	$1.67\text{-}7.07 \times 10^{-7}$
Folic acid		$2.3\text{-}5.2 \times 10^{-8}$	$1.6\text{-}2 \times 10^{-8}$
Fructose		$0\text{-}5 \times 10^{-5}$	$7\text{-}8 \times 10^{-5}$
Furosemide glucuronide		-	$1\text{-}400 \times 10^{-6}$
Galactose (children)		-	$<2 \times 10^{-4}$
Gastric inhibitory peptide (GIP)		-	$<1.25\text{-}4.0 \times 10^{-10}$
Gastrin		-	
"	mean	-	7×10^{-11}
"	maximum	-	$<20 \times 10^{-11}$

...(Contd.)

1	2	3	4
Globulin	-	-	total $2.2\text{-}4 \times 10^{-2}$
alpha-1-Globulin		-	$1\text{-}4 \times 10^{-3}$
alpha-2-Globulin		-	$4\text{-}10 \times 10^{-3}$
beta globulin		-	$5\text{-}12 \times 10^{-3}$
gamma globulin		-	$6\text{-}17 \times 10^{-3}$
Glucagon		-	
"	range	-	$5\text{-}15 \times 10^{-11}$
"	mean	-	$7.1\text{-}7.9 \times 10^{-11}$
Glucosamine	-	-	-
"	fetus	$4\text{-}6 \times 10^{-4}$	$4.2\text{-}5.5 \times 10^{-4}$
"	child	$5\text{-}7 \times 10^{-4}$	$5.2\text{-}6.9 \times 10^{-4}$
"	adult	$6\text{-}8 \times 10^{-4}$	$6.1\text{-}8.2 \times 10^{-4}$
"	aged	$7\text{-}9 \times 10^{-4}$	$7.0\text{-}8.9 \times 10^{-4}$
Glucose		-	
"	newborn	$2\text{-}3 \times 10^{-4}$	-
"	adult	$6.5\text{-}9.5 \times 10^{-4}$	$7\text{-}10.5 \times 10^{-4}$
"	diabetic	$14\text{-}120 \times 10^{-4}$	-
Glucuronic acid		$4.1\text{-}9.3 \times 10^{-5}$	$8\text{-}11 \times 10^{-6}$
Glutamic acid	-	$2\text{-}28 \times 10^{-6}$	
Glutamine		-	$4.6\text{-}10.6 \times 10^{-5}$
Glutathione, reduced		$2.5\text{-}4.1 \times 10^{-4}$	0
Glycerol, free		-	$2.9\text{-}17.2 \times 10^{-6}$
Glycine		$1.7\text{-}2.3 \times 10^{-5}$	$8\text{-}54 \times 10^{-6}$
Glycogen		$1.2\text{-}16.2 \times 10^{-5}$	0
Glycoprotein, acid		-	$4\text{-}15 \times 10^{-4}$
cGMP		-	$0.6\text{-}4.4 \times 10^{-9}$
Gonadotropic releasing hormone		-	$1\text{-}80 \times 10^{-12}$
Guanidine		$1.8\text{-}2.3 \times 10^{-6}$	-
Haptoglobin		-	$3\text{-}22 \times 10^{-4}$
Hæmoglobin		$1.2\text{-}1.75 \times 10^{-1}$	$1\text{-}4 \times 10^{-5}$
"	newborn	$1.65\text{-}1.95 \times 10^{-1}$	-
"	children, varies with age	$1.12\text{-}1.65 \times 10^{-1}$	-
"	adult, male	$1.4\text{-}1.8 \times 10^{-1}$	-
"	adult, female	$1.2\text{-}1.6 \times 10^{-1}$	-
"	inside erythrocyte	$\sim 3.3 \times 10^{-1}$	-
"	per red blood cell	27-32 picograms	-

...(Contd.)

1	2	3	4
Hexosephosphate P		$1.4\text{-}5 \times 10^{-5}$	$0\text{-}2 \times 10^{-6}$
Histamine		$6.7\text{-}8.6 \times 10^{-8}$	-
Histidine		$9\text{-}17 \times 10^{-6}$	$1.1\text{-}3.8 \times 10^{-5}$
Hydrogen ion(pH 7.4)		4×10^{-11}	-
beta-Hydroxybutyric acid		$1\text{-}6 \times 10^{-6}$	$1\text{-}9 \times 10^{-6}$
17-Hydroxycorticosteroids		$4\text{-}10 \times 10^{-8}$	-
17-Hydroxyprogesterone	-	-	-
"	male	-	$20\text{-}250 \times 10^{-11}$
"	female, follicular	-	$20\text{-}80 \times 10^{-11}$
"	female, luteal	-	$80\text{-}300 \times 10^{-11}$
"	female, postmenopausal	-	$4\text{-}50 \times 10^{-11}$
"	female, child	-	$20\text{-}140 \times 10^{-11}$
Antibodies	-	-	-
"	Immunoglobulin A (IgA)	-	$5\text{-}39 \times 10^{-4}$
"	Immunoglobulin D (IgD)	-	$0.5\text{-}8.0 \times 10^{-5}$
"	Immunoglobulin G (IgG)	-	$5.0\text{-}19 \times 10^{-3}$
"	Immunoglobulin M (IgM)	-	$3.0\text{-}30 \times 10^{-4}$
"	Immunoglobulin E (IgE)	-	$<5 \times 10^{-7}$
Indican	-	-	$8\text{-}50 \times 10^{-7}$
Inositol	-	-	$3\text{-}7 \times 10^{-6}$
Insulin	-	-	$2.0\text{-}8.4 \times 10^{-10}$
Insulin-like growth factor	-	-	$9.9\text{-}50 \times 10^{-8}$
Iodine, total	-	$2.4\text{-}3.2 \times 10^{-8}$	$4.5\text{-}14.5 \times 10^{-8}$
Iron, adult	-	$4\text{-}6 \times 10^{-4}$	$6\text{-}18 \times 10^{-7}$
Isoleucine	-	$9\text{-}15 \times 10^{-6}$	$1.2\text{-}4.2 \times 10^{-5}$
Ketone bodies	-	$2.3\text{-}10 \times 10^{-6}$	$1.5\text{-}30 \times 10^{-6}$
alpha-Ketonic acids, adult	-	$1\text{-}30 \times 10^{-6}$	-
L-Lactate	-	-	-
"	arterial	$<11.3 \times 10^{-5}$	$4.5\text{-}14.4 \times 10^{-5}$
"	venous	$8.1\text{-}15.3 \times 10^{-5}$	$4.5\text{-}19.8 \times 10^{-5}$
Lead	-	-	-
"	normal	$1\text{-}5 \times 10^{-7}$	$1\text{-}7.8 \times 10^{-8}$
"	toxic	$>6\text{-}10 \times 10^{-7}$	-
Lecithin	-	$1.1\text{-}1.2 \times 10^{-3}$	$1\text{-}2.25 \times 10^{-3}$
Leptin	-	1.2×10^{-8}	-

...(Contd.)

1	2	3	4
Leucine	-	1.4-2 × 10^{-5}	1.2-5.2 × 10^{-5}
Leukocytes (#/cm^3) Total:	-	-	-
"	total, birth	9.0-30.0 × 106	-
"	total, pediatric	4.5-15.5 × 106	-
"	total, adult, range	4.3-11.0 × 106	-
"	total, adult, median	7.0 × 106	-
"	Neutrophil granulocytes, birth	6.0-26.0 × 106	-
"	Neutrophils, pediatric	1.5-8.5 × 106	-
"	Neutrophils, adult, range	1.83-7.25 × 106	-
"	Neutrophils, adult, median	3.65 × 106	-
"	Eosinophil granulocytes birth	0.4 × 106	-
"	Eosinophils, pediatric	0.2-0.3 × 106	-
"	Eosinophils, adult, range	0.05-0.7 × 106	-
"	Eosinophils, adult, median	0.15 × 106	-
"	Basophil granulocytes, adult, range	0.015-0.15 × 106	-
"	Basophils, adult, median	0.03 × 106	-
"	Lymphocytes, birth	2-11 × 106	-
"	Lymphocyte, pediatric	1.5-8.0 × 106	-
"	Lymphocyte, adult, range	1.5-4.0 × 106	-
"	Lymphocyte, adult, median	2.5 × 106	-
"	Monocytes, birth, range	0.4-3.1 × 106	-
"	Monocytes, birth, median	1.05 × 106	-
"	Monocytes, pediatric	0.4 × 106	-
"	Monocytes, adult, range	0.21-1.05 × 106	-
"	Monocytes, adult, median	0.43 × 106	-
"	Phagocytes, birth, range	6-26 × 106	-
"	Phagocytes, birth, median	11 × 106	-
"	Phagocytes, pediatric, range	1.5-8.5 × 106	-
"	Phagocytes, pediatric, median	4.1 × 106	-
"	Phagocytes, adult, range	3.5-9.2 × 106	-
"	Phagocytes, CD4 cell count	0.5-1.5 × 106	-
Lipase P	-	1.2-1.4 × 10^{-4}	-
Lipids, total	-	4.45-6.1 × 10^{-3}	4-8.5 × 10^{-3}
Lipoprotein (Sr 12-20)	-	-	1-10 × 10^{-4}

...(Contd.)

1	2	3	4
Proinsulin	-	-	-
"	fasting	-	$0.5\text{-}5 \times 10^{-10}$
"	mean	-	$1.42\text{-}1.70 \times 10^{-10}$
Prolactin (male) $<20 \times 10^{-9}$	-	-	-
"	while awake	-	$1\text{-}7 \times 10^{-9}$
"	during sleep	-	$9\text{-}20 \times 10^{-9}$
Prolactin (female)	-	-	-
"	follicular	-	$<23 \times 10^{-9}$
"	luteal	-	$5\text{-}40 \times 10^{-9}$
Proline	-	-	$1.2\text{-}5.7 \times 10^{-5}$
Prostaglandins	-	-	-
"	PGE	-	$3.55\text{-}4.15 \times 10^{-10}$
"	PGF	-	$1.26\text{-}1.56 \times 10^{-10}$
"	15-keto-PGF2a	-	5×10^{-10}
"	15-keto-PGE2	-	$<5 \times 10^{-11}$
Protein, total	-	$1.9\text{-}2.1 \times 10^{-1}$	$6.0\text{-}8.3 \times 10^{-2}$
Protoporphyrin	-	$2.7\text{-}6.1 \times 10^{-7}$	-
Prostate specific antigen	-	$0\text{-}5 \times 10^{-9}$	-
Pseudoglobulin I	-	-	$8\text{-}19 \times 10^{-3}$
Pseudoglobulin II	-	-	$2\text{-}8 \times 10^{-3}$
Purine, total	-	$9.5\text{-}11.5 \times 10^{-5}$	-
Pyrimidine nucleotides	-	$2.6\text{-}4.6 \times 10^{-5}$	$2\text{-}12 \times 10^{-7}$
Pyridoxine (Vitamin B6)	-	-	$3.6\text{-}90 \times 10^{-9}$
Pyruvic acid	-	$3\text{-}10 \times 10^{-6}$	$3\text{-}12 \times 10^{-6}$
RANTES	-	-	7×10^{-11}
Relaxin	-	-	-
"	day <100 preparturition	-	$<2 \times 10^{-9}$
"	day 100 to 2 days preceding	-	$5\text{-}40 \times 10^{-9}$
"	day preceding parturition	-	$100\text{-}200 \times 10^{-9}$
"	day following parturition	-	$<2 \times 10^{-9}$
Retinol (Vitamin A)	-	-	$1\text{-}8 \times 10^{-7}$
Riboflavin (Vitamin B2)	-	$1.5\text{-}6 \times 10^{-7}$	$2.6\text{-}3.7 \times 10^{-8}$
RNA	-	$5\text{-}8 \times 10^{-4}$	$4\text{-}6 \times 10^{-5}$
Secretin	-	-	$2.9\text{-}4.5 \times 10^{-11}$
Serine	-	-	$3\text{-}20 \times 10^{-6}$

...(Contd.)

1	2	3	4
Serotonin (5-hydroxytryptamine)	-	1.55-1.81 × 10-77	$0.8\text{-}2.1 \times 10^{-7}$
Silicon	-	$1.4\text{-}2.95 \times 10^{-6}$	$2.2\text{-}5.7 \times 10^{-6}$
Sodium	-	-	$3.1\text{-}3.4 \times 10^{-3}$
Solids, total	-	$2\text{-}2.5 \times 10^{-1}$	$8\text{-}9 \times 10^{-2}$
Somatotropin (growth hormone)	-	-	$4\text{-}140 \times 10^{-10}$
Sphingomyelin	-	$1.5\text{-}1.85 \times 10^{-3}$	$1\text{-}4 \times 10^{-4}$
Succinic acid	-	-	5×10^{-6}
Sugar, total	-	$7\text{-}11 \times 10^{-4}$	-
Sulfates, inorganic	-	-	$8\text{-}12 \times 10^{-6}$
Sulfur, total	-	$3.8\text{-}5 \times 10^{-2}$	$3.1\text{-}3.8 \times 10^{-2}$
Taurine	-	-	$3\text{-}21 \times 10^{-6}$
Testosterone (male)	-	-	-
"	free	-	$5.6\text{-}10.2 \times 10^{-11}$
"	total	-	$275\text{-}875 \times 10^{-11}$
Testosterone (female)	-	-	-
"	free	-	$0.24\text{-}0.38 \times 10^{-11}$
"	total	-	$23\text{-}75 \times 10^{-11}$
"	pregnant	-	$38\text{-}190 \times 10^{-11}$
Thiamine (Vitamin B1)	-	$3\text{-}10 \times 10^{-8}$	$1\text{-}9 \times 10^{-8}$
Thiocyanate	-	$5\text{-}14 \times 10^{-6}$	-
"	nonsmoker	-	$1\text{-}4 \times 10^{-6}$
"	smoker	-	$3\text{-}12 \times 10^{-6}$
Threonine	-	$1.3\text{-}2 \times 10^{-5}$	$0.9\text{-}3.2 \times 10^{-5}$
Thyroglobulin (Tg)	-	-	$<5 \times 10^{-8}$
Thyroid hormones	-	-	$4\text{-}8 \times 10^{-8}$
Thyrotropin-releasing hormone	-	-	$5\text{-}60 \times 10^{-12}$
Thyroxine (FT4)	-	-	-
"	free	-	$8\text{-}24 \times 10^{-12}$
"	total	-	$4\text{-}12 \times 10^{-8}$
Thyroxine-binding prealbumin	-	-	$2.8\text{-}3.5 \times 10^{-4}$
Thyroxine-binding globulin	-	-	$1.0\text{-}3.4 \times 10^{-7}$
Tin	-	$0\text{-}4 \times 10^{-7}$	$0\text{-}1 \times 10^{-7}$
alpha-Tocopherol (Vitamin E)	-	-	$5\text{-}20 \times 10^{-6}$
Transcortin	-	-	-
"	male	-	$1.5\text{-}2 \times 10^{-5}$
"	female	-	$1.6\text{-}2.5 \times 10^{-5}$

...(Contd.)

1	2	3	4
Transferrin	-	-	-
"	newborn	-	1.3-2.75 × 10^{-3}
"	adult	-	2.2-4 × 10^{-3}
"	age >60 yrs	-	1.8-3.8 × 10^{-3}
Triglycerides	-	8.5-23.5 × 10^{-4}	2.5-30 × 10^{-4}
Triiodothyronine	-	-	-
"	free	-	2.3-6.6 × 10^{-12}
"	total (T3)	-	0.75-2.50 × 10^{-9}
Tryptophan	-	5-10 × 10^{-6}	9-30 × 10^{-6}
Tyrosine	-	8-14 × 10^{-6}	4-25 × 10^{-6}
Urea	-	2-4 × 10^{-4}	2.8-4 × 10^{-4}
Uric acid	-	6-49 × 10^{-6}	2-7 × 10^{-5}
"	child	-	3.5-7.2 × 10^{-5}
"	adult, male	-	2.6-6.0 × 10^{-5}
"	adult, female	2.0-5.5 × 10^{-5}	
Valine	-	2-2.9 × 10^{-5}	1.7-4.2 × 10^{-5}
Vasointestinal peptide (VIP)	-	-	6-16 × 10^{-12}
Vasopressin	-	-	-
"	hydrated	-	4.5 × 10^{-13}
"	dehydrated	-	3.7 × 10^{-12}
Zinc	-	5-13 × 10^{-6}	7-15 × 10^{-7}

The Elements of Blood and the Whole Blood (Diagrammatic Representation)

Whole Blood is the term used in transfusion medicine for human blood from a standard blood donation. The blood is typically combined with an anticoagulant during the collection process, but is generally otherwise unprocessed. In the US, the capitalized "Whole Blood" means a specific standardized product for transfusion or further processing, where "whole blood" is any unmodified collected blood.

Historically, blood was transfused as Whole Blood without further processing. Most blood banks now split the Whole Blood into two or more components, typically red blood cells and a plasma component such as Fresh Frozen Plasma. Platelets for transfusion can also be prepared from a unit of Whole Blood. Some blood banks have replaced this with platelets collected by Plateletpheresis because whole blood Platelets, sometimes called "random" platelets, must be pooled from multiple donors to get enough for a therapeutic dose.The collected blood is generally separated into components by one of three methods. A centrifuge can be used in a "hard spin" which separates Whole Blood into plasma and red cells or for a "soft spin" which separates it into plasma, buffy coat (used to make platelets), and red blood cells. The third method is sedimentation: the Whole Blood simply sits overnight and the red cells and plasma are separated by gravity. Whole Blood has similar risks to a transfusion of Red Blood Cells and must be cross-matched to avoid hemolytic transfusion reactions. Most of the indications for use are identical to those for RBCs, and Whole Blood is not used because the extra plasma can contribute to transfusion associated circulatory overload (TACO), a potentially dangerous complication. Whole Blood is sometimes "recreated" from stored red blood cells and FFP for neonatal transfusions. This is done to provide a final product with a very specific hematocrit (percentage of red cells) with type O red cells and type AB plasma to minimize the chance of complications.

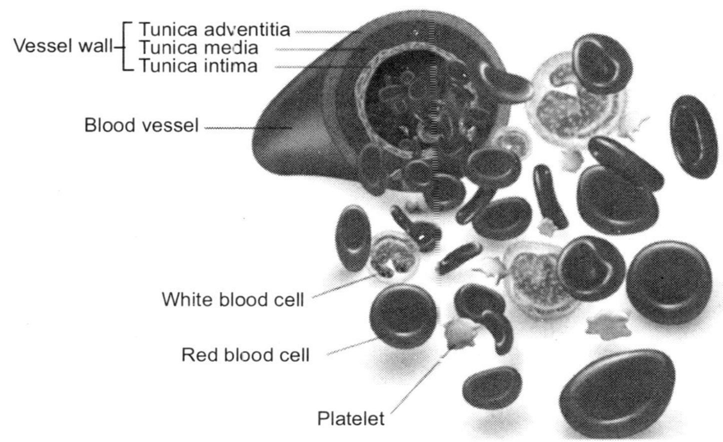

Fig. 8.2 : Showing Diagramatically the Biological Components of Blood

Whole Blood is typically stored under the same conditions as Red Blood Cells and can be kept up to 35 days if collected with CPDA-1 storage solution or 21 days with other common storage solutions such as CPD.

If the blood will be used to make platelets, it is kept at room temperature until the process is complete. This must be done quickly to minimize the warm storage of RBCs in the unit.

- *Red Blood Cells* carry oxygen and are needed by surgical patients or those with anemia or kidney disease
- *Platelet Rich Plasma* is an intermediate stage in the production of platelet concentrate and plasma components.
- *Frozen Red Blood Cells* have been frozen to extend their storage period. Very rare blood types are frozen to insure they are always on hand.
- *Washed Red Blood Cells* have been washed to prevent reactions in some sensitive patients.
- *Filtered Red Blood Cells* have been filtered to prevent reactions in some sensitive patients.
- *Plasma* is the yellow liquid portion of blood. It is also a source of proteins that stop bleeding by forming blood clots.
- *Platelet Concentrates* are fragile blood cells needed by leukemia and other cancer patients to control bleeding.

- *Frozen Plasma* to be manufactured into derivatives is stored and shipped in the frozen state.
- *Fresh Frozen Plasma* must be processed and frozen within eight hours of the whole blood donation to preserve the less stable clotting proteins. It is used mainly for people with bleeding complications.
- *Cryoprecipitate*, a part of the plasma, contains two important clotting factors found in whole blood. It is used to treat two common clotting disorders.
- *Fractionation* is the process of breaking plasma down into "fractions." When plasma gets mixed with certain chemicals, proteins can be separated and processed into products which are stable and easily stored. Some of the more common fractions are as follows:

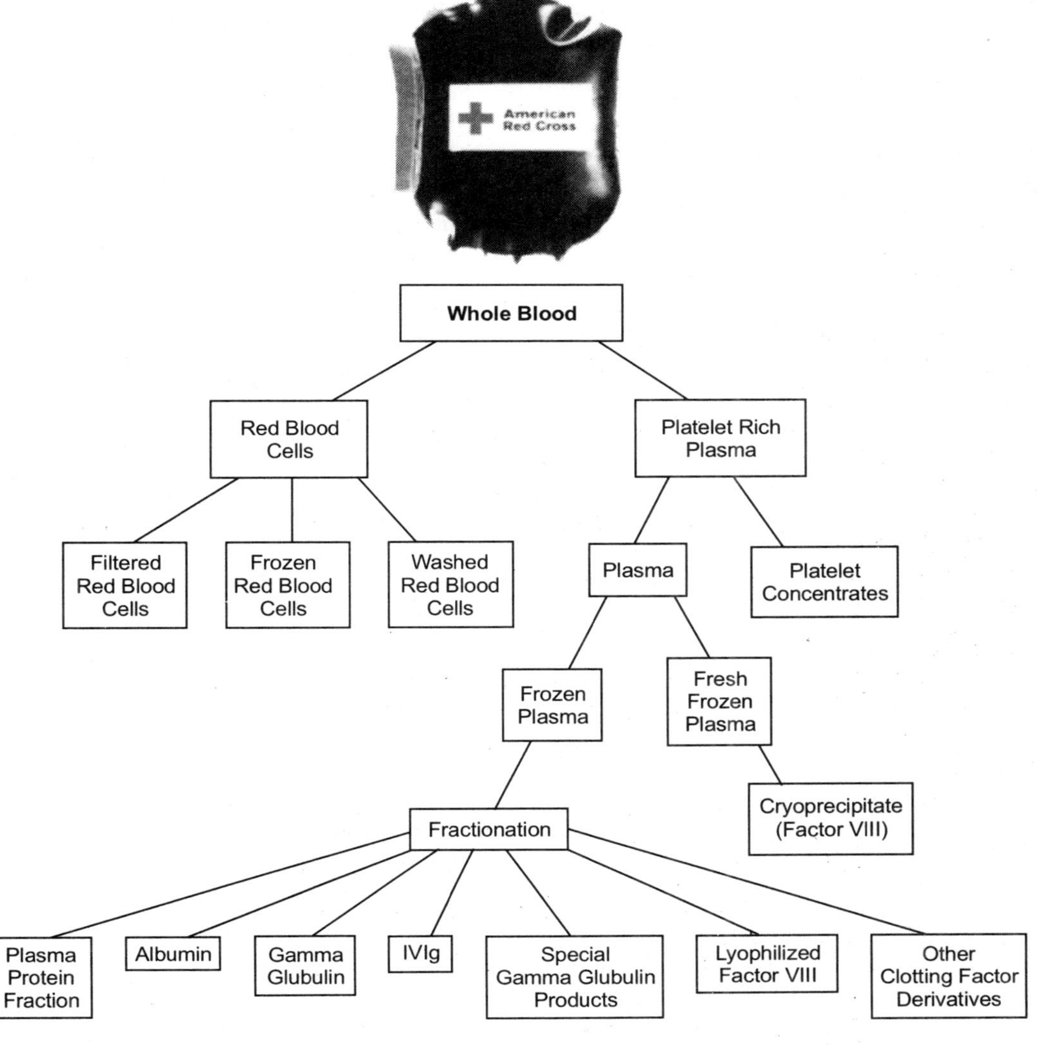

Albumin - used in the treatment of certain kidney and liver diseases. Because of the relative simplicity of storage and administration, it is also used for emergency cases, such as accident or shock victims, particularly where facilities for administering blood are not available or where time does not permit its use.

Plasma Protein Fraction - similar to albumin; the two products are often used interchangeably.

Gamma Globulin - contains the antibodies in plasma and is able to modify or prevent measles and other infectious diseases, such as some varieties of hepatitis.

HBIg, RHIg, VZIg - are gamma globulin products prepared from the plasma of donors with high levels of specific desirable antibodies.

Lyophilized Factor VIII - concentrates are used for the treatment of Hemophilia A.

Other Clotting Factor Derivatives - contain concentrated amounts of specific clotting factors and are used in certain less common clotting disorders.

Now we shall discuss the Biological Components of blood(can be treated as whole blood) in detail.

RBC

Normal Count: 4,500,000 to 6,000,000 for men 4,200,000 to 5,400,000 for Women
Life Span: 100-120 days

The first person to describe red blood cells was probably the young Dutch biologist Jan Swammerdam, who had used an early microscope in 1658 to study the blood of a frog. Unaware of this work, Anton van Leeuwenhoek provided another microscopic description in 1674, this time providing a more precise description of red blood cells, even approximating their size, "25,000 times smaller than a fine grain of sand".

In 1901 Karl Landsteiner published his discovery of the three main blood groups - A, B, and C (which he later renamed to O). Landsteiner described the regular patterns in which reactions occurred when serum was mixed with red blood cells, thus identifying compatible and conflicting combinations between these blood groups. A year later Alfred von Decastello and Adriano Sturli, two colleagues of Landsteiner, identified a fourth blood group - AB - the serum of which causes both A and B red cells to agglutineate.

In 1959, by use of X-ray crystallography, Dr. Max Perutz was able to unravel the structure of hæmoglobin, the red blood cell protein that carries oxygen.

Red blood cells (also referred to as **erythrocytes**) are the most common type of blood cell and the vertebrate organism's principal means of delivering oxygen (O_2) to the body tissues via the blood flow through the circulatory system. They take up oxygen in the lungs or gills and release it while squeezing through the body's capillaries. These cells' cytoplasm is rich in hæmoglobin, an iron-containing biomolecule that can bind oxygen and is responsible for the blood's red color.

In humans, mature red blood cells are flexible biconcave disks that lack a cell nucleus and most organelles. The cells develop in the bone marrow and circulate for about 100-120 days in the body before their components are recycled by macrophages. Each circulation takes about 20 seconds. Approximately a quarter of the cells in the human body are red blood cells.

Red blood cells are also known as **RBC's**, **red blood corpuscles** (an archaic term), **haematids**, **erythroid cells** or **erythrocytes** (from Greek *erythros* for "red" and *kyttaro*[1]" for cell". The capitalized term **Red Blood Cells** is the proper name in the US for erythrocytes in storage solution used in transfusion medicine.

Erythrocytes consist mainly of hæmoglobin, a complex metalloprotein containing hæme groups whose iron atoms temporarily bind to oxygen molecules (O_2) in the lungs or gills and release them throughout the body. Oxygen can easily diffuse through the red blood cell's cell membrane. Hæmoglobin in the erythrocytes also carries some of the waste product carbon dioxide back from the tissues; most of the carbon dioxide, however, is transported as bicarbonate (HCO_3^-) dissolved in the blood plasma. Myoglobin, a compound related to hæmoglobin, acts to store oxygen in muscle cells.

The colour of erythrocytes is due to the hæme group of hæmoglobin. The blood plasma alone is straw-coloured, but the red blood cells change colour depending on the state of the hæmoglobin: when combined with oxygen the resulting oxyhæmoglobin is scarlet, and when oxygen has been released the resulting deoxyhæmoglobin is darker, appearing bluish through the vessel wall and skin. Pulse oximetry takes advantage of this color change to directly measure the arterial blood oxygen saturation using colorimetric techniques.

The sequestration of oxygen carrying proteins inside specialized cells (rather than having them dissolved in body fluid) was an important step in the evolution of vertebrates as it allows for less viscous blood, higher concentrations of oxygen, and better diffusion of oxygen from the blood to the tissues. The size of erythrocytes varies widely among vertebrate species; erythrocyte width is on average about 25% larger than capillary diameter and it has been hypothesized that this improves the oxygen transfer from erythrocytes to tissues.

The only known vertebrates without erythrocytes are the crocodile icefishes (family Channichthyidae); they live in very oxygen rich cold water and transport oxygen freely dissolved in their blood. While they don't use hæmoglobin anymore, remnants of hæmoglobin genes can be found in their genome.

Nucleus

Erythrocytes in mammals are *anucleate* when mature, meaning that they lack a cell nucleus. In comparison, the erythrocytes of other vertebrates have nuclei; the only known exceptions are salamanders of the *Batrachoseps* genus and fish of the *Maurolicus* genus with closely related species.

Secondary Functions

When erythrocytes undergo shear stress in constricted vessels, they release ATP which causes the vessel walls to relax and dilate so as to promote normal blood flow.

When their hæmoglobin molecules are deoxygenated, erythrocytes release S-nitrosothiols which also acts to dilate vessels, thus directing more blood to areas of the body depleted of oxygen.

Erythrocytes also play a part in the body's immune response: when lysed by pathogens such as bacteria, their hæmoglobin releases free radicals that break down the pathogen's cell wall and membrane, killing it.

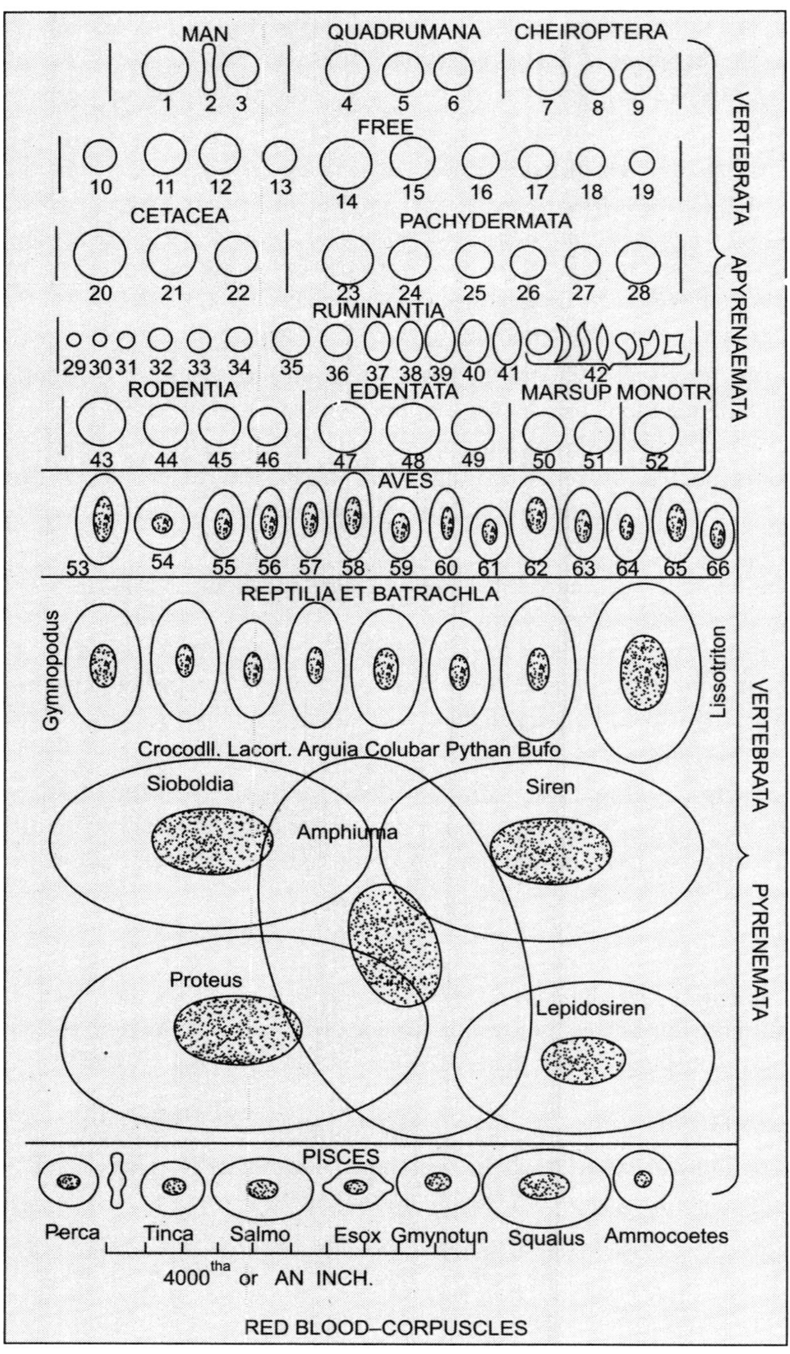

Fig. 8.3 : The Different Types of Erythrocytes in Vertibrates

Human Erythrocytes

A typical human erythrocyte has a disk diameter of 6-8 μm and a thickness of 2 μm, being much smaller than most other human cells. These cells have a volume of about 90 fL with a surface of about 136 μm^2, and can swell up to a sphere shape containing 150 fL, without membrane distension.

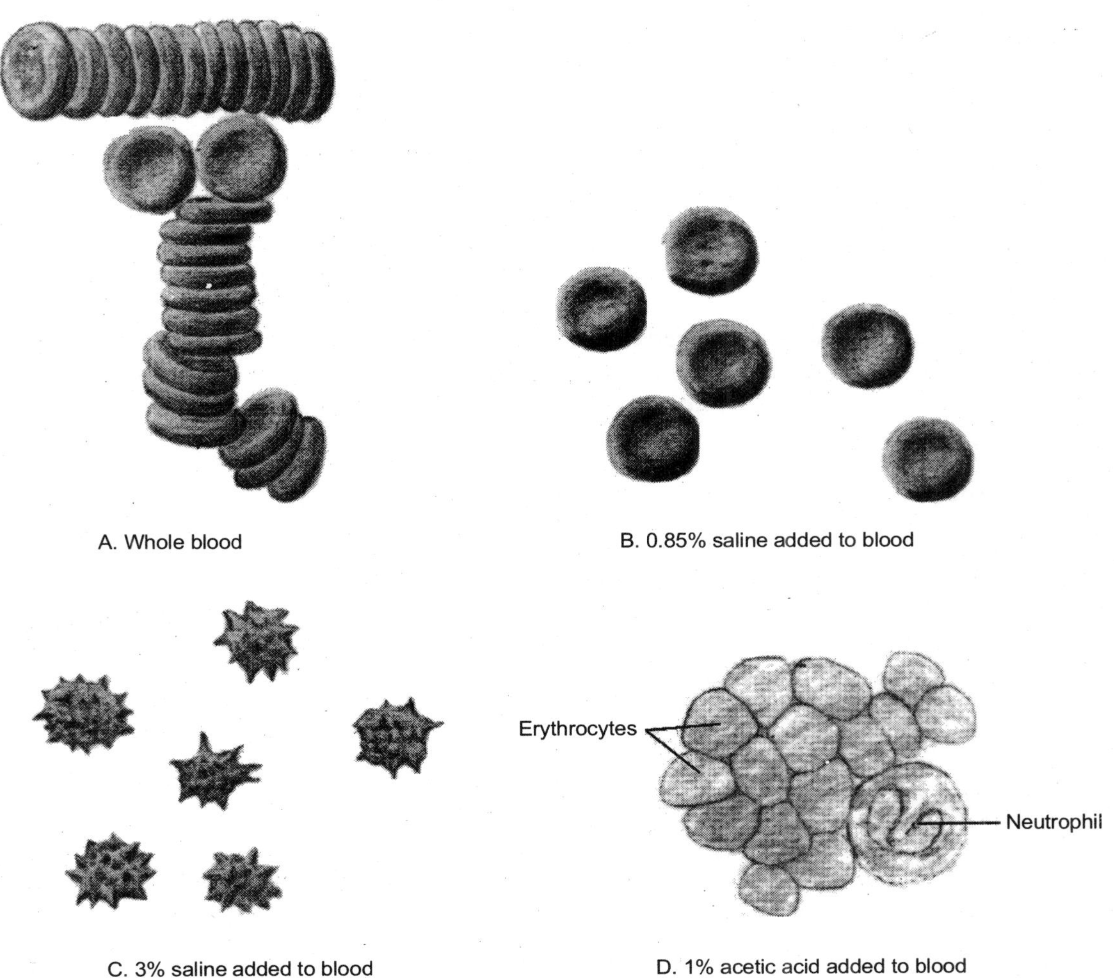

Fig. 8.4 : The Different Erythrocytes

Adult humans have roughly 2-3 × 10^{13} (20-30 trillion) red blood cells at any given time, comprising approximately one quarter of the total human body cell number (women have about 4 to 5 million erythrocytes per microliter (cubic millimeter) of blood and men about 5 to 6 million; people living at high altitudes with low oxygen tension will have more). Red blood cells are thus much more common than the other blood particles: there are about 4,000-11,000 white blood cells and about 150,000–400,000 platelets in each microliter of human blood.

Human red blood cells take on average 20 seconds to complete one cycle of circulation. As red blood cells contain no nucleus, protein biosynthesis is currently assumed to be absent in these cells, although a recent study indicates the presence of all the necessary biomachinery in human red blood cells for protein biosynthesis.

The blood's red colour is due to the spectral properties of the hemic iron ions in hæmoglobin. Each human red blood cell contains approximately 270 million of these hæmoglobin biomolecules, each carrying four hæme groups; hæmoglobin comprises about a third of the total cell volume. This protein is responsible for the transport of more than 98% of the oxygen (the remaining oxygen is carried dissolved in the blood plasma). The red blood cells of an average adult human male store collectively about 2.5 grams of iron, representing about 65% of the total iron contained in the body.

Life-cycle

Human erythrocytes are produced through a process named erythropoiesis,developing from committed stem cells to mature erythrocytes in about 7 days. When matured, these cells live in blood circulation for about 100-120 days. At the end of their life-span, they become senescent, and are removed from circulation.

Erythropoiesis

Erythropoiesis is the development process in which new erythrocytes are produced, through which each cell matures in about 7 days. Through this process erythrocytes are continuously produced in the red bone marrow of large bones, at a rate of about 2 million per second in a healthy adult. (In the embryo, the liver is the main site of red blood cell production.) The production can be stimulated by the hormone erythropoietin (EPO), synthesised by the kidney. Just before and after leaving the bone marrow, the developing cells are known as reticulocytes; these comprise about 1% of circulating red blood cells.

Functional Lifetime

This phase lasts about 100-120 days, during which the erythrocytes are continually moving by the blood flow push (in arteries), pull (in veins) and squeezing through microvessels such as capillaries as they compress against each other in order to move.

Senescence

The aging erythrocyte undergoes changes in its plasma membrane, making it susceptible to selective recognition by macrophages and subsequent phagocytosis in the reticuloendothelial system (spleen, liver and bone marrow), thus removing old and defected cells and continually purging the blood. This process is termed eryptosis, or erythrocyte programmed cell death. This process normally occurs at the same rate of production by erythropoiesis, balancing the total circulating red blood cell count. Much of the resulting important breakdown products are recirculated in the body. The hæme constituent of hæmoglobin are broken down into Fe^{3+} and biliverdin. The biliverdin is reduced to bilirubin, which is released into the plasma and recirculated to the liver bound to albumin. The iron is released into the plasma to be recirculated by a carrier protein called transferrin. Almost all erythrocytes are removed in this manner from the circulation before they are old enough to hemolyze. Hemolyzed hæmoglobin is bound to a protein in plasma called haptoglobin which is not excreted by the kidney.

Membrane Composition

The membranes of red blood cells play many roles that aid in regulating their surface deformability, flexibility, adhesion to other cells and immune recognition. These functions are highly dependent on its composition, which defines its properties. The red blood cell is composed of 3 layer: the glycocalyx on the exterior, which is rich in carbohydrates; the lipid bilayer which contains many transmembrane proteins, besides its lipidic main constituents; and the membrane skeleton, a structural network of proteins located on the inner surface of the lipid bilayer. The proteins of the membrane skeleton are responsible for the deformability, flexibility and durability of the red blood cell, enabling it to squeeze through capillaries less than half the diameter of the erythrocyte (7-8 μm) and recovering the discoid shape as soon as these cells stop receiving compressive forces, in a similar fashion to an object made of rubber.[27]

Membrane Lipids

The erythrocyte cell membrane is a typical lipid bilayer, similar to what we can find in virtually all human cells. Simply said, it is composed by cholesterol and phospholipids in equal proportions by weight.

Unlike cholesterol which is evenly distributed between the inner and outer leaflets, the 5 major phospholipids are asymmetrically disposed, as shown below:

Outer monolayer

- Phosphatidylcholine (PC);
- Sphingomyelin.

Inner monolayer

- Phosphatidylethanolamine (PE);
- Phosphatidylserine (PS);
- Phosphoinositide (PI) constituents (small amounts).

This asymmetric phospholipid distribution among the bilayer is the result of the function of several energy-dependent and energy-independent phospholipid transport proteins. Proteins called "Flippases" move phospholipids from the outer to the inner monolayer while others called "floppases" do the opposite operation, against a concentration gradient in an energy dependent manner. Additionally, there are also "scramblase" proteins that move phospholipids in both directions at the same time, down their concentration gradients in an energy independent manner. There is still considerable debate ongoing regarding the identity of these membrane maintenance proteins in the red cell membrane.

The maintenance of an asymmetric phospholipid distribution in the bilayer (such as an exclusive localization of PS and phosphoinositides in the inner monolayer) is critical due to several reasons:

- Macrophages recognize and phagocytose red cells that expose PS at their outer surface. Thus the confinement of phosphatidylserine in the inner monolayer is essential if the cell is to survive its frequent encounters with macrophages of the reticuloendothelial system, especially in the spleen.
- Premature destruction of thallassemic and sickle red cells has been linked to disruptions of lipid asymmetry leading to exposure of PS on the outer monolayer.
- An exposure of PS can potentiate adhesion of red cells to vascular endothelial cells, effectively preventing normal transit through the microvasculature. Thus it is important that PS is maintained only in the inner leaflet of the bilayer to ensure normal blood flow in microcirculation.
- Both PS and phosphatidylinositol-4,5-bisphosphate (PIP2) can regulate membrane mechanical function, due to their interactions with skeletal proteins such as spectrin and protein 4.1R. Recent studies have shown that binding of spectrin to PS promotes membrane mechanical stability. PIP2 enhances the binding of protein band 4.1R to glycophorin C but decreases its interaction with protein band 3, and thereby may modulate the linkage of the bilayer to the membrane skeleton.

The presence of specialized structures named "lipid rafts" in the erythrocyte membrane have been described by recent studies. These are structures enriched in cholesterol and sphingolipids associated with specific membrane proteins, namely flotillins, stomatins (band 7), G-proteins, and β adrenergic receptors. Lipid rafts that have been implicated in cell signaling events in nonerythroid cells have been shown in erythroid cells to mediate β2-adregenic receptor signaling and increase cAMP levels, and thus regulating entry of malarial parasites into normal red cells.

Membrane Proteins

There are currently more than 50 known membrane proteins, which can be exist in a few hundred to a million copies per erythrocyte. Approximately 25 of these membrane proteins carry the various blood group antigens, such as the A, B and Rh antigens, among many others. These membrane proteins can perform a wide diversity of functions, such as transporting ions and molecules across the red cell membrane, adhesion and interaction with other cells such as endothelial cells, as signaling receptors, as well as other currently unknown functions. The blood types of humans are due to variations in surface glycoproteins of erythrocytes. Disorders of the proteins in these membranes are associated with many disorders, such as hereditary spherocytosis, hereditary elliptocytosis, hereditary stomatocytosis, and paroxysmal nocturnal hæmoglobinuria.

The red blood cell membrane proteins organized according to their function:

Transport

- Band 3 - Anion transporter, also an important structural component of the erythrocyte cell membrane, makes up to 25% of the cell membrane surface, each red cell contains approximately one million copies. Defines the Diego Blood Group;

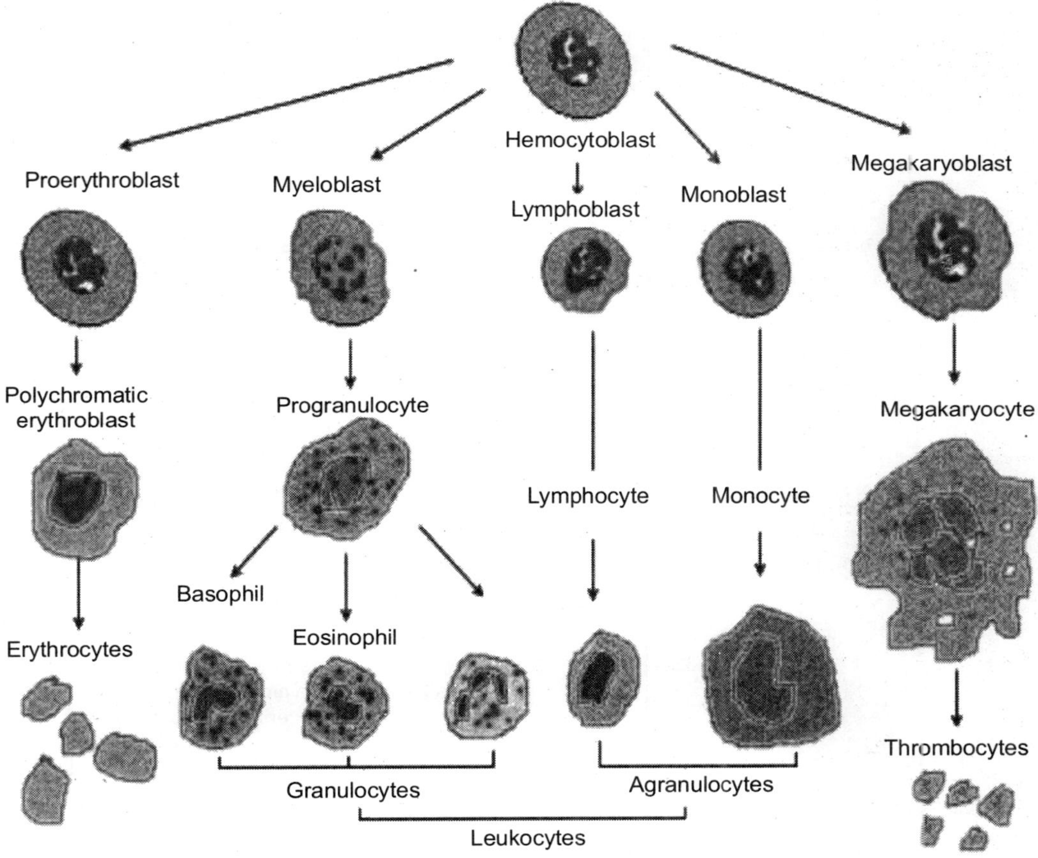

There are several different types of white blood cells. They all have many things in common, but are all different. A major distinguishing feature of some leukocytes is the presence of granules; white blood cells are often characterized as granulocytes or agranulocytes:

- **Granulocytes** (polymorphonuclear leukocytes): leukocytes characterised by the presence of differently staining granules in their cytoplasm when viewed under light microscopy. These granules are membrane-bound enzymes which primarily act in the digestion of endocytosed particles. There are three types of granulocytes: neutrophils, basophils, and eosinophils, which are named according to their staining properties.
- **Agranulocytes** (mononuclear leucocytes): leukocytes characterized by the apparent absence of granules in their cytoplasm. Although the name implies a lack of granules these cells do contain non-specific azurophilic granules, which are lysosomes. The cells include lymphocytes, monocytes, and macrophages.

Overview Table

Type	Diagram	Approx. % in adults	Diameter (μm)	Main targets	Nucleus	Granules	Lifetime
1	2	3	4	5	6	7	8
Neutrophil		54-62%	10-12	• bacteria • fungi	multilobed	fine, faintly pink (H&E) Stain)	6 hours-few days (days in spleen and other tissue)
Eosinophil		1-6%	10-12	• parasites • in allergic reactions	bi-lobed	full of pink-orange (H&E Stain)	8-12 days (circulate for 4-5 hours)
Basophil		<1%	12-15	• in allergic reactions	bi-lobed or tri-lobed	large blue	
Lymphocyte		25-33%	7-8	• B cells: various pathogens • T cells: CD4+ (helper): extracellular bacteria broken down into peptides presented by MHC class 2 molecule. • CD8+ cytotoxic T cells: virus-infected and tumor cells. • Natural killer cells: virus-infected and tumor cells	deeply staining, eccentric	NK-cells and Cytotoxic (CD8+) T-cells	weeks

...(Contd.)

1	2	3	4	5	6	7	8
Monocyte		2-8%	14-17	Monocytes migrate from the bloodstream to other tissues and differentiate into tissue resident macrohages or dendritic cells	kidney shaped	none	hours to days
Macrophage			21 (human)	Phagocytosis (engulfment and digestion) of cellular debris and pathogens, and stimulation of lymphocytes and other immune cells that respond to the pathogen			activated days immature: months to years
Dendritic cells				Main function is as an antigen-presenting cell (APC) that activates T lymphocytes			similar to macrop hages

Neutrophil

Normal Count : - 3000 - 7000 (54 - 62 %) is a standard normal range

Life Span: - 1-2 days

Neutrophil granulocytes, generally referred to as **neutrophils**, are the most abundant type of white blood cells in mammals and form an essential part of the innate immune system. They form part of the polymorphonuclear cell family (PMNs) together with basophils and eosinophils. (For an overview of neutrophils and their function. The name, neutrophil, derives from staining characteristics on hematoxylin and eosin (H&E) histological or cytological preparations. Whereas basophilic white blood cells stain dark blue and eosinophilic white blood cells stain bright red, neutrophils stain a neutral pink. Normally neutrophils contain a nucleus divided into 2-5 lobes.Neutrophils are normally found in the blood stream. However, during the beginning (acute) phase of inflammation, particularly as a result of bacterial infection and some cancers, neutrophils are one of first-responders of inflammatory cells to migrate toward the site of inflammation, firstly through the blood vessels, then through interstitial tissue, following chemical signals (such as Interleukin-8 (IL-8), Interferon-gamma (IFN-gamma), and C5a) in a process called chemotaxis. They are the predominant cells in pus, accounting for its whitish/yellowish appearance. Neutrophils are recruited to the site of injury within minutes following trauma and are the hallmark of acute inflammation.

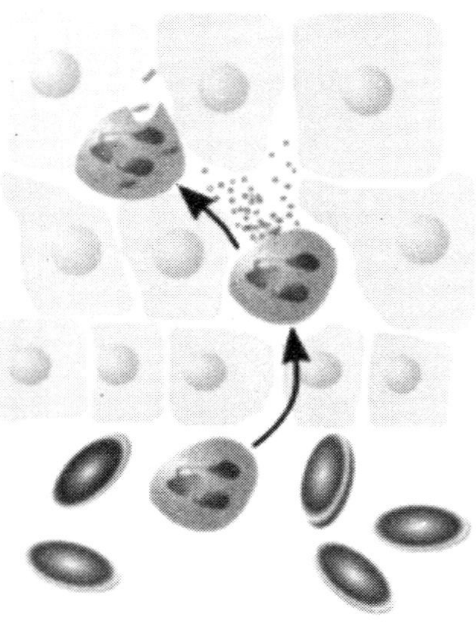

Fig. 8.5 : The Neutrophil granulocyte migrates from the blood vessel to the matrix, sensing proteolytic enzymes, in order to determine intercellular connections (to the improvement of its mobility) and envelop bacteria through phagocytosis

Measurement of Neutrophils

With the eosinophil and the basophil, they form the class of *polymorphonuclear cells*, named for the nucleus's characteristic multilobulated shape (as compared to lymphocytes and monocytes, the other types of white cells). Neutrophils are the most abundant white blood cells in humans (approximately 10^{-11} are produced daily) ; they account for approximately 70% of all white blood cells (leukocytes).The stated normal range for human blood counts varies between laboratories, but. People of African and Middle Eastern descent may have lower counts, which are still normal.

A report may divide neutrophils into segmented neutrophils and bands.

Life-span

The average half-life of non-activated neutrophils in the circulation is about 12 hours. Upon activation, they marginate (position themselves adjacent to the blood vessel endothelium), and undergo selectin-dependent capture followed by integrin-dependent adhesion in most cases, after which they migrate into tissues, where they survive for 1-2 days.

Neutrophil granulocytes have an average diameter of 12-15 micrometers (μm) in peripheral blood smears.

Neutrophils are much more numerous than the longer-lived monocyte/macrophage phagocytes. A pathogen (disease-causing microorganism or virus) is likely to first encounter a neutrophil. Some experts hypothesize that the short lifetime of neutrophils is an evolutionary adaptation. The short lifetime of neutrophils minimizes propagation of those pathogens that parasitize phagocytes, the more time such parasites spend outside a host cell, the more likely they will be destroyed by some component of the body's defenses. Also, because neutrophil antimicrobial products can also damage host tissues, their short life limits damage to the host during inflammation.

Chemotaxis

Neutrophils undergo a process called chemotaxis, which allows them to migrate toward sites of infection or inflammation. Cell surface receptors allow neutrophils to detect chemical gradients of molecules such as interleukin-8 (IL-8), interferon gamma (IFN-gamma), and C5a, which these cells use to direct the path of their migration.

Anti-Microbial Function

Being highly motile, neutrophils quickly congregate at a focus of infection, attracted by cytokines expressed by activated endothelium, mast cells, and macrophages. Neutrophils express and release cytokines, which in turn amplify inflammatory reactions by several other cell types.

In addition to recruiting and activating other cells of the immune system, neutrophils play a key role in the front-line defence against invading pathogens. Neutrophils have three strategies for directly attacking micro-organisms: phagocytosis (ingestion), release of soluble anti-microbials (including granule proteins) and generation of neutrophil extracellular traps (NETs)

Phagocytosis

Neutrophils are phagocytes, capable of ingesting microorganisms or particles. They can internalize and kill many microbes, each phagocytic event resulting in the formation of a phagosome into which reactive oxygen species and hydrolytic enzymes are secreted. The consumption of oxygen during the generation of reactive oxygen species has been termed the "respiratory burst," although unrelated to respiration or energy production.

The respiratory burst involves the activation of the enzyme NADPH oxidase, which produces large quantities of superoxide, a reactive oxygen species. Superoxide dismutates, spontaneously or through catalysis via enzymes known as superoxide dismutases (Cu/ZnSOD and MnSOD), to hydrogen peroxide, which is then converted to hypochlorous acid HOCl, by the green hæme enzyme myeloperoxidase. It is thought that the bactericidal properties of HOCl are enough to kill bacteria phagocytosed by the neutrophil, but this may instead be step necessary for the activation of proteases.

Degranulation

Neutrophils also release an assortment of proteins in three types of granules by a process called degranulation:

Granule type	Protein
specific granules (or "secondary granules")	Lactoferrin and Cathelicidin
azurophilic granules (or "primary granules")	myeloperoxidase, bactericidal/permeability increasing protein (BPI), Defensins and the serine proteases neutrophil elastase and cathepsin G
tertiary granules	cathepsin and gelatinase

Neutrophil Extracellular Traps (NETs)

Zychlinsky and colleagues described a new striking observation that activation of neutrophils causes the release of web-like structures of DNA; this represents a third mechanisms for killing bacteria. These neutrophil extracellular traps (NETs) comprise a web of fibers composed of chromatin and serine proteases that trap and kill microbes extracellularly. It is suggested that NETs provide a high local concentration of antimicrobial components and bind, disarm, and kill microbes independent of phagocytic uptake. In addition to their possible antimicrobial properties, NETs may serve as a physical barrier that prevents further spread of pathogens. Trapping of bacteria may be a particularly important role for NETs in sepsis, where NET are formed within blood vessels. Recently, NETs have been shown to play a role in inflammatory diseases, as NETs could be detected in preeclampsia, a pregnancy related inflammatory disorder in which neutrophils are known to be activated.

Role in Disease

Low neutrophil counts are termed *neutropenia*. This can be congenital (genetic disorder) or it can develop later, as in the case of aplastic anemia or some kinds of leukemia. It can also be a side-effect of medication, most prominently chemotherapy. Neutropenia makes an individual highly susceptible to infections. Neutropenia can be the result of colonization by intracellular neutrophilic parasites.

Functional disorders of neutrophils are often hereditary. They are disorders of phagocytosis or deficiencies in the respiratory burst (as in chronic granulomatous disease, a rare immune deficiency, and myeloperoxidase deficiency).

In alpha 1-antitrypsin deficiency, the important neutrophil enzyme elastase is not adequately inhibited by alpha 1-antitrypsin, leading to excessive tissue damage in the presence of inflammation - most prominently pulmonary emphysema.

In Familial Mediterranean fever (FMF), a mutation in the *pyrin* (or *marenostrin*) gene, which is expressed mainly in neutrophil granulocytes, leads to a constitutively active acute phase response and causes attacks of fever, arthralgia, peritonitis, and - eventually - amyloidosis.

Neutrophils display highly directional amoeboid motility in infected footpad and phalanges. Intravital imaging was performed in the footpad path of LysM-eGFP mice 20 min after infection with LM. Public Library of Science.

Eosinophil

Normal Count: - 50 – 500 (1 – 3%)
Life Span: -8–12 days (circulate for 4–5 hours)

Eosinophil granulocytes, usually called *eosinophils* (or, less commonly, *acidophils*), are white blood cells that are one of the immune system components responsible for combating multicellular parasites and certain infections in vertebrates. Along with mast cells, they also control mechanisms associated with allergy and asthma. They are granulocytes that develop during haematopoiesis in the bone marrow before migrating into blood.

These cells are eosinophilic or 'acid-loving': Normally transparent, they appear brick-red after staining with eosin, a red dye, using the Romanowsky method. The staining is concentrated in small granules within the cellular cytoplasm, which contain many chemical mediators, such as histamine and proteins such as eosinophil peroxidase, ribonuclease (RNase), deoxyribonucleases, lipase, plasminogen, and major basic protein. These mediators are released by a process called degranulation following activation of the eosinophil, and are toxic to both parasite and host tissues.

In normal individuals eosinophils make up about 1-6% of white blood cells, and are about 12-17 micrometers in size. They are found in the medulla and the junction between the cortex and medulla of the thymus, and, in the lower gastrointestinal tract, ovary, uterus, spleen, and lymph nodes, but not in the lung, skin, esophagus, or some other internal organs under normal conditions. The presence of eosinophils in these latter organs is associated with disease. Eosinophils persist in the circulation for 8-hours, and can survive in tissue for an additional 8-12 days in the absence of stimulation.

Eosinophil Development, Migration and Activation

Eosinophils develop and mature in bone marrow. They differentiate from myeloid precursor cells in response to the cytokines interleukin 3 (IL-3), interleukin 5 (IL-5), and granulocyte macrophage colony-stimulating factor (GM-CSF). Eosinophils produce and store many secondary granule proteins prior to their exit from the bone marrow. After maturation, eosinophils circulate in blood and migrate to inflammatory sites in tissues, or to sites of helminth infection in response to chemokines like CCL11 (eotaxin-1), CCL24 (eotaxin-2), CCL5 (RANTES), and certain leukotrienes like leukotriene B4 (LTB4). At these infectious sites, eosinophils are activated by Type 2 cytokines released from a specific subset of helper T cells (T_h2); IL-5, GM-CSF, and IL-3 are important for eosinophil activation as well as maturation.

Functions of Eosinophils

Following activation, eosinophils effector functions include production of:

- cationic granule proteins and their release by degranulation.
- reactive oxygen species such as superoxide, peroxide, and hypobromite (hypobromous acid, which is preferentially produced by eosinophil peroxidase).
- lipid mediators like the eicosanoids from the leukotriene (e.g., LTC_4, LTD_4, LTE_4) and prostaglandin (e.g., PGE_2) families. enzymes, such as elastase.
- growth factors such as TGF beta, VEGF, and PDGF.
- cytokines such as IL-1, IL-2, IL-4, IL-5, IL-6, IL-8, IL-13, and TNF alpha.

In addition, eosinophils play a role in fighting viral infections, which is evident from the abundance of RNAses they contain within their granules, and in fibrin removal during inflammation. Eosinophils along with basophils and mast cells, are important mediators of allergic responses and asthma pathogenesis and are associated with disease severity. They also fight helminth (worm) colonization and may be slightly elevated in the presence of certain parasites. Eosinophils are also involved in many other biological processes, including postpubertal mammary gland development, oestrus cycling, allograft rejection and neoplasia. They have also recently been implicated in antigen presentation to T cells.

Eosinophil Granule Proteins

Following activation by an immune stimulus, eosinophils degranulate to release an array of cytotoxic granule cationic proteins that are capable of inducing tissue damage and dysfunction. These include:

- major basic protein (MBP)
- eosinophil cationic protein (ECP)
- eosinophil peroxidase (EPO)
- eosinophil-derived neurotoxin (EDN)

Major basic protein, eosinophil peroxidase, and eosinophil cationic protein are toxic to many tissues. Eosinophil cationic protein and eosinophil-derived neurotoxin are ribonucleases with antiviral activity. Major basic protein induces mast cell and basophil degranulation, and is implicated in peripheral nerve remodelling. Eosinophil cationic protein creates toxic pores in the membranes of target cells allowing potential entry of other cytotoxic molecules to the cell, can inhibit proliferation of T cells, suppress antibody production by B cells, induce degranulation by mast cells, and stimulate fibroblast cells to secrete mucus and glycosaminoglycan. Eosinophil peroxidase forms reactive oxygen species and reactive nitrogen intermediates that promote oxidative stress in the target, causing cell death by apoptosis and necrosis.

Eosinophilia

An increase in eosinophils, i.e., the presence of more than 500 eosinophils/microlitre of blood is called an eosinophilia, and is typically seen in people with a parasitic infestation of the intestines, a collagen vascular disease (such as rheumatoid arthritis), malignant diseases such as Hodgkin's disease, extensive skin diseases (such as exfoliative dermatitis), Addison's disease, in the squamous epithelium of the esophagus in the case of reflux esophagitis, eosinophilic esophagitis, and with the use of certain drugs such as penicillin. In 1989, contaminated L-tryptophan supplements caused a deadly form of eosinophilia known as eosinophilia-myalgia syndrome.

Treatment

Treatments used to combat autoimmune diseases and conditions caused by eosinophils include:

- corticosteroids- promote apoptosis. Numbers of eosinophils in blood are rapidly reduced

- monoclonal antibody therapy against IL-5 - promote apoptosis
- antagonists of leukotriene synthesis or receptors
- Gleevec (STI571)- inhibits PDGF-BB in hypereosinophilic leukemia

Basophil Granulocytes
Normal Count: 0 – 50 (0 – 0.75%)
Life Span:

Basophil granulocytes, sometimes referred to as basophils, are the least common of the granulocytes, representing about 0.01% to 0.3% of circulating white blood cells.

The name comes from the fact that these leucocytes are basophilic, i.e., they are susceptible to staining by base dyes, as shown in the picture.

Basophils contain large cytoplasmic granules which obscure the cell nucleus under the microscope. However, when unstained, the nucleus is visible and it usually has 2 lobes. The mast cell, a cell in tissues, has many similar characteristics. For example, both cell types store histamine, a chemical that is secreted by the cells when stimulated in certain ways (histamine causes some of the symptoms of an allergic reaction). Like all circulating granulocytes, basophils can be recruited out of the blood into a tissue when needed.

Function

Reference ranges for blood tests of white blood cells, comparing basophil granulocyte amount (shown in violet) with other cells.

Basophils appear in many specific kinds of inflammatory reactions, particularly those that cause allergic symptoms. Basophils contain anticoagulant heparin, which prevents blood from clotting too quickly. They also contain the vasodilator histamine, which promotes blood flow to tissues. They can be found in unusually high numbers at sites of ectoparasite infection, e.g., ticks. Like eosinophils, basophils play a role in both parasitic infections and allergies[1]. They are found in tissues where allergic reactions are occurring and probably contribute to the severity of these reactions. Basophils have protein receptors on their cell surface that bind IgE, an immunoglobulin involved in macroparasite defense and allergy. It is the bound IgE antibody that confers a selective response of these cells to environmental substances, for example, pollen proteins or helminth antigens. Recent studies in mice suggest that basophils may also regulate the behavior of T cells and mediate the magnitude of the secondary immune response.

Secretions

When activated, basophils degranulate to release histamine, proteoglycans (e.g. heparin and chondroitin), and proteolytic enzymes (e.g. elastase and lysophospholipase). They also secrete lipid mediators like leukotrienes, and several cytokines. Histamine and proteoglycans are pre-stored in the cell's granules while the other secreted substances are newly generated. Each of these substances contributes to inflammation. Recent evidence suggests that basophils are an important source of the cytokine, interleukin-4, perhaps more important than T cells. Interleukin-4 is considered one of the critical cytokines in the development of allergies and the production of IgE antibody by the immune system. There are other substances that can activate basophils to secrete which suggests that these cells have other roles in inflammation.

Basopenia (a low basophil count) is difficult to demonstrate as the normal basophil count is so low; it has been reported in association with autoimmune urticaria (a chronic itching condition). Basophilia is also uncommon but may be seen in some forms of leukaemia or lymphoma.

Lymphocyte

Life Span: - weeks to several years

A lymphocyte is a type of white blood cell in the vertebrate immune system.

By their appearance under the light microscope, there are two broad categories of lymphocytes, namely the large granular lymphocytes and the small lymphocytes. Functionally distinct subsets of lymphocytes correlate with their appearance. Most, but not all large granular lymphocytes are more commonly known as the natural killer cells (NK cells). The small lymphocytes are the T cells and B cells. Lymphocytes play an important and integral role in the body's defenses .

The three major types of lymphocyte are T cells, B cells and natural killer (NK) cells

Natural Killer Cells

The discovery of NK cells occurred in the early 1970s during research on the well-characterized ability of T-lymphocytes to lyse tumor cells against which they had been previously immunized. During these experiments, investigators consistently observed what was termed a *natural* reactivity, that is, a certain population of cells seemed to be able to lyse tumor cells without having been previously sensitized to them. As these discoveries were incompatible with established model at the time, many of these observations were initially considered artifacts.

However, by 1973, 'natural killing' activity was established across a wide variety of species, and the existence of a separate lineage of cells possessing this ability was postulated. Through the use of monoclonal antibodies, *natural killing* ability was mapped to the subset of large, granular lymphocytes known today as NK cells.

The cells were named "natural killer" because of the initial notion that they do not require activation in order to kill cells that are "missing self" recognition ("missing-self" recognition is a term used to describe cells with low levels of MHC class I cell surface marker molecules—a situation that could arise due to viral infection, or in tumors under strong selection pressure of killer T cells).

With the discovery of activating receptors almost two decades after the discovery of the inhibitory receptors these cells continue to be called by the same name, though "natural" no longer means the same thing. The term "natural killer" continues to be justified by:

- a morphology characteristic of activated cytotoxic lymphocytes, e.g., large size, high protein synthesis activity in the abundant endoplasmic reticulum (ER), and preformed granules
- the mature state (does not require much new protein synthesis and remodelling before starting to kill)
- the rapid killing activity observed in freshly-isolated NK cells.

Natural killer cells (or **NK cells**) are a type of cytotoxic lymphocyte that constitute a major component of the innate immune system. NK cells play a major role in the rejection of tumors and cells infected by viruses. The cells kill by releasing small cytoplasmic granules of proteins called perforin and granzyme that cause the target cell to die by apoptosis.

NK cells are defined as large granular lymphocytes (LGL) and constitute the third kind of cells differentiated from the common lymphoid progenitor generating B and T lymphocytes.[1] They do not express T-cell antigen receptors (TCR) or Pan T marker CD3 or surface immunoglobulins (Ig) B cell receptor but they usually express the surface markers CD16 (FcγRIII) and CD56 in humans, and NK1.1/NK1.2 in certain strains of mice. Up to 80% of NK cells also express CD8.

They were named "natural killers" because of the initial notion that they do not require activation in order to kill cells that are missing "self" markers of major histocompatibility complex (MHC) class I.

Given their strong cytolytic activity and the potential for auto-reactivity, Natural Killer cell activity is tightly regulated. Natural Killer cells must receive an activating signal, which can come in a variety of forms, the most important of which are listed below.

- *Cytokines*

  The cytokines play a crucial role in NK cell activation. As these are stress-molecules, released by cells upon viral infection, they serve to signal to the NK cell the presence of viral pathogens.

- *Fc Receptor*

  NK cells, along with macrophages and several other cell types, express the FcR molecule, an activating biochemical receptor that binds the Fc portion of antibodies. This allows Natural Killer cells to target cells against which a humoral response has been mobilized and to lyse cells through Antibody-dependent cellular cytotoxicity (ADCC).

- *Activating and inhibitory receptors*

  Aside from the Fc receptor, Natural Killer cells express a variety of receptors that serve to either activate or suppress their cytolytic activity. These receptors bind to various ligands on target cells, both endogenous and exogenous, and have an important role in regulating the NK cell response.

Mechanism

NK cells are cytotoxic; small granules in their cytoplasm contain proteins such as perforin and proteases known as granzymes. Upon release in close proximity to a cell slated for killing, perforin forms pores in the cell membrane of the target cell through which the granzymes and associated molecules can enter, inducing apoptosis. The distinction between apoptosis and cell lysis is important in immunology: lysing a virus-infected cell would only release the virions, whereas apoptosis leads to destruction of the virus inside.

NK cells are activated in response to interferons or macrophage-derived cytokines. They serve to contain viral infections while the adaptive immune response is generating antigen-specific cytotoxic T cells that can clear the infection. Patients deficient in NK cells prove to be highly susceptible to early phases of herpes virus infection.

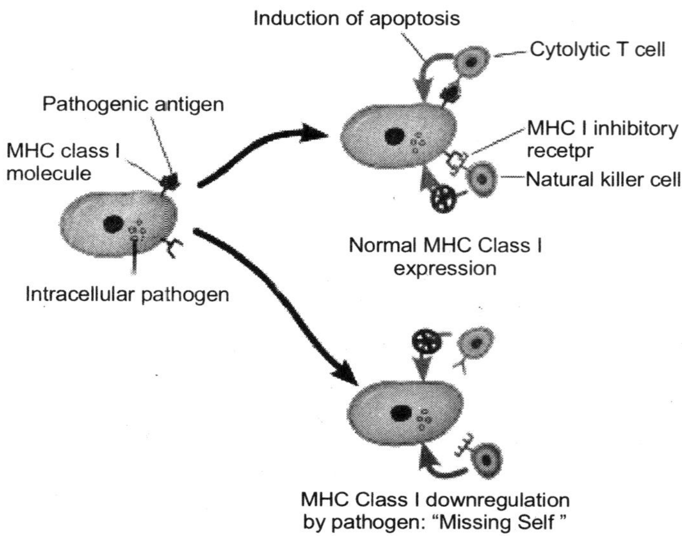

Fig. 8.6 : The Schematic Diagram Indicating the Complementary Activities of Cytotoxic T-cells and NK Cells

In order for NK cells to defend the body against viruses and other pathogens, they require mechanisms that enable the determination of whether a cell is infected or not. The exact mechanisms remain the subject of current investigation, but recognition of an "altered self" state is thought to be involved. To control their cytotoxic activity, NK cells possess two types of surface receptors: *activating receptors* and *inhibitory receptors*. Most of these receptors are not unique to NK cells and can be present in some T cell subsets as well.

These inhibitory receptors recognize MHC class I alleles, which could explain why NK cells kill cells possessing low levels of MHC class I molecules. This inhibition is crucial to the role played by NK cells. MHC class I molecules consist of the main mechanism by which cells display viral or tumor antigens to cytotoxic T-cells. A common evolutionary adaption to this seen in both intracellular microbes and tumours is a chronic down-regulation of these MHC I molecules, rendering the cell impervious to T-cell mediated immunity. It is believed that NK cells, in turn, evolved as an evolutionary response to this adaption, as the loss of the MHC would deprive these cells of the inhibitory effect of MHC and render these cells vulnerable to NK cell mediated apoptosis.

Receptor Types

NK cell receptor types (with inhibitory as well as some activating members) are differentiated by structure:

1. **CD94 : NKG2** (heterodimers) — a C-type lectin family receptor, conserved in both rodents and primates and identifies non-classical (also non-polymorphic) MHC I molecules like HLA E. Though indirect, this is a way to survey the levels of classical (polymorphic) HLA molecules, however, because expression of HLA-E at the cell surface is dependent upon the presence of classical MHC class I leader peptides.
2. **Ly49** (homodimers) — a relatively ancient, C-type lectin family receptor; are of multigenic presence in mice, while humans have only one pseudogenic Ly49; the receptor for classical (polymorphic) MHC I molecules.
3. **KIR** (Killer-cell immunoglobulin-like receptors) — belong to a multigene family of more recently-evolved Ig-like extracellular domain receptors; are present in non-human primates; and are the main receptors for both classical MHC I (HLA-A, HLA-B, HLA-C) and also non-classical HLA-G in primates. Some KIRs are specific for certain HLA subtypes.
4. **ILT** or **LIR** (leukocyte inhibitory receptors) — are recently-discovered members of the Ig receptor family.

NK cell receptors can also be differentiated based on function. Natural cytotoxicity receptors directly induce apoptosis after binding to ligands that directly indicate infection of a cell. The MHC dependent receptors (described above) use an alternate pathway to induce apoptosis in infected cells. NK cells are a part of innate immune system and play a major role in defending the host from both tumors and virally infected cells. NK cells distinguish infected cells and tumors from normal and uninfected cells by recognizing level changes of a surface molecule called MHC (major histocompatibility complex) class I. NK cells are activated in response to a family of cytokines called interferons. Activated NK cells release cytotoxic (cell-killing) granules which then destroy the altered cells.[2] They were named "natural killer" because of the initial notion that they do not require prior activation in order to kill cells which are missing MHC class I.

T Cells and B Cells

T cells and B cells are the major cellular components of the adaptive immune response. T cells are involved in cell-mediated immunity whereas B cells are primarily responsible for humoral immunity (relating to antibodies). The function of T cells and B cells is to recognize specific "non-self" antigens, during a process known as antigen presentation. Once they have identified an invader, the cells generate specific responses that are tailored to maximally eliminate specific pathogens or pathogen infected cells. B cells respond to pathogens by producing large quantities of antibodies which then neutralize foreign objects like bacteria and viruses. In response to pathogens some T cells, called *helper T cells*, produce cytokines that direct the immune response while other T cells, called *cytotoxic T cells*, produce toxic granules that induce the death of pathogen infected cells. Following activation, B cells and T cells leave a lasting legacy of the antigens they have encountered, in the form of *memory cells*. Throughout the lifetime of an animal these memory cells will "remember" each specific pathogen encountered, and are able to mount a strong response if the pathogen is detected again.B cells are not only produced in the **bone marrow** but also mature there.

However, the precursors of T cells leave the bone marrow and mature in the **thymus** (which accounts for their designation).

Each B cell and T cell is **specific** for a particular **antigen**. What this means is that each is able to **bind to** a particular molecular structure.

The specificity of binding resides in a **receptor** for antigen:

- the B cell receptor (**BCR**) for antigen and
- the T cell receptor (**TCR**) respectively.

Both BCRs and TCRs **share** these properties:

- They are integral membrane proteins.
- They are present in thousands of identical copies exposed at the cell surface.
- They are made before the cell ever encounters an antigen.
- They are encoded by genes assembled by the recombination of segments of DNA.
- They have a unique binding site.
- This site binds to a portion of the antigen called an antigenic determinant or **epitope**.
- The binding, like that between an **enzyme** and its **substrate** depends on complementarity of the surface of the receptor and the surface of the epitope.
- The binding occurs by non-covalent forces (again, like an enzyme binding to its substrate).
- Successful binding of the antigen receptor to the epitope, **if accompanied by additional signals**, results in:
  - stimulation of the cell to leave G_0 and enter the cell cycle.
  - Repeated mitosis leads to the development of a **clone** of cells bearing the same antigen receptor; that is, a clone of cells of the identical specificity.

BCRs and TCRs **differ** in:

- their structure;
- the genes that encode them;
- the type of epitope to which they bind.

B Cells

- BCRs bind soluble antigens (like diphtheria toxoid, the protein introduced into your body in the DTP vaccine).
- The bound antigen molecules are engulfed into the B cell by receptor-mediated endocytosis.
- The antigen is digested into fragments
- which are then displayed at the cell surface nestled inside a class II histocompatibility molecule.
- Helper T cells specific for this structure (i.e., with complementary TCRs) bind the B cell and

- secrete **lymphokines** that:
  - stimulate the B cell to enter the cell cycle and develop, by repeated mitosis, into a **clone** of cells with identical BCRs;
  - switch from synthesizing their BCRs as integral membrane proteins to a soluble version;
  - differentiate into **plasma cells** that secrete these soluble BCRs, which we now call **antibodies**.

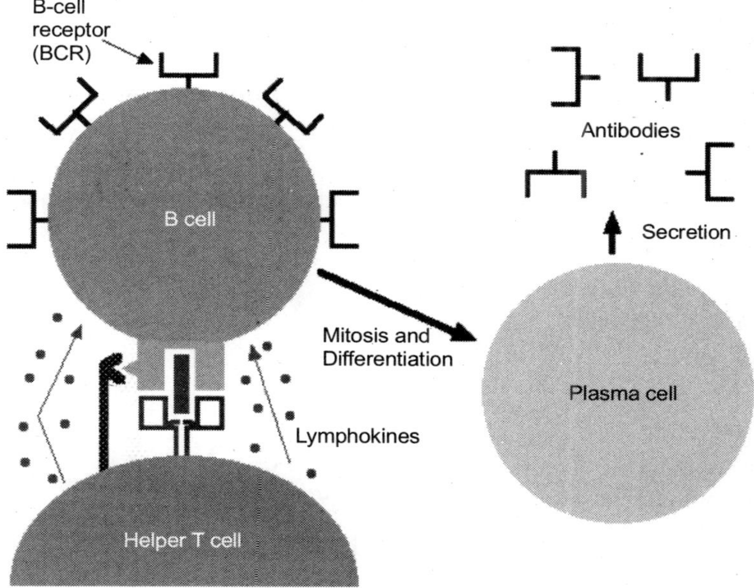

Fig. 8.7 : The function of T Cell 1

T Cells

The surface of each T cell also displays thousands of identical **T cell receptors (TCRs)**.

There are two types of T cells that differ in their TCR:

- **alpha/beta** (αβ) T cells. Their TCR is a heterodimer of an alpha chain with a beta chain. Each chain has a variable (V) region and a constant (C) region. The V regions each contain 3 **hypervariable regions** that make up the antigen-binding site.
- **gamma/delta** (γβ) T cells. Their TCR is also a heterodimer of a gamma chain paired with a delta chain.

The discussion that follows now concerns alpha/beta T cells. Gamma/delta T cells, which are less well understood, are discussed at the end.

The TCR (of alpha/beta T cells) binds a bimolecular complex displayed at the surface of **some other cell** called an **antigen-presenting cell (APC)**. This complex consists of:

- a fragment of an antigen lying within the groove of a
- histocompatibility molecule

The complex has been compared to a “hot dog in a bun”.

Most of the T cells in the body belong to one of two subsets. These are distinguished by the presence on their surface of **one or the other** of two glycoproteins designated:

- **CD4**
- **CD8**

Which of these molecules is present determines what types of cells the T cell can bind to.

- **CD8$^+$** T cells bind epitopes that are part of **class I histocompatibility molecules**. Almost all the cells of the body express class I molecules.
- **CD4$^+$** T cells bind epitopes that are part of **class II histocompatibility molecules**. Only specialized antigen-presenting cells express class II molecules. These include:
  - dendritic cells
  - phagocytic cells like **macrophages** and
  - **B cells**!

CD8$^+$ T cells

The best understood CD8$^+$ T cells are **cytotoxic T lymphocytes** (**CTL**s). They secrete molecules that destroy the cell to which they have bound.

This is a very useful function if the target cell is infected with a virus because the cell is usually destroyed before it can release a fresh crop of viruses able to infect other cells.

An example will show the beauty and biological efficiency of this mechanism.

Every time you get a virus infection, say influenza (flu), the virus invades certain cells of your body (in this case cells of the respiratory passages). Once inside, the virus subverts the metabolism of the cell to make more virus. This involves synthesizing a number of different macromolecules encoded by the viral genome.

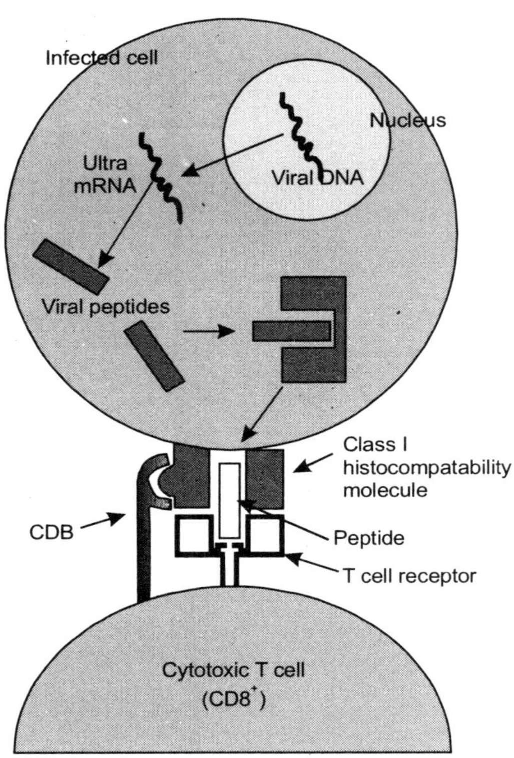

Fig . 8.8 : The CD8 T cells

In due course, these are assembled into a fresh crop of virus particles that leave the cell (often killing it in the process) and spread to new target cells.

Except while in transit from their old homes to their new, the viruses work inside of your cells safe from any antibodies that might be present in blood, lymph, and secretions.

But early in the process, infected cells display fragments of the viral proteins in their surface class I molecules. CTLs specific for that antigen will be able to bind to the infected cell and often will be able to destroy it before it can release a fresh crop of viruses.

In general, the role of the $CD8^+$ T cells is to monitor all the cells of the body, ready to destroy any that express foreign antigen fragments in their class I molecules.

$CD4^+$ T cells

$CD4^+$ T cells bind an epitope consisting of an antigen fragment lying in the groove of a class II histocompatibility molecule. $CD4^+$ T cells are essential for both the cell-mediated and antibody-mediated branches of the immune system:

- **cell-mediated immunity**

  These $CD4^+$ cells bind to antigen presented by antigen-presenting cells (APCs) like phagocytic macrophages and dendritic cells. The T cells then release **lymphokines** that attract other cells to the area. The result is **inflammation**: the accumulation of cells and molecules that attempt to wall off and destroy the antigenic material (an abscess is one example, the rash following exposure to poison ivy is another).

- **antibody-mediated immunity**

  These $CD4^+$ cells, called **helper T cells**, bind to antigen presented by B cells (as shown above). The result is the development of **clones** of **plasma cells** secreting antibodies against the antigenic material.

AIDS Patients lose their $CD4^+$ T Cells

AIDS provides a vivid and tragic illustration of the importance of $CD4^+$ T cells in immunity. The human immunodeficiency virus (**HIV**) binds to CD4 molecules and thus is able to invade and infect $CD4^+$ T cells. As the disease progresses, the number of $CD4^+$ T cells declines below its normal level of about 1000 per microliter (μl). (A partial explanation for this may be the unceasing efforts of the patient's $CD8^+$ T cells to destroy the infected $CD4^+$ cells. However, it turns out that only a small fraction of the patients $CD4^+$ T cells are infected at any given time. How **uninfected** $CD4^+$ cells may be induced to commit suicide is discussed in the page on apoptosis.)

When the number of $CD4^+$ T cells drops below 400 per microliter, the ability of the patient to mount an immune response declines dangerously. Not only does the patient become hypersusceptible to pathogens that give all of us grief but also to microorganisms, especially viruses and fungi, that normally inhabit our tissues without harming us. Eventually the patient dies of these **opportunistic infections**.

Building the T-cell Repertoire

T cells have receptors (TCRs) that bind to antigen fragments nestled in MHC molecules. But

- all cells express class I MHC molecules containing fragments derived from self proteins;
- many cells express class II MHC molecules that also contain self peptides.

This presents a risk to the animal of the T cells recognizing these self-peptide/self-MHC complexes and mounting an autoimmune attack against them. Fortunately, this is usually avoided by a process of selection that goes on in the thymus (where all T cells develop).

The process works like this:

- The **precursors** of T cells — like all blood cells — are formed in the bone marrow.
- These cells then migrate to the cortex of the thymus. At this time they have neither a complete TCR nor either CD4 or CD8 (thus are called "double-negative" or **DN** cells).
- In the cortex of the thymus, they
  - begin to form a TCR;
  - synthesize both CD4 and CD8 (so now they are "double-positive" or **DP** cells).
  - The cortical cells of the thymus express a wide variety of small molecules, usually a peptide of 6–8 amino acids derived from body proteins; that is, "self" proteins such as
    - proteins within the cytosol
    - serum proteins; i.e., proteins circulating in the blood and lymph

    nestled in a histocompatibility molecule (encoded by the MHC).
  - Most of the cells (~97%) will produce a TCR that does not bind to any of the peptide-MHC molecules present on the surface of the cortical cells. Unless they can try again with a new TCR, these cells die by "neglect" (by apoptosis, actually).
  - Those remaining cells whose TCR
    - has bound a peptide antigen presented in **class II** MHC molecule stop expressing CD8 and become $CD4^+$ T cells. It is these cells that will go on to become
      - **Th1** cells in cell-mediated immune responses;
      - Th1 helper cells for cytotoxic T lymphocytes (CTLs);
      - **Th2** helper cells for **B cells** [View].
    - has bound a peptide antigen presented in **class I** MHC molecule stop expressing CD4 and become $CD8^+$ T cells.
  - Both sets of cells are said to have undergone **positive selection**.
- After positive selection, these cells migrate to the **medulla** of the thymus.
- There those cells whose TCR binds very strongly to complexes of self-peptide and self-MHC are destroyed (again by apoptosis).

- This process of **negative selection** is important as it eliminates cells that might otherwise mount an autoimmune attack. It is one of the ways in which **tolerance** to self antigens is achieved.
- The cells whose TCRs bind antigen at an affinity below the threshold that triggers apoptosis are free to leave the thymus and migrate throughout the immune system (lymph nodes, spleen, etc.)
- It is this population which we depend on to mount immune responses against foreign antigens. A TCR that binds self-peptide/self-MHC with low affinity may well bind a foreign-peptide in self MHC with high affinity.

Gamma/Delta (γδ) T Cells

Gamma/delta T cells differ from their alpha/beta cousins in several ways:

- Their TCR is encoded by different gene segments.
- Their TCR binds to antigens that can be
  - intact proteins (just as antibodies do) as well as a variety of other types of organic molecules (often containing phosphorus atoms).
  - **not** "presented" within class I or class II histocompatibility molecules;
  - not presented by "professional" **a**ntigen-**p**resenting **c**ells (APCs) like dendritic cells.
- Most of these T cells have neither CD8 nor CD4 on their surface. This makes sense because they have no need to recognize class I and class II histocompatibility molecules.
- Gamma/Delta T cells, like alpha/beta T cells, develop in the thymus. However, they migrate from there into body tissues, especially epithelia (e.g., intestine, skin, lining of the vagina), and don't recirculate between blood and lymph nodes (they represent no more than 5% of the T cells in the blood and are even rarer in lymph nodes). They encounter antigens on the surface of the epithelial cells that surround them rather than relying on the APCs found in lymph nodes.

What is the Function of γδ T cells?

That is still something of a mystery.

Situated as they are at the interfaces between the external and internal worlds, they may represent a first line of defense against invading pathogens. Their response does seem to be quicker than that of αβ T cells.

Curiously, many of the antigens to which γδ T cells respond are found not only on certain types of invaders (e.g., **Mycobacterium tuberculosis**, the agent of tuberculosis) but also in host cells that are under attack by pathogens.

Knockout mice that cannot make γδ T cells are slower to heal injuries to their skin. They are also much more susceptible to skin cancers than normal mice. Perhaps, immune surveillance is one of the functions of γδ T cells.

Lymphocyte Development

Mammalian stem cells differentiate into several kinds of blood cell within the bone marrow (see Fig. 8.5) This process is called haematopoiesis. All lymphocytes originate, during this process, from a common lymphoid progenitor before differentiating into their distinct lymphocyte types. The differentiation of lymphocytes follows various pathways in a hierarchical fashion as well as in a more plastic fashion. The formation of lymphocytes is known as lymphopoiesis. B cells mature into B lymphocytes in the bone marrow, while T cells migrate to and mature in a distinct organ, called the thymus. Following maturation, the lymphocytes enter the circulation and peripheral lymphoid organs (e.g. the spleen and lymph nodes) where they survey for invading pathogens and/or tumor cells.

The lymphocytes involved in adaptive immunity (i.e. B and T cells) differentiate further after exposure to an antigen; they form effector and memory lymphocytes. Effector lymphocytes function to eliminate the antigen, either by releasing antibodies (in the case of B cells), cytotoxic granules (cytotoxic T cells) or by signaling to other cells of the immune system (helper T cells). Memory cells remain in the peripheral tissues and circulation for an extended time ready to respond to the same antigen upon future exposure.

They live weeks to several years, which is very long compared to other leukocytes.

Characteristics

Microscopically, in a Wright's stained peripheral blood smear, a normal lymphocyte has a large, dark-staining nucleus with little to no eosinophilic cytoplasm. In normal situations, the coarse, dense nucleus of a lymphocyte is approximately the size of a red blood cell (about 7 micrometres in diameter).[3] Some lymphocytes show a clear perinuclear zone (or halo) around the nucleus or could exhibit a small clear zone to one side of the nucleus. Polyribosomes are a prominent feature in the lymphocytes and can be viewed with an electron microscope. The ribosomes are involved in protein synthesis allowing the generation of large quantities of cytokines and immunoglobulins by these cells.

Typical Recognition Markers for Lymphocytes

Lymphocyte Class	Function of Lymphocyte	Proportion	Phenotypic Marker(s)
NK cells	Lysis of virally infected cells and tumour cells	7% (2-13%)	CD16 CD56 but not CD3
Helper T cells	Release cytokines and growth factors that regulate other immune cells	46% (28-59%)	TCRαβ, CD3 and CD4
Cytotoxic T cells	Lysis of virally infected cells, tumour cells and allografts	19% (13-32%)	TCRαβ, CD3 and CD8
γδ T cells	Immunoregulation and cytotoxicity		TCRγδ and CD3
B cells	Secretion of antibodies	23% (18-47%)	MHC class II, CD19 and CD21

It is impossible to distinguish between T cells and B cells in a peripheral blood smear.[3] Normally, flow cytometry testing is used for specific lymphocyte population counts. This can be used to specifically determine the percentage of lymphocytes that contain a particular combination of specific cell surface proteins, such as immunoglobulins or cluster of differentiation (CD) markers or that produce particular proteins (for example, cytokines using intracellular cytokine staining (ICCS)). In order to study the function of a lymphocyte by virtue of the proteins it generates, other scientific techniques like the ELISPOT or secretion assay techniques can be used.

In the circulatory system they move from lymph node to lymph node. This contrasts with macrophages, which are rather stationary in the nodes.

Lymphocytes and Disease

A lymphocyte count is usually part of a peripheral complete blood cell count and is expressed as percentage of lymphocytes to total white blood cells counted.

An increase in lymphocyte concentration is usually a sign of a viral infection (in some rare case, leukemias are found through an abnormally raised lymphocyte count in an otherwise normal person). A general increase in the number of lymphocytes is known as lymphocytosis whereas a decrease is lymphocytopenia.

A low normal to low absolute lymphocyte concentration is associated with increased rates of infection after surgery or trauma.

Once basis for low T cell lymphocytes occurs when the human immunodeficiency virus (HIV) infects and destroys T cells (specifically, the $CD4^+$ subgroup of T lymphocytes). Without the key defense that these T cells provide, the body becomes susceptible to opportunistic infections that otherwise would not affect healthy people. The extent of HIV progression is typically determined by measuring the percentage of $CD4^+$ T cells in the patient's blood. The effects of other viruses or lymphocyte disorders can also often be estimated by counting the numbers of lymphocytes present in the blood.

Platelets

Normal Count: 150,000 and 450,000 per μl (microlitre) of blood (150–450 × 10^9/L)
Life Span: between 8 and 12 days

Brewer traced the history of the discovery of the platelet. Although red blood cells had been known since van Leeuwenhoek, it was the German anatomist Max Schultze (1825-1874) who first offered a description of the platelet in his newly-founded journal *Archiv für mikroscopische Anatomie*]. He describes "spherules" to be much smaller than red blood cells that are occasionally clumped and may participate in collections of fibrous material. He recommends further study of the findings.

Giulio Bizzozero (1846-1901), building on Schultze's findings, used "living circulation" to study blood cells of amphibians microscopically *in vivo*. He is especially noted for discovering that platelets clump at the site of blood vessel injury, a process that precedes the formation of a blood clot. This observation confirmed the role of platelets in coagulation

Platelets, or **thrombocytes** (from Greek *Thrombos*— «clot» and *kyttaro* — «cell»), are small, irregularly-shaped anuclear cells (i.e. cells that do not have a nucleus containing DNA), 2-3 μm in diameter, which are derived from fragmentation of precursor megakaryocytes. The average lifespan of a platelet is between 8 and 12 days. Platelets play a fundamental role in hemostasis and are a natural source of growth factors. They circulate in the blood of mammals and are involved in hemostasis, leading to the formation of blood clots.

If the number of platelets is too low, excessive bleeding can occur. However, if the number of platelets is too high, blood clots can form (thrombosis), which may obstruct blood vessels and result in such events as a stroke, heart attack, pulmonary embolism or the blockage of blood vessels to other parts of the body, such as the extremities of the arms or legs. An abnormality or disease of the platelets is called a thrombocytopathy which could be either a low number of platelets (thrombocytopenia), a decrease in function (thrombasthenia), or an increase in the number of (thrombocytosis). There are disorders that reduce the number of platelets, such as heparin-induced thrombocytopenia (HIT) or thrombotic thrombocytopenic purpura (TTP) that typically cause thromboses, or clots, instead of bleeding.

Platelets release a multitude of growth factors including Platelet-derived growth factor (PDGF), a potent chemotactic agent, and TGF beta, which stimulates the deposition of extracellular matrix. Both of these growth factors have been shown to play a significant role in the repair and regeneration of connective tissues. Other healing-associated growth factors produced by platelets include basic fibroblast growth factor, insulin-like growth factor 1, platelet-derived epidermal growth factor, and vascular endothelial growth factor. Local application of these factors in increased concentrations through Platelet-rich plasma (PRP) has been used as an adjunct to wound healing for several decades

- Platelets are produced in blood cell formation (thrombopoiesis) in bone marrow, by budding off from megakaryocytes.
- The physiological range for platelets is 150-400 $\times$ 10^9 per litre.
- Around 1 $\times$ 10^{11} platelets are produced each day by an average healthy adult.
- The lifespan of circulating platelets is 7 to 10 days.
- Megakaryocyte and platelet production is regulated by thrombopoietin, a hormone usually produced by the liver and kidneys.
- Each megakaryocyte produces between 5,000 and 10,000 platelets.
- Old platelets are destroyed by phagocytosis in the spleen and by Kupffer cells in the liver.
- A reserve of platelets are stored in the spleen and are released when needed by sympathetically-induced splenic contraction.

Thrombus Formation

The function of platelets is the maintenance of haemostasis. This is achieved primarily by the formation of thrombi, when damage to the endothelium of blood vessels occurs. On the converse, thrombus formation must be inhibited at times when there is no damage to the endothelium.

Activation

The inner surface of blood vessels is lined with a thin layer of endothelial cells that, in normal hemostasis, acts to inhibit platelet activation by producing nitric oxide, endothelial-ADPase, and PGI_2. Endothelial-ADPase clears away the platelet activator, ADP.

Endothelial cells produce a protein called von Willebrand factor (vWF), a cell adhesion ligand, which helps endothelial cells adhere to collagen in the basement membrane. Under physiological conditions, collagen is not exposed to the bloodstream. vWF is secreted constitutively into the plasma by the endothelial cells, and is stored in granules within the endothelial cell and in platelets. When the endothelial layer is injured, collagen, vWF and tissue factor from the subendothelium is exposed to the bloodstream. When the platelets contact collagen or vWF, they are activated. They are also activated by thrombin (formed with the help of tissue factor). They can also be activated by a negatively-charged surface, such as glass. Platelet activation further results in the scramblase-mediated transport of negatively-charged phospholipids to the platelet surface. These phospholipids provide a catalytic surface (with the charge provided by phosphatidylserine and phosphatidylethanolamine) for the tenase and prothrombinase complexes. Calcium ions are essential for binding of these coagulation factors.

Shape Change

Activated platelets change in shape to become more spherical, and pseudopods form on their surface. Thus they assume a stellate shape.

Granule Secretion

Platelets contain alpha and dense granules. Activated platelets excrete the contents of these granules into their canalicular systems and into surrounding blood. There are two types of granules:

- dense granules (containing ADP or ATP, calcium, and serotonin).
- α-granules (containing platelet factor 4, transforming growth factor-β1, platelet-derived growth factor, fibronectin, B-thromboglobulin, vWF, fibrinogen, and coagulation factors V and XIII).

Thromboxane A2 Synthesis

Platelet activation initiates the arachidonic acid pathway to produce TXA_2. TXA_2 is involved in activating other platelets and its formation is inhibited by CO_x inihibitors, such as aspirin.

Adhesion and Aggregation

Platelets aggregate, or clump together, using fibrinogen and vWF as a connecting agent. The most abundant platelet aggregation receptor is glycoprotein IIb/IIIa (gpIIb/IIIa); this is a calcium-dependent receptor for fibrinogen, fibronectin, vitronectin, thrombospondin, and von Willebrand factor (vWF). Other receptors include GPIb-V-IX complex (vWF) and GPVI (collagen).

Activated platelets will adhere, via glycoprotein (GP) Ia, to the collagen that is exposed by endothelial damage. Aggregation and adhesion act together to form the platelet plug. Myosin and actin filaments in platelets are stimulated to contract during aggregation, further reinforcing the plug.

Platelet aggregation is stimulated by ADP, thromboxane, and α2 receptor-activation, but inhibited by other inflammatory products like PGI2 and PGD2. Platelet aggregation is enhanced by exogenous administration of anabolic steroids.

Wound Repair

The blood clot is only a temporary solution to stop bleeding; vessel repair is therefore needed. The aggregated platelets help this process by secreting chemicals that promote the invasion of fibroblasts from surrounding connective tissue into the wounded area to form a scar. The obstructing clot is slowly dissolved by the fibrinolytic enzyme, plasmin, and the platelets are cleared by phagocytosis.

Role in Disease (High and Low Counts)

A normal platelet count in a healthy individual is between 150,000 and 450,000 per µl (microlitre) of blood ($150–450 \times 10^9$/L). Ninety-five per cent of healthy people will have platelet counts in this range. Some will have statistically abnormal platelet counts while having no demonstrable abnormality. However, if it is either very low or very high, the likelihood of an abnormality being present is higher.

Both thrombocytopenia and thrombocytosis may present with coagulation problems. In general, low platelet counts increase bleeding risks; however there are exceptions. For example, immune heparin-induced thrombocytopenia and thrombocytosis (high counts) may lead to thrombosis, although this is mainly when the elevated count is due to myeloproliferative disorder.

Low platelet counts are, in general, not corrected by transfusion unless the patient is bleeding or the count has fallen below 5×10^9/L. Transfusion is contraindicated in thrombotic thrombocytopenic purpura (TTP), as it fuels the coagulopathy. In patients undergoing surgery, a level below 50×10^9/L is associated with abnormal surgical bleeding, and regional anaesthetic procedures such as epidurals are avoided for levels below 80-100.

Normal platelet counts are not a guarantee of adequate function. In some states, the platelets, while being adequate in number, are *dysfunctional*. For instance, aspirin irreversibly disrupts platelet function by inhibiting cyclooxygenase-1 (COX1), and hence normal hemostasis. The resulting platelets have no DNA and are unable to produce new cyclooxygenase. Normal platelet function will not return until the use of aspirin has ceased and enough of the affected platelets have been replaced by new ones, which can take over a week. Ibuprofen, another NSAID, does not have such a long duration effect, with platelet function usually returning within 24 hours, and taking ibuprofen before aspirin will prevent the irreversible effects of aspirin[]. Uremia, a consequence of renal failure, leads to platelet dysfunction that may be ameliorated by the administration of desmopressin.

Monocytes

Normal Count : 0-9%
Life Span : –59 days

Monocyte is a type of white blood cell, part of the human body's immune system. Monocytes have two main functions in the immune system: (1) replenish resident macrophages and dendritic cells under normal states, and (2) in response to inflammation signals, monocytes can move quickly (approx. 8-12 hours) to sites of infection in the tissues and divide/differentiate into macrophages and dendritic cells to elicit an immune response. Half of them are stored in the spleen. Monocytes are usually identified in stained smears by their large bilobate nucleus.

Monocytes are produced by the bone marrow from haematopoietic stem cell precursors called monoblasts. Monocytes circulate in the bloodstream for about one to three days and then typically move into tissues throughout the body. They constitute between three to eight percent of the leukocytes in the blood. Half of them are stored as a reserve in the spleen in clusters in the red pulp's Cords of Billroth. In the tissues monocytes mature into different types of macrophages at different anatomical locations.

Monocytes are responsible for phagocytosis (ingestion) of foreign substances in the body. Monocytes can perform phagocytosis using intermediary (opsonising) proteins such as antibodies or complement that coat the pathogen, as well as by binding to the microbe directly via pattern-recognition receptors that recognize pathogens. Monocytes are also capable of killing infected host cells via antibody, termed antibody-mediated cellular cytotoxicity. Vacuolization may be present in a cell that has recently phagocytized foreign matter.

Monocytes which migrate from the bloodstream to other tissues will then differentiate into tissue resident macrophages or dendritic cells. Macrophages are responsible for protecting tissues from foreign substances but are also suspected to be the predominant cells involved in triggering atherosclerosis. They are cells that possess a large smooth nucleus, a large area of cytoplasm and many internal vesicles for processing foreign material.

Monocyte Subpopulations

There are two types of monocytes in human blood:

(*a*) the classical monocyte, which is characterized by high level expression of the CD14 cell surface receptor (CD14++ monocyte) and

(*b*) the non-classical, pro-inflammatory monocyte with low level expression of CD14 and with additional co-expression of the CD16 receptor (CD14+CD16+ monocyte).

The CD14+CD16+ monocytes develop from the CD14++ monocytes, i.e. they are a more mature version. After stimulation with microbial products the CD14+CD16+ monocytes produce high amounts of pro-inflammatory cytokines like tumor necrosis factor and interleukin-12.

Diagnostic Use

A *monocyte count* is part of a complete blood count and is expressed either as a ratio of monocytes to the total number of white blood cells counted, or by absolute numbers. Both may be useful in determining or refuting a possible diagnosis.

Monocytosis

Monocytosis is the state of excess monocytes in the peripheral blood. It may be indicative of various disease states. Examples of processes that can increase a monocyte count include:

- chronic inflammation
- stress response
- hyperadrenocorticism
- immune-mediated disease
- infectious mononucleosis
- pyogranulomatous disease
- necrosis
- red cell regeneration
- Viral Fever.

A high count of CD14+CD16+ monocytes is found in severe infection (sepsis) and a very low count of these cells is found after therapy with immuno-suppressive glucocorticoids

Monocytopenia

Monocytopenia is a form of leukopenia associated with a deficiency of monocytes.

Dendritic Cells

Monocytes can be used to generate dendritic cells in vitro by adding cytokines like Granulocyte Monocyte Colony Stimulating Factor (GMCSF) and IL-4.

Plasma

Blood plasma is the yellow liquid component of blood, in which the blood cells in whole blood would normally be suspended. It makes up about 55% of the total blood volume. It is mostly water (90% by volume) and contains dissolved proteins, glucose, clotting factors, mineral ions, hormones and carbon dioxide (plasma being the main medium for excretory product transportation). Blood plasma is prepared by spinning a tube of fresh blood in a centrifuge until the blood cells fall to the bottom of the tube. The blood plasma is then poured or drawn off. Blood plasma has a density of approximately 1025 kg/m^3, or 1.025 kg/l.

Blood serum is blood plasma without fibrinogen or the other clotting factors (i.e., whole blood minus both the cells *and* the clotting factors).

Plasmapheresis is a medical therapy that involves blood plasma extraction, treatment, and reintegration.

"Fresh frozen plasma" (FFP) is prepared from a single unit of blood or by apheresis, drawn from a single person. It is frozen to –40 °C (–104°F) after collection and can be stored for ten years from date of collection. The term "FFP" is sometimes used informally to mean any frozen transfusable plasma product, including products which do not meet the standards for FFP. The FFP contains all of the coagulation factors and proteins present in the original unit of blood. It is used to treat coagulopathies from warfarin overdose, liver disease, or dilutional coagulopathy. Other transfusable plasma is identical except that the coagulation factors are no longer considered completely viable. This is particularly important for Factor VIII and hemophilia, but these have been mostly replaced by more specific Factor VIII concentrates in the developed world and true FFP is rarely used for that indication.

Plasma used as a source of Cryoprecipitate (Plasma, Cryoprecipitate Reduced) cannot be used for treatment of some coagulation problems but is still acceptable for many uses.

Dried Plasma

"Dried plasma" was developed and first used in World War II. Prior to the United States' involvement in the war, liquid plasma and whole blood were used. The "Blood for Britain" program during the early 1940s was quite successful (and popular in the United States) based on Dr. Charles Drew's contribution. A large project was begun in August of the year 1940 to collect blood in New York City hospitals for the export of plasma to Britain. Dr. Drew was appointed medical supervisor of the "Plasma for Britain" project. His notable contribution at this time was to transform the test tube methods of many blood researchers, including himself, into the first successful mass production techniques.Nonetheless, the decision was made to develop a dried plasma package for the armed forces as it would reduce breakage and make the transportation, packaging, and storage much simpler. The resulting Army-Navy dried plasma package came in two tin cans containing 400 cc bottles. One bottle contained enough distilled water to completely reconstitute the dried plasma contained within the other bottle. In about three minutes, the plasma would be ready to use and could stay fresh for around four hours. Following the "Plasma for Britain" invention, Dr. Drew was named director of the Red Cross blood bank and assistant director of the National Research Council, in charge of blood collection for the United States Army and Navy. Dr. Drew argued against the armed forces directive that blood/plasma was to be separated by the race of the donor. Dr. Drew argued that there was no racial difference in human blood and that the policy would lead to needless deaths as soldiers and sailors were required to wait for "same race" blood. By the end of the war the American Red Cross had provided enough blood for over six million plasma packages. Most of the surplus plasma was returned to the United States for civilian use. Serum albumin replaced dried plasma for combat use during the Korean War.

Composition

Water	91.5%
Protein	7.0%
Inorganic Ions	1.5%

Inorganic Ions

Without sodium, the body will lack extracellular fluid and might affect the blood pressure, leading to insufficient circulation of oxygen causing drowsiness, nausea etc. Lack of potassium will result in muscle abnormalities and weakness, leading to vomiting and diarrhoea. Hence all the inorganic ions facilitate in a very important role as do other cells within the blood.

Ion	Symbol	Concentration (Mmol/L)
Sodium	Na^+	135-146
Potassium	K^+	3.5-5.2
Calcium	Ca^{++}	2.1-2.7
Chloride	Cl^-	98-108
Hydrogen Carbonate	HCO_3^-	23-31
Phosphate	PO_4^-	0.7-1.4

Plasma Proteins

Plasma proteins are made up of 3 main types.

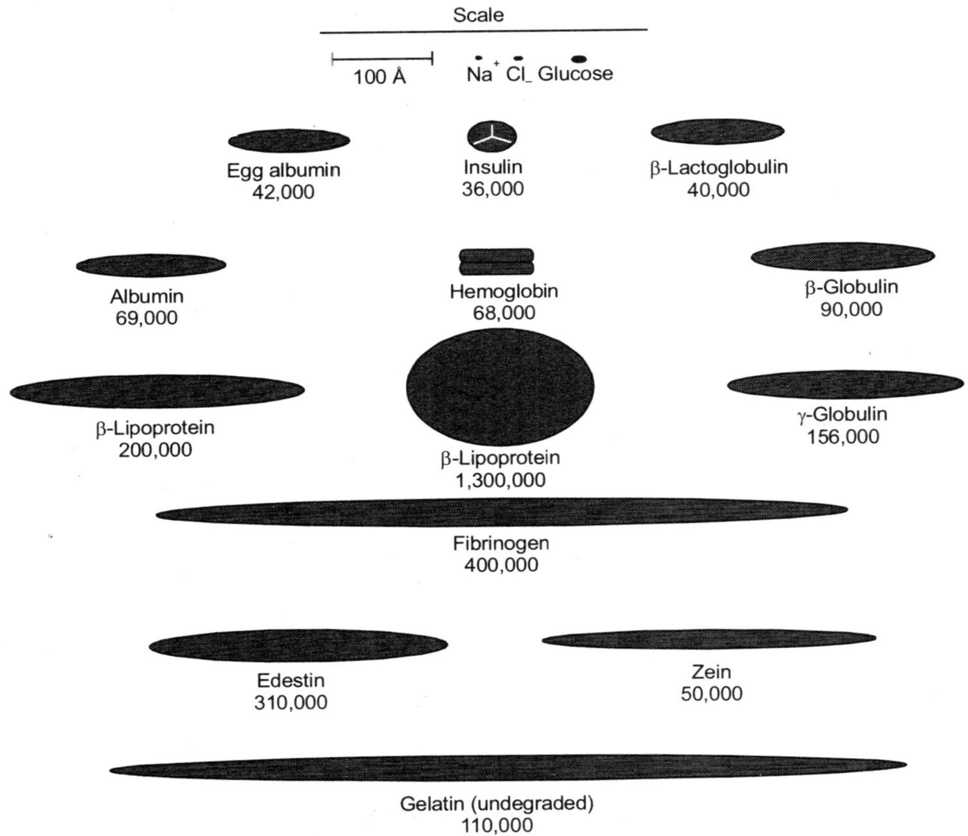

Fig. 8.9 : The Dimensions of the Plasma Protein

- Serum Albumin
- Serum Globulin
- Fibrinogen
- Prothrombin.

The total protein content of whole blood amounts on the average to some 22% and represents a bit more than 3% of protein nitrogen as contrasted with an average of only about 32 mg% of non-protein nitrogen in blood.

The hæmoglobin of the red cells represent about two-thirds of the mass of blood proteins and the plasma proteins about one-third. The proteins of stroma, leucocytes and platelets represents insignificant quantities of the total protein of the blood. Although the single protein hæmoglobin makes up most of the protein content of the red cells, the plasma proteins represent an exceedingly complex mixture of dozens of proteins. These include " simple" proteins lipoproteins and glycoproteins and tratces of large number of proteins which functions as immune bodies (antibodies) enzymes and hormones, The relative dimensions of some of the blood proteins has been diagrammatically represented in Fig. 8.9.

Plasma proteins are generally reffered to as the "albumins" " globulins" and " fibrinogen" and these type s include the complexity of proteins reffered to as above. The fraction of plasma proteins precipitated or salted out by the half saturation with ammonium sulphate was designated as " albumin fraction"

These albumin and globin fractions correspond in solubility and salting out characteristics to the classical definitions of albumins and globulins. Fibrinogen is precipitated with the globulin fractions but because it is a very special protein concerned with the formation of fibrin in the coagulation of blood, it has received a separate designation.

Many of the plasma proteins are synthesized in the liver, which includes albumin and α and β-globulins lipoproteins and the proteins of blood coagulations. Their breadth of function as antibodies and are made in B lymphocytes.

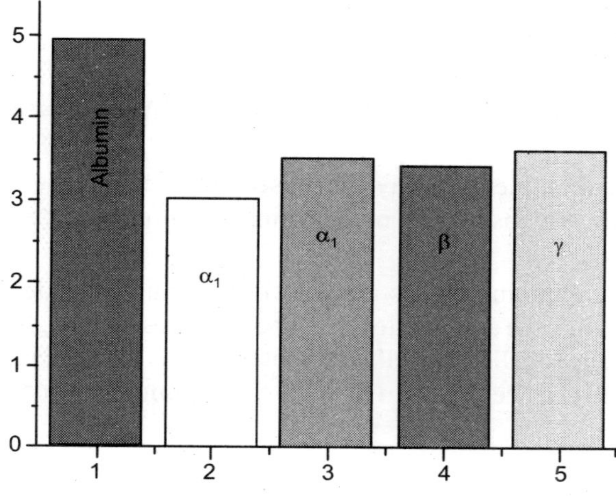

Fig. 8.10 : The Plasma Proteins

The Plasma proteins are classified accordingly from the serum of the blood by carrying out the electrophoresis on cellulose acetate or agrose gel electrophoresis on cellulose acetate or agrose gel. The major classes are pre albumin, albumin, β-globulin and γ-globulin shown in the Fig. 8.10.

Albumin

The Plasma Protein Electrophoretic analysis is of particular use in the diagnosis of nephritic syndrome, lypogam maglobluanemia, cirrhosis α_1 antitypsin deficiency and paraproteinæmia.

Albumin (Latin: albus, white) refers generally to any protein with water solubility, which is moderately soluble in concentrated salt solutions, and experiences heat coagulation (protein denaturation). Substances containing albumin, such as egg white, are called *albuminoids*. It has a molecular weight of about 65kD (65,000 Atomic mass units) and consists of 584 amino acids and contains no carbohydrate.

Types

Serum Albumin

The most well-known type of albumin is serum albumin. It is most common in the blood or serum (providing its name) but it can also appear in other fluid compartments (providing the basis for the CSF/serum albumin ratio, for example.)

Serum albumin is the most abundant blood plasma protein and is produced in the liver and forms a large proportion of all plasma protein. The human version is human serum albumin, and it normally constitutes about 60% of human plasma protein.

Serum albumins are important in regulating blood volume by maintaining the oncotic pressure (also known as colloid osmotic pressure) of the blood compartment. They also serve as carriers for molecules of low water solubility this way isolating their hydrophobic nature, including lipid soluble hormones, bile salts, unconjugated bilirubin, free fatty acids (apoprotein), calcium, ions (transferrin), and some drugs like warfarin, phenobutazone, clofibrate and phenytoin. For this reason, it's sometimes referred as a molecular "taxi". Competition between drugs for albumin binding sites may cause drug interaction by increasing the free fraction of one of the drugs, thereby affecting potency.

Specific types include:

- human serum albumin
- bovine serum albumin (cattle serum albumin) or BSA, often used in medical and molecular biology labs.

Low albumin (hypoalbuminemia) may be caused by liver disease, nephrotic syndrome, burns, protein-losing enteropathy, malabsorption, malnutrition, late pregnancy, artefact, genetic variations and malignancy.

High albumin (hyperalbuminemia) is almost always caused by dehydration. In some cases of retinol (Vitamin A) deficiency the albumin level can become raised to High-normal values (ex: 4.9g/dL). This is because retinol causes cells to swell with water (this is also the reason too much Vitamin A is toxic). In lab experiments it has been shown that All-trans retinoic acid down regulates human albumin production.

Normal range of human serum albumin in adults (> 3 y.o.) is 3.5 to 5 g/dL. For children less than three years of age, the normal range is broader, 2.5-5.5 g/dL.

Albumin binds to the cell surface receptor Albondin.

Other Types

Other types include the storage protein ovalbumin in egg white, and different storage albumins in the seeds of some plants.

- Note that the protein 'albumin' is spelled with an "i", while "albumen" with an "e", is the white of an egg which contains (among other things) several dozen types of albumin (with an 'i'), mostly ovalbumin.

Albumin Measurement

Plasma albumin is a component of the Liver Function Tests (LFTs) but may be ordered separately. Albumin can be measured in Serum (yellow-top tube), plain tube with no additives (red-top tube) or heparin plasma (green-top tube). The reference interval is 36 - 52 g/L. (note upper limit increased from 47 g/L on the 15th June 2007) One of the method used is Bromocresol green on a Roche Modular or Olympus AU2700 analyser.

Alpha 1-Antitrypsin or **α_1-antitrypsin (A1AT)** is a glycoprotein and generally known as **serum trypsin inhibitor**. Alpha 1- antitrypsin is also referred to as alpha-1 proteinase inhibitor (A1PI) because it is a serine protease inhibitor (serpin), inhibiting a wide variety of proteases. It protects tissues from enzymes of inflammatory cells, especially elastase, and has a reference range in blood of 1.5 - 3.5 gram/liter (in US the reference range is generally expressed as mg/dL or micromoles), but the concentration can rise manyfold upon acute inflammation. In its absence, elastase is free to break down elastin, which contributes to the elasticity of the lungs, resulting in respiratory complications such as emphysema, or COPD (chronic obstructive pulmonary disease) in adults and cirrhosis in adults or children.

Function

A1AT is a 52-kDa serpin (serine protease inhibitor), and, in medicine, it is considered the most prominent serpin; the terms *α1-antitrypsin* and *protease inhibitor* (P_i) are often used interchangeably.

Most serpins inactivate enzymes by binding to them covalently, requiring very high levels to perform their function. In the acute phase reaction, a further elevation is required to "limit" the damage caused by activated neutrophil granulocytes and their enzyme elastase, which breaks down the connective tissue fiber elastin.

Like all serine protease inhibitors, A1AT has a characteristic secondary structure of beta sheets and alpha helices. Mutations in these areas can lead to non-functional proteins that can polymerise and accumulate in the liver (infantile hepatic cirrhosis).

Role in Disease

Disorders of this protein include alpha 1-antitrypsin deficiency, a hereditary disorder in which a deficiency of alpha 1-antitrypsin leads to a chronic uninhibited tissue breakdown. This causes the degradation especially of lung tissue, and eventually leads to characteristic manifestations of pulmonary emphysema.

Evidence has shown that cigarette smoke can lead to oxidation of methionine 358 of α_1-antitrypsin, a residue essential for binding elastase; this is thought to be one of the primary mechanisms by which cigarette smoking (or second-hand smoke) can lead to emphysema. Because A1AT is created in the liver, certain mutations in the DNA code for the enzyme can cause misfolding and impaired secretion of the enzyme, which can lead to liver cirrhosis.

A remarkable form of P_i, termed $P_{iPittsburgh}$, functions as an antithrombin (a related serpin), due to a mutation (Met358Arg). One patient with this abnormality has been described; he died of a lethal bleeding diathesis.

Liver biopsy will show abundant PAS-positive globules within periportal hepatocytes.

Nomenclature

The protein was originally named "antitrypsin" because of its ability to covalently bind and irreversibly inactivate the enzyme trypsin in vitro. Trypsin, a type of peptidase, is a digestive enzyme active in the duodenum and elsewhere.

The term *alpha-1* refers to the protein's behavior on protein electrophoresis. On electrophoresis, the protein component of the blood is separated by electric current. There are several *clusters*, the first being albumin, the second being the *alpha*, the third *beta* and the fourth *gamma* (immunoglobulins). The non-albumin proteins are referred to as globulins.

The *alpha* region can be further divided into two sub-regions, termed "1" and "2". Alpha 1-antitrypsin is the main enzyme of the alpha-globulin 1 region.

Another name used is *alpha-1 proteinase inhibitor* (α_1-PI).

Genetics

The gene is located on the long arm of the fourteenth chromosome (14q32.1).

Over 100 different variants of α_1-antitrypsin have been described in various populations. North-Western Europeans are most at risk for carrying one of the most common mutant forms of A1AT, the Z mutation.

Biochemical Properties

A1AT is a single-chain glycoprotein consisting of 394 amino acids in the mature form. The three N-linked glycosylations sites are mainly equipped with so-called diantennary N-glycans. However, one particular site shows a considerable amount of heterogeneity since tri- and even tetra-antennary N-glycans can be attached to the Asparagine 107 (ExPASy amino acid nomenclature). These glycans carry different amounts of negatively-charged sialic acids, this causes the heterogeneity observed on normal A1AT when analysed by isoelectric focussing. In addition, the fucosylated triantennary N-glycans were shown to have the fucose as part of a so-called Sialyl Lewis × epitope, which could confer this protein particular protein-cell recognition properties. The single cysteine residue of A1AT in position 256 (ExPASy nomenclature) is found to be covalently linked to a free single cysteine by a disulfide bridge.

Analysis

As protein electrophoresis is imprecise, A1AT is analysed by isoelectric focusing (IEF) in the pH range 4.5-5.5, where the protein migrates in a gel according to its isoelectric point or charge in a pH gradient.

Normal A1AT is termed *M*, as it is migrates toward the center of such an IEF gel. Other variants are less functional, and are termed A-L and N-Z, dependent on whether they run proximal or distal to the M band. The presence of deviant bands on IEF can signify the presence of alpha 1-antitrypsin deficiency. Since the number of identified mutations has exceeded the number of letters in the alphabet, subscripts have been added to most recent discoveries in this area, as in the Pittsburgh mutation described above.

As every person has two copies of the A1AT gene, a heterozygote with two different copies of the gene may have two different bands showing on electrofocusing, although heterozygote with one null mutant that abolishes expression of the gene will only show one band.

In blood test results, the IEF results are notated as in P_iMM, where P_i stands for protease inhibitor and "MM" is the banding pattern of that patient.

Other detection methods include use of enzyme-linked-immuno-sorbent-assays in vitro and radial immunodiffusion.

Alpha 1-antitrypsin levels in the blood depend on the genotype. Some mutant forms fail to fold properly and are, thus, targeted for destruction in the proteasome, whereas others have a tendency to polymerise, being retained in the endoplasmic reticulum. The serum levels of some of the common genotypes are:

- PiMM: 100% (normal)
- PiMS: 80% of normal serum level of A1AT
- PiSS: 60% of normal serum level of A1AT
- PiMZ: 60% of normal serum level of A1AT
- PiSZ: 40% of normal serum level of A1AT
- PiZZ: 10-15% (severe alpha 1-antitrypsin deficiency)
- PiZ is caused by a glutamate to lysine mutation at position 342
- PiS is caused by a glutamate to valine mutation at position 264

Other rarer forms have been described; in all there are over 80 variants.

Therapeutic Use

Recombinant alpha 1-antitrypsin is not yet commercially available, but is under investigation as a therapy for alpha 1-antitrypsin deficiency.

Therapeutic concentrates are prepared from the blood plasma of blood donors. The US FDA has approved the use of three alpha 1-antitrypsin products derived from a human plasma: Prolastin, Zemaira, and Aralast. A recent study analyzed and compared the three FDA-approved products in terms of their primary structure and glycosylation. All three products showed minor differences compared to the normal human plasma A1AT, and are introduced during the specific purifications procedures. However, these detected differences are not believed to have any negative implications to the patients.

Aerosolized-augmented A1AT therapy is under study. This involves inhaling purified human A1AT into the lungs and trapping the A1AT into the lower respiratory tract. This method proves more successful than intravenous-augmented A1AT therapy because intravenous use of A1AT results in only 10%-15% of the A1AT reaching the lower respiratory tract, whereas 25%-45% of A1AT can reach the lower respiratory tract through inhalation. However, inhaled A1AT may not reach the elastin fibers in the lung where elastase injury actually occurs. Further study is clearly warranted.

The hematopoietic System of Blood

The majority of blood cells have a relatively short life span, and new cells must be continually produced to replace them. This process depends on a highly regulated series of events in which immature precursors progress through several stages of growth and differentiation, resulting in the formation of mature blood cells of a particular type. Blood cell development is known as hemopoiesis, and this occurs in cavities within the bones. The soft tissue that occupies the cavities is the bone marrow, and it is composed of the various hæmopoietic populations and the supporting cells that form the environment in which blood cell production occurs. These supporting tissues are collectively referred to as the stroma, and they play an important role in the regulation of blood cell production.

Blood Cell Types and Their Functions

Classification of Blood Cells

Hæmopoietic cells are classified into two broad categories: myeloid cells and lymphoid cells. Myeloid populations include red blood cells, or erythrocytes, as well as members *or* the granulocytic, monocytic, and megakaryocytic lineages. The two principle lymphoid cells are Band T lymphocytes. Additional terms, such as leukocytes or white blood cells, are also used and refer only to granulocytes, monocytes, and lymphocytes. In any case, mature myeloid and lymphoid cells are easily detected upon examination of the marrow, and some of their distinguishing characteristics are described below. In addition to the fully differentiated cells, immature progenitors of each of the different blood cell types are also present.

Lymphoid Cells

Lymphocytes are found in the blood and various lymphoid tissue, such as the spleen and lymph nodes. These mononuclear, nonphagocytic white blood cells are key components of the body's immunological defense system and are classified as T cells or B cells.

Several sub-populations of T cells have now been identified based on functional criteria and their expression of cell-surface antigenic determinants; these include helper T cells and cytotoxic T cells. Helper cells are important in the regulation of the immune response and can affect the function of other T cells or B lymphocytes through direct contact with those cells or via the release of soluble factors. Cytotoxic T cells function to eliminate tumor cells or virus-infected cells. Some investigators believe that a third type of suppressor T cell exists, but this is controversial. T-cell development occurs in the thymus, and precursors present in the marrow are believed to migrate to that tissue and provide a continual source of immature cells from which thymocytes are generated. The maturational state of these marrow emigrants is unknown, but they could be a prothymocyte committed to T cell development or the pluripotent hæmopoietic stem cell itself.

The other main type of lymphocyte is the B cell. B lymphocytes are produced in the bone marrow and are the class of lymphocyte responsible for the production of immunoglobulin (Ig) or antibody molecules. Each newly produced B cell expresses an immunoglobulin on its cell surface that can specifically recognize a particular foreign material known as an antigen. Following binding of the antigenic substance to the Ig molecule, the B cell is stimulated to synthesize identical Igs, which are secreted. All of the immunoglobulin molecules secreted by anyone B cell or its progeny have the same antigenic specificity. Whether or not, analogous to T cells, sub populations of B cens exist is an active area of interest. Some investigators believe that those murine B cells that express the Lyl cell surface antigen or human B cells that express the CDS cell-surface antigen constitute a separate B lymphocyte lineage.

Natural killer (NK) cells are an additional population best considered with the lymphocytes. These cells are present in the spleen, lymph nodes, peripheral blood, and peritoneal cavity and are capable of killing a target, usually a tumor or virus-infected cell, without prior sensitization. NK cells are also known as large granular lymphocytes due to the presence of granules in their cytoplasm. To what lineage NK cells belong is unclear, and this has been an area of considerable controversy. NK cells originate in the bone marrow from hæmopoietic precursors

Myeloid Cells

The majority of blood cells can be classified as myeloid cells and include monocytes-macrophages, several different types of granulocytes, megakaryocytes, and erythrocytes. Macrophages are mononuclear cells derived from monocytes and are specialized for phagocytosis. These cells can ingest particulate antigens or microorganisms and digest these materials with hydrolytic enzymes they produce. Macrophages vary in size and shape depending on their state of activation. In general, however, they have a cell surface from which projects numerous membrane processes, a spherical or oval nucleus, and a cytoplasm with granules that contain numerous hydrolytic enzymes. Macrophages play a key role in the immune response by presenting antigen to lymphocytes and through the secretion of various factors that can modulate the function of lymphocytes and other cells in the body.

The granulocytes are a second category of myeloid cells. These cells receive their name because of the presence in their cytoplasm of granules that contain various enzymes or active proteins. Based on their appearance after hematologic staining, granulocytes are classified as neutrophils, eosinophils, and basophils. Neutrophils can be recognized by their multi-lobed nucleus and numerous cytoplasmic granules. The latter are classified as primary granules, which develop first during neutrophil differentiation and contain hydrolases, lysozymes, myeloperoxidase, and cationic proteins, and secondary granules, which constitute up to three-quarters of the granules and contain lysozymes and lactoferrin. These. agents function to destroy or break down substances phagocytized by the cell. Neutrophils are sensitive to chemotactic agents and can migrate to the site of an inflammatory response at which they phagocytize and kill bacteria and other infectious agents. Eosinophils are bilobed cells that are characterized by the presence of large reddish orange staining granules that contain peroxidase and other enzymes. Because their numbers increase during parasitic infections, they may be involved in defense against those organisms. Eosinophils may also play a role in various allergic reactions. Basophils have a constricted nucleus and a cytoplasm that contains coarse bluish black staining granules. The granules in these cells contain heparin, histamine, and serotonin. These substances are released following stimulation of the cells and contribute to such allergic responses as smooth muscle contraction and increased blood vessel permeability. Basophils have also been implicated in defense against parasites. Megakaryocytes or thrombocytes are huge cells that can have large nuclei that form as a result of polyploidy. The function of megakaryocytes is to produce platelets, and this occurs as a result of budding of cytoplasmic fragments from the cells. Plate lets function in coagulation, hemostasis, and thrombus formation.

Erythrocytes, or red blood cells, are non-nucleated biconcave disks that are able to transport oxygen to the tissues. This is because the cells contain hæmoglobin, a protein that combines readily with oxygen.

Introduction to Hæmopoietic Hierarchy

If a sample of bone marrow is smeared on a slide and stained with an appropriate hematologic reagent, a marked heterogeneity of cell types can be observed. The mature cells described above can be easily recognized as well as other populations not readily identifiable. These are hæmopoietic cells of various lineages and at different stages of development. The maturational state of these cells can be determined using a variety of criteria that include their histologic appearance, expression of defined cell-surface or cytoplasmic antigenic determinants, responsiveness to growth and differentiation factors, and proliferative and differentiative potential. All of these parameters have provided useful information that has permitted a schæme of hemopoiesis to be formulated.

The Pluripotent Hæmopoietic Stem Cell

At the head of the hæmopoietic hierarchy is the pluripotent hæmopoietic stem cell (PHSC). This is defined as a cell that has the ability to divide and generate additional stem cells (a property known as self-renewal) as well as to differentiate and give rise to the blood cell types described above. The PHSC is a rare cell and may account for <0.1% of the nucleated cells in the bone marrow. Their incidence is even lower in other tissues such as the spleen. Despite this, these cells clearly exist.

One study that provided evidence for the existence of a multi potential stem cell analyzed blood cells in women who were heterozygous for the enzyme glucose-6-phosphate dehydrogenase (G-6-PD). Each cell and all of its progeny, in these individuals expresses only one of the two forms of the enzyme. Therefore, all blood cells that derive from anyone pluripotent hæmopoietic stem cell would express one or the other G-6-PD type. When cells from G-6-PD heterozygous females who had developed chronic myelogenous leukemia (a disease that originates in a single cell) were analyzed, cells from most of the hæmopoietic lineages were shown to express the same G-6-PD type. This provided strong evidence that these populations were the progeny of a common stem cell that exhibited a marked differentiative capability. Another classic study that demonstrated the existence of the PHSC involved the induction of chromosomal markers inbone marrow cells by subjecting them to a low, sub-lethal dose of ionizing irradiation. This induces random chromosomal translocations in a high proportion of the surviving cells, and because each is unique, it can be used as a cytogenetic marker to detect cells derived from the marked cells. This cell population (called clonal) was used to repopulate X-irradiated mice, in which the bone marrow is destroyed. At various times following reconstitution, cytogenetic analysis was performed on the myeloid, B, and T cells which derived from the cells used to repopulate the animal. The study showed that in some of the mice myeloid, B, and T cells all bore the same unique chromosomal marker, providing very strong evidence that they were all progeny of a single pluripotent stem cell in which the marker had been created.

Another series of experiments that demonstrated the existence of the PHSC repopulated mice with bone marrow cells that were infected with a retroviral vector carrying the bacterial neomycin resistance gene (which allows for the selection of the cells carrying the gene). The retrovirus genome integrates randomly into the genome of the marrow cells, and the viral integration site creates in each cell a unique change (called restriction fragment-length polymorphism), which can be recognized by preparing DNA from the cells and cutting it with an appropriate restriction enzyme. These fragments can be visualized by a suitable laboratory technique. At appropriate periods after reconstitution of irradiated mice, myeloid, B, and T cells were isolated from the animals, and the cells' DNA was analyzed. These studies demonstrated that neomycin resistant B cells, T cells, and myeloid cells could be detected in which the retroviral vector had integrated at the same site. The most straightforward explanation for these results was that these cells were the progeny of a pluripotent hæmopoietic stem cell that had been infected by the retrovirus.

A clonal analysis of hæmopoietic cells in the above mice was also performed at various times after reconstitution. This investigation showed that some hæmopoietic stem cells stably contributed to all hæmopoietic lineages in a reconstituted mouse over time, while others underwent temporal changes. For example, in some animals, myeloid and lymphoid progenies of a particular stem cell were observed at the initial sampling time, but only one or the other cell type was present at a later point. In additional animals, new stem cell clones were observed to have emerged, based on the observation that lymphoid and myeloid cells that bore the same unique viral integration site were detected at the time of a second sampling but not an earlier point following reconstitution. These studies are of importance, because they document that the contribution of a pluripotent stem cell to the various lineages can vary with time.

The above experiments have provided evidence for the existence of the pluripotent hæmopoietic stem cell, but they do not allow it to be directly studied in isolation. A major goal of experimental hematologists is to isolate and grow pure stem cell populations. This would present great advantages for studying how the growth and differentiation of that population is regulated and for identifying and characterizing genes that are temporally expressed during blood cell development. A variety of approaches have been used to purify stem cells from the marrow based on their physical characteristics and expression of cell-surface determinants. Recent work has indicated that murine stem cells have special surface characteristics by which they can be isolated. Efforts are now underway to identify markers expressed on human pluripotent hæmopoietic stem cells. If limited than that of a pluripotent stem cell.

Therefore, some investigators prefer to designate these cells as multipotential precursors to clearly distinguish them from the pluripotent stem cell. Two types of multi potential precursors have been described. A multipotential myeloid-precursor is defined as a cell that can generate all hæmopoietic cells except lymphocytes. The evidence that these restricted populations exist is based on several criteria. First. in the above experiments using retrovirally marked cells, patterns of repoptllation were observed in which only myeloid but not lymphoid cells bore the same viral integration site. A second piece of evidence is based on classic experiments that demonstrated that bone marrow cells will form macroscopic colonies on the spleen 8-13 days following their injection into lethally irradiated mice. The cell that initiates these colonies is known as the spleen colony-forming unit (CFU-S; S stands for spleen) and the colonies are called spleen colonies.

Based on the observation that lymphoid cells are rarely, if ever, detected as progeny of the CFU-S and the colonies primarily contain myeloid cells, the assay is generally accepted as detecting a multi-potential myeloid-restricted precursor. Spleen colonies form at various times after injection of bone marrow cells into the irradiated mice. Those that appear on day 13 are generally thought to be derived from a multipotential myeloid-restricted precursor. Spleen colonies also appear at earlier times, but they may be derived from more differentiated progenitor cells described below.

A common lymphoid precursor is a second type of restricted stem cell. By definition, such a cell would have the potential to generate both Band T lymphocytes but not myeloid cells. Several laboratories have provided indirect evidence for the existence of a common lymphoid precursor, but to date, a pure population of such cells has not been isolated. Whether lymphocytes are the progeny of a common lymphoid precursor committed to the B-cell and T-cell differentiative pathway or are direct descendants of the PHSC remains an unanswered question.

Progenitor Cells

Progenitor cells have a limited proliferative capability, and a restricted differentiative potential. For each of the blood cell populations described above, a progenitor cell committed to that particular linkeage exists. There are two principle means by which progenitor cells can be detected. The classical method takes advantage of the ability of progenitors to form lineage-restricted hæmopoietic colonies in semisolid medium. The growth of these colonies depends on the presence of a growth factor that potentiates their formation. Progenitor cells can also be identified by their expression of lineage-specific antigens.

Myeloid Progenitors

Investigators in Israel and Australia were the first to demonstrate that bone marrow or spleen cells could form colonies in a semisolid medium such as agar following culture under appropriate conditions. The colonies that grew contained granulocytes and/or macrophages, and subsequent analyses indicated they had arisen from a single progenitor cell. Distinct progenitor cells for most of the myeloid lineages have now been identified, and their formation of colonies in semisolid medium depends on the presence of a particular lineage specific growth and/or differentiation factor. For example, as shown in Fig. 8.4, it is now known that a bipotential granulocyte-macrophage progenitor cell exists that can differentiate into cells of one or both of these lineages.

Progenitor cells are not necessarily of the same precise maturational state, and multiple cells may be included in a progenitor cell compartment. This point is best illustrated by analysis of cells of the erythroid lineage. Two stages of erythropoiesis have been defined based on the ability of erythroid progenitors to form colonies that appear at different times following initiation of cultures. An erythroid progenitor known as the erythroid burst-forming unit (BFU-E) is an immature erythroid progenitor and the erythroid colony-forming unit (CFU-E) is a more differentiated cell. Both the BFU-E and the CFU-E meet the definition of progenitor cells, because they are lineage-restricted in their differentiative potential.

Lymphoid Progenitors

The bone marrow contains precursors that can generate B and T cells, but the development of these two types of lymphocytes occurs in different sites. B lymphopoiesis proceeds in the marrow of adult mammals, while primary T lymphocyte development occurs in the thymus.

Many researchers now believe that T-cell production depends on the migration of a bone marrow derived T-cell precursor to the thymus. Whether this is a lymphoid stem cell, a progenitor committed to the T-cell developmental pathway (the prothymocyte) and/or the pluripotent stern cell itself is not entirely clear. In any case, once the T-cell precursor enters that tissue, it interacts with the thymic stroma and a series of events occur that result in the expression of the T cell-receptor and T cell-surface antigens. Major efforts are underway to identify the anatomical sites within the thymus in which various differentiative events occur, to define the lineage relationships, which result in the production of the various T-cell subpopulations, and to further clarify the molecular events that result in the expression of the genes that encode the T cell receptor. It is generally accepted that a cellular intermediate that expresses both the CD4 and CD8 antigens develops during the intrathymic stage of differentiation, and this gives rise to T-cell subpopulations that include the CD4+ T helper cells and the CD8 + cytotoxic T cells. These newly produced T cells express the heterodimeric T-cell receptor, which most commonly consists of an *a*- and a, a-chain, although it may also be formed by a *y*-and a 8-chain. These chains are expressed during T-cell differentiation as a result of a complex process in which genes that encode segments of these molecules re-arrange and become expressed in developing thymocytes. The T-cell receptor on each newly produced T lymphocyte has a unique site at its amino terminus, known as the variable region that recognizes antigen in association with cell-surface determinants encoded by the major histocompatibility locus.

B-cell development occurs in the bone marrow, and maturation of cells in this lineage results in the expression of the Ig. molecule as well as other non-immunoglobulin cell-surface antigens. Distinct stages of B lymphopoiesis have been defined through the use of antibodies that recognize these molecules (Fig. 8.9). For example, the B220 molecule is ~22000-molecular weight glycoprotein present on murine B-cell progenitors. Similar molecules may be present on human B-cell progenitors.

The Ig or antibody molecule is formed by the assembly of two identical heavy-chain and two identical light-chain glycoproteins. The heavy-chain genes are expressed first, and following translation of heavy-chain messenger RNA, heavy chain of the immunoglobulin M class is expressed in the

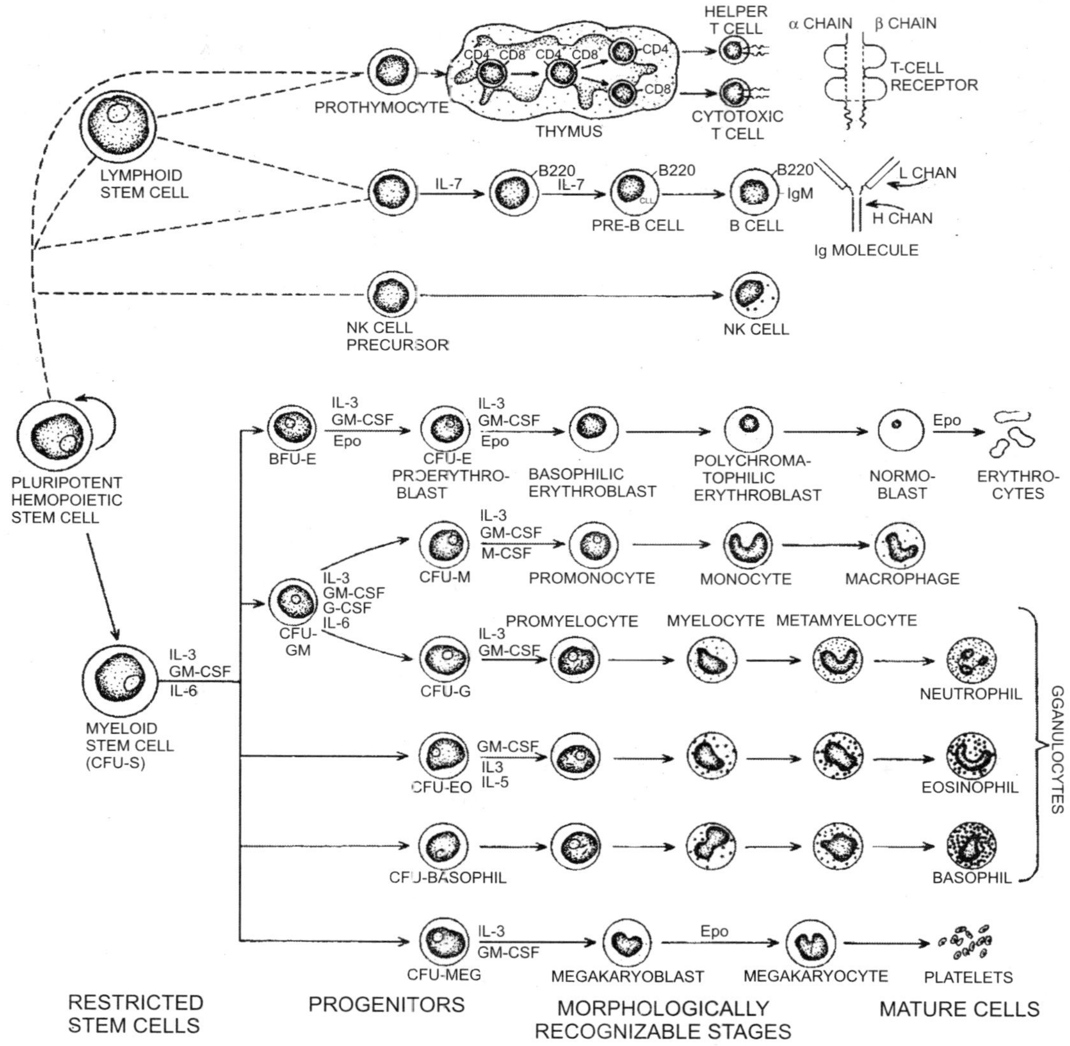

Fig. 8.11 : A scheme of the cells of the hæmopoietic system showing the various lineages and maturational states that have been defined. Broken lines indicate pathways in which lineage relationships remain unresolved. The arrow on the pluripotent hæmopoietic stem cell indicates that the cell is capable of self-renewal. Some of the cytokines that regulate the growth and/or diffe,rentiation of blood cells are indicated. Their placement in the diagram is intended to indicate the point at which they act, but this is only an approximation. The reader must consult primary literature for the precise, up to date effects of these and other cytokines.

cytoplasm of a cell population termed a pre-B cell. Light-chain genes are subsequently expressed, and as light-chain protein appears, the complete Ig molecule is assembled in the cytoplasm of the pre-B cell and immediately appears on the surface of what is now termed a newly pmduced B lymphocyte. As with the T-cell receptor, expression of the Ig molecule depends on the rearrangement of gene segments that encode the heavy- and light-chain molecules. The amino terminus of the heavy and light chains form a unique site specialized for recognition of an antigen.

Colony assays for B-cell progenitors, analogous to those for erythroid and myeloid populations, are a recent development that provides an additional means to identify B cell precursors.

Morphologic Identification of End Cell Populations

Before the development of colony assays or the availability of antibodies that recognize antigens expressed by hæmopoietic precursors, morphology was the principle means by which immature cells in the various lineages were identified. However, by the time a cell of a particular lineage is histologically recognizable, it is more mature than progenitors that are detected in colony assays. For example, it is possible to distinguish immature erythroid cells such as the erythroblast. This cell, which is thought to correspond to a mature CFU-E, has a basophilic cytoplasm and a nucleus containing multiple nucleoi. More differentiated progeny of the erythro-blast can also be defined based on cell size, the nuclear chromatin pattern, and cytoplasmic baso-philia (Fig. 8.11). Similarly, various stages of myelopoiesis can be recognized under the light microscope. One of the earliest recognizable precursors of granulocytes is the promyelocyte. This cell, which is certainly more differentiated than the progenitors that initiate colonies in semisolid medium has a large nucleus with a nucleolus. Peroxidase-positive granules are present in the cytoplasm, and a well-developed Golgi apparatus is present. From the promyelocyte stage on, cells in the granulocytic series are distinguished based on the formation of specific granules in their cytoplasm, and once these appear, the cells are classified as myelocytes. Cells are classified as metamyelocytes following the development of the distinguishing nuclear characteristics. Some appreciation of the appearance of the various hæmopoietic cells remains important. This is particularly true for the clinician, because diagnosis of certain types of hematologic dyscrasias such as leukemia is still based on the morphological appearance of the cells present in the marrow or peripheral blood.

Organization of Hæmopoietic Tissues

Blood cell development takes place in association with a fixed framework of supporting elements known as stromal cells. An understanding of the location of these stromal populations and the architecture of the marrow cavity is best appreciated in the context of blood cell circulation to the bone.

The principle blood supply to a typical long bone is the nutrient artery. This enters at midshaft through the nutrient foramen and travels in the central portion of the medullary cavity in parallel with. the longitudinal axis of the bone. This centrally running artery sends off branches, which radiate toward the inner or endosteal surface of the medullary cavity, and at that point these branches anastomose with tributaries from other arteries that supply the bone. These subendosteal vessels become continuous with a series of venous sinusoids that deliver their contents toward a central sinus in the center of the marrow cavity. Blood exits the bone from the central sinus via the nutrient foramen and enters the systemic circulation. Stromal cells are present in the intersinusoidal spaces and form the framework on which hemopoiesis proceeds. These cells, which have been referred to as reticular cells by morphologists, have numerous cytoplasmic projections that form extensive connections with one another and the processes of stromal cells in the adjacent intersinu-soidal spaces. Stromal cell processes also have extensive contact with the endothelial cells, which form the lining of the sinuses; Developing blood cells in the inter sinusoidal spaces are in close contact with the stromal cells, and this association led early morphologists to propose that the stroma can regulate blood cell production. As blood cells mattire, they must ultimately exit the medullary cavity and enter the circulation. They do so by passing froin the intersinusoidal spaces and through the endothelial cells to enter the venous sinusoids.

Under normal, steady-state conditions, the bone marrow is the only site in which the simultaneous production of myeloid cells and B lymphocytes occurs. This suggests that local tissue influences critical for that process are operative in the medullary cavity. It is generally accepted that the various stromal cell types and their products are one source of these regulatory signals, and the term hæmopoietic micro-environment has been used to describe the localized sites in the medullary cavity in which blood cell production proceeds. Some investigators consider that components of the hæmopoietic micro-environment are able to induce a pluripotent hæmopoietic stem cell to differentiate along a particular lineage. However, this is not universally accepted, and it remains unclear if the micro-environment acts directly to induce the commitment and differentiation of multi-potential stem cells or, instead, provides the permissive signals that allow the further development of a precursor that has become randomly committed to a particular lineage through its own intrinsic genetic programming. These controversies aside, it is evident that a number of growth and differentiation signals exist that affect hemopoiesis.

Regulation of Blood Cell Production Introduction

The identification of the environmental signals that regulate blood cell production and the characterization of their effects on hæmopoietic targets remain challenging areas of investigation. Such knowledge is of relevance to the basic understanding of a fundamental developmental process and has clinical relevance as well. When regulatory controls do not function properly, a number of hæmopoietic abnormalities can occur. Much of what is known about the regulation of hemopoiesis is based on studies of murine tissues and cells, and it is reasonable to expect that similar regulatory controls are operative in human marrow.

The environmental cues a differentiating hæmopoietic cell receives can be delivered via interactions with other cells, through the association of the cell with components of the extracellular matrix, or by the binding of a soluble mediator to a receptor on the cell membrane.

Cell-Cell Interactions

Morphologic examination of marrow sections reveals that hæmopoietic cells are tightly clustered in the intersinusoidal spaces in association with each other and the supporting stromal elements. This obviously creates the potential for direct cell-cell interactions between these populations and has suggested that hæmopoietic and stromal cells might associate with one another via recognition molecules that permit direct membrane binding. Such molecules are now being identified by several laboratories that have prepared monoclonal antibodies to stromal cells. These antibodies apparently block the interactions between stromal cells and blood cells. These reagents will be of particular value in functional studies and in purifying and characterizing cell interaction molecule(s).

The endothelial cells that line the marrow sinusoids also play a role in regulating the passage of cells from the marrow into the systemic circulation. A number of elegant ultramicroscopic analyses have demonstrated that blood cells actually pass through the endothelial cell wall on their way into the sinusoids. What signals may be transmitted from the endothelial cell to the migrating blood cells remain to be defined, but the potential for this exists, because endothelial cells are known to secrete various factors with effects on hemopoiesis.

An important means of regulating passage of cells from the marrow may be mediated by the reticular and endothelial cells. The processes of reticular cells are associated with the abluminal surface of the endothelial cells and the degree of association can be variable. It has been hypothesized that an extensive reticular-endothelial cell association may serve to retain cells in the marrow, while a retraction of reticular cell process from the' endothelium may facilitate blood cell, emigration.

Cell-Matrix Interactions

The extracellular matrix (ECM) is a meshwork of polysaccharides and proteins that form an organized framework in the tissues. Fibroblasts in the tissues secrete much of ECM. The two main ECM components include polysaccharide glycosaminoglycans that are linked to a protein backbone and fibrous proteins. The latter include structural components such as collagen and elastin and adhesive proteins such as fibronectin and laminin. The matrix was initially believed to serve as a packing material in the tissues, but studies now indicate that it plays an important role in the development, migration, proliferation, shape, and metabolic function of cells.

Hæmopoietic cells have been reported to have receptors for various ECM components to which they may bind. What particular events may be signalled by such binding remains to be determined. Also, evidence indicates that hæmopoietic growth factors can bind to ECM components and that the ECM thereby compartmentalizes these cytokines and helps to present them to developing blood cells.

Soluble Mediators

Experiments in which a conditioned medium was used to stimulate the formation of hæmopoietic colonies in agar demonstrated that some soluble factors could regulate the growth and! or differentiation of hæmopoietic cells. It is now known that the molecules that potentiate the formation of the granulocyte and macrophage colonies are members of a family of cytokines known as the colony-stimulating factors. Additional factors that can affect the growth, differentiation, survival, and activation of hæmopoietic cells have now been identified and include erythropoietin, which acts on erythroid cells and, more recently, a variety of interleukins with effects on lymphoid and myeloid cells (Table 8.1). The considerable progress that has been made in the identification and characterization of the actions of these multiple cytokines is due in part to the application of molecular approaches to cytokine analysis. Techniques for the cloning of the genes that encode these mediators has become relatively routine, and because of this, high concentrations of essentially pure recombinant factors are now available for *in vitro* and *in vivo* experiments.

Among the best characterized factors are the family of glycoproteins known as the colony-stimulating factors (CSFs). Three of the four CSFs were named according to their initially described effects on the growth and differentiation of a particular progenitor cell population and include macrophage (M-CSF), granulocyte (G-CSF), and granulocyte-macrophage (GM-CSF). While the primary targets of these mediators are indicated by their name, their actions are not restricted to those lineages. For example, GM-CSF targets include eosinophils, mega-karyocytes, and multipotential myeloid precursors. A fourth CSF is known as multi-CSF or interleukin-3 (IL-3). IL-3 targets include granulocytes, macrophages, eosinophils, megakaryocytes, and multi potential precursors. The various hæmopoietic cells noted above each express several hundred CSF receptors, and occupancy of as few as 10 per cent of them can stimulate a biological response by the cells. The CSFs have a high specific activity and are active at concentrations of 10^{-10}–10^{-12} molar. While a major action of CSFs is the stimulation of cell proliferation, they also have effects on the survival, differentiation, maturation, and activation of their targets.

Table 8.1 : Selected Cytokines and their Effects on Hemopoietic Cells[a]

Cytokine	Source	Effects
IL-1	Macrophages, monocytes, fibro-blasts, endothelial cells	Synergy with CSFs, stimulation of CSF and IL-6 section by stromal cells
IL-2	T cells	Stimulates thymocyte and T-cell growth
IL-3	T cells	Stimulation of proliferation-differentiation of myeloid-restricted stem cells and erythroid, monocytic, granulocytic, and megakaryocytic progenitors
IL-4	T cells, stromal cells	Pre-B-cell maturation, potentites responsiveness of myeloid pregenitors to CSFs
IL-5	T cells	Stimulation of eosinophil progenitors
IL-6	T cells, macrophages, stromal ceils	Acts as CSF to stimulate proliferation of myeloid progenitors
IL-7	Stromal cells	Stimulation of B cell progenitor-proliferation; stimulates thymocyte proliferation; augments proliferation of mature T cells to other cytokines
M-CSF	Macrophages, endothelial cells, stromal cells	Stimulates proliferation-differentiation of macrophage progenitors
G-CSF	T cells, stromal cells	Stimulates proliferation-differentiation of granulocyte progenitors
GM-CSF	T cells, stromal cells	Stimulates proliferation-differentiation of granulocyte-macrophage progenitors

[a]The cellular sources of a particular factor and the effects are not restricted to the parameters noted. This is only a selected tabulation of sources and actions of these cytokines. In addition, these are not the only cytokines with effects on hemopoietic progenitors.

Red blood cell production depends on the hormone erythropoietin. This is produced in the kidney, and its level of production is sensitive to the oxygen levels in the blood. A lack of oxygen or shortage of red blood cells results in an increase in erythropoietin synthesis and a stimulation of red blood cell production in the bone marrow.

Another class of molecules with effects on the growth and differentiation of hæmopoietic cells is the interleukins. This name was chosen to indicate that the source of these cytokines is leukocytes. In addition to IL-3 described above, at least nine other interleukin molecules have now been described. Interleukin-1 (IL-I) is primarily a macrophage-monocyte product, although other tissue cells such as fibroblasts and endothelial cells can produce it. IL-I exhibits potent synergistic effects with other CSFs. Other interleukins with hæmopoietic effects include interleukins-4, -5, -6, and -7. For example, IL-5 aCts as a differentiation factor for eosinophils. IL-6 is a recently described interleukin that was originally defined by its ability to promote the growth of Ig-secreting B cells. This molecule is now recognized to act as a CSF to stimulate granulocyte progenitors and multipotent myeloid precursors. IL-7 is the first cytokine available in recombinant form that has been shown to affect B-cell precursors in the bone marrow. The gene for IL-7 was cloned from a bone marrow stromal cell line and encodes a 14.9-kD molecule. IL-7 can stimulate the proliferation of B-cell progenitors. Its effects are not restricted to B-lineage cells, as both thymocytes and peripheral T cells are sensitive to its actions. IL-7 apparently does not have any differentiative function, and this would suggest that additional molecules with such actions on B-lineage cells exist.

Cytokine biology is a dynamic and complex area, and it is difficult to discuss all of the subtleties of this field. However, several important facts regarding the actions of these regulatory molecules mustbe appreciated and can be summarized as follows:

1. The effects of anyone cytokine are pleiotropic (i.e., multiple). For example, the interleukins are critical to the mediation of the immune, response and many were initially described by their, effects on mature lymphocytes. The fact that most of these. factors have potent effects on the growth and differentiation of hæmopoietic cells of various lineages was a secondary discovery.
2. The same factor may exhibit differential effects on cells within the same lineage. IL-4 has been proposed to stimulate or potentiate the differentiation of mature B cells while acting as an inhibitor of the proliferation-differentiation of early B-cell precursors. These dichotomous effects may be due to different concentrations of the factor.
3. The potential exists for additive and synergistic interactions between various cytokines. IL-1 and the CSFs are known to synergize and IL-4 has been shown to potentiate the response of myeloid progenitor cells to the actions of CSFs. Precisely how the interactions between two cytokines result in an augmented cellular response remains to be defined. If a cell expresses receptors for both cytokines, then the result of dual binding to these could easily explain an additive response. The explanation for a synergistic response is not as straight forward. In the case of IL-I-CSF synergy, it is possible that one molecule, such as IL-I, could stimulate an increase in the number of CSF receptors expressed by the cell and, thus, potentiate the response to that cytokine.
4. The effects of any particular hæmopoietic factor are not necessarily mediated through a direct action on the hæmopoietic target. Many of these molecules indirectly stimulate bystander cells to secrete a second molecule, which in turn acts on the hæmopoietic progenitor. Numerous cells such as fibroblasts, bone marrow stromal cells, and endothelial cells are known to express

IL-I receptors. One effect of IL-1 binding to them is that they are stimulated to secrete a number of other mediators that include the CSFs. These in turn may then act on the hæmopoietic target alone or synergistically with the IL-1 molecules that originally induced their production.

5. Certain hæmopoietic cells can secrete cytokines to which they respond. For example, macrophages are sensitive to the effects of M-CSF and may also produce that mediator. This raises the possibility that autocrine (self-regulating) and paracrine (regulating neighboring cells) regulatory mechanisms are operative.
6. It is becoming clear that control of cell production in the hæmopoietic system is also governed by various negative regulatory molecules that can have inhibitory effects on the growth and differentiation of cells. One such molecule, transforming growth factor-13, has been shown to affect the production of lymphoid and myeloid lineage cells. There have also been suggestive reports that the positive effects of many cytokines may be modulated via negative feedback circuits. In this case, a cytokine could stimulate a particular hæmopoietic cell while inducing the secretion of a negative regulator by another cell.

It is important to note that virtually nothing is known about factors that 'regulate the self-renewal and differentiation of pluripotent hæmopoietic stem ~ cells. Some of the factors described above and in Table I might act on stem cells, but it is difficult to obtain a pure population on which to test this. A factor known as leukemia inhibitory factor (LIP) has been shown to inhibit the differentiation and promote the growth in culture of a cell known as an embryonic stem cell. These cells are isolated from the inner cell mass and have the capability to generate all cells and tissues of an embryo. Whether or not LIF acts to promote the self-renewal and expansion of the hæmopoietic stem-cell compartment is unknown. Nevertheless, the fact that such molecules with effects on undifferentiated embryonic cells exist suggests that comparable mediators with actions on the hæmopoietic stem cell will be described.

Interface between Basic and Clinical Studies

Because hæmopoietic cells are easily obtained and manipulated, the hæmopoietic system is an excellent model for investigating the parameters that regulate cell growth and differentiation. The development of culture systems that facilitate these studies has been a major technological advance, and systems that allow long-term, stromal cell-dependent murine myelopoiesis or lymphopoiesis have been described and are in widespread use. These cultures are being used to study the growth of human hæmopoietic cells as well. The accessibility of bone marrow, the relative ease in manipulating it, and the development of techniques for its transplantation have contributed to great interest in applying the basic knowledge obtained about hæmopoietic cells to clinical medicine. For example, the experiments in which retroviral vectors were used to infect stem cells have potential use far beyond simply establishing lineage relationships. There is now considerable interest in using this approach to correct genetic defects that affect blood cell development or function. Thus, hæmopoietic stem cells could be harvested from an individual and infected with a retrovirus that contains a normal copy of the defective gene. These cells would then be used to repopulate that individual's hæmopoietic system with "normal" cells. This therapy depends on perfecting techniques for retroviral infection of stem cells and for ensuring that the exogenous gene is expressed in a regulated manner.

Another example of how the basic studies of the hæmopoietic system have rapidly overlapped with clinical applications is evident from the rapid progress in cloning the genes for the various cytokines. Virtually unlimited quantities of factors such as erythropoietin, GM-CSF, and G-CSF are now available and are being administered to patients with promising results. While considerably more basic research in the mechanisms of action of these factors remains to be conducted, information regarding their actions is being obtained from these clinical trials. Thus, not only can basic information about the hæmopoietic system be applied to human medicine, but the results obtained from these applications can contribute to the fundamental understanding of blood cell production.

HEMOGLOBIN AND MYOGLOBIN

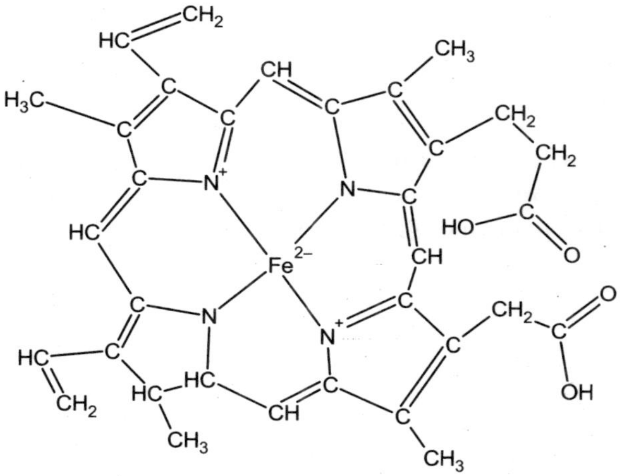

Hemoglobin is the protein of the red blood cells that allows vertebrates to transport oxygen from the lungs to the tissues and helps the return transport of carbon dioxide from the tissues back to the lungs. Like all proteins, hæmoglobin is made of small organic molecules called amino acids, strung together in a linear sequence called a polypeptide chain. There are 20 different amino acids, and their sequence in the chain is genetically determined. The hemoglobin molecule is made up of four polypeptide chains: two α chains of 141 amino acid residues each and two β chains of 146 residues each. The α and β chains have different sequences of amino acids, but fold up to form similar three-dimensional structures. Each chain harbors one hæme, which gives blood its red color. The hæme consists of a ring of carbon, nitrogen, and hydrogen atoms called porphyrin, with an atom of iron at its centre (Fig. 8.10). A single polypeptide chain combined with a single hæme is called a subunit of hæmoglobin, or a monomer of the molecule. In the complete molecule four subunits are closely joined, as in a three-dimensional jigsaw puzzle, to form a tetramer.

Fig. 8.12 : The Hæme Molecule the Iron protophorpyrin

The hemoglobin molecule contains about 36 negatively charged and 36 positively charged groups.

The hemoglobin molecules is the combination of two molecules the hæme and the globin: the hæme group appears to be attached to the globin surface at three places, the two propionic acid groups of the hæme are attached with the two strongly basic groups of globin and the iron of hæme is linked via histidine molecule to the globin protein; this establishment is clearly visible in Fig. 8.13 next page:

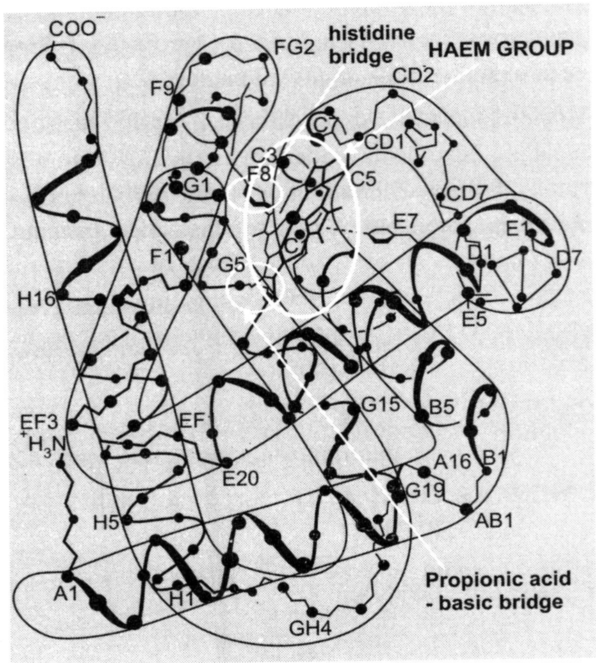

Fig. 8.13 : The Hemoglobin Molecule

In the hæmoglobin solutions nin the absence of O_2 valence is attached to H_2O ad the oxygenation of hæmoglobin involves displacement of this H_2O by O_2

$$HHb.H_2O \rightleftharpoons HHbO_2 + H_2O$$

The six linkages of the iron atom in the hæmoglobin appear to be quite analogous to the linkages in the ferrocyanide ion

$$\left[Fe^{4-}\left(N^{+}\equiv C^{-}\right)_6 \right]$$

In ferrocyanide all the bonds are equivalent on account of resonance and the group as a whole has a net negative charge of four of the cyanide groups may be pictured as lying in the place of the paper and one as lying above and other below this plane.

In this hæmoglobin formula the four N atoms of the protoporphyrin ring may be considered to be connected to iron by linkage 1, 2, 3 and 4 in the plane of the paper, while linkage 5 to the imidoazole N is below this plane and linkage 6, which in ferrohæmoglobin unites with O_2 and CO_2 is above this plane, In ferrohæmoglobin the hæme complex possesses no net charge, but in ferrohæmoglobin there is a net positive charge of one, which holds a negative group such as hydroxyl. In passing from ferro – to ferrohæmoglobin, the molecule loses its capacity to combine with O_2 and CO.

The conjugation of the iron of the hæme group with the globin via histidine may be used to keep the iron in ferrous state and fortunately giving rise to another linkage of the carbonic acid, this linkage helps the iron of hæme to get autoxidize to ferric state when hæme gets detached from globin molecule in the presence of acid or alkali and oxygen.

Hæmoglobin with iron is called ferrohæmoglobin and this is strongly paramagnetic and possesses four unpaired electrons. The oxygen molecule, O_2 contains two unpaired electrons, when O_2 adds to ferrohæmoglobin to form oxyferrohæmoglobin (oxyhæmoglobin) the unpaired electrons of both the iron and oxygen become paired, and oxyhæmoglobin diamagnetic. The manner in which the O_2 molecule is held in oxyhæmoglobin is probably related to its failure to convert the hæmoglobin to ferric state.

The best approximate molecule of hæmoglobin appears to be a spheroidial or egg shaped structure 54Å 52Å thick and 72Å long. There is a relatively deep grove around the centre of the molecule along the greater axis.

The globin molecule (molecular weight 66,000) of normal human hæmoglobin (hæmoglobin A) is made up of four polypeptide chains, two α-and two β chains, both types having valine as the N-terminal amino acids. The N-terminal amino acid sequences of the α-chains is valine leucine and that of the β chains is valine–histidine leucine.

Normal human hæmoglobin, or hæmoglobin A is represented by $\alpha_2{}^A\beta_2{}^A$ indication of the presence of two and two β chains of the A hæmoglobin in the molecule.

Fetal hæmoglobin, or hæmoglobin F, usually makes up to 70% to 90% of the hæmoglobin of a full term infant, the rest of the hæmoglobin being of the adult or A type, fetal hæmoglobin of the infant generally entirely replaced by adult hæmoglobin, A by the first year if the life.

Fetal hæmoglobin differs from normal adult hæmoglobin in crystal form Electrophoretic mobility, pH, solubility resistance to alkali, absorption spectrum chromatographic behaviour, amino acid composition, N-terminal amino acids and immunologic properties. These differences are all due to differences in the compositions of the globin molecules of normal and fetal hæmoglobins, the hæme group of all the hæmoglobin being alike.

So it can be said that four hæme groups of all types of hæmoglobins are arranged symmetrically on the four globin peptide chain, this is because all the four hæme appear to possess the same chemical properties.

Hæmoglobin has four polypeptides each with its own hæme. Human have several types of hæmoglobin shown the table below:

Type	Subunit	Importance
Major Adult(HbA)	$\alpha_2\beta_2$	97% adult hæmoglobin
Minor adult(HbA_2)	$\alpha_2\delta_2$	2% to 3% of adult hæmoglobin
Fetal (HbF)		Major hæmoglobin in second and third trimesters of pregnancy

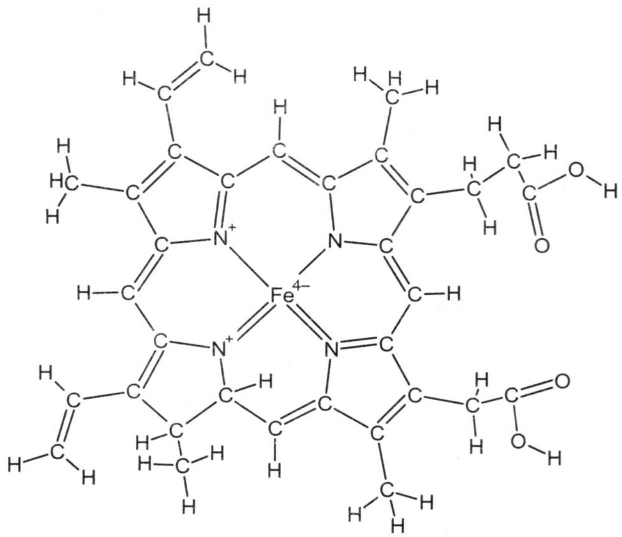

The hæmoglobin A (HbA) consisting of two α chains and two chains it is the major adult hæmoglobin. Also the minor adult hæmoglobin (HbA2) and fetal hæmoglobin (HbF) have two α-chains, but instead of the β chains, they have δ chains and γ chains respectively.

The α chains have 141 amino acids, and the β, γ and δ chains have 146. All these chains are structures related to each other. The α and β chains differ in 10 and the β and γ chains in 39 of their 146 amino acids see (Fig. 8.13).

Hæme is a tightly bound prosthetic group of many proteins including hæmoglobin myoglobin and cytochromes. It is bound to its appoproteins either non covalently as in hæmoglobin and myoglobin or by covalently in cytochrome C.

Hæme consists of porphyrin, known as protoporphyrin ix with an iron cheated in its centre. The porphyrin system consists of four pyrrole rings links by methane (–CH=) bridges.

Hæme (Fe protoporphyrin IX)

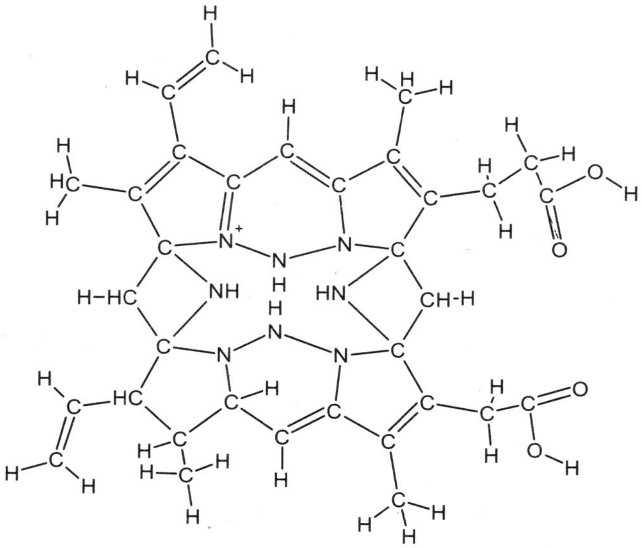

The conjugated (alternating) double bonds absorb visible light. Therefore, the hæme proteins are coloured. Porphyrinogen are important intermediates in hæme biosynthesis, biosynthesis. Their pyrrole rings are connected by methylene ($-CH_2-$) bridges, and they do not absorb light because the double bonds are not conjugated over the whole system.

Porphyrinogen are prone to non-enzymatic oxidation to the corresponding porphyrins. Although a small amount of dietary hæme is absorbed in the small intestine, essentially all the hæme in the human body is derived from endogenous synthesis. The next subhead will discuss the pathways of biosynthesis of hæme and also the diseases in which the pathways are disrupted.

Synthesis of Hæme

The 800 to 900 g of hæmoglobin in the human body contains 30 to 35mg of hæme 250 to 300mg of hæme have to synthesized in the red, bone marrow every day, most of this in the erythroblasts and proerythroblasts.

The bone marrow accounts for 70% to 80% of the total synthesis of the body

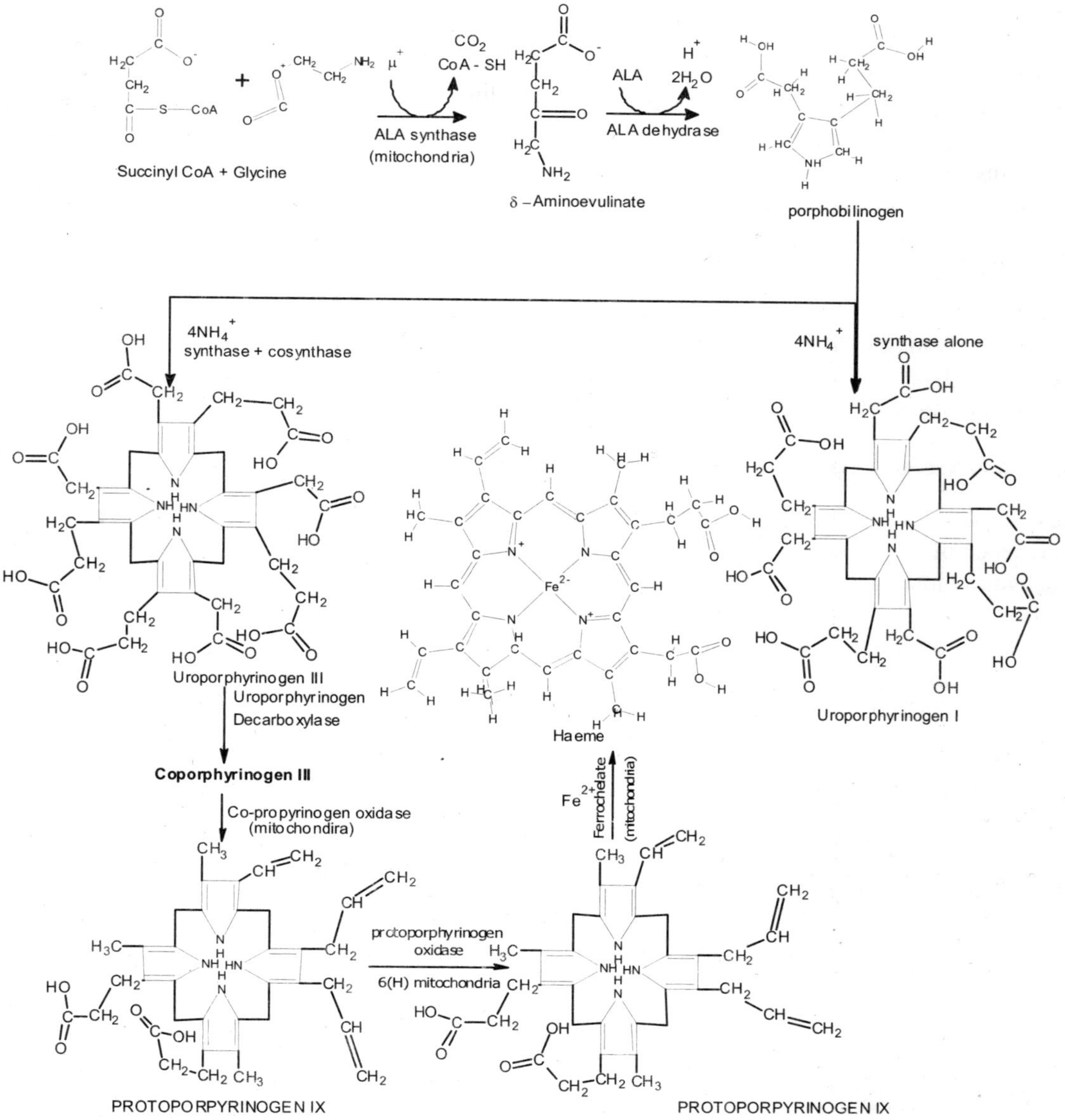

Succinyl CoA + Glycine

CO_2 CoA - SH

ALA synthase (mitochondria)

δ –Aminoevulinate

ALA H^+ $2H_2O$

ALA dehydrase

porphobilinogen

$4NH_4^+$ synthase + cosynthase

$4NH_4^+$ synthase alone

Uroporphyrinogen III

Uroporphyrinogen Decarboxylase

Haeme

Uroporphyrinogen I

Coporphyrinogen III

Co-propyrinogen oxidase (mitochondira)

Fe^{2+} Ferrochelate (mitochondria)

protoporphyrinogen oxidase 6(H) mitochondria

PROTOPORPYRINOGEN IX

PROTOPORPYRINOGEN IX

Biosynthesis of Hæme

The second most important site is the liver because of its high content of cytochrome P-450 and other hæme enzymes. These enzymes have for shorter half-lives then do hæmoglobin. Therefore hepatic hæme synthesis is quite productive accounting for approximately 15% of the total.

The hæme biosynthesis starts with the formation of δ–aminolevulinate (ALA) from succinyl CoA and glycine, catalyzed by the hæme containing vitamin B_6 dependant enzyme δ-aminolevulinate synthase. This is the committed step of the hæme synthesis as shown in the reaction equation above.

The second reaction catalyzed by ALA dehydrase, forms the pyrrole ring of porphobilinogen. Lead and other heavy metals can inhibit this enzyme by binding to sulphydryl groups in the enzyme protein.

The third enzyme uroporphobilinogen I synthase (also called porphobilinogen deaminases) Left on its own, this product can spontaneously however, a second protein uroporphyrinogen III cosynthase, channels the reaction into the formation of the isomer uroporphyrinogen III.

All naturally occurring porphyrins, including hæme processed to hæme as shown in the reaction equation above.

The first reaction and the last three reactions of the pathway are mitochondrial; the others are cytoplasmic, Aminolevulinic acid (ALA) is the regulated enzyme ALA synthase has an unusually short biological half-life of I to 3 hours in the liver and its synthesis is very effectively suppressed by hæme. Only free non-protein bound hæme acts as a feedback inhibitor.

The Chemical Properties of Hæme

Acids and alkalies act upon hæmoglobin to separate the hæme from globin, when the reaction occurs in the presence of oxygen the ferrohæme is promptly oxidized to ferrohæme. For example the action of dilute hydrochloric acid splits it into globin and ferrohæme the latter bring quickly oxidized to ferrihæme which holds a chlorine ion and is therefore called "acid hæmatin" or "hæmin"

Other acids act similarly to form acids hæmatin, the only variations consisting the differences in the acid anion held by the ferrihæme. Thus sulphuric acid hydroiodic acid form ferrihæme and iodide respectively. Alkalies split hæmoglobin into globin and ferrohæme, the latter then being oxidized to ferrihæme, which is combined with hydroxyl ion to give alkali hæmatin. If this reaction is carried out in the presence of a strong reducing agent such as sodium bisulphate $Na_2S_2O_4$, the ferrohæme is retained in the ferrous state.

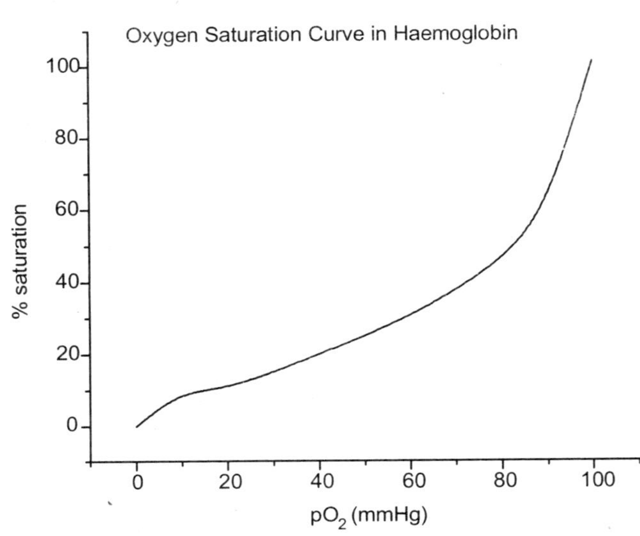

Fig. 8.14 : The Oxygen Saturation Curve of hæmoglobin

Globin may be coupled with hæme to form hæmoglobin (ferrohæmoglobin) by carrying out the reaction at the proper pH and in the presence of hydrosulphite to keep the hæme in the reduced state. Ferrihæme also may be combined with globin to form methæmoglobin (ferrihæmoglobin) Acids and alkalies directly split methæmoglobin into globin, acid and alkali hæmatin. In this case the hæme already is in the ferrihæme state. Now we are going to discuss about the gas transport mechanism in the blood. Hæmoglobin reacts reversely with the oxygen and the reaction equilibrium can be expressed in terms of the percentage of the oxygenated form, which will vary with the concentration of O_2. The relation has represented graphically in the fig. 8.14 where O_2 concentration is expressed as the partial pressure (pO_2) in mmHg.

According to Daltons Law, the total pressure of gas mixture is the sum of the partial pressure of each of the components. Also the partial pressure gases are proportional to the corresponding mole fractions of each has in the mixture, Finally, the concentration of the gas in the liquid phase of the gas liquid system is proportional to its partial pressure in the gas phase, the factor being the solubility of the gas in the liquid.

Since we are frequently concerned with air at 760 mm Hg, which contains O_2, N_2, CO_2 and water vapour. Therefore Daltons's Law can be expressed as follows

$$P_{tot} = pO_2 + pN_2 + pH_2O + pCO_2$$

The mole fraction of any gas, such as O_2 in the mixture is the following

$$\frac{pO_2}{p_{tot}} \text{ or } \frac{pO_2}{760}$$

The amount of O_2 dissolved in liquid at a given temperature is expressed in terms of a Bunsen co-efficient, the volume (millimeter) of a gas at the specified temperature and 760 mm Hg pressure that will dissolve in 1 ml of liquid. The Bunsen coefficient although corresponding to a particular temperature, is always corrected to the standard conditions of 760 mmHg pressure and 273°K. Under such conditions the volume of a gas is easily converted to a molar concentration, with 22.40 ml of an ideal gas being equivalent to 1 mmol.

The use of partial pressures of gasses is so common and important in problems of respiration, anesthesiology and any quantitative consideration involving gases that it is useful to illustrate the interrelationships of gas concentrations. Consider the concentrations of O_2 in blood plasma. The O_2 derived from the inspired air in which the pO_2 is about 158 mm Hg. As a result of some mixing with air already present in the lung as well as a result of its saturation with water vapour, the air in the alveolæ of the lung has a pO_2 of about 100 mmHg. Diffusion of the O_2 into the capillaries of the lungs result further lowering of the O_2 tension (pO_2) to about 90mm Hg. Therefore the arterial blood plasma can be considered in contact with a gas that has a pO_2 of 90 mmHg. The temperature of the body is close to 38°C at which temperature the Bunsen Coefficient for O_2 in blood plasma is 0.024. This means that 0.624 ml of O_2 at 760 mm Hg and 273K (0°C), will dissolve in 1 ml of plasma at 38°C when the pO_2 is 760mmHg, Since the amount of O_2 dissolved is proportional to its partial pressure in the alveolar air is 90 mmHg the actual volume of O_2 in arterial plasma at the lungs is the following

$$0.024 \times \frac{90}{760} O_2 / \text{ml plasma}$$

$$2.842105263 \times 10^{-3} \text{ml } O_2 / \text{ml plasma}$$

Since this volume of O_2, derived from the Bunsen co-efficient, gives the volume of the dissolved gas corrected to standard conditions the volume is converted to mill mole by dividing with 22.40

$$\frac{2.842105263 \times 10^{-3}}{22.4} \text{ml} O_2 / \text{ml plasma}$$

$$= 1.267857143 \times 10^{-4}$$

Since this is the amount of O_2 in 1 ml of plasma the concentration of dissolved O_2 is 0.127μmol/ml or 0.127mmol/L

Similar calculations can be made for any of the other gases in plasma, given their Bunsen co-efficient. If as with CO_2 they are not ideal gases the volume of 1 mmol of gas at 760mmHg and 273°K also must be known.

Considering again the oxygenation curves of hæmoglobin the saturation of hæmoglobin forms a sigmodial* curve see Fig. 8.14.

Through out the range of pO_2 values, the O_2 is bound more readily to myoglobin is independent of another molecule because there is only one O_2 binding state site of each monomeric subunit of hæmoglobin also binds O_2 but in the tetrameric quaternary structure the binding affinity of O_2 depends on whether an O_2 molecule is bound to one of the other subunits. The bindings of O_2 is said to be co-operative. The binding of first O_2 affect the binding of second O_2, these two O_2 affect that of the third one and so on. This situation reflects a co-operatively in the binding processes. In effect, there are four different key values for each of the binding sites with K_1 being less than K_2 and K_3, and the final equilibrium constant demonstrates that the last O_2 molecule is more tightly bound than the first

$$Hb_4 + O_2 \xrightleftharpoons{K_1} Hb_4(O_2) + O_2 \xrightleftharpoons{K_2} Hb_4(O_2)_2 + O_2 \xrightleftharpoons{K_3} Hb_4(O_2)_3 + O_2 \xrightleftharpoons{K_4} Hb_4(O_2)_4$$

$$K_1 < K_2 \approx K_3 < K_4$$

Explanation for the co-operative effects in hæmoglobin have been based on evidence obtained by X-Ray crystallography of oxy-hæmoglobin and deoxy-hæmoglobin. It is known that the confirmations of the A and β chains are altered when their respective hæme residues combine with O_2, The stepwise process may be described using several models unfortunately none of these does actually explains the real fact.

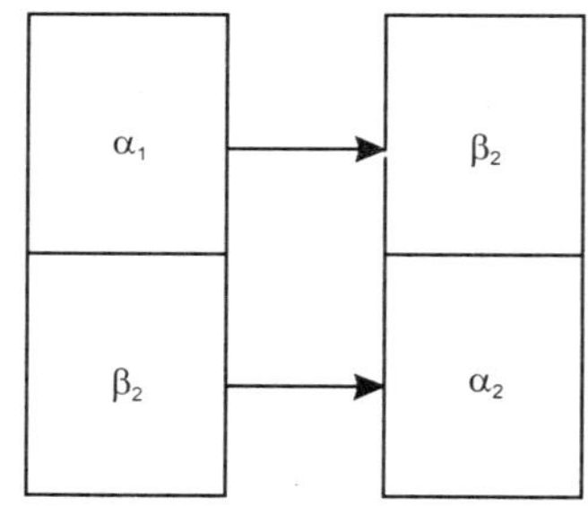

Fig. 8.15 : The relation of the Hæmoglobin Pattern

The $(\alpha\beta_2)$ tetramer of hæmoglobin has two types of contacts interface (1) between the α_1 and β_1 subunit and (2) between $\alpha_1\beta_2$

* Sigmodial Curve: The 'C' shaped curve detailed discussion later.

which is identical to α_{11}. This can be expressed in the flow diagram (fig. 8.15A).

When the hæmoglobin becomes oxygenated a large structural change occurs at the $\alpha_1\beta_2$ and $\alpha_2\beta_1$ interfaces and much less of a change occurs at the $\alpha_1\beta_1$ and α_{22} interfaces.

The binding of the first O_2 molecule can occur at α or β subunit, that is at two different binding states, hæmoglobin with three O_2 molecules can have two binding states and the fully oxygenated molecule has one state in which all subunits carry an O_2. Thus a total of nine binding states exist. The free energy levels of these nine states fall into three groups as represented in the Fig. 8.16 graphically.

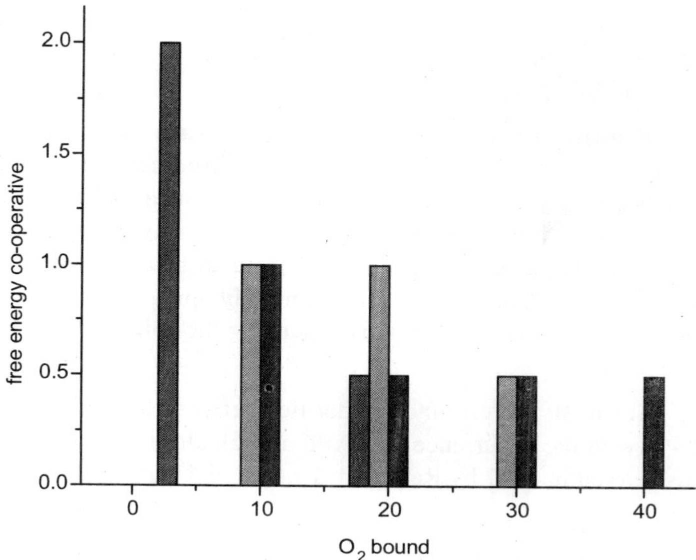

Fig. 8.16 : The Free Energy Levels for Tetramers of Hæmoglobin in various Binding States

These three energy states suggest that in the course of sequential binding, the hæmoglobin molecule switches into three molecular forms as proposed to only two, an oxy hæmoglobin and deoxy hæmoglobin confirmation. Several confirmations probably form the transitions between the two extremes of oxy and deoxy hæmoglobin, with some too fine to measure. Evidence shows that the switches in confirmation are not simple are found in dissociation studies of carboxy hæmoglobin and similarly bonded molecules proton nuclear magnetic spectroscopy and X-ray studies of partially oxygenated hæmoglobin.

These confirmation switches occur in response to oxygen binding and thus the affinity of the O_2 for the hæmoglobin states changes as was noted in the four equilibrium constant.

As a result a sigmodial saturation curve of Hb_4 for O_2 exists. The conformational switches occurring in the protein also are partly related to the slight change in position of the Fe(II) in the hæme when it binds O_2. The conformational changes produce displacement of an L-tyrosol residue from a pocket in the globular polypeptide. Along with this residue, the NH_2 terminal residue that has been in a salt linkage with nearby negativity charged groups is also displaced. The number of salt linkage broken in the Hb_4

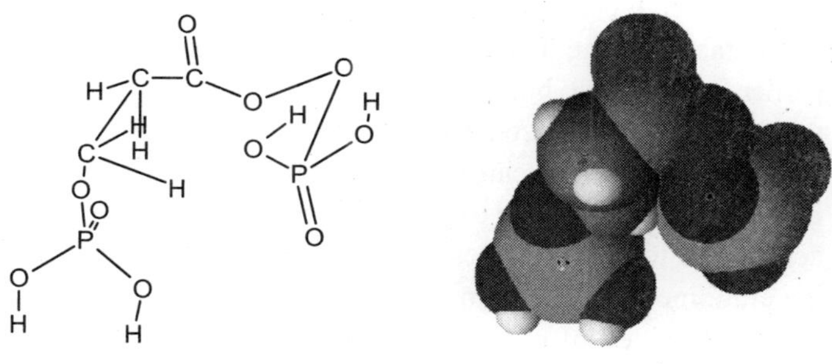

Fig. 8.17: The 2-3 biphospho-D-glycerate

depends on the number of O_2 molecule bound, which again is related to the different affinities for O_2 in the intermediate states. This change is called the Bohr Effect, and has been discussed in detail in the subhead following.

Binding of 2-3 biphospho-D-glycerate to Hæmoglobin (2,3-BPG, also known as 2,3-diphosphoglycerate or 2,3-DPG) is a three carbon isomer of the glycolytic intermediate 1,3-bisphosphoglycerate. 2,3-BPG is present in human red blood cells (RBC; erythrocyte) at approximately 5 mmol/L. It binds with greater affinity to deoxygenated hæmoglobin (e.g. when the red cell is near respiring tissue) than it does to oxygenated hæmoglobin (e.g. in the lungs). In bonding to partially deoxygenated hæmoglobin it allosterically upregulates the release of the remaining oxygen molecules bound to the hæmoglobin, thus enhancing the ability of RBCs to release oxygen near tissues that need it most.

Scientists have found similarities between low amounts of 2,3-BPG with the occurrence of HAPE at high altitudes. Its function was discovered in 1967 by Reinhold Benesch and Ruth Benesch.

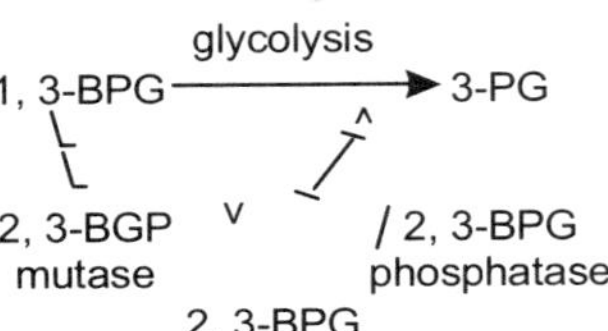

2,3-BPG is formed from 1,3-BPG by an enzyme called bisphosphoglycerate mutase. It is broken down by a phosphatase to form 3-phosphoglycerate. Its synthesis and breakdown are therefore a way around a step of glycolysis.

Effects of Binding

When 2,3-BPG binds deoxyhæmoglobin, it acts to stabilize the low oxygen affinity state (T state) of the oxygen carrier. It fits neatly into the cavity of the deoxy-conformation, exploiting the molecular symmetry and positive polarity by forming salt bridges with lysine and histidine residues in the four subunits of hæmoglobin. The R state, with oxygen bound to a hæme group, has a different conformation and does not allow this interaction.

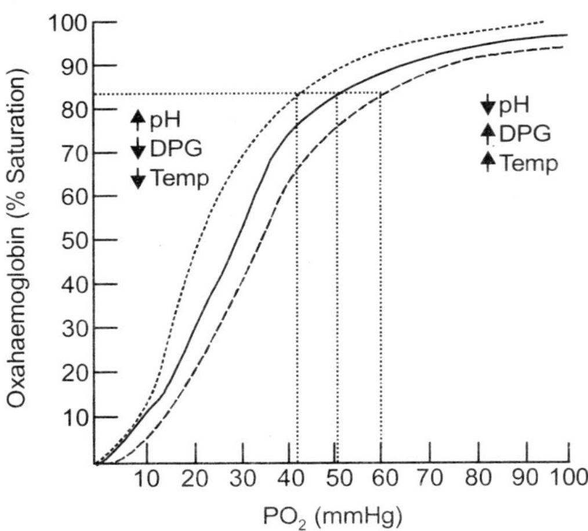

Fig. 8.18 : Oxygen-haemoglobin dissociation Curve

By selectively binding to deoxyhæmoglobin, 2,3-BPG stabilizes the T state conformation, making it harder for oxygen to bind hæmoglobin and more likely to be released to adjacent tissues. 2,3-BPG is part of a feedback loop that can help prevent tissue hypoxia in conditions where it is most likely to occur. Conditions of low tissue oxygen concentration such as high altitude (2,3-BPG levels are higher in those acclimated to high altitudes), airway obstruction, or congestive heart failure will tend to cause RBCs to generate more 2,3-BPG in their effort to generate energy by allowing more oxygen to be released in tissues deprived of oxygen. Ultimately, this mechanism increases oxygen release from RBCs under circumstances where it is needed most. This release is potentiated by the Bohr Effect in tissues with high energetic demands.

Deoxygenation and Carboxylation of Hæmoglobin

The carboxylation of hæmoglobin is mainly the combination of carbon dioxide with hæmoglobin to form carboxy hæmoglobin and thereby to interfere with the oxygenation of these substances. Carbon monoxide also combines with ferrous cytochrome oxidase one of the respiratory enzyme of tissue and prevents its action particularly at higher concentrations of the gas.

Carbon monoxide also adds on the hæme iron atoms exactly as does oxygen but with much more affinity, the overall reactions of O_2 and CO with hæmoglobin (Hb).

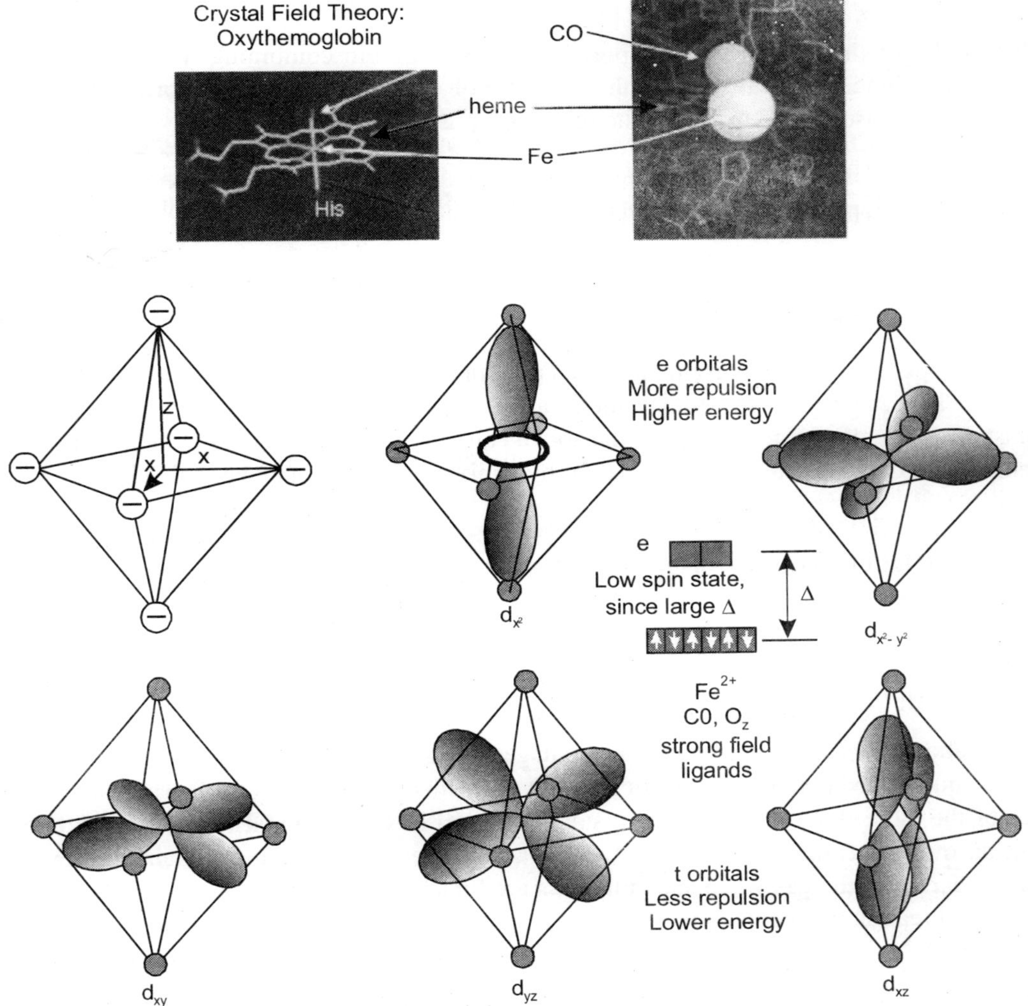

Fig. 8.19 : The crystalline properties of Hæme

$$O_2 + Hb \rightleftharpoons HbO_2$$
$$CO + Hb \rightleftharpoons HbCO$$

take place when hæmoglobin is exposed to a mixture of the gases there is competition for the hæmoglobin according to the equation.

$$HbO_2 + CO \rightleftharpoons HbCO + O_2$$

It is interesting to note that light promotes the dissociation of carboxy hæmoglobin which necessitates equibrating blood or hæmoglobin solutions with CO in the dark when studying reactions of the gas with hæmoglobin.

It has shown that the relative proportions of hæmoglobin combining with O_2 and CO when hæmoglobin solutions are saturated with a mixture of the two gases are proportional to the partial pressures of the gases in the mixture. The relationship is as follows:

$$\frac{[HbCO][O_2]}{[HbO_2][CO]} = K \text{ or } \frac{[HbCO]_p CO_2}{[HbO_2]CO} = K$$

These equations hold only when all the hæmoglobin is converted to carboxy hæmoglobin

or

$$\frac{[HbCO]}{[HbO_2]} = \frac{K_pCO}{pO_2}$$

In which pCO and pO_2 represent the partial pressure of CO and O_2 and K is the affinity constant of the action. The value of K for human blood at 38°C is about 210, which means that under these circumstances the affinity for O_2. From the above equation the following expressions for the fraction of hæmoglobin converted to HbCO and HbO_2 by a mixture of gases may be derived.

$$\frac{[HbCO]}{[HbCO]+[HbO_2]} = \frac{K_pCO}{pO_2 + KpCO}$$

and

$$\frac{[HbO_2]}{[HbCO]+[HbO_2]} = \frac{pO_2}{pO_2 + KpCO}$$

Table above gives the approximate percentages of carboxyhæmoglobin formed in human blood when it is equilibrated in the dark with air plus CO and with oxygen plus CO at 760 mm and 380°C. The figures of the table show that when very small amounts of CO are present in air, a relatively large proportion of the hæmoglobin is converted to carboxyhaemoglobin. This explains the exceedingly poisonous action of the gas when such mixtures are breathed. The figures also show that if a person poisoned by breathing air containing CO is given pure oxygen, the conversion of carboxyhaemoglobin to oxyhæmoglobin will be promoted. In such a case the high oxygen concentration in the blood tends to displace CO from carboxyhaemoglobin by mass action and the CO is irreversibly exhaled by the lungs:

$$HbCO + O_2 \rightarrow HbO_2 + CO$$

CO Per Cent	Per Cent HbCO	
	Air 760 mm	Oxygen 760 mm
0.04	52	13
0.08	68	22
0.12	77	30
0.16	81	36
0.20	84	42
0.24	86	46
0.28	88	50

Haldane pointed out that the anoxic symptoms caused by CO poisoning greatly surpass in severity those of anemia when the blood oxygen content is the same in the two conditions. In such cases of CO poisoning some of the iron atoms of the haemoglobin molecule hold CO and some hold O_2 molecules. At the low oxygen tensions present in the tissues and capillaries these mixed molecules give up their oxygen much less readily than do haemoglobin molecules holding the same quantity of oxygen alone. Consequently, the presence of CO with oxygen in the haemoglobin molecule may cause severe tissue hypoxia, whereas the same oxygen content in the absence of CO may prove adequate. Carbon monoxide thus exerts a double action. It prevents the uptake of normal quantities of oxygen in the lungs and interferes with the unloading of oxygen in the tissues.

Carboxyhaemoglobin is strikingly distinguished from haemoglobin and oxy-hæmoglobin by its cherry red colour. The skin and tissues of victims of CO poisoning are tinged with this colour. The absorption spectrum of carboxy-haemoglobin shows bands with centers at 570 and 535 mμ, respectively, which are close to the bands of oxyhæmoglobin. Various chemical tests for carboxyhaemoglobin in blood have been devised. A simple test is to treat the suspected blood with a little NaOH, dilute it greatly, and then compare it with normal blood similarly treated. The normal blood shows a greenish hue, while the CO blood remains pink. Simple dilution without adding alkali also shows the CO blood to be pink, as compared with the more brownish red of normal blood.

The Formation Different Types of Hæmoglobin

Methaemoglobin or Ferrihaemoglobin

The oxidation of the ferrous iron of haemoglobin to the ferric state results in the formation of methaemoglobin or ferrihaemoglobin. Methaemoglobin is dark brown in color, in contrast to the purple of haemoglobin, the dark red of oxyhaemoglobin, and the cherry red of carboxyhaemoglobin. It is readily reduced back to haemoglobin by strong reducing agents, such as hydrosulfite.

The iron of methaemoglobin carries an extra positive charge and holds a hydroxyl or other negative group. Since the haemoglobin molecule contains four hæme groups with their ferrous iron atoms, the oxidation of haemoglobin to methaemoglobin proceeds in stages, so that intermediate molecules containing both ferrous and ferric iron atoms must be formed in the oxidation process:

$$Hb_4(Fe^{++})_3(Fe^{+++}\text{-}OH),\ Hb_4(Fe^{++})_2(Fe^{+++}\text{-})_2(Fe^{+++}\text{-}OH)_2 \text{ etc}$$

Conversion of ferrous to ferric iron in haemoglobin destroys its capacity to combine with O_2 and CO, and methaemoglobin consequently is useless in the transport of O_2. It appears that the intermediate compounds of haemoglobin oxidation, in which some ferrous iron is present, combine with O_2 and in proportion serve to transport it.

Various oxidizing agents convert haemoglobin to methaemoglobin, ferricyanide and nitrite being most commonly employed in the laboratory. Molecular O_2 rather slowly converts haemoglobin of blood to methaemoglobin upon standing.

Neill and Hastings showed that fully oxygenated haemoglobin is resistant to oxidation, and that conversion of haemoglobin to methaemoglobin proceeds through oxidation of the deoxygenated form.

The oxidation of haemoglobin to met

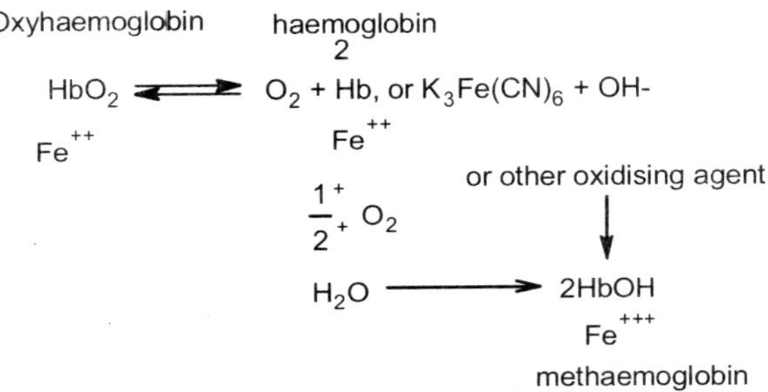

Haemoglobin in blood or in haemoglobin solutions may be represented as follows, where Hb indicates the amount of haemoglobin corresponding to one iron atom:

When the reversible reaction:

$$HbO_2 \rightleftharpoons O_2 + Hb$$

is made irreversible by oxidizing the Hb to methaemoglobin which cannot hold O_2, the oxygen then is liberated and escapes. This fact is utilized in determining the O_2 of blood and haemoglobin solutions in the gas apparatus of Van Slyke.

Normally, the conversion of hæmoglobin to methæmoglobin takes place in the circulating blood, but the reducing systems present in red cells tend to prevent the accumulation of any appreciable quantity of methæmoglobin. The amount present in human blood has been estimated to range from 1.1 to 1.4 per cent of the total hæmoglobin, with an average value of 1.7. It is interesting to note that glutathione, which is present in high concentration in the erythrocytes, reduces methæmoglobin to hæmoglobin, which may represent an important function of the compound. Cox and Wendel found that high concentrations of methæmoglobin in dog erythrocytes are reduced at a rate of about 10 to 12 per cent per hour, which decreases with decreasing methæmoglobin concentrations. Many different chemical agents have been found to convert hæmoglobin to methæmoglobin. These include oxidizing agents, such as chlorates, nitrites (both inorganic. and organic), ferricyanides, quinones, peroxides, and various oxidizing dyes; reducing agents, such as hydroquinone, pyrogallol, hydroxylamine, and hydrazobenzene; and compounds which are neither pronounced oxidizing nor reducing agents, such as aminophenol, azoxybenzene, nitrobenzene, aniline, acetanilide, and the sulfonamide drugs. It appears that agents which are not capable directly of oxidizing hæmoglobin to methæmoglobin are first converted in the blood to such agents or to substances which catalyze the oxidation of hæmoglobin to methæmoglobin. Thus it seems that the action of aromatic amino and nitro compounds may be the result of their conversion to aminophenol and then into corresponding quinones which are oxidizing agents. Another mechanism by which some of these agents may promote methæmoglobin accumulation in blood is by inhibition of the systems which reduce the methæmoglobin normally formed in blood.

Although most of the agents listed above presumably may cause methæmoglobinemia in circulating blood, this is not true in all cases. For example, ferricyanide does not penetrate the red cell and consequently may be absorbed and excreted without forming methæmoglobin. Nitrite, on the other hand, rapidly penetrates the erythrocytes and produces methæmoglobinemia.

Clinically, methæmoglobinemia is most often observed in factory workers who have absorbed through the skin or lungs aromatic nitro and amino compounds, such as nitro-benzene, nitro phenols, and aniline, and in patients taking large amounts of drugs such as acetanilide and sulfonamides. Dennig, from experiments on dogs poisoned with acetanilide, concluded that the conversion of about two-thirds of the hæmoglobin to methæmoglobin causes fatal results. In nonfatal cases he observed an increase in blood methæmoglobin for several hours after giving the drug orally, followed by reversion to normal hæmoglobin, which was complete in 48 hours. The hypoxia due to methæmoglobinemia, like that due to carbon monoxide poisoning, is more severe than the hypoxia due simply to anemia giving the same total blood oxygen content. This is related to the fact that the intermediate ferrous-ferrihæmoglobin oxygen complexes release O_2 less readily than does an equivalent amount of oxyhæmoglobin, a situation quite analogous to conditions in carbon monoxide poisoning.

The chief symptoms observed in severe methæmoglobinemia are cyanosis (blue skin and membranes) and dyspnea (labored breathing). The cyanosis is due to the dark brown color of methæmoglobin, and the dyspnea to the diminution in the oxygen transported to the tissues by the hæmoglobin. The other symptoms are those commonly associated with tissue anoxia.

When carefully neutralized cyanide is added to methæmoglobin solutions, the color changes from brown to bright red, due to the formation of cyan methæmoglobin which shows an absorption spectrum maximum at 540 mμ. The compound was first crystallized by von Zeynek, who showed the presence of one cyanide group for each iron atom. The importance of this compound is its effectiveness in combating cyanide poisoning. Cyanide poisons primarily by combining with the cytochrome oxidase enzyme of tissues, thereby blocking intracellular oxidations. The conversion of a part of the hæmoglobin of blood to methæmoglobin permits the latter to withdraw the cyanide from tissue cytochrome oxidase, which then allows cellular oxidations to proceed. The cyanide is then detoxified by conversion to thiocyanate. Sodium nitrite is generally given intravenously in order to provide the requisite methæmoglobin.

Methæmoglobin also forms complexes with many other substances. These include nitric oxide, azide, hydrogen sulfide, hydrogen peroxide, cyanate, thiocyanate, and fluoride.

Sulphæmoglobin

Hoppe-Seyler observed the production of a greenish hæmoglobin derivative in the reaction of hydrogen sulfide on oxyhæmoglobin which has been shown to be sulphæmoglobin. This substance appears to be formed from hæmoglobin only in the presence of oxygen, indicating that oxyhæmoglobin reacts with H_2S. Sulphæmoglobin contains one more sulfur atom than does hæmoglobin. Lemberg has suggested that the most probable mechanism of reaction between oxyhæmoglobin and H_2S is as follows:

$$Hb(XH_2)O_2 + H_2S \rightarrow Hb(X)S + 2H_2O$$

Oxyhæmoglobin Sulphæmoglobin X = group attached to 2H, in globin molecule.

The exact linkage of the sulfur atom in the molecule is uncertain. Complete conversion of hæmoglobin to sulphæmoglobin has never been accomplished. In vitro preparations are generally mixed with a green pigment, choleglobin.

Sulphæmoglobin does not combine reversibly with O_2 but may be oxidized to sulph-methæmoglobin. However, it does combine with carbon monoxide to form carboxysulph-hæmoglobin.

The presence of sulphæmoglobin in blood has been reported in a number of conditions. These include therapy with sulphonamides and aromatic amine drugs such as phenacetin and acetanilide, dosage with sulfur, severe constipation, and cases of severe bacteremia due to Clostridium welchii. The H_2S for sulphæmoglobin production arises in the intestine from bacterial action upon sulfur compounds. Normally, it is rapidly oxidized to sulfate after absorption into the body, but when produced in excessive quantities and when certain catalytic agents are present, it may lead to the formation of sulphæmoglobin. However, blood seldom contains more than 10 per cent sulphæmoglobin. It is of interest to note that sulphæmoglobin, unlike methæmoglobin, is not converted back to functional hæmoglobin in blood but persists in the corpuscles throughout their life span of several months.

Verdohæmoglobin or Green Hæmoglobin

Verdohæmoglobin appears to be the substance formed in the first stage of the conversion of hæmoglobin to bile pigments. According to Watson, verdohæmoglobin is formed by removing the α-methylidyne bridge and opening the hæme ring of hæmoglobin, with the iron remaining attached. The exact structure of the resulting iron-containing group is unknown. Barkan and Walker state that the iron of verdohæmoglobin is more easily split off than is the iron of hæmoglobin. These workers found the blood pigment of normal persons to contain about 5 per cent of this type of easily split-off iron. Havemann, by the use of a direct method, estimated that as much as 8 per cent of the total hæmoglobin of normal persons is in the form of verdohæmoglobin.

As we have seen before that when CO combines with hæmoglobin competitively at the same site and as does O_2 which will be better discussed in the. Bohr Effect, CO binding is 200 times stronger that O_2 binding. The amount of O_2 that can be carried by the remaining untreated Fe(II) of the hæmoglobin molecule is reduced if CO is present at one end of the hæme residues. Moreover, the O_2 that is

	right shift	left shift
temperature	high	low
2.3-DPG	high	low
$p(CO_2)$	high	low
p(CO)	low	high
pH (Bohr effect)	low (acidosis)	high (alkalosis)
type of haemoglobin	adult haemoglobin	foetal haemoglobin

Left shift of the curve is a sign of haemoglobin's increased affinity for oxygen (e.g. at the lungs). Similarly, right shift shows decreased affinity, as would appear with an increase in body temperature, hydrogen ion, 2,3-diphosphoglycerate (also known as bisphosphoglycerate) or carbon dioxide concentration (the Bohr effect).

Carbon monoxide has a much higher affinity for haemoglobin than oxygen does. In carbon monoxide poisoning, oxygen cannot be transported and released to body tissues thus resulting in hypoxia. With foetal haemoglobin, the shift facilitates diffusion of oxygen across the placenta. The oxygen dissociation curve for myoglobin exists even further to the left. The causes of shift to right can be remembered using the mnemonic, "**CADET**, face Right!" for **C**O2, **A**cid, 2,3-**D**PG, **E**xercise and **T**emperature.

Factors that Affect the Standard Dissociation Curve

The strength with which oxygen binds to hæmoglobin is affected by several factors. In effect these factors shift or reshape the oxyhæmoglobin curve ("the standard curve"). The standard curve is shifted to the right by an increase in temperature, 2,3-DPG, or pCO_2, or a decrease in pH.

A rightward shift indicates that the hæmoglobin under study has a decreased affinity for oxygen. This makes it more difficult for hæmoglobin to bind to oxygen (requiring a higher partial pressure of oxygen to achieve the same oxygen saturation), but it makes it easier for the hæmoglobin to release oxygen bound to it. The effect of this rightward shift of the curve increases the partial pressure of oxygen in the tissues when it is most needed, such as during exercise, or hemorrhagic shock.

In contrast, the curve is shifted to the left by the opposite of these conditions. This leftward shift indicates that the hæmoglobin under study has an increased affinity for oxygen so that hæmoglobin binds oxygen more easily, but unloads it more reluctantly.

Variation of the Hydrogen ion Concentration

This changes the blood's pH. A decrease in pH shifts the standard curve to the right, while an increase shifts it to the left. Effects of carbon dioxide

Carbon dioxide affects the curve in two ways: first, it influences intracellular pH (the Bohr effect), and second, CO_2 accumulation causes carbamino compounds to be generated through chemical interactions, which bind to hæmoglobin forming Carbaminohaemoglobin. Low levels of carbamino compounds have the effect of shifting the curve to the right, while higher levels cause a leftward shift. However, this isn't the overriding effect of CO_2 accumulation. Only about 5-10% of the total CO_2 content of blood is transported as carbamino compounds.

Most of the CO_2 content (80-90%) is transported as bicarbonate ions. The formation of a bicarbonate ion will release a proton into the plasma. Hence, the elevated CO_2 content creates a respiratory acidosis and shifts the oxygen-haemoglobin dissociation curve to the right.

Effects of 2,3-DPG

2,3-Disphosphoglycerate (also 2,3 Bisphosphoglycerate or BPG), or 2,3-DPG, is an organophosphate, which is created in erythrocytes during glycolysis. The production of 2,3-DPG is likely an important adaptive mechanism, because the production increases for several conditions in the presence of diminished peripheral tissue O_2 availability, such as hypoxemia, chronic lung disease, anæmia, and congestive heart failure, among others. High levels of 2,3-DPG shift the curve to the right, while low levels of 2,3-DPG cause a leftward shift, seen in states such as septic shock and hypophosphatemia. Temperature

Temperature does not have such a dramatic effect compared to the previous factors, but **hyperthermia causes a rightward shift, while hypothermia causes a leftward shift.**

Respiratory Control of Blood pH

The pO_2 of blood can be changed quite rapidly by the rate and/ or depth of breathing (pulmonary exchange). Slow, shallow breathing (hypoventilation) results in a building up of pCO_2 and a reduction in this diffusion of CO_2 from the blood into the pulmonary alveolar gas phase. This increases the pCO_2 in the plasma and lower the blood pH. Hyperventilation has the effect. These two effects can be illustrated by simple example of a normal, healthy person who develops hiccups. Knowing that such an affect can usually be stopped by increasing the level of CO_2 in the blood the individual holds his or her breadths with closing the nose and mouth (i.e. inhales air with high pH CO_2 of 58 mm Hg and total CO_2 content of 29 mmol/L. At this time the pH of the blood can be calculated.

$$pH = 6.1 + \log_{10}\frac{29 - 0.0301 \times 58}{0.0301 \times 58}$$

$$pH = 6.1 + \log_{10}\frac{29 - 1.7458}{1.7458}$$

$$pH + 6.1 + \log_{10} 15.6160$$

$$pH = 6.1 + 1.19$$

$$pH = 7.2$$

To eliminate the excess CO_2, the individual breaths rapidly and deeply for a few mins (hyperventilation). The blood plasma then contains 25.5 mmol/L $[HCO_3]^-$ and the pCO_2 is 35 mmHg. By how much the blood pH has changed.

$$pH = 6.1 + \log_{10}\frac{25.5}{0.0301 \times 35}$$

$$pH = 6.1 + \log_{10}\frac{25.5}{1.0535}$$

$pH + 6.1 + \log_{10} 24.205$

$pH = 6.1 + 1.38390$

or 7.5

Thus pH has moved from the acid side of normal to the alkaline side changing 0.29 pH units by going from a state of hypoventilation to one of hyperventilation.

Transportation of Gases in Blood

Consider the circulation of blood through the lungs where the pO_2 is approximately 90 mmHg and the pCO_2 about 30mmHg. One can see from the fig 284 that in this condition the hæmoglobin is almost 100% saturated. As this blood moves to the peripheral capillary bed, the pCO_2 is reduced and then increased because the cells takes up and utilize the O_2 to oxidize nutrients to CO_2 and they release the CO_2 that is formed back into the blood. This increases the blood pCO_2, which in turn produces a lowered blood pH. The amount of O_2 combined with hæmoglobin at a given pCO_2 because of the decrease in pH, permitting more O_2 to be taken up the tissues, Additional release of O_2 in the capillary bed is caused by the effect

Of 2, 3 bisphosphate-D-glycerate in the erythrocyte. The overall function of the erythrocyte is therefore to bind the maximum amount of O_2 in the lungs and to release a portion of it to the tissues. Release of O_2 to the tissues in enhanced by the presence 2-3 bisphosphate-D-glycerate.

The CO_2 that is produced by catabolism must be returned to the lungs, where it is to be exhaled. About 10% is transported in the red blood cells as carboamino hæmoglobin. In this form the CO_2 is linked covalently to the NH_2 terminal value residues of hæmoglobin subunits.

$$CO_2 + HbNH_2 \leftrightarrows HbNH^- COO^- + H^+$$

The reaction is rapidly reversible, and probably not catalyzed by any enzyme.

Sohydric Transport of CO_2

Most of the CO_2 is transported in the plasma as bicarbonates which is produced in the erythrocytes from CO_2 (see Fig. 8.22).

Several reactions occur in this transport process.

1. Since oxyhæmoglobin is a stronger acid that deoxyhæmoglobin the following reactions take place.

$$HbH_4^+ + O_2 \rightleftharpoons HbO_2 + xH^+$$
$$(pK_a = 7.71)(pK_a = 7.13)$$

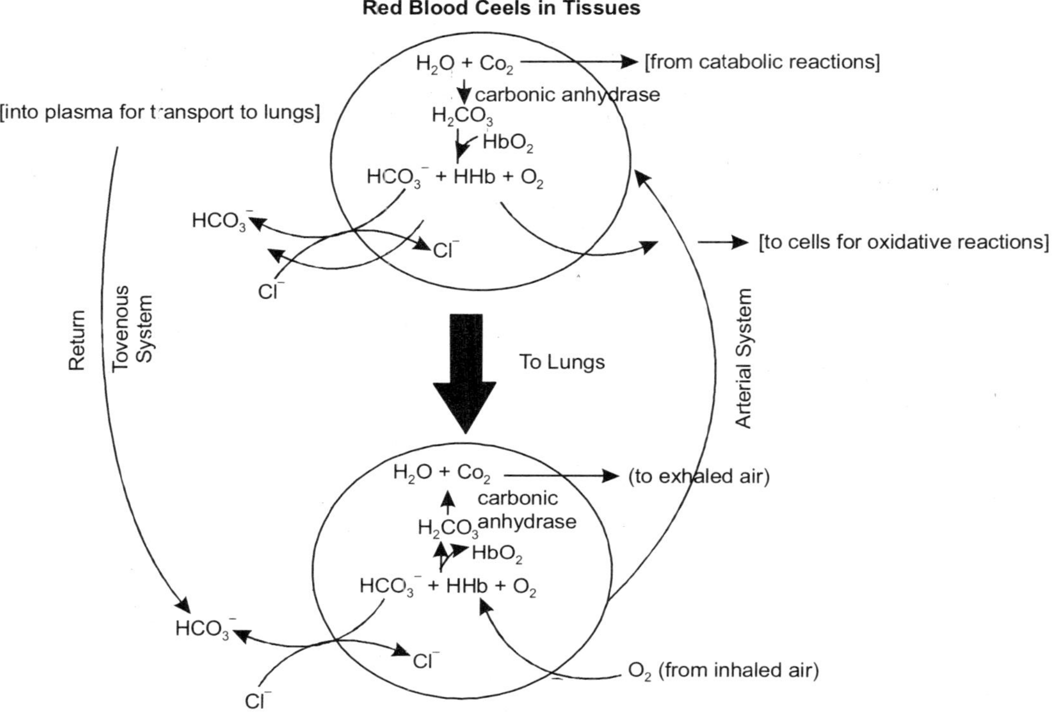

Fig. 8.22 : The gas exchange between the Erythrocytes and Lungs

The pK_a values of these two weak acids are such that $x = 0.68$ approx. This property of the hæmoglobins demonstrates their buffering effect. It also explains the transport of an appreciable quantity of the CO_2 released from the tissues without change in pH the so called iso hydric transport of CO_2. The oxyhæmoglobin, arriving in the red blood cells at the capillary bed, dissociates to release O_2, a process that is facilitated by the state of the tissues relative to incoming arterial blood: low pO_2, lower pH and higher pCO_2

$$0.68H^+ + HbO_2 \rightleftharpoons HbH^+_{0.68} + O_2$$

The H^+ in this reaction is derived from the discussion of H_2CO_3, which is formed from the CO_2 diffusing from the tissue into the plasma and finally into the erythrocytes (rbc)

$$H_2CO_3 \leftrightarrows H^+ + HCO_3^-$$

The update of H^+ by hæmoglobin buffers the effect of the carbonic acid. The HCO_3^- formed in the erythrocyte diffuses into the plasma and is carried back by the venous blood to the lungs, where reduced hæmoglobin is oxygenated. This results in the release of H^+, which reacts with HCO_3^- to give H_2CO_3. The concentration in the erythrocyte is therefore being reduced so that HCO_3^- from the plasma eturns to the cell. The buffering effect reduces the pH change as a result of the oxygenation of HbH^+, Also the H_2CO_3 dehydrates (catalyzed by carbonic anhydrase) to form CO_2 for exhalation the lungs. The reactions of hæmoglobin with CO_2 and O_2 of hæmoglobin with CO_2 and O_2 occur in support of each other and of proper handling of the H^+ load.

2. Added to this H^+ load is the H^+ from the formation of carboamino hæmoglobin the erythrocyte to maintain electro neutrality. Thus in the lung the movement of ion move from the cell into the plasma to take its place. This part is played by Cl^-, to so that the $[Cl^-]$ is higher and the $[HCO_3^-]$ lower in arterial than in venous blood, conversely in the capillary bed CO_2 diffuses into the plasma and then into the erythrocyte, HCO_3^- is formed (as explained in the discussion of the first reaction of the transport process) and moves to the plasma. Its place in the erythrocyte is taken by Cl^- from the plasma, so that venous blood has a lower $[Cl^-]$ and higher $[HCO_3^-]$ in the plasma than does arterial blood. This point can easily be demonstrated in vitro by blowing O_2 over a sample of venous blood.

   Not all the H^+ produced by the first two reactions is accommodated by the protonation of deoxyhæmoglobin. Therefore the venous blood plasma is more acidic than arterial blood, state reflected by a higher pCO_2

Myoglobin

Myoglobin is a single-chain globular protein of 153 amino acids, containing a hæme (iron-containing porphyrin) prosthetic group in the center around which the remaining apoprotein folds. It has eight alpha helices and a hydrophobic core. It has a molecular weight of 16,700 daltons, and is the primary oxygen-carrying pigment of muscle tissues. Unlike the blood-borne hæmoglobin, to which it is structurally related, this protein does not exhibit cooperative binding of oxygen, since positive cooperativity is a property of multimeric/oligomeric proteins only. Instead, the binding of oxygen by myoglobin is unaffected by the oxygen pressure in the surrounding tissue. Myoglobin is often cited as having an "instant binding tenacity" to oxygen given its hyperbolic oxygen dissociation curve. High concentrations of myoglobin in muscle cells allow organisms to hold their breaths longer.

In 1958, John Kendrew and associates successfully determined the structure of myoglobin by high-resolution X-ray crystallography. For this discovery, John Kendrew shared the 1962 Nobel Prize in chemistry with Max Perutz. Despite being one of the most studied proteins in biology, its true physiological function is not yet conclusively established: mice genetically

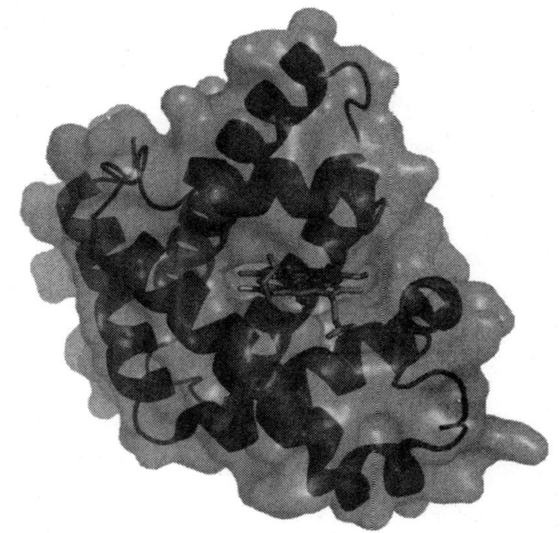

Fig. 8.23 : The structure of Myoglobin

engineered to lack myoglobin are viable, but showed a 30% reduction in cardiac systolic output. They adapted to this deficiency through hypoxic genetic mechanisms and increased vasodilation.

Myoglobin forms pigments responsible for making meat red. The colour that meat takes is partly determined by the charge of the iron atom in myoglobin and the oxygen attached to it. When meat is in its raw state, the iron atom is in the +2 oxidation state, and is bound to a dioxygen molecule (O_2). Meat cooked well done is brown because the iron atom is now in the +3 oxidation state, having lost an electron, and is now coordinated by a water molecule. Under some conditions, meat can also remain pink all through cooking, despite being heated to high temperatures. If meat has been exposed to nitrites, it will remain pink because the iron atom is bound to NO, nitric oxide (true of, e.g., corned beef or cured hams). Grilled meats can also take on a pink "smoke ring" that comes from the iron binding a molecule of carbon monoxide to give metmyoglobin. Raw meat packed in a carbon monoxide atmosphere also shows this same pink "smoke ring" due to the same coordination chemistry. Notably, the surface of the raw meat also displays the pink color, which is usually associated in consumers' minds with fresh meat. This artificially-induced pink color can persist in the meat for a very long time, reportedly up to one year. Hormel and Cargill are both reported to use this meat-packing process, and meat treated this way has been in the consumer market since 2003. Myoglobin is found in Type I muscle, Type II A and Type II B, but most texts consider myoglobin not to be found in smooth muscle.

Role in Disease

Myoglobin is released from damaged muscle tissue (rhabdomyolysis), which has very high concentrations of myoglobin. The released myoglobin is filtered by the kidneys but is toxic to the renal tubular epithelium and so may cause acute renal failure.

Myoglobin is a sensitive marker for muscle injury, making it a potential marker for heart attack in patients with chest pain. However, elevated myoglobin has low specificity for Acute Myocardial Infarction (AMI) and thus CK-MB, cTnT, ECG, and clinical signs should be taken into account to make the diagnosis.

Structure and Bonding

Myoglobin contains a porphyrin ring with an iron center. There is a *proximal histidine* group attached directly to the iron center, and a *distal histidine* group on the opposite face, not bonded to the iron.

Many functional models of myoglobin have been studied. One of the most important is that of *picket fence porphyrin* by James Collman. This model was used to show the importance of the distal prosthetic group. It serves three functions:

1. To form hydrogen bonds with the dioxygen moiety, increasing the O_2 binding constant
2. To prevent the binding of carbon monoxide, whether from within or without the body. Carbon monoxide binds to iron in an end-on fashion, and is hindered by the presence of the distal histidine, which forces it into a bent conformation. CO binds to hæme 23,000 times better than O_2, but only 200 times better in hæmoglobin and myoglobin. Oxygen binds in a bent fashion, which can fit with the distal histidine.
3. To prevent irreversible dimerization of the oxymyoglobin with another deoxymyoglobin species.

At one time or another, everyone has experienced the momentary sensation of having to stop, to "catch one's breath," until enough O_2 can be absorbed by the lungs and transported through the blood stream. Imagine what life would be like if we had to rely only on our lungs and the water in our blood to transport oxygen through our bodies.

O_2 is only marginally soluble (< 0.0001 *M*) in blood plasma at physiological pH. If we had to rely on the oxygen that dissolved in blood as our source of oxygen, we would get roughly 1% of the oxygen to which we are accustomed. (Consider what life would be like if the amount of oxygen you received was equivalent to only one breath every 5 min, instead of one breath every 3 s.) The evolution of forms of life even as complex as an earthworm required the development of a mechanism to actively transport oxygen through the system. Our blood stream contains about 150 g/L of the protein known as **hæmoglobin** (Hb), which is so effective as an oxygen-carrier that the concentration of O_2 in the blood stream reaches 0.01 *M*—the same concentration as air. Once the Hb-O_2 complex reaches the tissue that consumes oxygen, the O_2 molecules are transferred to another protein **myoglobin** (Mb) which transports oxygen through the muscle tissue

At the center of the hæme is an Fe(II) atom. Four of the six coordination sites around this atom are occupied by nitrogen atoms from a planar **porphyrin** ring. The fifth coordination site is occupied by a nitrogen atom from a histidine side chain on one of the amino acids in the protein. The last coordination site is available to bind an O_2 molecule. The hæme is therefore the oxygen-carrying portion of the hæmoglobin and myoglobin molecules. This raises the question: What is the function of the globular protein or "globin" portion of these molecules?

The structure of myoglobin suggests that the oxygen-carrying hæme group is buried inside the protein portion of this molecule, which keeps pairs of hæmes group from coming too close together. This is important, because these proteins need to bind O_2 reversibly and the Fe(II) hæme, by itself, cannot do this. When there is no globin to protect the hæme, it reacts with oxygen to form an oxidized Fe(III) atom instead of an Fe(II)-O_2 complex.

Hæmoglobin consists of four protein chains, each about the size of a myoglobin molecule, which fold to give a structure that looks very similar to myoglobin. Thus, hæmoglobin has four separate hæme groups that can bind a molecule of O_2. Even though the distance between the iron atoms of adjacent hæmes in hæmoglobin is very large between 250 and 370 nm the act of binding an O_2 molecule at one of the four hæmes in hæmoglobin leads to a significant increase in the affinity for O_2 binding at the other hæmes.

This **cooperative interaction** between different binding sites makes hæmoglobin an unusually good oxygen-transport protein because it enables the molecule to pick up as much oxygen as possible once the partial pressure of this gas reaches a particular threshold level, and then give off as much oxygen as possible when the partial pressure of O_2 drops significantly below this threshold level. The hæmes are much too far apart to interact directly. But, changes that occur in the structure of the globin that surrounds a hæme when it picks up an O_2 molecule are mechanically transmitted to the other globins in this protein. These changes carry the signal that facilitates the gain or loss of an O_2 molecule by the other hæmes.

Drawings of the structures of proteins often convey the impression of a fixed, rigid structure, in which the side-chains of individual amino acid residues are locked into position. Nothing could be further from the truth. The changes that occur in the structure of hæmoglobin when oxygen binds to the hæmes are so large that crystals of deoxygenated hæmoglobin shatter when exposed to oxygen. Further evidence for the flexibility of proteins can be obtained by noting that there is no path in the crystal structures of myoglobin and hæmoglobin along which an O_2 molecule can travel to reach the hæme group. The fact that these proteins reversibly bind oxygen suggests that they must undergo simple changes in their conformation changes that have been called **breathing motions**. that open up and then close down the pathway along which an O_2 mclecule travels as it enters the protein. Computer simulations of the motion within proteins suggests that the interior of a protein has a significant "fluidity," with groups moving within the protein by as much as 20 nm.

The Differences of the Hæmoglobin and Myoglobin

Because myoglobin and hæmoglobin each bind oxygen, we can assume that there should be some similarities between their structures. However, there should also be some dissimilarities because of their different roles as oxygen binding proteins. One obvious difference between the two proteins is that myoglobin is a single polypeptide chain while vertebrate hæmoglobins are tetrameric.

Myoglobin

Myoglobin comes form the Greek word Myos (muscle) it is the muscle hæmoglobin, it works in the muscle.

Human myoglobin has 153 amino acid residues.

Myoglobin has a highly folded and compact structure that has 8 separate and distinct alpha helical secondary structures.

Located at the surface are the polar, hydrophilic residues. The animation will cut slices through myoglobin, allowing you to "slab" through the protein, allowing you to see that the interior of globular proteins such as myoglobin is comprised of mostly hydrophobic residues, while the water-exposed exterior is predominantly hydrophilic.

Hæmoglobin

In human hæmoglobin, the overall structure is made up of four individual polypeptide subunits.

Human hæmoglobin is comprised of two alpha chains made up of 144 residues and two beta chains made up of 146 residues. The alpha and beta subunits associate more strongly with each other than similar subunits (alpha-alpha or beta-beta). For this reason, hæmoglobin is sometimes referred to as a "dimer of dimers".

This is interesting in light of the fact that these three polypeptide residues have only a 27% identity in their amino acid sequence.

alpha subunit—beta subunit

The subunits of hæmoglobin are arranged in a tetrahedral array with a tight spherical overall appearance and, like myoglobin, each individual polypeptide is folded in such a manner to maximize polar residues being on the exposed surface and non-polar interactions being internal, making this large protein water soluble. The arrangement of polypeptides is held together by hydrogen bonding, hydrophobic interactions and multiple ionic interactions, that take place at the contact points between subunits. These subunit interactions play a critical role in the binding of oxygen to hæmoglobin.

Oxygen Binding in Myoglobin and Hæmoglobin

The hæme is a small but important non-protein molecule, or prosthetic group, that is associated with these oxygen binding proteins. The hæme is comprised of protoporphyrin IX (an organic ring structure) which has a chelated (bound) iron atom.

The hæme resides in a small hydrophobic cleft within each polypeptide. The hydrophobic residues will be shown in green, while hydrophilic are colored red.

There is one hæme associated with myoglobin.

There are four hæmes found in hæmoglobin, one in each of the subunits.

Therefore, myoglobin binds reversibly to only one oxygen molecule, while and hæmoglobin binds four. These differences help to explain the different biological functions of hæmoglobin and myoglobin. Hæmoglobin is used to transport oxygen over large distances (from the lungs to all tissues), whereas myoglobin is present in muscle and behaves as a local "storage" reservoir of oxygen.

Also, due to the intrinsic nature of hæmoglobin, it can more precisely control the binding and release of oxygen in the appropriate areas of the body. The internal arrangement of hæme in this cleft is very similar in myoglobin and hæmoglobin. The hæme prosthetic group is made up of a protoporphyrin IX ring structure that is chelated to an iron atom in the ferrous (+2) oxidation state. Incidentally, when a metal ion binds to a porphyrin, which has a great number of conjugated double bonds, it will lead to a compound that is highly colored. This is why blood is red. In plants, the porphyrin is chelated with a magnesium ion, giving chlorophyll a greenish-blue colour.

This prosthetic group is physically held in the cleft by hydrophobic interactions as well as a coordinate covalent bond between the iron atom and a nearby nitrogen atom, belonging to the side chain of a histidine residue.

This histidine residue is often referred to as the proximal histidine. On the side of the hæme group opposite the proximal histidine, where oxygen reversibly binds, is another histidine residue called the distal histidine. This residue serves two very important functions in the polypeptide. First, it prevents oxidation of the iron ion to a higher oxidation state by any number of possible nearby oxidizing agents. At a higher oxidation state, the iron (Fe^{+++}) is unable to bond to oxygen. Secondly, this histidine residue acts to reduce carbon monoxide (CO) binding to the hæme. If the distal histidine was absent, even low levels of CO would greatly interfere with oxygen binding. The hæme group alone (in the absence of the surrounding protein of myoglobin or hæmoglobin) has a much greater bonding affinity for carbon monoxide than for oxygen.

The physical process of binding and releasing oxygen by hæmoglobin has a marked effect on its three-dimensional structure. Hæmoglobin in its deoxygenated state has a low affinity for oxygen compared to myoglobin. When oxygen is bound to the first subunit of hæmoglobin it leads to subtle changes to the quaternary structure of the protein. This in turn makes it easier for a subsequent molecule of oxygen to bind to the next subunit. Thus, with the initial oxygen binding to a subunit, the remaining unbound subunits become more receptive to oxygen. This phenomenon is called an allosteric (through space) interaction and is clearly illustrated in the sigmoidal curve shown above for oxygen binding to hæmoglobin.

The mechanism for this effect is initiated when upon binding to oxygen, the iron (Fe^{++}) atom is pulled into the plane of the ring. (In the unbound state, the iron (Fe^{++}) atom lies just to one side of the porphyrin ring.) Since this iron atom is attached to the proximal histidine, this causes the local helix to move also. This movement within the polypeptide subunit causes subtle but substantive changes localized in the regions between subunits. These changes include breaking and reforming hydrogen bonding and ionic linkages. The alpha1-beta2 and alpha2-beta1 subunit junctions are the segments most affected.. In the following depictions shown with Chime, this subunit motion is illustrated by the differences between oxyhæmoglobin and deoxyhæmoglobin.

Deoxy Hb—Oxy Hb

The binding of oxygen to hæmoglobin can be dramatically altered by a small group of substances called allosteric effectors. Hydrogen ions (protons), carbon dioxide, and 2,3-bisphosphoglycerate are effectors that can promote the release of oxygen by favoring the deoxygenated form of hæmoglobin. Since these allosteric effectors bind to sites that are specific to each kind of compound, their effects are cumulative.

Hydrogen ions and carbon dioxide are found in high concentrations around actively metabolizing tissues. In the capillaries, the environment favors the release of oxygen from hæmoglobin and the binding of these allosteric effectors. The overall result is to facilitate oxygen release into blood plasma and subsequent uptake of oxygen by the high affinity myoglobin in the tissues. The specific reactions of the hydrogen ions and carbon dioxide with hæmoglobin causing the release of additional oxygen is called the Bohr effect.

The reactions of the Bohr effect are reversible. When deoxygenated hæmoglobin returns to the lungs, the concentration of the hydrogen ions and the partial pressure of carbon dioxide is low. This causes these compounds to be released from hæmoglobin. The carbon dioxide is expelled out of the body through expired air. In effect hæmoglobin not only carries oxygen to the cells but it also carries waste products from the cells to the lungs, eventually to be eliminated out of the body.

In addition to hydrogen ions and carbon dioxide, a very important allosteric effector is 2,3-bisphosphoglycerate (2,3-BPG). It is a small organic molecule with a large overall negative charge due to the presence of a number of phosphate groups. This species is present along with hæmoglobin in red blood cells.

BPG affects oxygen binding affinity by binding in a small cavity at the center axis of deoxygenated hæmoglobin. It is held in place by a number of electrostatic interactions with positively charged residues in hæmoglobin. In oxygenated hæmoglobin, this cavity is too small to effectively accommodate 2,3-

BPG. It is only in the deoxygenated state that this molecule has the ability to bind into this cavity. When bound, 2,3-BPG greatly diminishes the binding of oxygen to hæmoglobin and facilitates oxygen unloading to actively respiring tissues.

Some points about Myoglobin

- Myoglobin is a haemoglobin-like, iron-containing pigment found in muscle fibres.
- It consists of a single alpha polypeptide chain and binds only one oxygen molecule (as opposed to haemoglobin which binds 4 oxygen molecules).
- The oxygen dissociation curve for myoglobin is hyperbolic (as opposed to the sigmodial curve for haemoglobin) and is to the left of that for haemoglobin.
- Myoglobin takes up oxygen from the haemoglobin in the blood and stores oxygen within the muscle itself

HAPTOGLOBIN

Haptoglobin (abbreviated as Hp) is a protein that in humans is encoded by the *Hp* gene. In blood plasma, haptoglobin binds free hemoglobin (Hb) released from erythrocytes with high affinity and thereby inhibits its oxidative activity. The haptoglobin-hemoglobin complex will then be removed by the reticuloendothelial system (mostly the spleen). In clinical settings, the haptoglobin assay is used to screen for and monitor intravascular hemolytic anemia. In intravascular hemolysis free hemoglobin will be released into circulation and hence haptoglobulin will bind the Hb. This causes a decline in Hp levels. Conversely, in extravascular hemolysis the reticuloendothelial system, especially splenic monocytes, phagocytose the erythrocytes and hemoglobin is not released into circulation and hence haptoglobulin levels are normal. This gene encodes a preproprotein that is processed to yield both alpha and beta chains, which subsequently combine as a tetramer to produce haptoglobin. Haptoglobin functions to bind free plasma hemoglobin, which allows degradative enzymes to gain access to the hemoglobin while at the same time preventing loss of iron through the kidneys and protecting the kidneys from damage by hemoglobin.

Haptoglobin is synthesized predominantly in the liver cells; in addition, it is formed also in the spleen and in the lymphatic system. The extrahepatical synthesis is not significant in physiological conditions. About 30-50% of the intravascular haptoglobin is synthesized daily. According to the characteristic genetic polymorphism, 3 haptoglobin types can be distinguished: Hp 1-1, Hp 2-1, Hp 2-2. Type Hp 1-1 is the most frequent one in Africa, in North- and in South-America; type Hp 2-2 is the predominant one in Asia, while in Central-Europe the Hp 2-1 type is the most prevalent one (about half of the population). The molecular mass of the type Hp 1-1 is cca 100 000 dalton; in the case of the rest of the phenotypes, there is a significant tendency for polymerization. The main physiological function of haptoglobin is hemoglobin binding. It is capable to bind only the globin part of hemoglobin, not the heme part. The linkage means a non-covalent bond involving the β-chain of haptoglobin and the α-chain of the globin molecule. Four binding sites were shown on the haptoglobin molecule. However, hemoglobin saturation does not develop in all cases; complexes involving only one or two hemoglobin molecule(s) also occur. The haptoglobin-hemoglobin complex quickly disappears from the circulation (T/2 value of the unsaturated complex is 9 min, that of native haptoglobin is 4 days).

The complex undergoes degradation primarily in the liver; in addition, it is broken down also in the spleen and in the bone marrow. According to recent observations, rather the hepatocytes and not the Kupffer-cells are involved in the degradation of the complex. Haptoglobin present in the human blood plasma is capable to bind approx. Three g hemoglobin, which equals about 10 ml red blood cell or 12 mg iron. This amount is half of the total daily turnover of hemoglobin. Thus, haptoglobin continuously prevents loss of iron, and at the same time, it protects the integrity of the kidney tubules.

Haptoglobin – as an acute phase protein – participates also in the preventive mechanisms of the organism. It was shown to inhibit the action of some proteases (e.g. cathepsin B), that are released in significant quantities during tissue damage. Haptoglobin, in its simplest form, consists of two α- and two β-chains, connected by disulfide bridges. The chains originate from a common precursor protein, which is proteolytically cleaved during protein synthesis.

Hp exists in two allelic forms in the human population, so-called *Hp1* and *Hp2*, the latter one having arisen due to the partial duplication of *Hp1* gene. Three genotypes of Hp, therefore, are found in humans: Hp1-1, Hp2-1, and Hp2-2. Hp of different genotypes have been shown to bind hemoglobin with different affinities, with Hp2-2 being the weakest binder.

Hp has been found in all mammals studied so far, some birds, e.g., cormorant and ostrich but also, in its simpler form, in bony fish, e.g., zebrafish. It is interesting to note that Hp is absent in at least some amphibians (*Xenopus*) and neognathous birds (chicken and goose).

Mutations in this gene and/or its regulatory regions cause ahaptoglobinemia or hypohaptoglobinemia. This gene has also been linked to diabetic nephropathy, the incidence of coronary artery disease in type 1 diabetes, Crohn's disease, inflammatory disease behavior, primary sclerosing cholangitis, susceptibility to idiopathic Parkinson's disease, and a reduced incidence of Plasmodium falciparum malaria.

Since the reticuloendothelial system will remove the haptoglobin-hemoglobin complex from the body, haptoglobin levels will be decreased in hemolytic anemias. In the process of binding hemoglobin, haptoglobin sequesters the iron within hemoglobin, preventing iron-utilizing bacteria from benefiting from hemolysis. It is theorized that, because of this, haptoglobin has evolved into an acute-phase protein. HP has a protective influence on the hemolytic kidney.

Some studies associate the HP with the risk of developing schizophrenia.

Haptoglobin is ordered whenever a patient exhibits symptoms of anemia, such as pallor, fatigue, or shortness of breath, along with physical signs of hemolysis, such as jaundice or dark-coloured urine. The test is also commonly ordered as a hemolytic anemia battery, which also includes a reticulocyte count and a peripheral blood smear. It can also be ordered along with a Direct Antiglobulin Test when a patient is suspected of having a transfusion reaction or symptoms of autoimmune hemolytic anemia. Also, it may be ordered in conjunction with a bilirubin.

A decrease in haptoglobin can support a diagnosis of hemolytic anemia, especially when correlated with a decreased RBC count, Hemoglobin, and Hematocrit, and also an increased reticulocyte count.

If the reticulocyte count is increased, but the haptoglobin level is normal, this may indicate that cellular destruction is occurring in the spleen and liver, which may indicate a drug-induced hemolysis, or a red cell dysplasia. The spleen and liver recognize an error in the red cells (either Drug coating the red cell membrane or a dysfunctional red cell membrane), and destroy the cell. This type of destruction does not release hemoglobin into the peripheral blood, so the haptoglobin cannot bind to it. Thus, the haptoglobin will stay normal.

If there are symptoms of anemia but both the reticulocyte count and the haptoglobin level are normal, the anemia is most likely not due to hemolysis, but instead some other error in cellular production, such as aplastic anemia

Haptoglobin levels that are decreased but do not accompany signs of anemia may indicate liver damage, as the liver is not producing enough haptoglobin to begin with.

As haptoglobin is indeed an acute-phase protein, any inflammatory process (infection, extreme stress, burns, major crush injury, allergy, etc.) may increase the levels of plasma haptoglobin.

Binding of Free Plasma and Cooperativity

Cooperative binding is a special case of allostery. Cooperative binding requires that the macromolecule have more than one binding site, since cooperativity results from the interactions between binding sites. If the binding of ligand at one site increases the affinity for ligand at another site, the macromolecule exhibits **positive cooperativity**. Conversely, if the binding of ligand at one site *lowers* the affinity for ligand at another site, the protein exhibits **negative cooperativity**. If the ligand binds at each site independently, the binding is **non-cooperative**. The Hill coefficient n provides a quantitative method for characterizing binding cooperativity. The macromolecule is assumed to bind to n ligands simultaneously (where n is to be determined)

$$P + \eta L \leftrightarrow C$$

to form the complex C. Hence the dissociation constant equals

$$K_d = \frac{[P][L]^n}{[C]}$$

The variable θ represents the fraction of binding sites that are occupied on the macromolecule. Therefore, 1 – θ represents the fraction of binding sites that are *not* occupied, giving the ratio

$$\frac{\theta}{1-\theta} = \frac{[C]}{[P]} = \frac{[L]}{K_d}$$

Taking the logarithm yields an equation linear in n

$$\log\left[\frac{\theta}{1-\theta}\right] = n\log[L] - \log K_d$$

Hence, the slope of this line yields n, whereas its intercept is determined by log K_d.

More generally, plotting $\log\left[\frac{\theta}{1-\theta}\right]$ versus log[L] and taking the slope gives the effective number of ligands n that are binding cooperatively at a particular ligand concentration [L]. In a non-cooperative system such as myoglobin, the plot is a straight line with slope $n = 1$ at all ligand concentrations. By contrast, in a system with positive cooperativity such as hemoglobin, the plot begins as a line with slope $n = 1$, then ramps up to a new line (also with slope $n = 1$) that is offset upwards. The degree of cooperativity is characterized by the maximum slope n in the "ramping up" region, which is ~2.8 for hemoglobin; thus, at its most cooperative, hemoglobin effectively binds three ligands in concert. The "ramping up" corresponds to an increase in the affinity (decrease in K_d) that occurs as the amount of bound ligand increases. Such plots are sometimes characterized as "sigmoid" due to their subtle "S"-shape.

Mechanisms of Cooperativity

Two models were hypothesized to account for the binding cooperativity observed in proteins, the MWC model and the KNF model.

The **Monod-Wyman-Changeux (MWC) model** was advanced by Jacques Monod, Jeffries Wyman and Jean-Pierre Changeux in 1965. It posits that the protein has only two states, a low-affinity state T and a high-affinity state R, where the T state is thermodynamically favored. Hence, at low amounts of bound ligand, the protein prefers the low-affinity T state; however, as the amount of bound ligand increases, the protein comes to prefer the high-affinity state. Structural studies have supported the MWC model and elucidated the R and T states; however, the model cannot explain *negative* cooperativity.

An alternative model is the sequential or "induced fit" model of Daniel Koshland, George Némethy and Filmer (KNF model), in which ligand binding at one site causes a local conformational change ("induced fit") that causes small conformational changes at nearby binding sites, affecting their affinity for the ligand. Thus, according to the KNF model, the protein has many slightly different conformational states, corresponding to all possible modes of ligand binding.

In non-cooperative binding, the way the affinity depends on the concentration of ligand in solution often is described as "hyperbolic," because a graph of this dependence traces a hyperbola.It is often the case that biochemical systems require tighter control than can be achieved with the sort of simple affector

$$K_4 = \frac{[Hb][O_2]}{[HbO_2]}$$

$$[HbO_2] = \frac{[Hb][O_2]}{K_d}$$

$$[Hb]_{total} = [Hb] + [HbO_2]$$

$$\frac{[HbO_2]}{[Hb]_{total}} = \frac{[HbO_2]}{[Hb]+[HbO_2]} = \frac{\frac{[Hb][O_2]}{K_4}}{[Hb]+\frac{[Hb][O_2]}{K_d}}$$

Simplifying:

$$\frac{[HbO_2]}{[Hb]_{total}} = \frac{\frac{[O_2]}{K_d}}{1+\frac{[O_2]}{K_d}} = \frac{[O_2]}{K_d+[O_2]}$$

system that is based on just a single linear binding event. Often one needs a response more like a switch—little effect at low concentration then a rapid rise to maximal activity. A classic example has to do with the protein hemoglobin which binds oxygen in your lungs and carries it to other tissues of your body. The problem is simply this, in the lungs you want hemoglobin to bind oxygen tightly so that a large fraction of the hemoglobin will pick up an oxygen before cycling back through the other tissues. However, in the other tissues, one wants hemoglobin to bind oxygen loosely so that it can be easily relinquished for use in respiration. How do we control the binding? Obviously the oxygen concentration in the blood near the lungs is going to be higher than it is in tissues were oxygen is rapidly being used up. But the difference is not large enough to cause nearly complete binding in the lungs and nearly complete oxygen release in the other tissues. A curve describing the fraction of hemoglobin molecules bound to oxygen as a function of the oxygen concentration is calculated assuming the sort of simple equilibrium binding model we discussed before is as follows: where Kd is the dissociation constant, [Hb] is the free hemoglobin concentration, and [Hb]t is the total hemoglobin concentration. The final equation should look very familiar to you. The dissociation constant, Kd, is just a backwards equilibrium constant. What we can see is that the curve is not very sharp. That is, it takes a relatively large change in the oxygen concentration in the blood to cause a significant uptake or release of oxygen from Hb. We would like a finer switch, where a small change in oxygen concentration resulted in going from almost all Hb in the unbound state to almost all Hb in the bound state. Nature has come up with an interesting solution to this problem which is called cooperativity. Hb is actually a tetramer of four subunits each of which can bind an oxygen. This binding is cooperative, that is, the binding of the first oxygen makes binding of the second more favorable which makes binding of the third even more favorable which makes binding of the fourth even more favorable. Lets draw a picture of this:

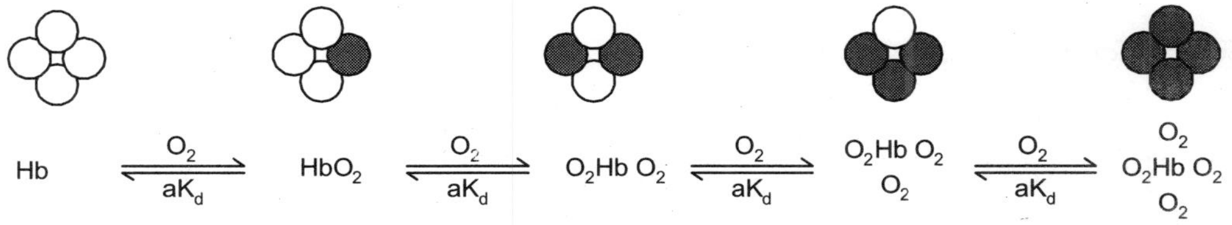

$$\text{Hb} \underset{aK_d}{\overset{O_2}{\rightleftharpoons}} \text{HbO}_2 \underset{aK_d}{\overset{O_2}{\rightleftharpoons}} \text{O}_2\text{Hb O}_2 \underset{aK_d}{\overset{O_2}{\rightleftharpoons}} \begin{matrix}\text{O}_2\text{Hb O}_2\\ \text{O}_2\end{matrix} \underset{aK_d}{\overset{O_2}{\rightleftharpoons}} \begin{matrix}\text{O}_2\\ \text{O}_2\text{Hb O}_2\\ \text{O}_2\end{matrix}$$

Fig. 8.24 : The Binding of the Hæmoglobin Cooperativity

Four successive oxygen molecules are bound to hemoglobin and each time one binds the dissociation constant for the next one decreases by a factor a. Remember that a lower dissociation constant means tighter binding. What does this do to the fraction bound vs. oxygen concentration curve? Let's work through it first with a simpler case. Consider (right side):

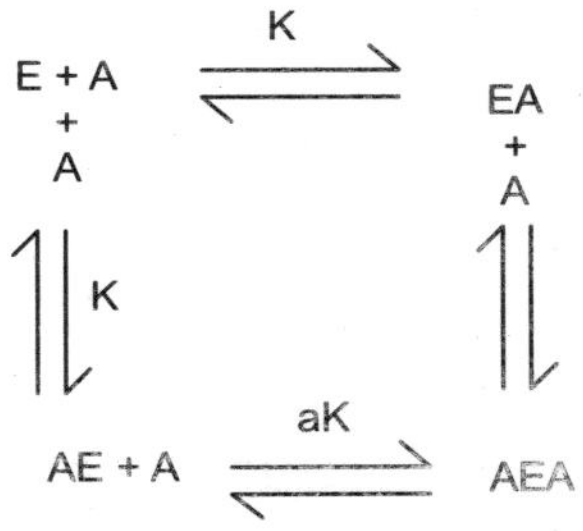

$$\begin{matrix} \text{E + A} & \overset{K}{\rightleftharpoons} & \text{EA + A} \\ K \updownarrow & & \updownarrow \\ \text{AE + A} & \overset{aK}{\rightleftharpoons} & \text{AEA} \end{matrix}$$

We want to know how many A's are bound per E (the occupancy of A on E).

$$N = \frac{A_{sound}}{E_r} = \frac{[EA]+[AE]+2[AEA]}{[E]+[EA]+[AE]+[AEA]}$$

Here N is the number of A's bound per enzyme. This is equal to all bound forms of A (remember there are two possible sites for A on E that are distinguishable resulting in AE and EA). The AEA term is multiplied by 2 because it contains two bound A molecules. Et is the total enzyme concentration. It equals all forms of both the free enzyme and enzyme bound to A. In this case AEA term is not multiplied by two since we are interested in the number of enzyme molecules represented in the AEA complex.

$$K = \frac{[E][A]}{[EA]} = \frac{[E][A]}{[AE]}$$

Thus

$$[EA] = [AE] = \frac{1}{K}[E][A]$$

Remember that K defined in this way is a dissociation constant, the inverse of a normal equilibrium constant. It is often written as K_s or K_d to avoid confusion, but I will just refer to it as K here. Cooperativity is represented by multiplying the dissociation constant by the factor a when considering binding of the second A to E. If a=1 there is no cooperativity. If a<1 then binding of the second A is stronger than the first. If a>1 then binding of the second A is weaker than the first. We can see that

$$aK = \frac{[EA][A]}{[AEA]}$$

$$[AEA] = \frac{1}{aK}[AE][A]$$

$$= \frac{1}{aK^2}[E][A]^2$$

Solving for the occupancy of A on E:

$$\frac{[AE]+[EA]+2[AEA]}{[E]+[AE]+[EA]+[AEA]} = \frac{\frac{[A][E]}{K_d} + \frac{[A][E]}{K_4} + 2\frac{[A]^2[E]}{aK_d}}{[E] + \frac{[A][E]}{K_d} + \frac{[A][E]}{K_d} + \frac{[A]^2[E]}{aK_d}}$$

$$= \frac{2\frac{[A]}{K_d} + 2\frac{[A]^2}{aK_d}}{1 + 2\frac{[A]}{K_d} + \frac{[A]^2}{aK_d}} = 2\frac{\frac{[A]}{K_d} + \frac{[A]^2}{aK_d}}{1 + 2\frac{[A]}{K_d} + \frac{[A]^2}{aK_d}}$$

Now if a = 1 (no cooperativity) this reduces to

$$\frac{2[A]}{K+[A]}$$

which is essentially the same as the simple equation for Hb binding to one O_2 except that it is multiplied by two because two A's can bind per E.

What the cooperativity does, however, is it makes the terms with $[A]^2$ much more important. For positive cooperativity, a<1 and therefore the terms with a in the denominator become larger. This makes the curve in Fig. 1 look more like a parabola at low [A] instead of a straight line. The result is a sigmoidal curve with a sharper dependence on the concentration of A.

If we now consider a more realistic situation with hemoglobin which has four total binding sites, the situation becomes a bit more complex. We must again remember that the four binding sites are distinguishable, and therefore we must consider all forms of singly, doubly, triply and quadruplely bound hemoglobin (Fig. 8.25):

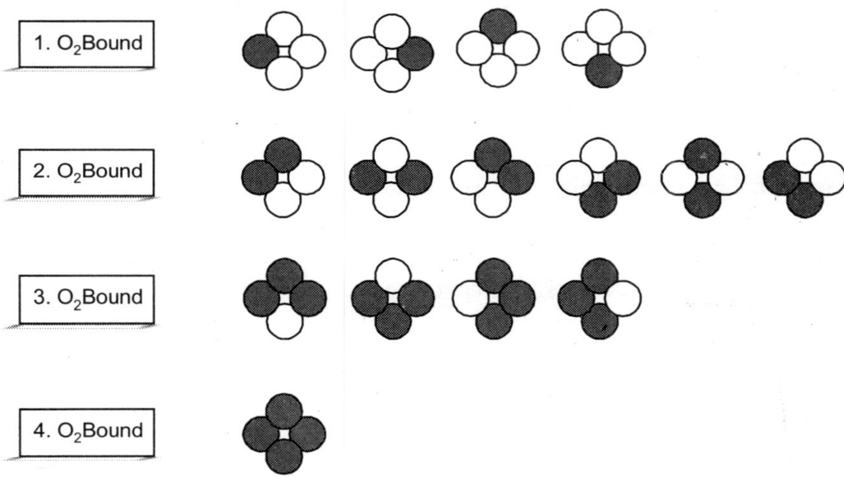

Fig. 8.25 : The Detail Cooperativity

Note that there are four possible ways for one O_2 to be bound, six ways for two, four ways for three and only one way for four O_2's to be bound (Fig. 8.25). As before, we can write the concentration of one bound, two bound, three bound or four bound species in terms of the concentration of Hb and O_2 by using the dissociation constant, K, and the cooperativity factor, a:

$$[HbO_2]_f = \frac{4}{K}[Hb][O_2]$$

$$\left[Hb(O_2)_2\right]_f = \frac{6}{aK^2}[Hb][O_2]^2$$

$$\left[Hb(O_2)_3\right]_f = \frac{4}{a^3K^3}[Hb][O_2]^3$$

$$\left[Hb(O_2)_4\right]_f = \frac{4}{a^4K^4}[Hb][O_2]^4$$

This is more complicated than the case for two ligands binding, but you can see the pattern is the same. The expression for N, the number of O2's bound per hemoglobin (the occupancy) is:

$$N = \frac{\left[Hb(O_2)\right]_f + 2\left[Hb(O_2)_2\right]_f + 3\left[Hb(O_2)_3\right]_f + 4\left[Hb(O_2)_4\right]_f}{[Hb] + \left[Hb(O_2)\right]_f + \left[Hb(O_2)_2\right]_f + \left[Hb(O_2)_3\right]_f \left[Hb(O_2)_4\right]_f}$$

substituting the expressions for the different bound forms of hemoglobin into the above equation yields (after some simplification):

$$N = \frac{\frac{4}{K}[O_2] + \frac{12}{aK^2}[O_2]^2 + \frac{12}{a^3K^3}[O_2]^3 + \frac{4}{a^6K^4}[O_2]^4}{1 + \frac{4}{K}[O_2] + \frac{6}{aK^2}[O_2]^2 + \frac{4}{a^3K^3}[O_2]^3 + \frac{1}{a^6K^4}[O_2]^4}$$

For a = 1 (no cooperativity) this factors into:

$$N = 4\frac{[O_2]}{K + [O_2]}$$

as one might expect for a four binding site molecule with no cooperativity. We can also consider what happens if a is significantly less than 1. In this case, only the higher order terms become significant and we have:

$$N = \frac{\frac{4}{a^6K^4}[O_2]^4}{1 + \frac{1}{a^6K^4}[O_2]^4} = \frac{[O_2]^4}{a^6K^4 + [O_2]^4}$$

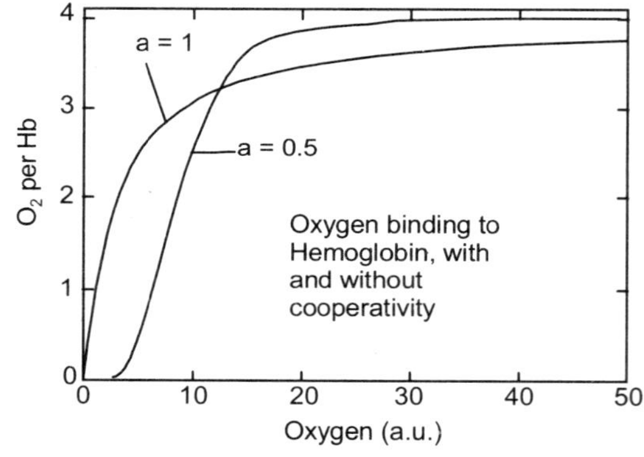

Fig. 8.26 : The Cooperativity Curve

The resulting plot is shown as the dashed line in Fig. 8.26. As you can see, there is now a much smaller change in oxygen concentration required to go from mostly bound to mostly unbound. Positive cooperativity of substrate binding is a very common method for causing the control of a reaction or process to be much more sensitive to the substrate concentration.

As practice please do the following problem:

The repressor of lambda phage is a dimer in its active form (if you want to read about lambda repressor you can look it up in any biochem text book). This is true of a number of regulatory proteins. Consider the chemical interactions between a regulator and an operator diagrammed below (if you don't know what an operator is, it doesn't really matter, just consider it as something that R binds to):

Case 1:

$$R + O \xrightleftharpoons{K} RO$$

Case 2:

$$R + R \xrightleftharpoons{K_1} RR$$

$$RR + O \xrightleftharpoons{K_2} RRO$$

where K, K_1 and K_2 are all equilibrium constants for the reactions as shown, R is the regulator which is functional as a monomer in case 1 and as a dimer in case 2, and O is the operator that the regulator binds to. For each of the two cases, derive an expression for the fraction of the operator sites that have regulator bound to them in terms only of the equilibrium constants shown and the concentration of the regulator. Remember that in case 2 only the dimer can bind to the operator. The monomer does not. To get you started, realize that the fraction of the operator with regulator bound in case 1 is just:

$$\frac{[RO]}{[O]+[RO]}$$

The definition of the fraction of the operator with dimer regulator bound in case 2 is:

$$\frac{[RRO]}{[O]+[RRO]}$$

Make a sketch of these two functions showing how the binding curve differs in the case of a dimer regulator from that of a monomer. Why is this difference important to the gene regulation of bacteriophage lambda?

TRANSFERRIN

Transferrin is a blood plasma protein for iron ion delivery that, in humans, is encoded by the *TF* gene. Transferrin is a glycoprotein that binds iron very tightly but reversibly. Although iron bound to transferrin is less than 0.1% (4 mg) of the total body iron, it is the most important iron pool, with the highest rate of turnover (25 mg/24 h). Transferrin has a molecular weight of around 80 kiloDaltons and contains 2 specific high-affinity Fe(III) binding sites. The affinity of transferrin for Fe(III) is extremely high (10^{23} M^{-1} at pH 7.4) but decreases progressively with decreasing pH below neutrality.

When not bound to iron, it is known as "apo-transferrin" (see also apoprotein).

When a transferrin protein loaded with iron encounters a transferrin receptor on the surface of a cell (e.g., to erythroid precursors in the bone marrow), it binds to it and, as a consequence, is transported into the cell in a vesicle. The pH of the vesicle is reduced by hydrogen ion pumps (H^+ ATPases), causing transferrin to release its iron ions. The receptor (with its ligand, transferrin, bound) is then transported through the endocytic cycle back to the cell surface, ready for another round of iron uptake. Each transferrin molecule has the ability to carry two iron ions in the ferric form (Fe^{3+}).

The gene coding for transferrin in humans is located in chromosome band 3q21. Research on king snakes by Dessauer and Zwiefel in 1981 revealed that the inheritance of transferrin is a codominant trait.

Medical professionals may check serum transferrin level in iron deficiency, hemochromatosis, and other iron overload disorders.

Structure

In humans, transferrin consists of a poly peptide chain containing 679 amino acids. It is a complex composed of alpha helices and beta sheets to form two domains (the first situated in the N-terminus and the second in the C-terminus). The N- and C-terminal sequences are represented by globular lobes and between the two lobes is an iron-binding site.

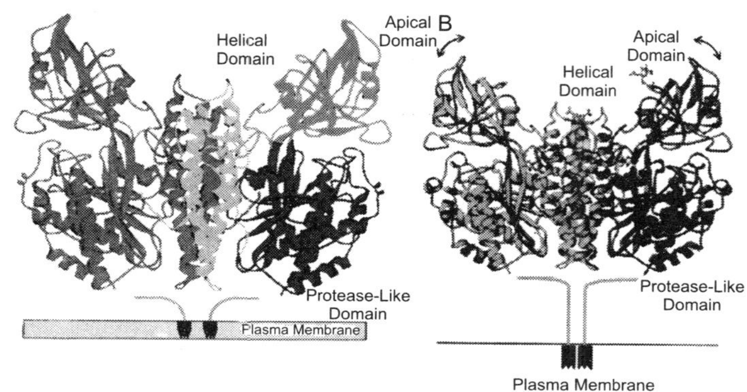

STRUCTUE OF TRANSFERRIN

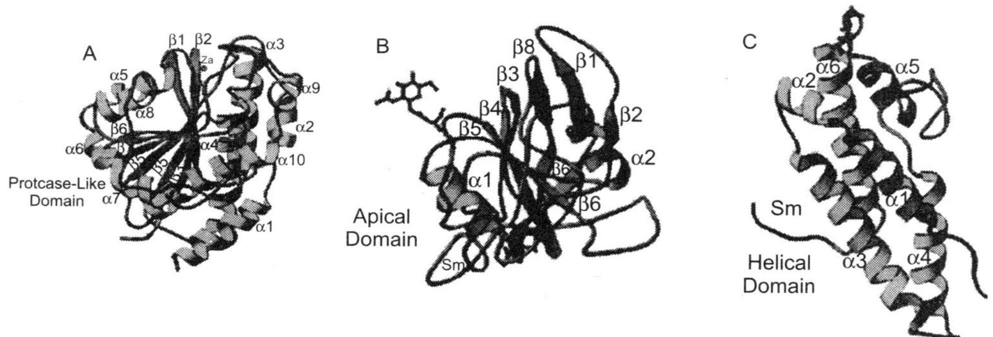

Fig. 8.27 : The Structure of Tranferrin

The liver is the main source of manufacturing transferrin, but other sources such as the brain also produce this molecule. The main role of transferrin is to deliver iron from absorption centres in the duodenum and red blood cell macrophages to all tissues. Predominantly, transferrin plays a key role where erythropoiesis and active cell division occur. In order for iron ion to be introduced into the cell a carrier protein is used, known as a transferrin receptor. The receptor helps maintain iron homeostasis in the cells by controlling iron concentrations. Transferrin's primary protein structure is made up of about 700 amino acids (80 kDa). Transferrin has a combination of alpha helices and beta sheets to form two different lobes: N- and C- terminus. These two domains are held together by a short peptide and create a deep hyrophobic site. The amino acids that bind the Ferric iron ion are the same for both lobes: two tyrosine residues, one aspartic acid, and one histotine. The binding of iron also needs an anion which is usually carbonate (CO_3^{2-}). The 3-charge, contributed by the two tyrosine and one aspartic acid, balances the 3+ charge of Ferric iron. The charge on the anion is balanced by the adjacent positive charge on the protein (in transferrins, this postive charge comes from the arginine side chain and the N-termins of an alpha helix).

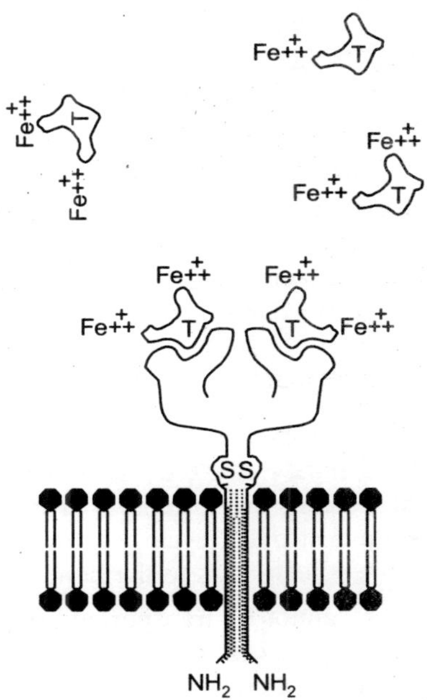

Fig. 8.28 : The binding Affinity of Transferrin

The transferrin receptor is a transmembrane homodimer consisting of two identical monomers. These monomers are able to bind up to two molecules of transferrin. The monomers are joined by two disulfide bonds at Cys89 and Cys89. This structure contains a short NH_2 terminal cytoplasmic region (residues 1 to 67), a single transmembrane pass (residues 68 to 88), and a large extracellular ectodomain (residues 89 to 760). The extracellular portion bears a trypsin-sensitive region and contains a binding site for transferrin. The transferrin receptor is butterfly-like in shape with three distinct domains: apical, protease-like, and helical. The membrane stalk probably involves disulfide bonded residues.

The binding affinity of diferric Tf for TfR is very high. Research concludes that how and where Tf bind to TfR is not fully understood. It was that the primary receptor is mainly in the C-lobe of Rf. However, in recent studies, both the N- and C- lobes are shown to play a role in Tf binding. In the TfR, the helical domain is thought to be the major player in Tf binding.

The Effects of Transferrin in the Metabolism

Transferrin is also associated with the innate immune system. Transferrin is found in the mucosa and binds iron, thus creating an environment low in free iron, where few bacteria are able to survive. The levels of transferrin decreases in inflammation, seeming contradictory to its function.

Transferrin imbalance can have serious health effects for those with low or high serum transferrin levels. A patient with an increased serum transferrin level often suffers from iron deficiency anemia. A patient with decreased plasma transferrin can suffer from iron overload diseases and protein malnutrition. An absence of transferrin in the body creates a rare genetic disorder known as atransferrinemia; a condition characterized by anemia and hemosiderosis in the heart and liver that leads to many complications including heart failure. Most recently, transferrin and its receptor have been tested to diminish tumour cells by using the receptor to attract antibodies. As research continues and additional information is collected possibly life saving, is being discovered through the molecular biology and protein biochemistry worlds.

The metal binding properties of transferrin have a great influence on the biochemistry of plutonium in humans. Transferrin has a bacteriocidal effect on bacteria, in that it makes Fe^{3+} unavailable to the bacteria. A deficiency is associated with atransferrinemia. The amino acids which bind the iron ion to the transferrin are identical for both lobes; two tyrosines, one histidine, and one aspartic acid. In order for the iron ion to bind an anion is required, preferably carbonate (CO_3^{2-}).

Transferrin also has a transferrin iron-bound receptor; it is a disulfide-linked homodimer. In humans, each monomer consists of 760 amino acids. It enables ligand bonding to the transferrin, as each monomer can bind to one or two molecules of iron. Each monomer consists of three domains: the protease domain, the helical domain, and apical domain. The shape of transferrin receptor resembles a butterfly-like complex, due to the three clearly shaped domains.

Fe Transport

It is obvious that iron must be transported around the human body. Firstly, it must be transported from the food in the gut to the places where it is required. Mostly, iron is required in the bone marrow, where red blood cells are formed. Red blood cells have a finite lifetime of about only four months, and old cells are destroyed, usually in the spleen. Iron from the destruction of these cells is then transported from the spleen back to the bone marrow to be recycled.

Iron cannot be transported around the body's circulation system as free iron, since it would be susceptible to chelation by siderophores, or may precipitate as iron(III) oxide, or may form iron(II). Therefore, a specific transport protein is required, called transferrin. (In fact, a whole class of transferrin-like proteins is involved in iron transport.) Transferrin is a medium-sized protein with a relative molecular mass of about 80,000. The crystal structures of transferrin with and without iron have been obtained, and the overall structure is shown in Figure 8.29, The structure without any iron (Figure 8.29a) shows that transferrin can be considered as two very similar polypeptides back-to-back, with each of the polypeptides having a large cleft. The apex of each cleft coordinates one iron atom; each transferrin molecule is therefore capable of transporting two iron atoms. Also, on the binding of iron, there is a significant change in the higher-order structure of the protein, such that the two sides of the cleft corne together and incarcerate the iron atoms (Figure 8.29b). Both iron atoms are now buried deep within the protein structure. (It is not fully clear why the iron atoms are buried in this way, but it may help in protecting the iron atom from microbial siderophores.)

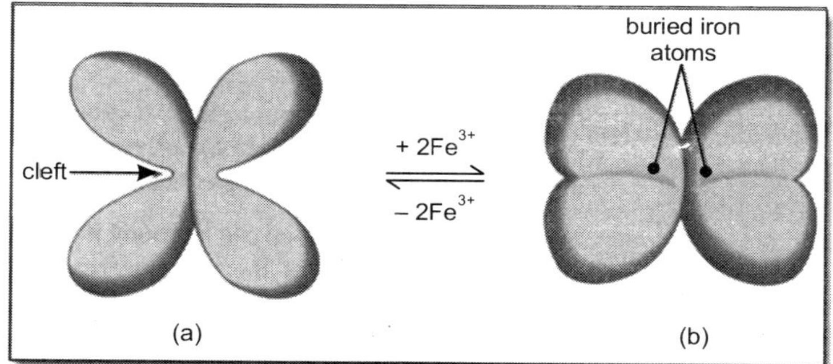

Fig. 8.29 : Schematic diagram of transferrin protein; (b) proposed higher-order structure change on complexation of iron(III)

The iron binding site is rich in hard ligands, which are suitable for binding iron (III) in a stable complex (Figure 8.30a). When the iron atom enters the active site (Figure 8.35b) it is coordinated by one η^1-aspartyl, one histidyl and two tyrosinate side-chains; a non-protein ligand also coordinates to it. This external ligand is a carbonate, CO_3^{2-}, which is held in place within the protein via hydrogen bonds to the protein backbone.

What is the mode of bonding of the CO_3^{2-} group?

The carbonate coordinates to the iron in an η^2 fashion; in other words, it is a chelating ligand.

Once the carbonate is held in place, we can see another example of ligand preorganisation, where an octahedral environment of hard/borderline ligands is ready to receive the iron(III) ion. The carbonate binding appears to facilitate the iron binding by the protein and vice versa, and so the system is said to be *synergistic*. It is not clear why this unusual synergistic binding of iron and carbonate occurs in transferrin, but it may have something to do with the way iron is released from transferrin.

Since all the coordinating ligands in transferrin can be considered as hard or 'borderline' ligands, it is no surprise that transferrin forms a very stable complex with iron(III). The stability constant of the Fe(III)–transferrin complex is *c*. 10^{20} mol^{-1} l. This is high enough to protect the iron(III) against the low concentration of any siderophores present. Indeed, the transferrins show mild antibacterial properties, in which their method of operation is to prevent extensive iron chelation by siderophores (see below the Iron in Human Milk.

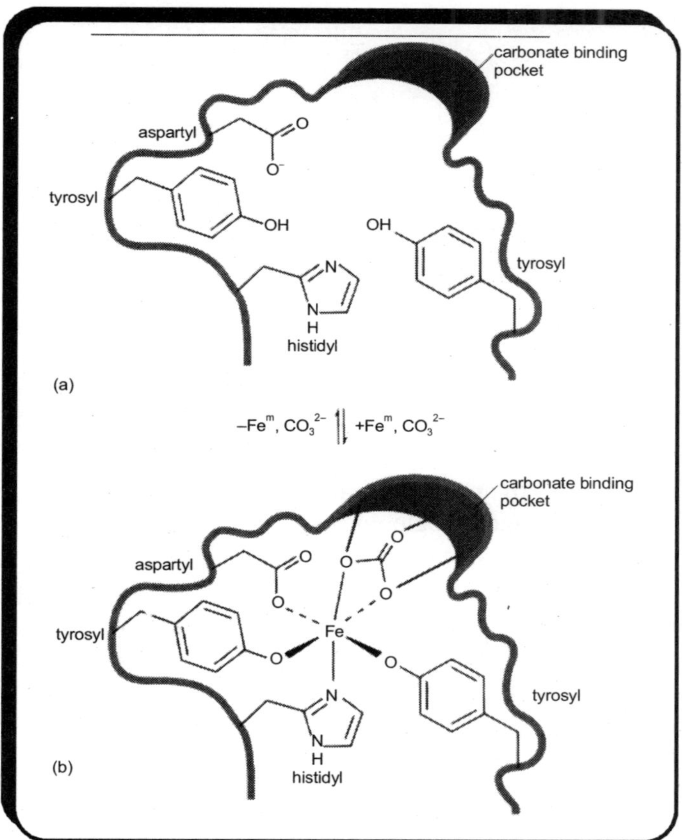

Fig. 8.30: (*a*) The iron binding site in transferrin; (*b*) the six-coordinate iron site; the coordination geometry is distorted octahedral. The carbonate is held in place by hydrogen bonds (green lines) to amino acid side-chains inside a small cleft

Iron in Human Milk

It has been known for some time that bottle-fed babies are more likely to suffer from gastric infections than breast-fed babies; this may be despite strict hygiene standards. The reason for this probably lies in the availability of iron within the baby's feed. Breast milk is known to contain a transferrin-like protein called lactoferrin. The lactoferrin chelates all the iron in the mother's milk, and prevents iron chelation by microbial siderophores. Formula milk, on the other hand, does not contain human lactoferrin, so the iron in the feed is more available for chelation by siderophores secreted by bacteria

Transferrin also forms relatively stable complexes with other hard metals (Table 8.2). Current thinking suggests that these other metals are transported by transferrin into cells, where they are potentially detrimental. One element in particular, aluminium, is of concern because it is used widely in cooking utensils.

Table 8.2 : Stability constants of Metal-transferrin complexes

Metal	Log_{10}(stability constant)
cadmium(II)	5.95
zinc(II)	7.80
aluminium(III)	13.50
iron(III)	22.80

Therefore, we can see from the structure and function of transferrin that the transport of iron within (and without) the body is very carefully managed, so as not to allow any free soluble iron to form. How, then, is iron stored? After all, we must store iron, since we need a reservoir of it for the synthesis of iron-containing proteins, most notably haemoglobin and myoglobin. As with iron transport, the iron storage systems need to ensure that free, soluble iron is not formed. In humans, iron is stored mainly in the bone marrow, spleen and liver. About 10 per cent of all the iron in the body is in storage. Two proteins are involved in iron storage; these are called ferritin and haemosiderin (they also occur in other organisms). We shall only study the better characterised (and simpler!) ferritin.

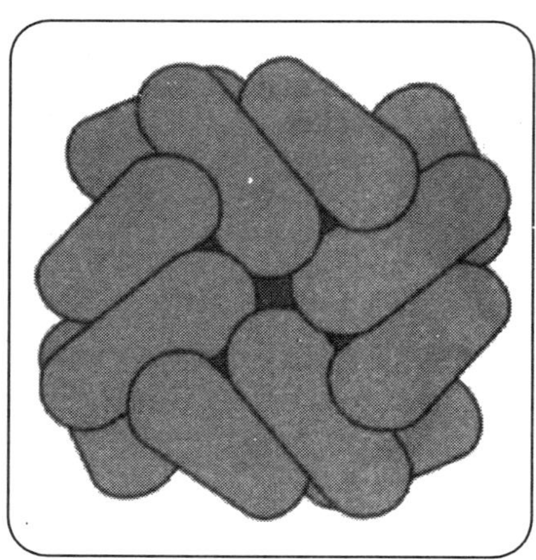

Fig. 8.31 : Sub-unit assembly of ferritin. Each sub-unit (shaped like a sausage) is made up of four parallel, α-helical polypeptide chains. The channel at the centre of the structure is clearly visible (dark green area); it lies on a fourfold axis of symmetry

Each ferritin molecule can store iron up to about 20 per cent of its total mass. This is a very high percentage, considering that less than 0.2 per cent of the total mass of proteins like transferrin and myoglobin is iron. Ferritin is a large protein with a relative molecular mass of 440 000. The crystal structure of ferritin with *no iron* is shown in Figure 8.31. The overall structure shows that ferritin is a huge, hollow protein, with a wall mostly made up of α-helical peptide chains. The structure is quite symmetrical, being roughly dodecahedral, and is one of the outstanding examples of symmetry in chemistry. The wall contains channels, which lead from the inside to the outside of the hollow 'sphere'. The channels are rich in amino acids with carboxylate side-chains, which are capable of chelating iron.

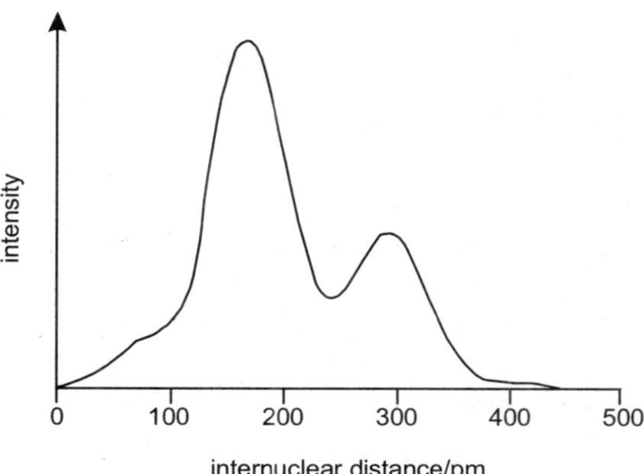

Fig. 8.32 : Fe-EXAFS radial distribution plot of iron-containing ferritin. Notice that there are two peaks, the first at 160 pm corresponding to a sphere of oxygen atoms, and the second at 290 pm corresponding to a sphere of iron atoms. (The peak due to the iron atoms is smaller than the peak due to the oxygen atoms; this is not in accord with the relative number of electrons in oxygen and iron atoms, The reasons for this are complex, but involve other factors beside the number of electrons in the intensity of back-scatterring. Also the actual structure of the hydrated iron(III) oxide in ferritin is not 'perfect', in that there are incomplete 'shells' of iron atoms, and poor long-range crystal order.)

The crystal structure of iron-containing ferritin is not known. However, some clues as to its structure have been obtained from EXAFS studies. EXAFS gives information about the direct coordination environment of a particular atom in terms of the number and type of its coordinated atoms (although no angular information is usually available). EXAFS studies on iron-containing ferritin showed that each iron atom is surrounded by an inner shell of six or seven oxygen atoms at a distance of about 160 pm, and by a second shell of seven or eight iron atoms at a distance of about 290 pm (Figure 8.33). This was a very strange result. How could seven or eight iron atoms be packed around each iron atom? The problem was solved when it was noticed that the EXAFS data were very similar to that of a hydrated iron(III) oxide mineral called ferrihydrite, $5Fe_2O_3.9H_2O$. From this result it was clear that ferritin stored iron partly as a crystalline, hydrated iron(III) oxide. Further studies showed that the inorganic crystalline part was within the hollow sphere of the protein (Figure 296).

Therefore, ferritin stores iron as crystalline, hydrated iron(III) oxide within its structure.

How will this affect the availability of the iron?

As the iron(III) oxide is very insoluble, it is unavailable to microbial iron-chelating ligands.

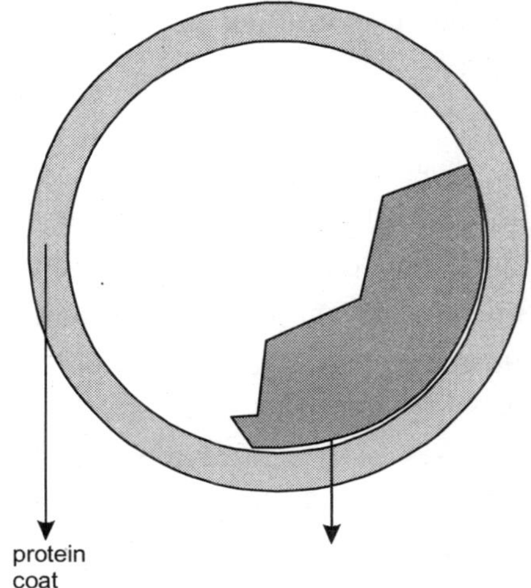

Fig. 8.33 : Schematic diagram showing growth of iron(III) oxide within a ferritin macromolecule. The full ferritin contains about 4500 iron atoms

The inorganic iron(III) oxide core is also protected from chelators by the outer protein coat. Moreover, this is a very space-efficient method of storing iron; each ferritin protein macromolecule can store a maximum of 4,500 iron atoms.

Iron is delivered to ferritin (after having been transported by transferrin) where it migrates through the carboxylate-rich channels in the surface of the protein to the interior. The inner side of the protein sphere is also rich in carboxylate residues. It is thought that these carboxylate residues coordinate iron atoms, such that the iron atoms are held in a regular array. This regular array has the correct spacing of iron atoms to encourage the growth of crystals of iron(III) oxide, and in this way an iron(III) oxide phase grows within the ferritin core. The growth of iron(III) oxide in ferritin is an example of biomineralisation.

So it is somewhat ironic that the formation of highly insoluble iron(III) oxides and hydroxides, which causes such a problem in the availability of iron in the environment, is the method by which iron is stored in ferritin!

We have seen in this unit that, despite having a high natural abundance, iron is in very short supply because of the insolubility of its oxides and hydroxides. A result of this is that organisms have developed methods for the uptake, transport and storage of iron. Bacteria, in particular, secrete very powerful iron chelators known as siderophores. Of all the iron–siderophore complexes, the iron(III)–enterobactin complex has the exceptionally high stability constant of 10^{49} mol^{-1}.

Other organisms, partly as a defence against siderophores and the need to avoid free iron in solution, have biochemical methods to transport and store iron. The protein most associated with iron transport is transferrin. Iron storage, in mammals, including humans, is achieved by ferritin, which stores iron as a hydrated iron(III) oxide; this is an example of biomineralisation.

The key points of this unit are:

1. The concentration of free Fe^{3+}(aq) at pH 7 is very small.
2. Free Fe^{2+}(aq) in an organism is dangerous, because it can react to produce the superoxide radical anion.
3. Iron(III) tends to be coordinated by hard ligands such as those containing oxygen or nitrogen.
4. The chelate effect confers extra stability on iron(III) complexes with polydentate ligands.
5. Siderophores are powerful microbial iron chelators.
6. Enterobactin is a particularly powerful iron chelator It acts as a preorganised hexadentate ligand for iron(III).
7. Esterases catalyse the hydrolysis of the triserine ring in enterobactin, thus enabling iron(III) from the enterobactin complex to be released.
8. Iron transport in mammals is carried out by transferrin, which holds iron(III) in a hexacoordinate site, one of the ligands being η^2-carbonate.
9. Ferritin is an iron storage protein capable of storing up to 4 500 iron atoms per protein macromolecule. The iron is stored as hydrated iron(III) oxide.

THE COAGULATION OF BLOOD HÆMOSTASIS

Initial Discoveries

Theories on the coagulation of blood have existed since antiquity. Physiologist Johannes Müller (1801-1858) described fibrin, the substance of a thrombus. Its soluble precursor, fibrinogen, was thus named by Rudolf Virchow (1821-1902), and isolated chemically by Prosper Sylvain Denis (1799-1863). Alexander Schmidt suggested that the conversion from fibrinogen to fibrin is the result of an enzymatic process, and labeled the hypothetical enzyme "thrombin" and its precursor "prothrombin". Arthus discovered in 1890 that calcium was essential in coagulation. Platelets were identified in 1865, and their function was elucidated by Giulio Bizzozero in 1882.

The theory that thrombin is generated by the presence of tissue factor was consolidated by Paul Morawitz in 1905.[7] At this stage, it was known that *thrombokinase/thromboplastin* (factor III) is released by damaged tissues, reacting with *prothrombin* (II), which, together with calcium (IV), forms *thrombin*, which converts fibrinogen into *fibrin* (I).

Coagulation is a complex process by which blood forms clots. It is an important part of hemostasis (the cessation of blood loss from a damaged vessel), wherein a damaged blood vessel wall is covered by a platelet and fibrin-containing clot to stop bleeding and begin repair of the damaged vessel. Disorders of coagulation can lead to an increased risk of bleeding (hemorrhage) or clotting (thrombosis).

Coagulation is highly conserved throughout biology; in all mammals, coagulation involves both a cellular (platelet) and a protein (coagulation factor) component. The system in humans has been the most extensively researched and, therefore, the best-understood.

Factors of Coagulation of Blood

The remainder of the biochemical factors in the process of coagulation were largely discovered in the 20th century.A first clue as to the actual complexity of the system of coagulation was the discovery of *proaccelerin* (initially and later called Factor V) by Paul Owren (1905-1990) in 1947. He also postulated its function to be the generation of accelerin (Factor VI), which later turned out to be the activated form of V (or Va); hence, VI is not now in active use. Factor VII (also known as *serum prothrombin conversion accelerator* or *proconvertin*, precipitated by barium sulfate) was discovered in a young female patient in 1949 and 1951 by different groups.Factor VIII turned out to be deficient in the clinically recognised but etiologically elusive hemophilia A; it was identified in the 1950s and is alternatively called *antihemophilic globulin* due to its capability to correct hemophilia A. Factor IX was discovered in 1952 in a young patient with hemophilia B named Stephen Christmas (1947-1993). His deficiency was described by Dr. Rosemary Biggs and Professor R.G. MacFarlane in Oxford, UK. The factor is, hence, called Christmas Factor. Christmas lived in Canada, and campaigned for blood transfusion safety until succumbing to transfusion-related AIDS at age 46. An alternative name for the factor is *plasma thromboplastin component*, given by an independent group in California. Hageman factor, now known as factor XII, was identified in 1955 in an asymptomatic patient with a prolonged bleeding time named of John Hageman. Factor X, or Stuart-Prower factor, followed, in 1956. This protein was identified in a Ms. Audrey Prower of London, who had a lifelong bleeding tendency. In 1957, an American group identified the same factor in a Mr. Rufus Stuart. Factors XI and XIII were identified in 1953 and 1961, respectively.

The view that the coagulation process is a "cascade" or "waterfall" was enunciated almost simultaneously by MacFarlane in the UK and by Davie and Ratnoff in the USA, respectively.

Nomenclature

The usage of Roman numerals rather than eponyms or systematic names was agreed upon during annual conferences (starting in 1955) of hemostasis experts. In 1962, consensus was achieved on the numbering of factors I-XII.[11] This committee evolved into the present-day International Committee on Thrombosis and Hemostasis (ICTH). Assignment of numerals ceased in 1963 after the naming of Factor XIII. The names Fletcher Factor and Fitzgerald Factor were given to further coagulation-related proteins, namely prekallikrein and high-molecular-weight kininogen, respectively. Factors III and VI are unassigned, as thromboplastin was never identified, and actually turned out to consist of ten further factors, and accelerin was found to be activated Factor V. Table Showing the Coagulation factors

Coagulation Factors and Related Substances

Number and/or name	Function
1	2
I (fibrinogen)	Forms clot (fibrin)
II (prothrombin)	Its active form (IIa) activates I, V, VII, VIII, XI, XIII, protein C, platelets
Tissue factor	Co-factor of VIIa (formerly known as factor III)
Calcium	Required for coagulation factors to bind to phospholipid (formerly known as factor IV)
V (proaccelerin, labile factor)	Co-factor of X with which it forms the prothrombinase complex
VI	Unassigned - old name of Factor Va
VII (stable factor)	Name: Pro Convertin - Activates IX, X
VIII (Anti Hemophilic factor A)	Co-factor of IX with which it forms the tenase complex
IX (Anti Hemophilic Factor B or Christmas factor)	Activates X: forms tenase complex with factor VIII
X (Stuart-Prower factor)	Activates II: forms prothrombinase complex with factor V
XI (plasma thromboplastin antecedent)	Activates IX

1	2
XII (Hageman factor)	Activates factor XI and prekallikrein
XIII (fibrin-stabilizing factor)	Crosslinks fibrin
von Willebrand factor	Binds to VIII, mediates platelet adhesion
prekallikrein	Activates XII and prekallikrein; cleaves HMWK
high-molecular-weight kininogen (HMWK)	Supports reciprocal activation of XII, XI, and prekallikrein
fibronectin	Mediates cell adhesion
antithrombin III	Inhibits IIa, Xa, and other proteases;
heparin cofactor II	Inhibits IIa, cofactor for heparin and dermatan sulfate ("minor antithrombin")
protein C	Inactivates Va and VIIIa
protein S	Cofactor for activated protein C (APC, inactive when bound to C4b-binding protein)
protein Z	Mediates thrombin adhesion to phospholipids and stimulates degradation of factor X by ZPI
Protein Z-related protease inhibitor (ZPI)	Degrades factors X (in presence of protein Z) and XI (independently)
plasminogen	Converts to plasmin, lyses fibrin and other proteins
alpha 2-antiplasmin	Inhibits plasmin
tissue plasminogen activator (tPA)	Activates plasminogen
urokinase	Activates plasminogen
plasminogen activator inhibitor-1 (PAI1)	Inactivates tPA & urokinase (endothelial PAI)
plasminogen activator inhibitor-2 (PAI2)	Inactivates tPA & urokinase (placental PAI)
cancer procoagulant	Pathological factor X activator linked to thrombosis in cancer

Factors of Blood Coagulation a Discussion

Fibrin (also called **Factor Ia**) is a fibrous protein involved in the clotting of blood, and is non globular. It is a fibrillar protein that is polymerised to form a "mesh" that forms a hemostatic plug or clot (in conjunction with platelets) over a wound site.

Fibrin is made from **fibrinogen**, a soluble plasma glycoprotein that is synthesised by the liver. Processes in the coagulation cascade activate the zymogen prothrombin to the serine protease thrombin, which is responsible for converting fibrinogen into fibrin. Fibrin is then cross linked by factor XIII to form a clot. FXIIIa stabilizes fibrin further by incorporation of the fibrinolysis inhibitors alpha-2-antiplasmin and TAFI (thrombin activatable fibrinolysis inhibitor, procarboxypeptidase B), and binding to several adhesive proteins of various cells. Both the activation of Factor XIII by thrombin and plasminogen activator (t-PA) are catalyzed by fibrin. Fibrin specifically binds the activated coagulation factors factor Xa and thrombin and entraps them in the network of fibers, thus functioning as a temporary inhibitor of these enzymes which stay active and can be released during fibrinolysis. Recent research has shown that fibrin plays a key role in the inflammatory response and development of rheumatoid arthritis.

Fibrin is involved in the following biological processes: signal transduction, blood coagulation, platelet activation, and protein polymerization. Excessive generation of fibrin due to activation of the coagulation cascade leads to thrombosis, while ineffective generation or premature lysis of **fibrin** predisposes to hemorrhage. Dysfunction or disease of the liver can lead to a decrease in fibrinogen production or the production of abnormal fibrinogen molecules with reduced activity (dysfibrinogenaemia). Hereditary abnormalities of fibrinogen (the gene is carried on chromosome 4) are of both quantitative and qualitative in nature and include; afibrinogenaemia, hypofibrinogenaemia, dysfibrinogenaemia, and hypodysfibrinogenaemia.

Consequences of reduced, absent, or dysfunctional fibrin is likely to render patients as hemophiliacs. Congenital deficiency (afibrinognenemia) or disturbed function of fibrinogen has been described in a few cases.

It can lead to either bleeding, thromboembolic complications or is clinically without pathological findings. More common are acquired deficiency stages which can be detected by laboratory tests in blood plasma or in whole blood by means of thrombelastometry. Acquired deficiency is found after hemodilution, blood losses and/or consumption such as in trauma patients, during some phases of disseminated intravascular coagulation (DIC), and also in sepsis. In patients with fibrinogen deficiency, the correction of bleeding is possible by infusion of fresh frozen plasma (FFP), cryoprecipitate (a fibrinogen rich plasma fraction) or by fibrinogen concentrates. There is increasing evidence that correction of fibrinogen deficiency or fibrinogen polymerization disorders is very important in patients with bleeding.

Fibrinogen levels can be measured in venous blood. Normal levels are about 1.5-3.0 g/L, depending on the method which is used. Typically fibrinogen is measured in citrated plasma samples in the laboratory, however the analysis of whole blood samples by use of thrombelastometry (platelet function is inhibited with cytochalasin D) is also possible. Higher levels are, amongst others, associated with cardiovascular disease (>4.6 g/L). It may be elevated in any form of inflammation, as it is an acute phase protein; for example, it is especially apparent in human gingival tissue during the initial phase of periodontal disease.

It is used in veterinary medicine as an inflammatory marker: in horses a level above the normal range of 1.0-4.0 g/L suggests some degree of systemic inflammatory response.

Low levels of fibrinogen can indicate a systemic activation of the clotting system, with consumption of clotting factors faster than synthesis. This excessive clotting factor consumption condition is known as Disseminated Intravascular Coagulation or "DIC." DIC can be difficult to diagnose, but a strong clue is low fibrinogen levels in the setting of prolonged clotting times (PT or PTT), in the context of acute critical illness such as sepsis or trauma. Besides low fibrinogen level, fibrin polymerization disorders which can be induced by several factors, including plasma expanders, can also lead to severe bleeding problems. Fibrin polymerization disorders can be detected by viscoelastic methods such as thrombelastometry.

Fibrinogen (also called factor I) is a 340 KDa glycoprotein synthesised in the liver by hepatocytes and megakaryocytes. The concentration in blood plasma is 1.5-4.0 g/L (normally measured using the Clauss method) or about 7 μM. In its natural form, fibrinogen can form bridges between platelets, by binding to their GpIIb/IIIa surface membrane proteins; however its major function is as the precursor to fibrin.

Fibrinogen, the principal protein of vertebrate blood clotting is a hexamer containing two sets of three different chains (α, β, and γ), linked to each other by disulfide bonds. The N-terminal sections of these three chains contain the cysteines that participate in the cross-linking of the chains. The C-terminal parts of the α, β and γ chains contain a domain of about 225 amino-acid residues, which can function as a molecular recognition unit. In fibrinogen as well as in angiopoietin this domain is implicated in protein-protein interactions. In lectins, such as mammalian ficolins and invertebrate tachylectin 5A, the fibrinogen C-terminal domain binds carbohydrates. On the fibrinogen α and β chains, there is a small peptide sequence (called a fibrinopeptide). It is these small peptides that prevent fibrinogen spontaneously forming polymers with itself.

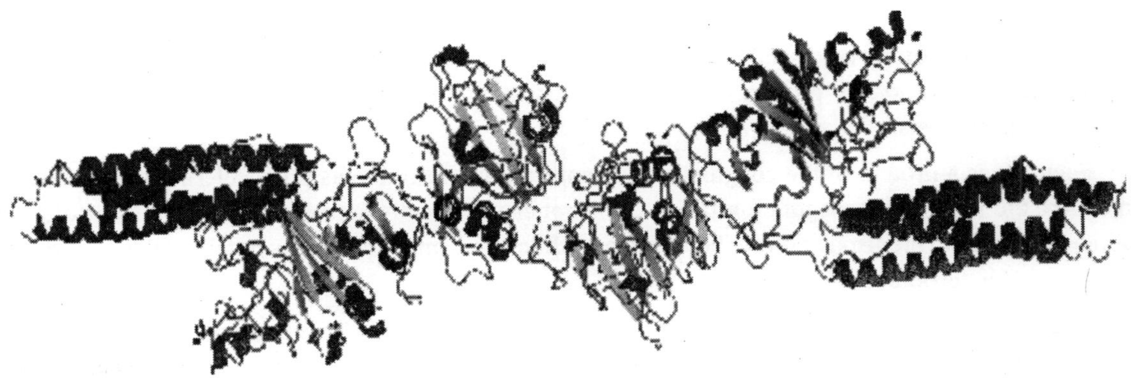

Fig. 8.34 : The Molecule of Fibrin

The image at the left is a crystal structure of the double-d fragment from human fibrin with two bound ligands. The experimental method used to obtain the image was X-ray diffraction, and it has a resolution of 2.30 Å. The structure is mainly made up of single alpha helices shown in red and beta sheets shown in yellow. The two blue structures are the bound ligands. The chemical structures of the ligands are Ca^{+2} ion, alpha-D-mannose ($C_6H_{12}O_6$), and D-glucosamine ($C_8H_{15}NO_6$).

Factor V (pronounced factor 5) Proaccelerin or Labile Factor

Until the discovery of factor V, coagulation was regarded as a product of four factors: calcium (IV) and thrombokinase (III) together acting on prothrombin (II) to produce fibrinogen (I); this model had been outlined by Paul Morawitz in 1905.

The suggestion that an additional factor might exist was made by Dr Paul Owren (1905-1990), a Norwegian physician, during his investigations into the bleeding tendency of a lady called Mary (1914-2002). She had suffered from nosebleeds and menorrhagia (excessive menstrual blood loss) for most her life, and was found to have a prolonged prothrombin time, suggesting either vitamin K deficiency or chronic liver disease leading to prothrombin deficiency. However, neither were the case, and Owren demonstrated this by correcting the abnormality with plasma from which prothrombin had been removed. Using Mary's serum as index, he found that the "missing" factor, which he labelled V (I-IV having been used in Morawitz' model), had particular characteristics. Most investigations were performed during the World War II, and while Owren published his results in Norway in 1944, he could not publish them internationally until the war was over. They appeared finally in *The Lancet* in 1947.

The possibility of an extra coagulation factor was initially resisted on methodological grounds by Drs Armand Quick and Walter Seegers, both world authorities in coagulation. Confirmatory studies from other groups led to their final approval several years later.

Owren initially felt that factor V (labile factor or proaccelerin) activated another factor, which he named VI. VI was the factor that accelerated the conversion from prothrombin to thrombin. It was later discovered that factor V was "converted" (activated) by thrombin itself, and later still that factor VI was simply the activated form of factor V.

The complete amino acid sequence of the protein was published in 1987. In 1994 factor V Leiden, resistant to inactivation by protein C, was described; this abnormality is the most common genetic cause for thrombosis.

Factor V is a protein of the coagulation system, rarely referred to as proaccelerin or labile factor. In contrast to most other coagulation factors, it is not enzymatically active but functions as a cofactor. Deficiency leads to predisposition for hemorrhage, while some mutations (most notably factor V Leiden) predispose for thrombosis.

The gene for factor V is located on the first chromosome (1q23). It is genomically related to the family of multicopper oxidases, and is homologous to coagulation factor VIII. The gene spans 70 kb, consists of 25 exons, and the resulting protein has a relative molecular mass of approximately 330000.

Factor V circulates in plasma as a single-chain molecule with a plasma half-life of about 12 hours. Half-lives up to 36 hours have been reported, though.

Factor V is able to bind to activated platelets and is activated by thrombin. On activation, factor V is spliced in two chains (heavy and light chain with molecular masses of 110000 and 73000, respectively) which are noncovalently bound to eaeh other by calcium. Factor V is active as a cofactor of the thrombinase complex. The activated factor X (FXa) enzyme requires Ca^{++} and activated factor V to convert prothrombin to thrombin on the cell surface membrane.

Factor Va is degraded by activated protein C, one of the principal physiological inhibitors of coagulation. In the presence of thrombomodulin, thrombin acts to decrease clotting by activating Protein C; therefore, the concentration and action of protein C are important determinants in the negative feedback loop through which thrombin limits its own activation.

Various hereditary disorders of factor V are known. Deficiency is associated with a rare mild form of hemophilia (termed parahemophilia or Owren parahemophilia), the incidence of which is about 1:1,000,000. It inherits in an autosomal recessive fashion.

Other mutations of factor V are associated with venous thrombosis. They are the most common hereditary causes for thrombophilia (a tendency to form blood clots). The most common one of these, factor V Leiden, is due to the replacement of an arginine residue with glutamine at amino acid position 506 (R506Q). All prothrombotic factor V mutations (factor V Leiden, factor V Cambridge, factor V Hong Kong) make it resistant to cleavage by activated protein C ("APC resistance"). It therefore remains active and increases the rate of thrombin generation.

Factor VII (formerly known as proconvertin) is one of the central proteins in the coagulation cascade. It is an enzyme (EC 3.4.21.21) of the serine protease class.

The main role of factor VII (FVII) is to initiate the process of coagulation in conjunction with tissue factor (TF). Tissue factor is found on the outside of blood vessels - normally not exposed to the bloodstream. Upon vessel injury, tissue factor is exposed to the blood and circulating factor VII. Once bound to TF, FVII is activated to FVIIa by different proteases, among which are thrombin (factor IIa), factor Xa, IXa, XIIa, and the FVIIa-TF complex itself. The most important substrates for FVIIa-TF are Factor X and Factor IX.

The action of the factor is impeded by tissue factor pathway inhibitor (TFPI), which is released almost immediately after initiation of coagulation. Factor VII is vitamin K dependent; it is produced in the liver. Use of warfarin or similar anticoagulants impairs its function.

The gene for factor VII is located on chromosome 13 (13q34).

Deficiency is rare (congenital proconvertin deficiency) and inherits recessively. Factor VII deficiency presents as a hemophilia-like bleeding disorder. It is treated with recombinant factor VIIa (NovoSeven).

Recombinant human factor VIIa (NovoSeven, eptacog alfa [activated], ATC code B02BD08) has been introduced for use in uncontrollable bleeding in hemophilia patients (with Factor VIII or IX deficiency) who have developed inhibitors against replacement coagulation factor.

It is being increasingly used in uncontrollable hemorrhage. The first report of its use was in an Israeli soldier with uncontrollable bleeding in 1999. The rationale for its use in hemorrhage is, that it will only induce coagulation in those sites where tissue factor (TF) is also present.

According to a 2005 study, recombinant human factor VII improves outcomes in acute intracerebral hemorrhage. Factor VII has been shown to interact with Tissue factor.

Factor IX (or **Christmas factor**) is one of the serine proteases (EC 3.4.21.22) of the coagulation system; it belongs to peptidase family S1. Deficiency of this protein causes hemophilia B. It was discovered after a young boy named Stephen Christmas was found to be lacking this exact factor, leading to hemophilia, in 1952.

Factor IX is inactive unless activated by factor XIa (of the contact pathway) or factor VIIa (of the tissue factor pathway). When activated into **factor IXa**, in the presence of Ca^{2+}, membrane phospholipids, and a Factor VIII cofactor, it hydrolyses one arginine-isoleucine bond in factor X to form factor Xa.

The gene for factor IX is located on the X chromosome (Xq27.1-q27.2). It was first cloned in 1982 by Kotoku Kurachi and Earl Davie.

Polly, a transgenic cloned Poll Dorset sheep carrying the gene for factor IX, was produced by Dr Ian Wilmut at the Roslin Institute in 1997.

Deficiency of factor IX causes Christmas disease (hemophilia B). Over 100 mutations of factor IX have been described; some cause no symptoms, but many lead to a significant bleeding disorder.

Some rare mutations of factor IX result in elevated clotting activity, and can result in clotting diseases, such as deep vein thrombosis.

Factor VIII (FVIII) is an essential blood clotting factor also known as anti-hemophilic factor (AHF). In humans, Factor VIII is encoded by the *F8* gene. Defects in this gene results in hemophilia A, a well known recessive X-linked coagulation disorder.

Factor VIII participates in blood coagulation; it is a cofactor for factor IXa which, in the presence of Ca^{+2} and phospholipids forms a complex that converts factor X to the activated form Xa. The Factor VIII gene produces two alternatively spliced transcripts. Transcript variant 1 encodes a large glycoprotein, isoform a, which circulates in plasma and associates with von Willebrand factor in a noncovalent complex. This protein undergoes multiple cleavage events. Transcript variant 2 encodes a putative small protein, isoform b, which consists primarily of the phospholipid binding domain of factor VIIIc. This binding domain is essential for coagulant activity.

The gene for Factor VIII is located on the X chromosome (Xq28). The gene for factor VIII presents an interesting primary structure, as another gene is embedded in one of its introns.

FVIII is a glycoprotein procofactor. It has been found to be synthesized and released into the bloodstream by the vascular, glomerular, and tubular endothelium, and the sinusoidal cells of the liver, though there is still considerable ambiguity as to what the primary site of release in humans is. In the circulating blood, it is mainly bound to von Willebrand factor to form a stable complex. Upon activation by thrombin, (Factor IIa), it dissociates from the complex to interact with Factor IXa in the coagulation cascade. It is a cofactor to Factor IXa in the activation of Factor X, which, in turn, with its cofactor Factor Va, activates more thrombin. Thrombin cleaves fibrinogen into fibrin which polymerizes and crosslinks (using Factor XIII) into a blood clot.

No longer protected by vWF, activated FVIII is proteolytically inactivated in the process (most prominently by activated Protein C and Factor IXa) and quickly cleared from the blood stream.

Factor VIII is not affected by liver disease. In fact, levels usually are elevated in such instances.

FVIII concentrated from donated blood plasma (Aafact), or alternatively recombinant FVIII can be given to hemophiliacs to restore hemostasis.

The transfer of a plasma byproduct into the blood stream of a patient with hemophilia often led to the transmission of diseases such as hepatitis B and C and HIV before purification methods were improved. Antibody formation to Factor VIII can also be a major concern for patients receiving therapy against bleeding; the incidence of these inhibitors is dependent of various factors, including the Factor VIII product itself.

In the 1980s, some pharmaceutical companies such as Bayer sparked controversy by continuing to sell contaminated factor VIII after new heat-treated versions were available. In the early 1990s, pharmaceutical companies began to produce recombinant synthesized factor products, which now prevent nearly all forms of disease transmission during replacement

Factor XI *or* ***plasma thromboplastin antecedent*** is the zymogen form of factor XIa, one of the enzymes of the coagulation cascade. Like many other coagulation factors, it is a serine protease. In humans, Factor XI is encoded by the *F11* gene.

Factor XI (FXI) is produced by the liver and circulates as a homo-dimer in its inactive form. The plasma half-life of FXI is approximately 52 hours. The zymogen factor is activated into *factor XIa* by factor XIIa (FXIIa), thrombin, and it is also autocatalytic, and FXI is a member of the "contact pathway" due to activation by FXIIa (with includes HMWK, prekallikrein, factor XII, factor XI and factor IX).

Factor XIa activates factor IX by selectively cleaving arg-ala and arg-val peptide bonds. Factor IXa, in turn, activates factor X.

Inhibitors of factor XIa include protein Z-dependent protease inhibitor (ZPI, a member of the serine protease inhibitor/serpin class of proteins), which is independent of protein Z (its action on factor X, however, is protein Z-dependent, hence its name).

Although synthesized as a single polypeptide chain, FXI circulates as a homodimer. Every chain has a relative molecular mass of approximately 80000. Typical plasma concentrations of FXI are 5 mg/l, corresponding to a plasma concentration (of FXI dimers) of approximately 30nM. The FXI gene is 23kb in length, has 15 exons, and is found on chromosome 4q32-35.

Deficiency of factor XI causes the rare Haemophilia C; this mainly occurs in Ashkenazi Jews and is believed to affect approximately 8% of that population, of both sexes. The condition has been described in other populations at around 1% of cases. It is an autosomal recessive disorder. There is little spontaneous bleeding, but surgical procedures may cause excessive blood loss, and prophylaxis is required.

Low levels of factor XI also occur in many other disease states, including Noonan syndrome.

High levels of factor XI have been implicated in thrombosis, although it is uncertain what determines these levels and how serious the procoagulant state is.

Tissue plasminogen activator (abbreviated tPA or PLAT) is a protein involved in the breakdown of blood clots. Specifically, it is a serine protease (EC 3.4.21.68) found on endothelial cells, the cells that line the blood vessels. As an enzyme, it catalyzes the conversion of plasminogen to plasmin, the major enzyme responsible for clot breakdown. Because it works on the clotting system, tPA is used in clinical medicine to treat only embolic or thrombolytic stroke. Use is contraindicated in hemorrhagic stroke and head trauma.

tPA may be manufactured using recombinant biotechnology techniques. tPA created this way may be referred to as **recombinant tissue plasimogen activator** or **rtPA**.

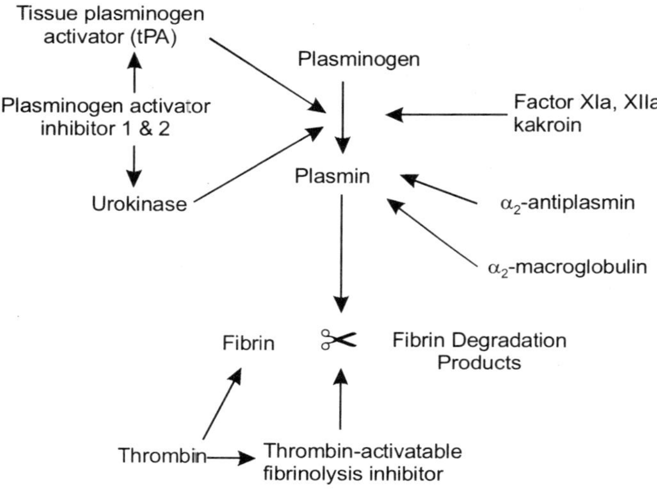

Fig. 8.35 : The tPA in action 1

A simplified illustration demonstrates clot breakdown (fibrinolysis), with blue arrows denoting stimulation, and red arrows inhibition.

The classic role of tPA is in the clotting system. Specifically, tPA catalyzes the conversion of plasminogen into plasmin. It does so by cleaving the single-chained plasminogen into two chains. These two chains are linked by a disulfide bond and the resulting molecule is called plasmin.

Increased enzymatic activity causes hyperfibrinolysis, which manifests as excessive bleeding. Decreased activity leads to hypofibrinolysis which can result in thrombosis or embolism.

Tissue plasminogen activator also plays a role in cell migration and tissue remodelling.

Tissue plasminogen activator is a protein encoded by the *PLAT* gene, which is located on chromosome 8. The primary transcript produced by this gene undergoes alternative splicing, producing three distinct messenger RNAs.

tPA is used in diseases that feature blood clots, such as pulmonary embolism, myocardial infarction and stroke, in a medical treatment called thrombolysis. To be effective, tPA must be administered within the first three hours of the event to be given intravenously, or within six hours to be administered through an arterial catheter directly to the site of occlusion. The guideline in Ontario, Canada hospitals for ischemic strokes is that tPA must be given within three hours of the onset of symptoms. Because of this, only about 3% of patients qualify for this treatment tPA appears to show benefit not only for large artery occlusions but also for lacunar strokes. Since tPA dissolves blood clots, there is risk of hemorrhage with its use.

The tPA has been used to dissolve thrombi associated with ischemic strokes and brain injury.

In addition, people with frostbite who were treated with tPA had fewer amputations than those who were not.

In tPA overdose, aminocaproic acid works as an antidote.

Tissue plasminogen activator has been shown to interact with Fibrinogen alpha chain, LRP1 and SERPINI1.

Von Willebrand factor (vWF) is a blood glycoprotein involved in hemostasis. It is deficient or defective in von Willebrand disease and is involved in a large number of other diseases, including thrombotic thrombocytopenic purpura, Heyde's syndrome, and possibly hemolytic-uremic syndrome.

vWF is a large multimeric glycoprotein present in blood plasma and produced constitutively in endothelium (in the Weibel-Palade bodies), megakaryocytes (α-granules of platelets), and subendothelial connective tissue.

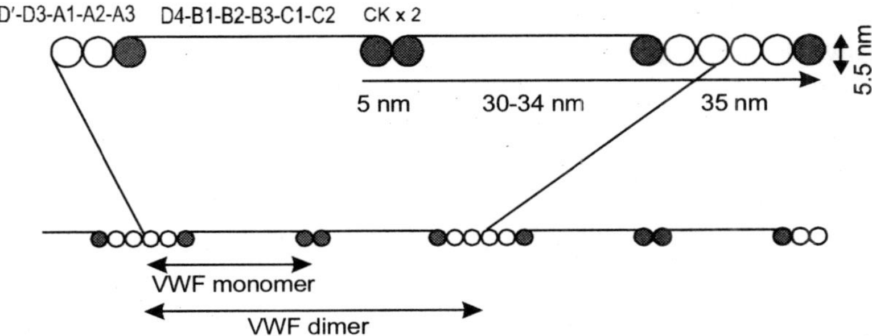

Fig. 8.36 : Von Willebrand Factor

The basic vWF monomer is a 2050 amino acid protein. Every monomer contains a number of specific domains with a specific function; elements of note are:

- the D'/D3 domain, which binds to Factor VIII
- the A1 domain, which binds to:
  - platelet GPIb-receptor
  - heparin
  - possibly collagen
- the A3 domain, which binds to collagen
- the C1 domain, in which the RGD domain binds to platelet integrin $\alpha_{IIb}\beta_3$ when this is activated

- the "cysteine knot" domain (at the C-terminal end of the protein), which vWF shares with platelet-derived growth factor (PDGF), transforming growth factor-β (TGFβ) and β-human chorionic gonadotropin (βHCG, of pregnancy test fame).

Monomers are subsequently N-glycosylated, arranged into dimers in the endoplasmic reticulum and into multimers in the Golgi apparatus by crosslinking of cysteine residues via disulfide bonds. With respect to the glycosylation, vWF is one of the few proteins that carry ABO blood group system antigens.

Multimers of vWF can be extremely large, >20,000 kDa, and consist of over 80 subunits of 250 kDa each. Only the large multimers are functional. Some cleavage products that result from vWF production are also secreted but probably serve no function.

Mechanism of Coagulation or Coagulation Cascade

Coagulation begins almost instantly after an injury to the blood vessel has damaged the endothelium (lining of the vessel), this releases phospholipid components called tissue factor and fibrinogen that initiate a chain reaction. Platelets immediately form a plug at the site of injury; this is called *primary hemostasis*. *Secondary hemostasis* occurs simultaneously: Proteins in the blood plasma, called *coagulation factors* or *clotting factors*, respond in a complex cascade to form fibrin strands, which strengthen the platelet plug.Damage to blood vessel walls exposes subendothelium proteins, most notably von Willebrand factor (vWF), present under the endothelium. vWF is a protein secreted by healthy endothelium, forming a layer between the endothelium and underlying basement membrane. When the endothelium is damaged, the normally-isolated, underlying vWF is exposed to blood and recruits Factor VIII, collagen, and other clotting factors. Circulating platelets bind to collagen with surface collagen-specific glycoprotein Ia/IIa receptors. This adhesion is strengthened further by additional circulating proteins vWF), which forms additional links between the platelets glycoprotein Ib/IX/V and the collagen fibrils. These adhesions activate the platelets.Activated platelets release the contents of stored granules into the blood plasma. The granules include ADP, serotonin, platelet-activating factor (PAF), vWF, platelet factor 4, and thromboxane A_2 (TXA_2), which, in turn, activate additional platelets. The granules' contents activate a G_q-linked protein receptor cascade, resulting in increased calcium concentration in the platelets' cytosol. The calcium activates protein kinase C, which, in turn, activates phospholipase A_2 (PLA_2). PLA_2 then modifies the integrin membrane glycoprotein IIb/IIIa, increasing its affinity to bind fibrinogen. The activated platelets change shape from spherical to stellate, and the fibrinogen cross-links with glycoprotein IIb/IIIa aid in aggregation of adjacent platelets.

The blood clothing system or coagulation pathway, like the complement system, is a proteolytic cascade. Each enzyme of the pathway is present in the plasma as a zymogen, in other words in an inactive form, which on activation undergoes proteolytic cleavage to release the active factor from the precursor molecule. The coagulation pathway functions as a series of positive and negative feedback loops which control the activation process. The ultimate goal of the pathway is to produce thrombin, which can then convert soluble fibrinogen into fibrin, which forms a clot. The generation of thrombin can be divided into three phases, the *intrinsic* and *extrinsic pathways* that provide alternative routes for the generation of factor X, and the final *common pathway* which results in thrombin formation. The intrinsic pathway is activated when blood comes into contact with sub-endothelial connective tissues or with negatively charged surface that are exposed as a result of tissue damage. Quantitatively it is the most important of the two pathways, but is slower to cleave fibrin than the extrinsic pathway. The *Hageman factor* (factor XII), factor XI, prekallikrein, and *high molecular weight kininogen* (HMWK) are involved in this pathway of activation. Thus this pathway provides a further of the interrelationship

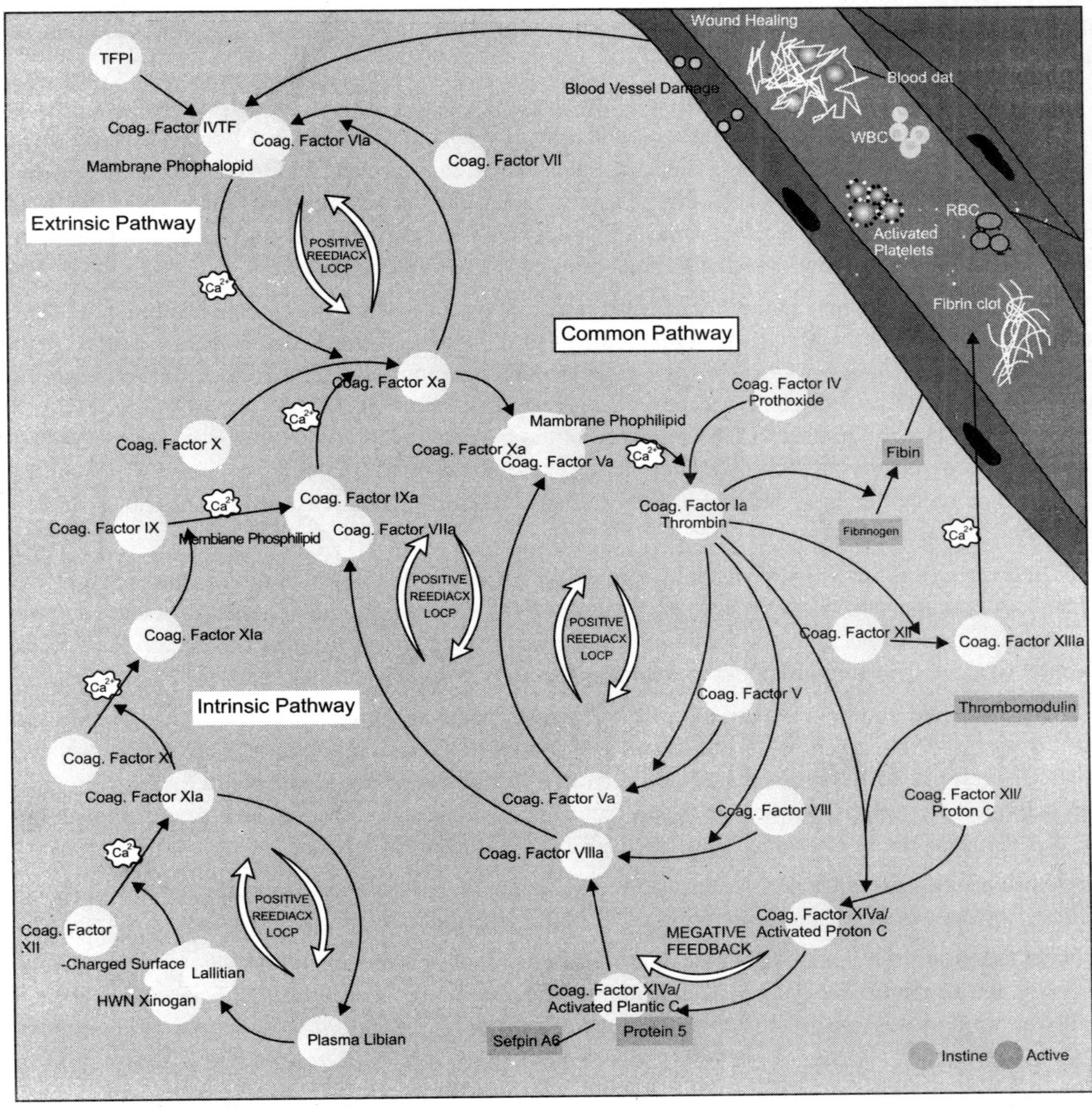

Fig. 8.37 : The Blood Coagulation Cascade

between the various enzyme cascade systems in plasma. The first step is the binding of Hageman factor to a sub-endothelial surface exposed by an injury. A complex of prekallikrein and HMWK also interacts with the exposed surface in close proximity to the bound factor XII, which becomes activated. During activation, the single chain protein of the native Hageman factor is cleaved into two chains (50 and 28 kDa), that remain linked by a disulphide bond. The light chain (28kDa) contains the active site and the molecule is referred to as activated Hageman factor (factor XIIa). There is evidence that the Hageman factor can autoactivate, thus the pathway is self-amplifying once triggered (compare with the alternative pathway of complement).

Activated Hageman factor in turn activates prekallikrein. The *kallikrein* produced can then also cleave factor XII, and a further amplification mechanism is triggered. The activated factor XII remains in close contact with the activating surface, such that it can activate factor XI, the next step in the intrinsic pathway which, to proceed efficiently, requires **Ca^{2+}** Also involved at this stage is HMWK, which binds to factor XI and facilitates the activation process. Activated factors XIa, XIIa, and kallikrein are all serine proteases, like many of the enzymes of the complement system.

Eventually the intrinsic pathway activates factor X, a process that can also be brought about by the extrinsic pathway. Factor X is the first molecule of the *common pathway* and is activated by a complex of molecules containing activated factor IX, factor VIII, calcium, and phospholipid which is provided by the *platelet surface*, where this reaction usually takes place. The precise role of factor VIII in this reaction is not clearly understood. Its presence in the complex is obviously essential, as evidenced by the serious consequences of factor VIII deficiency experienced by haemophiliacs. Factor VIII is modified by thrombin, a reaction that results in greatly enhanced factor VIII activity, promoting the activation of factor X.

The extrinsic pathway is an alternative route for the activation of the clothing cascade. It provides a very rapid response to tissue injury, generating activated factor X almost instantaneously, compared to the seconds or even minutes required for the intrinsic pathway to activate factor X. The main function of the extrinsic pathway is to *augment* the activity of the intrinsic pathway.

There are two components unique to the extrinsic pathway, tissue factor or factor III, and factor VII. *Tissue factor* is present in most human cells bound to the cell membrane. The activation process for tissue factor is not entirely clear. Once activated, tissue factor binds rapidly to factor VII which is then activated to form a complex of tissue factor, activated factor VII, calcium, and a phospholipid, and this complex then rapidly activates factor X.

The intrinsic and extrinsic systems converge at factor X to a single common pathway which is ultimately responsible for the production of thrombin (factor IIa).

Clot formation. The end result of the clotting pathway is the production of *thrombin* for the conversion of fibrinogen to fibrin. *Fibrinogen* is a dimer soluble in plasma. Exposure of fibrinogen to thrombin results in rapid proteolysis of fibrinogen and the release of *fibrinopeptide A*. The loss of small peptide A is not sufficient to render the resulting fibrin molecule insoluble, a proces that is required for clot formation, but it tends to form complexes with adjacent fibrin and fibrinogen molecules. A second peptide, *fibrinopeptide B*, is then cleaved by thrombin, and the fibrin monomers formed by this second proteolytic cleavage polymerize spontaneously to form an insoluble gel. The polymerized *fibrin*, held together by noncovalent and electrostatic forces, is stabilized by the transamidating enzyme factor XIIIa, produced by the action of thrombin on factor XIII. These insoluble fibrin aggregates (clots), together with aggregated platelets (thrombi), block the damaged blood vessel and prevent further bleeding.

There is an *interrelationships between the coagulation pathway and other plasma enzyme* systems. Contact activation of the coagulation pathway, in addition to promoting blood clotting, results in the generation of plasminogen activator activity, which is involved in *fibrinolysis* or clot removal. Activated Hageman factor and its peptides can also initiate the formation of kallikrein from plasma prekallikrein, and this triggers the release of bradykinin from kininogens in the plasma. *Kinins* are responsible for dilating small blood vessels, inducing a fall in blood presssure, triggering smooth muscle contraction, and increasing the permeability of vessel walls. In addition, activation of the coagulation pathway produces a vascular permeability factor, as well as chemotactic peptides for professional phagocytes.

The Anti Coagulants

An **anticoagulant** is a substance that prevents coagulation; that is, it stops blood from clotting. A group of pharmaceuticals called anticoagulants can be used *in vivo* as a medication for thrombotic disorders. Some chemical compounds are used in medical equipment, such as test tubes, blood transfusion bags, and renal dialysis equipment.

Anticoagulants are given to people to stop thrombosis (blood clotting inappropriately in the blood vessels). This is useful in primary and secondary prevention of deep vein thrombosis, pulmonary embolism, myocardial infarctions and strokes in those who are predisposed.

The oral anticoagulants are a class of pharmaceuticals that act by antagonizing the effects of vitamin K. Examples include warfarin. It is important to note that it takes at least 48 to 72 hours for the anticoagulant effect to develop fully. In cases when any immediate effect is required, heparin must be given concomitantly. Generally, these anticoagulants are used to treat patients with deep-vein thrombosis (DVT), pulmonary embolism (PE), atrial fibrillation (AF), and mechanical prosthetic heart valves.

Patients aged 80 years or more may be especially susceptible to bleeding complications with a rate of 13 bleeds per 100 person-years. These oral anticoagulants are used widely as poisons for mammalian pests, especially rodents. (For details, see rodenticide and warfarin.) Depletion of vitamin K by coumarine therapy increases risk of arterial calcification and heart valve calcification, especially if too much vitamin D is present.

Heparin is a biological substance, usually made from pig intestines. It works by activating antithrombin III, which blocks thrombin from clotting blood. Heparin can be used *in vivo* (by injection), and also *in vitro* to prevent blood or plasma clotting in or on medical devices. Vacutainer brand test tubes containing heparin are usually colored green. Low molecular weight heparin is a more highly processed product that is useful as it does not require monitoring of the APTT coagulation parameter (it has more predictable plasma levels) and has fewer side effects.

- Fondaparinux is a synthetic sugar composed of the five sugars (pentasaccharide) in heparin that bind to antithrombin. It is a smaller molecule than low molecular weight heparin.
- Idraparinux

In March 2008 major recalls of Heparin were announced by pharmaceuticals due to a suspected and unknown contamination of the raw Heparin stock imported from China. The U.S. Food and Drug Administration was quoted as stating that at least 19 deaths were believed linked to a raw Heparin ingredient imported from the People's Republic of China, and that they had also received 785 reports of serious injuries associated with the drug's use. According to the *New York Times*: 'Problems with heparin reported to the agency include difficulty breathing, nausea, vomiting, excessive sweating and rapidly falling blood pressure that in some cases led to life-threatening shock'.

Another type of anticoagulant is the direct thrombin inhibitor. Current members of this class include argatroban, lepirudin, bivalirudin, and dabigatran. An oral direct thrombin inhibitor, ximelagatran (Exanta) was denied approval by the Food and Drug Administration (FDA) in September 2004 and was pulled from the market entirely in February 2006 after reports of severe liver damage and heart attacks.

Laboratory instruments, test tubes, blood transfusion bags, and medical and surgical equipment will get clogged up and become nonoperational if blood is allowed to clot. Chemicals can be added to stop blood clotting. Apart from heparin, most of these chemicals work by binding calcium ions, preventing the coagulation proteins from using them.

- EDTA is denoted by mauve or purple caps on Vacutainer brand test tubes. This chemical strongly and irreversibly binds calcium. It is in a powdered form.
- Citrate is usually in blue Vacutainer tube. It is in liquid form in the tube and is used for coagulation tests, as well as in blood transfusion bags. It gets rid of the calcium, but not as strongly as EDTA. Correct proportion of this anticoagulant to blood is crucial because of the dilution. It can be in the form of sodium citrate or ACD.
- Oxalate has a mechanism similar to that of citrate. It is the anticoagulant used in fluoride (grey top) tubes.

Available Anti Coagulating Agents

- Warfarin (Coumadin) *This is the main agent used in the U.S. and UK*
- Acenocoumarol and phenprocoumon *This is used more commonly outside the U.S. and the UK*
- Phenindione

Warfarin (Coumadin) (also known under the brand names Coumadin, Jantoven, Marevan, Lawarin, and **Waran**) is an anticoagulant. It was initially marketed as a pesticide against rats and mice and is still popular for this purpose, although more potent poisons such as brodifacoum have since been developed. A few years after its introduction, warfarin was found to be effective and relatively safe for preventing thrombosis and embolism (abnormal formation and migration of blood clots) in many disorders. It was approved for use as a medication in the early 1950s and has remained popular ever since; warfarin is the most widely prescribed anticoagulant drug in North America. Despite its effectiveness, treatment with warfarin has several shortcomings. Many commonly used medications interact with warfarin, as do some foods, and its activity has to be monitored by frequent blood testing for the international normalized ratio (INR) to ensure an adequate yet safe dose is taken.

Warfarin is a synthetic derivative of coumarin, a chemical found naturally in many plants, notably woodruff (*Galium odoratum*, Rubiaceae), and at lower levels in licorice, lavender, and various other species. Warfarin and related coumarins decrease blood coagulation by inhibiting vitamin K epoxide reductase, an enzyme that recycles oxidized vitamin K to its reduced form after it has participated in the carboxylation of several blood coagulation proteins, mainly prothrombin and factor VII. For this reason, drugs in this class are also referred to as vitamin K antagonists.

Heparin, a highly-sulfated glycosaminoglycan, is widely used as an injectable anticoagulant, and has the highest negative charge density of any known biological molecule. It can also be used to form an inner anticoagulant surface on various experimental and medical devices such as test tubes and renal dialysis machines. Pharmaceutical grade heparin is derived from mucosal tissues of slaughtered meat animals such as porcine (pig) intestine or bovine (cow) lung.

Although used principally in medicine for anticoagulation, the true physiological role in the body remains unclear, because blood anti-coagulation is achieved mostly by endothelial cell-derived heparan sulphate proteoglycans. Heparin is usually stored within the secretory granules of mast cells and released only into the vasculature at sites of tissue injury. It has been proposed that, rather than anticoagulation, the main purpose of heparin is in a defensive mechanism at sites of tissue injury against invading bacteria and other foreign materials. In addition, it is preserved across a number of widely different species, including some invertebrates that do not have a similar blood coagulation system.

Acenocoumarol is an anticoagulant that functions as a vitamin K antagonist (like warfarin). It is a derivative of coumarin and is marketed under the brand names **Sintrom** and **Sinthrome**.

Phenindione is an anticoagulant which functions as a Vitamin K antagonist.

THE FUNCTIONS OF BLOOD

Blood is a liquid tissue. This means that it contains cells suspended in a liquid. Red blood cells carry oxygen and help to carry carbon dioxide. White blood cells are involved in the body's defense mechanisms. Platelets are fragments of cells; they help blood to clot. The liquid is called plasma. It contains many important substances which must be carried around the body.

- Approximately 8% of an adult's body weight is made up of blood.
- Females have around 4-5 litres, while males have around 5-6 litres. This difference is mainly due to the differences in body size between men and women.
- Its mean temperature is 38 degrees Celcius.
- It has a pH of 7.35-7.45, making it slightly basic (less than 7 is considered acidic).
- Whole blood is about 4.5-5.5 times as viscous as water, indicating that it is more resistant to flow than water. This viscosity is vital to the function of blood because if blood flows too easily or with too much resistance, it can strain the heart and lead to severe cardiovascular problems.

- Blood in the arteries is a brighter red than blood in the veins because of the higher levels of oxygen found in the arteries.
- An artificial substitute for human blood has not been found.

The Major Functions of Blood

Blood has three main functions: transport, protection and regulation.

Transport

Blood transports the following substances:

- Gases, namely oxygen (O_2) and carbon dioxide (CO_2), between the lungs and rest of the body
- Nutrients from the digestive tract and storage sites to the rest of the body
- Waste products to be detoxified or removed by the liver and kidneys
- Hormones from the glands in which they are produced to their target cells
- Heat to the skin so as to help regulate body temperature

Protection

Blood has several roles in inflammation:

- Leukocytes, or white blood cells, destroy invading microorganisms and cancer cells
- Antibodies and other proteins destroy pathogenic substances
- Platelet factors initiate blood clotting and help minimize blood loss

Regulation

Blood helps regulate:

- pH by interacting with acids and bases
- Water balance by transferring water to and from tissues

Blood as a Transport System

The plasma carries several different substances ***in solution*** around the body.

Substance	Origin	Destination
1	2	3
carbon dioxide	body	lungs
digested food	gut	liver & rest of body
wastes (e.g. urea)	liver	kidneys
hormones	endocrine glands	"where needed" & rest of body

1	2	3
Plasma proteins - normally remain in blood		**Outline of function**
albumin	liver	keeping blood osmotic pressure and pH constant, and assisting in transport and release of other soluble components
fibrinogen	liver	complex reaction when wound occurs resulting in formation of insoluble fibrin, which is the basis of a clot which seals the wound
globulins (antibodies)	lymphatic system	when an infection occurs: stick to invading organisms and inactivate them

Red cells are specialized to carry one substance in particular, (*actually not strictly in solution*).

Hormones from the Glands in which they are Produced to their Target Cells

Most hormones are secreted into the general circulation to exert their effects on appropriate distant target tissues. There are important exceptions, however, such as self-contained portal circulations in which blood is directed to a specific area. A portal circulation begins in a capillary bed. As the capillaries extend away from the capillary bed, they merge to form a set of veins, which then divide to form a second capillary bed. Thus, blood collected from the first capillary bed is directed solely into the tissues nourished by the second capillary bed. Two portal circulations in which hormones are transported are present in the human body. One system, the hypothalamic-hypophyseal portal circulation, collects blood from capillaries originating in the hypothalamus and, through a plexus of veins surrounding the pituitary stalk, directs the blood into the anterior pituitary gland. This allows the neurohormones secreted by the neuroendocrine cells of the hypothalamus to be transported directly to the cells of the anterior pituitary. These hormones are largely, but not entirely, excluded from the general circulation. In the second system, the hepatic portal circulation, capillaries originating in the gastrointestinal tract and the spleen merge to form the portal vein, which enters the liver and divides to form portal capillaries. This allows hormones from the islets of Langerhans of the pancreas, such as insulin and glucagon, as well as certain nutrients absorbed from the intestine, to be transported into the liver before being distributed through the general circulation. In serum, many hormones exist both as free, unbound hormone and as hormone bound to a serum carrier or transport protein. These proteins, which are produced by the liver, bind to specific hormones in the serum. Transport proteins include sex hormone-binding globulin, which binds estrogens and androgens; corticosteroid-binding globulin, which binds cortisol; and growth hormone-binding.

Transport Mechanism of Thyroid Hormone

Thyroid hormones (TH) are essential for normal development, differentia- tion growth and metabolism of every cell in the body. The pro-hormone thyrox- ine (T4) is synthesized by the thyroid follicles together with a small amount of the biologically active hormone triiodothyronine (T3), which derives mainly from tissue T4 deiodination. Approximately 0.03% of total T4 and 0.3% of total T3 in serum are circulating in a free or unbound form while the major part of TH is bound to circulating plasma proteins. These plasma proteins are responsible for the maintenance of the large extrathyroidal pool of TH, but their function is otherwise not quite clear, since wide differences in their concentrations do not influence the thyroid functional status of the individual to any large degree.

Thyroid Hormone Transport

Transport in the Blood

More than 99% of the circulating thyroid hormone is bound to plasma pro- teins but can be liberated with great rapidity for entry into cells. The thyroid hormone-binding proteins are comprised of thyroxine-binding globulin (TBG), transthyretin (TTR or thyroxine-binding prealbumin), human serum albumin (HSA) and lipoproteins. Their functions are most probably to ensure a constant supply of TH to the cells and tissues by preventing urinary loss, protect the organism against abrupt changes in thyroid hormone production and degradation, protect against iodine deficiency and target the amount of TH delivery by ensuring a site-specific, enzymatic alteration of TBG. TBG has by far the highest affinity for T4, the result of which being that TBG binds 75% of serum T4, whereas TTR binds 20% and HSA 5% . Some of the properties of the binding proteins are displayed in table below:

	TBG	TTR	HSA
	1	2	3
Molecular weight, kDa	54*	55	66.5
Structure	monomer	tetramer	monomer
Carbohydrate content, %	20	—	—
Number of bindings sites for T4 and T3	1	2	several
Association constant, K_a (M-1)			
For T4	1×10^{10}	2×10^{8}**	1.5×10^{6}**
For T3	1×10^{9}	1×10^{6}	2×10^{5}
Concentration in serum (mean normal, mg/l)	16	250	40,000
Relative distribution of T4 and T3 in serum, %			
T4	75	20	5

	1	2	3
T3	75	<5	20
In vivo survival			
Half-life, days	5***	2	15
Degradation rate, mg/day	15	650	17,000

HSA = human serum albumin; TBG = Thyroxine-binding globulin; TTR=transthyretin.

*Apparent molecular weight on acrylamide gel electrophoresis 60 kDa.

**Value given is for the high affinity binding site only.

***Longer under the influence of estrogen.

Reproduced with permission from Hayashi and REfetoff: Molecular Endocrinology: Basic Concepts and Clinical Correlations, New York, Raven Press, 1995.

Thyroxine-Binding Globulin

TBG carries the major part of both circulating T4 and T3 (as well as reverse T3), and therefore quantitative or qualitative changes in TBG concentration have a high impact on total serum T4 and T3. The protein is encoded by a single gene on the X-chromosome and is produced and cleared by the liver. It has a single iodothyronine-binding site with a slightly higher affinity for T4 compared to T3. When it is fully saturated it carries approximately 200 g T4/l. The TBG concentration in serum is between 11 and 21 mg/l (180–350 nmol/l), present from 12th week of fetal life and 1.5 times higher in newborns and children until 2–3 years of age. Estrogen has a marked effect on TBG by prolonging the biological half-life from the normal 5 days, thus resulting in increased plasma concentrations of TBG and total TH while testosterone has the opposite effect. In children and adolescents this may have an implication in diseases with a severe sex hormone overproduction related to the age, as well as oral contraceptives and pregnancy in adolescent girls. Inherited TBG excess was first described in 1959, and several familial X-chromosome-linked TBG abnormalities have been described. A rare TBG abnormality is seen in carbohydrate-deficient glycoprotein syndrome, which is associated with severe mental and motor retardation. Acquired TBG abnormalities are mostly resulting in altered synthesis and/or degradation and caused by, e.g., severe terminal illness, hypo- and hyperthyroidism, severe liver disease and a variety of critical non-thyroidal illnesses. The latter may be mediated by interleukin-6 or other cytokines suppressing acute-phase reactants.

Transthyretin

TTR, previously called thyroxine-binding prealbumin binds only about

15–20% of the circulating TH and has a lower affinity for the hormones thus dissociating from them more rapidly and thus responsible for much of the immediate delivery of T4 and T3. Transthyretin is the major thyroid hormone-binding protein in cerebrospinal fluid. It is synthesized in the liver and the choroids plexus and secreted into the blood and cerebrospinal fluid, respec- tively. Only 0.5% of the circulating TTR is occupied by T4 and it has a rapid turnover of 2 days in plasma. Hence, acute reduction of the rate of synthesis results in a rapid decrease of its serum concentration. Acquired abnormali- ties in TTR include major illness, nephrotic syndrome, liver disease, cystic fibrosis, protein fasting and hyperthyroidism. However, changes in TTR concentrations have little effect on the serum concentrations of TH.

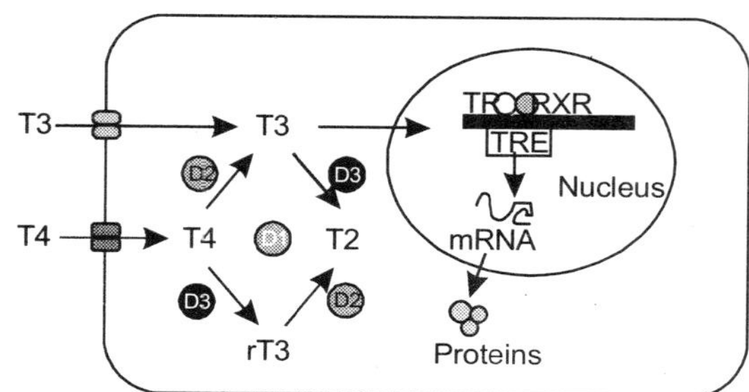

Fig. 8.38 : Thyroid hormone transport and metabolism in a 3,3 ,5-triiodothyronine (T3)

Albumin

HSA binds about 5% of the circulating T4 and T3. Its affinity for the hor- mones is even lower, and since HSA associates with a wide variety of sub- stances, including a number of different hormones and drugs, the association between TH and HSA can hardly be regarded specific. Even marked fluctuations in serum HSA concentrations have no effect on TH levels.

Lipoproteins

Lipoproteins transport a minor fraction of circulating T4 and to some extent T3. The binding site for TH on apolipoprotein A1 is distinct from that which binds to cellular protein receptors.

Consequences of Abnormal Binding Protein Concentrations

Abnormalities of the TH-binding proteins do not cause alterations in the metabolic state of the individual nor do they result in thyroid disease. Thus, abnormal concentrations of these binding proteins, due to changed synthesis, degradation or stability, result in maintaining normal free TH concentrations.

However, they do give rise to misinterpretation of most of the measurements of serum levels of TH by available techniques. Depending on the severity of the abnormality only total TH concentrations are affected, but also the measured free TH levels by automated currently used methods give rise to incorrect results . In such cases, it may be necessary to provide a free TH estimate by quantifying total hormone concentration with a subsequent estimate of the available binding places by use of a TH uptake test or direct measurement of TBG. Even better is measurement of free TH concentrations by equilibrium dialysis or ultrafiltration, but not many laboratories in the world perform these measurements anymore.

Transport Across the Cell Membrane

The deiodinases involved in T4 to T3 conversion and T4 and T3 degradation as well as the T3 receptors are located intracellularly. Therefore, both action and metabolism of thyroid hormones are intracellular events requiring transport of iodothyronines across the cell membrane. For a long time it was believed that TH diffused passively over the cell membrane, but recent years of research has made it increasingly clear that cellular transmembrane transport of TH is mediated by transporters, that these transporters determine the availability of iodothyronines to the intracellular sites for metabolism and action, and that the TH transport is energy dependent (Fig. 8.38). Recently, specific transporters (organic anion transporters and amino acid transporters) known to facilitate cellular thyroid hormone uptake have been identified. Hennemann and Visser have defined requirements for (patho)physiological significance of thyroid hormone plasma membrane transport in the terms that it should be specific, without significant diffusion, plasma membrane transport subject to regulation, transport rate limiting on subsequent metabolism, and changes in transport should be appropriate from the (patho)physiological point of view.

Organic Anion Transporters

These mediate uptake of iodothyronines and their sulphonated derivatives and they are members of the Na /taurocholate cotransporting polypeptide (NTCP) and the Na-independent organic anion transporting polypeptide (OATP) families. NTCP is only expressed on hepatocytes and is the major transporter of conjugated bile acids in the liver. The OATPs are a large family responsible for transmembrane transport of a number of compounds including TH. The most interesting OATP superfamily members in terms of TH transport are OATP1C1 and OATP14. The former has been demonstrated to be widely expressed both in human brain and the Leydig cells of testis. In the brain they seem to participate in maintaining the T3 concentration along with parallel changes in D2 expression. It has been demonstrated that the thyroid state modulates OATP1C1, and by doing so counteracts the effects of alterations in circulating T4 levels on brain T4 uptake. In humans, OATP1C1 is also expressed in the testis where also D2 expression has been demonstrated. This combination supports a role of TH in development, growth and differentiation of Leydig cells. In particular T3 is very important for testosterone biosynthesis and may therefore have an important role in male puberty. Other OATPs have been demonstrated in a number of other tissues and may exert a variety of effects, but this is not well clarified, and they are possibly less tissue-specific considering the widespread expression. Some characteristics of the transporters are shown in table.

Amino Acid Transporters

Iodothyronines are a particular class of amino acids built from two tyrosine residues implying transport by specific amino acid transporters, in particular the L and T type amino acid transporters, which therefore are involved in TH uptake into several tissues. Among those are members of the heterodimeric amino acid transporter (HAT) family. Their exact role is not clear, but it has been demonstrated that overexpression of the heterodimer L-type transporter in cells resulted in increased intracellular T3 availability and a consequent augmentation of T3 action. Evidence has also been presented to suggest a role of members from the HAT family in supplying the placenta and developing fetus with thyroid hormone.

The monocarboxylate transporter (MCT) family comprise to date 14 identified members in various tissues from different species. MCTs are dispersed over autosomal chromosomes, except MCT8, which is X-linked and a specific TH transporter. Compared to other TH transporters the rate of T3 and

Characteristics of Thyroid Hormone Transporters

Gene	Protein	Species code	Accession some	Chromo- distri- bution	Tissue transport	Iodothyroine
SLC10A1	NTCP	human	NP_003040	14q24.1	liver	T4, T3, rT3, T2
SLC10A1	NTCP	rat	NP_058743	6q24		
SLCO1A1	OATP1A1	rat	NP_058807	4q44	liver, kidney, CP	T4, RT3, T3, T2
SLCO1A2	OATP1A2	human	NP_602307	12p12	brain, kidney, liver	T3, T2, T4, rT3
SLCO1A4	OATP1A4	rat	NP_5711981	4	liver brain, retina	T4, T2, T3, rT3
SLCO1A5	OATP1A5	rat	NP_110465	4q44	kidney, retina, liver	T3, T4
SLCO1B1	OATP1B1	human	NP_006437	12p	liver	T3, T4
SLCO1B2	OATP1B2	rat	NP_113838	4q44	liver	T3, T4
SLCO1B3	OATP1B3	human	NP_062818	12p12	liver	T3, T4
SLCO1C1	OATP1C1	human	NP_059131	12p12.3	brain, cochlea	T3, rT3, T3
SLCO1C1	OATP1C1	rat	NP_445893	4q44	brain	T4, T3, T3
SLCO4A1	OATP1A1	human	NP_057438	20q13.33	multiple	T4, rT3, T3
SLCO4A1	OATP1A1	rat	NP_598292	3q43	multiple	T3 (T4, rT3 NT)
SLCO4C1	OATP4C1	human	NP_851322	5q21.2	kidney, other	T3, T4
SLCO4C1	OATP4C1	rat	AAQ04697	9		T3, (T4 NT)
SLCO6B1	OATP6B1	rat	NP_596903	9q36	testis	T4, T3
SLCO6C1	OATP6C1	rat	NP_775460	9q36	testis	T4, T3
SLC7A5	LAT1	human	NP_003477	16q24.3	multiple (not lover),	T2, rT3, T3, T4
SLC7A5	LAT1	rat	NP_059049	19q12	tumors	T2, rT3, T3, T4
SLC7A8	LAT2	human	NP_036376	14q11.2	mulpile,	
SLC7A8	LAT2	rat	NP_445894	15p13	tumors	
SLC16A2	MCT8	human	NP_006508	Xq13.2	brain, liver,	T3, rT2, T4, rT3
SLC16A2	MCT8	rat	NP_671749	Xq31	kidney, heart, thyroid, eye, pituitary, other	

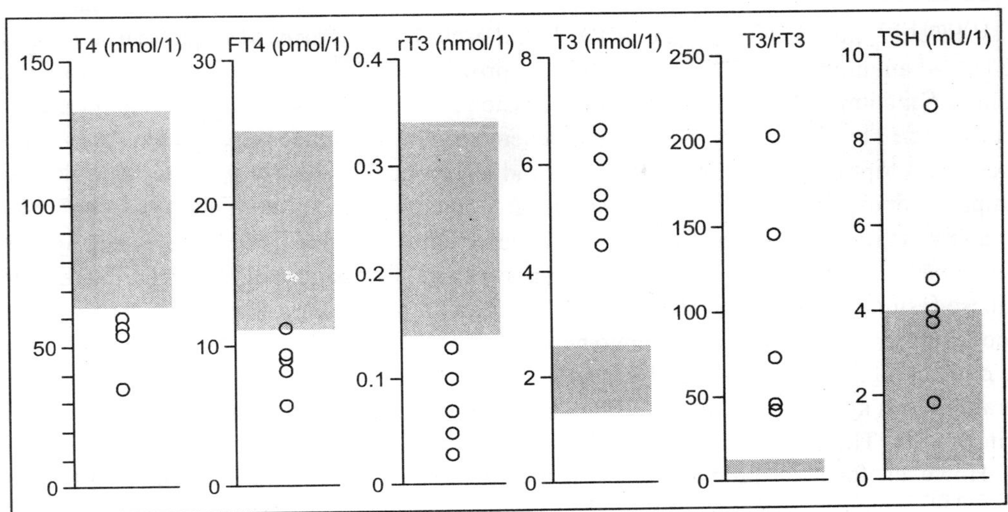

Fig. 8.39 : Thyroid hormone serum levels in patients with mutations in MCT8. Hatched areas indicate normal reference ranges for each analyte

T4 transport is much higher and follows the criteria set down for requirements of a transporter. The MCT8 gene is located in the region of the X-chromosome associated with X-linked diseases, and it was therefore hypothesized that a mutation in this gene would result in an X-linked form of thyroid hormone resistance. Indeed, this hypothesis was verified first in a 6-year-old boy with highly elevated serum T3 and severe psychomotor retardation of unknown origin, where a deletion of the first exon of the MCT8 gene was demonstrated. Since then the same group have described 5 unrelated young boys aged 1.5–6 years with mutations or deletions in the MCT8 gene. They all had a uniform type of severe psychomotor retardation of hitherto unknown origin. The described phenotype comprised symptoms such as severe proximal hypotonia with poor head control and lack of verticalization, absence of targeted grasping, severe mental retardation with only rudimentary commu- nicative skills and movement-induced increase in tone in the extremities. Concerning thyroid function variables, T3 was invariably strongly elevated in all the patients, T4 and free T4 were mildly increased while thyroid-stimulating hormone (TSH) was in the normal range for age in 4 patients and increased in one (fig. 8.40). The various mutations have been described in more detail in a recent review. All the mothers of the 5 patients were proven to be carriers, all of them with normal thyroid hormone levels and without psychomotor retardation. Another group has described two other

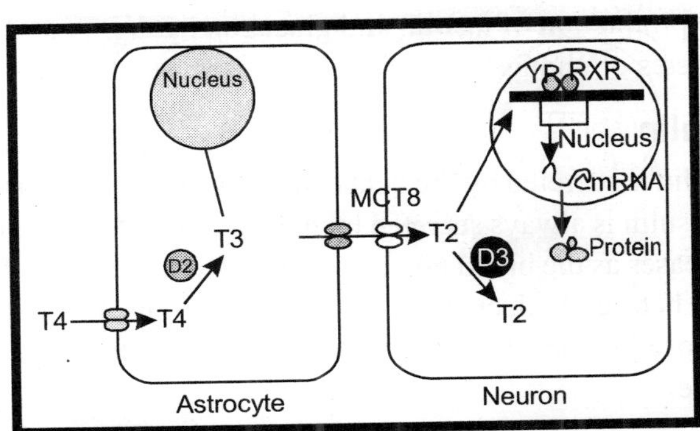

Fig. 8.40 : Role of MCT8 in the neuronal uptake of T3

cases with different mutations. By studying the complex clinical picture of these patients it was assumed that MCT8 had an important role in TH-dependent processes of brain development. To provide a clue to the cellular function of MCT8 in brain, the expression of MCT8 mRNA in the murine central nervous system was studied by in situ hybridization histochemistry. In addition to the choroid plexus structures, the highest transcript levels were found in neo- and allocortical regions (e.g. olfactory bulb, cerebral cortex, hippocampus, and amygdala), moderate signal intensities in striatum and cerebellum, and low levels in a few neuroen-docrine nuclei. Colocalization studies revealed that MCT8 was predominantly expressed in neurons. Together with the spatiotemporal expression pattern of MCT8 during the perinatal period, these results strongly indicated that MCT8 plays an important role for proper central nervous system development by transporting TH into neurons as its main target cells. Another hypothesis raised by these clinical pictures was that MCT8 must play an essential role in the supply of T3 to neurons in the central nervous system (Fig. 8.40). T3 binds to nuclear receptors in neurons, which are a primary action site for T3. The action of T3 is terminated by deiodination by D3, which is expressed in the neurons. However, for local production of T3 the neurons are dependent on neighboring astrocytes expressing D2, which is necessary for the local deiodination (Fig. 8.40). Inactivation of MCT8 by mutation in the gene will result in an impaired supply of T3 to the neuron, as well as a decrease in T3 clearance due to block of T3 access to D3 with a possible subsequent increase in serum T3, consequently stimulating a further expression of D1 in the liver and kidney. The resulting increase in conversion of T4 to T3 and breakdown of reverse T3 explains the serum thyroid hormone concentrations in these patients. The mutations in the MCT8 gene thus resulted in a severe hypothyroidism in the brain with the consequent phenotype, but other tissues and organs did not demonstrate signs of hypothyroidism e.g. bones and metabolism. It therefore seems that other tissues than the brain, are not dependent on MCT8 for uptake of TH. The elevated T3 did not exert any symptoms of hyperthyroidism in the patients, indicating that other yet unknown regulating mechanisms must be in place.

The Transport Mechanism of the Insulin Hormone in Blood

The human body requires that the blood glucose level is maintained in a very narrow range. Homeostasis is regulated by two hormones, insulin and glucagon which are both secreted by the endocrine pancreas. The production of insulin and glucagon by these pancreatic cells ultimately determines if a patient has diabetes or another related problem (see Fig. 8.41).

Insulin

Insulin is secreted by the beta cells of the pancreas in response to high blood sugar, although a low level of insulin is always secreted by the pancreas. After a meal, the amount of insulin secreted into the blood increases as the blood glucose rises. Likewise, as blood glucose falls, insulin secretion by the pancreatic islet beta cells decreases. In response to insulin, cells (muscle, red blood cells, and fat cells) take glucose in from the blood, which ultimately lowers the high blood glucose levels back to the normal range.

Glucagon

Glucagon is secreted by the alpha cells of the pancreas when blood glucose is low. Blood glucose is low between meals and during exercise. When blood glucose is high, no glucagon is secreted from the alpha cells. Glucagon has the greatest effect on the liver although it affects many different cells in the body. Glucagon's function is to cause the liver to release stored glucose from its cells into the blood. Glucagon also the production of glucose by the liver out of building blocks obtained from other nutrients found in the body, for example, protein.

Levels

Blood glucose is normally maintained between 70 mg/dl and 110 mg/dl. Blood glucose levels below 70mg/dl, denote the situation of "hypoglycemia". Although blood glucose levels of 110mg/dl can be normal, this is only the case if a meal has been taken within 2 to 3 hours. A blood glucose level of 180mg/dl or more, is termed "hyperglycemia." Diagnosis is made if blood glucose levels are above 200mg/dl after drinking a sugar-water drink (glucose tolerance test)].

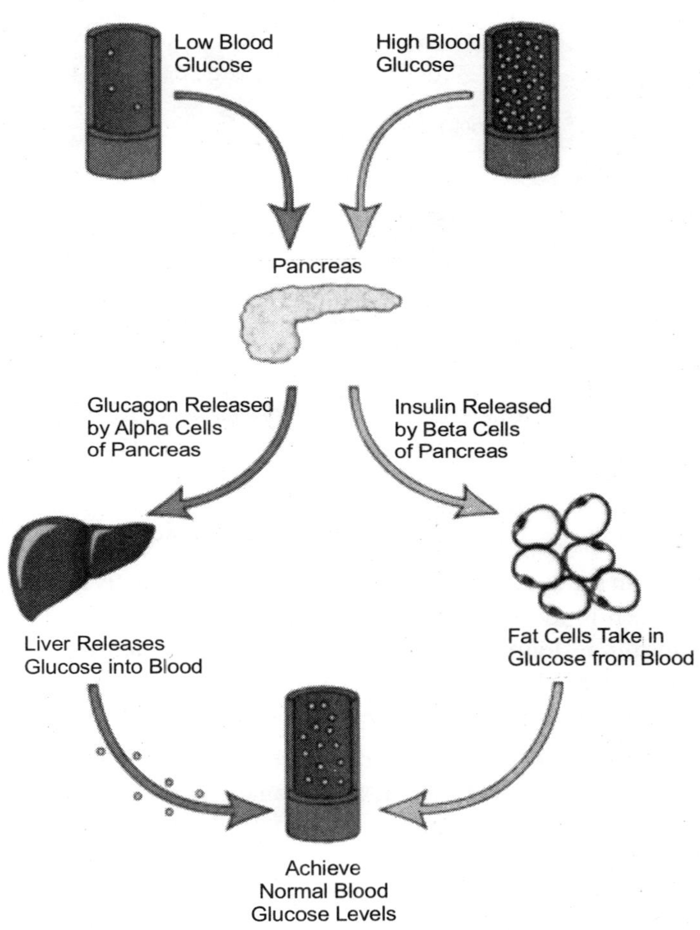

Fig. 8.41 : The Insulin Mechanism

Cell Types in the Endocrine Pancreas

Peptide hormone	A glucago	B Insulin	D Somatostatin	PP Pancreatic polypeptide
Molecular weight:	3500	5800	1500	4200
Number of aa :	29	51	14	36

It is the major actor in control of carbohydrate, lipid and protein metabolism. These aspects are well-discussed.

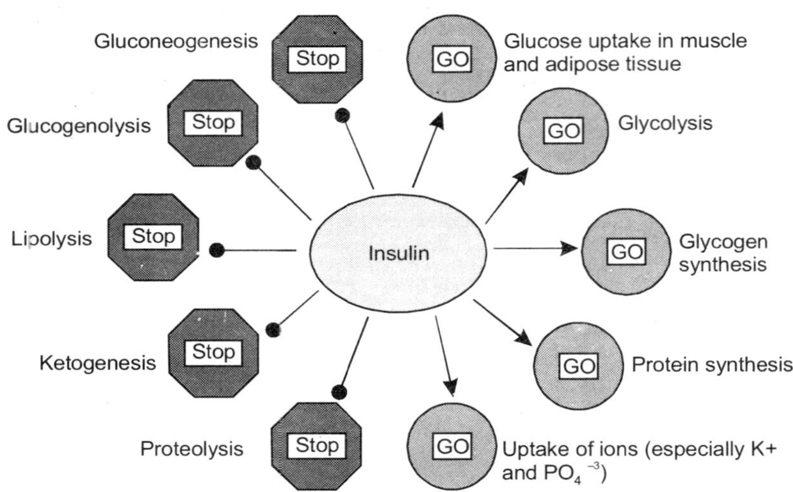

Fig. 8.42 : The Action of Insulin

Insulin signals a state of energy abundance, and activates glucose uptake, metabolism and storage as glycogen in muscle and fat tissue. These organs make up most of the body's mass. At the same time, insulin restrains processes that release stored energy; lipolysis and ketogenesis, glycogenolysis, proteolysis and gluconeogenesis. Insulin is necessary for uptake of amino acids to tissues and for protein synthesis. Insulin is the central actor in homeostasis; the stabilization of the internal milieu.

It has been previously pointed out that insulin and glucagon act together to balance metabolism In general we can say that insulin favors anabolic reactions; glucagon, catabolic reactions. Put more simply, insulin favors storing energy and production of proteins while glucagon activates release of stored energy in the form of glucose or fatty acids. The actions of these two hormones on individual metabolic processes are summarized in the following table.

Metabolic Actions of Insulin and Glucagon

Fatty acid uptake and release in fat	Insulin	Stimulates synthesis of triglycerides (TG) from free fatty acids (FFA); inhibits release of FFA from TG
	Glucagon	Stimulates release of FFA from TG.
Liver glycogen	Insulin	Increases synthesis and thereby glucose uptake and storage
	Glucagon	Stimulates glycogenolysis and glucose release.
Liver gluconeogenesis	Insulin	Inhibits, saves amino acids.
	Glucagon	Stimulates, glucose synthesized and released.
Glucose uptake, skeletal muscle	Insulin	Stimulates uptake, storage as glycogen and use in energy metabolism.
	Glucagon	No receptors, no effect

Glycogen, skeletal muscle	Insulin	Stimulates synthesis.
	Glucagon	No receptors, no effect.
Amino acid uptake	Insulin	Stimulates and is necessary for protein synthesis.
	Glucagon	No receptors, no effect.
Brain (hypothalamus)	Insulin	Reduces hunger through hypothalamic regulation.
	Glucagon	No effect.

Insulin action in Fat Cells

One of the primary actions of insulin is to control storage and release of fatty acids in and out of lipid depots . It does this through two mechanisms; regulation of several lipase enzymes and activation of glucose transport into the fat cell via recruitment of glucose-transport protein 4 (GLUT4).

Fat storage and synthesis of glycerol phosphate.

Splitting of triglycerides produces free fatty acids and glycerol. One might expect that the body would use this glycerol to aid in storage of fatty acids when required. However, that just does not occur. Adipocytes lack glycerokinase which is essential for synthesis of α-glycerol phosphate from glycerol. The glycerol released by lipolysis must be transported to the liver for further metabolism. Where does the glycerol phosphate that is needed to esterify fatty acids come from? How do adipocytes take up and store fat from the diet?

Insulin action in adipocytes and ketogenese in liver

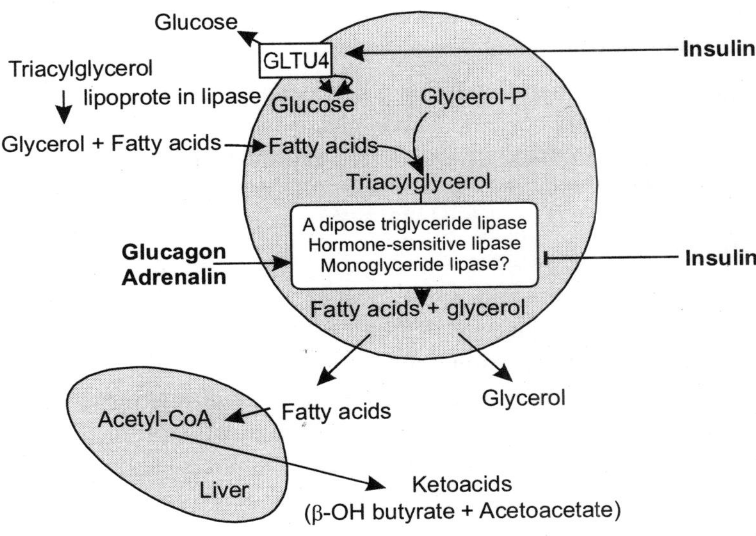

Fig. 8.43 : Action of Insulin in Adephocytes

The "backbone" of triglycerides is α-glycerol phosphate produced from glucose in fat cells. Storage of triglycerides after a meal is, therefore, dependent upon insulin-stimulated glucose uptake and glycolysis. Normally, fat cells take up fatty acids and glucose simultaneously. The fatty acids come from the action of lipoprotein lipase at the capillary wall. Glucose uptake is stimulated by insulin and occurs through the insulin-sensitive glucose transport protein GLUT4. The high blood lipid levels seen in diabetes are at least partially, due to reduced insulin-stimulation of glucose uptake at the fat cell membrane.

Splitting of triglycerides back to glycerol and fatty acids follows the actions of several lipases. Present (2008) knowledge suggests that triglycerides are split to diglycerides by adipose triglyceride lipase (ATGL) at the fat droplet-cytosol border.* Diglycerides are thereafter converted to monoglycerides by hormone-sensitive lipase (HSL). The monoglycerides are hydrolyzed by a cytosolic monoglyceride lipase. These three enzymes and their control elements (also proteins, especially perillipin A) are extensively phosphorylated, apparently by protein kinase A. Phosphorylation follows activation of adenyl cyclase, mainly through stimulation by glucagon and adrenalin. The cyclic AMP formed by the activated cyclase stimulates phosphorylation by protein kinase A (PKA) and activation of the various lipases. The important point to remember here is that the combined lipase activity increases and decreases in tact with cyclic AMP levels. Cyclic AMP synthesis and degradation is under strong hormonal control. Adrenalin and glucagon are the most important stimulating agents while insulin rapidly lowers cAMP levels through activation of phosphodiesterase. In effect, glucagon and adrenalin "turn on" lipolysis while insulin "turns off" breakdown of triglycerides in fat cells. This process is of major importance in our physiology and, in fact, is much more complex than I have described here. Please go to the following publications for more information:

1. Regulation of Lipolysis in Adipocytes,
2. Adipose Triglyceride Lipase, function, regulation by insulin and comparison with adiponutrin, Eating and working; how is blood glucose stabilized?

Mental awareness and the feeling of well-being is dependent upon a steady stream of fuel to the brain. This organ is normally entirely dependent upon glucose as its energy substrate. A twenty percent decline in blood sugar can lead to nausea, restlessness and other neurological symptoms. How are blood sugar levels controlled?

Resting Conditions

Let us first examine basal conditions, that is, the post-absorptive state in resting persons. Glucose uptake from the small intestine is not active and yet the level of blood glucose is held relatively constant. This occurs in spite of the fact that the resting body uses about 10 grams of glucose per hour. More than half of this goes to drive energy production in the brain. The brain uses 5-6 grams of glucose constantly; while sleeping, going for a walk or taking an examination in medical biochemistry! MR imaging has clearly demonstrated that various areas of the brain are specifically activated by specific tasks; some areas are quiescent, others active at various times. Never the less, the brain's total energy use is relatively constant, thus explaining the unvarying need for glucose.

* While this triglyceride lipase is called "adipose triglyceride lipase" or ATGL, it is found at considerable levels in many other tissues.

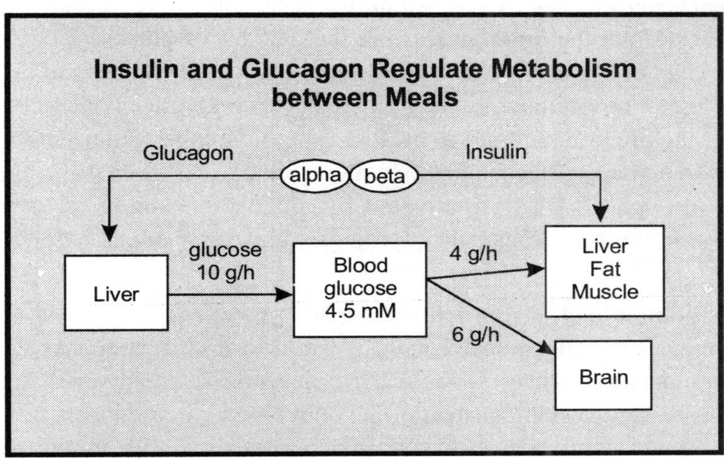

Fig. 8.44 : The Regulatory Action of Insulin and Glucagon between Meals

Between meals, the glucose that is taken up by the body's various tissues is replaced by the liver, either through glycogenolysis (use of stored glycogen reserves) or gluconeogenesis (synthesis of glucose from smaller metabolites). Regulation of these two processes requires the coordinated effects of insulin and glucagon. The molar ratio between the concentrations of these hormones controls metabolism. Secretion of hormones from the Islets of Langerhans is precisely adjusted to coordinate the various functions of insulin and glucagon and to stabilize blood sugar levels.

At Work

Skeletal muscle can increase its energy turnover 18-20 fold under exhaustive work loads, using fatty acids and glucose as substrates for aerobic metabolism and ATP production. However, hard-working

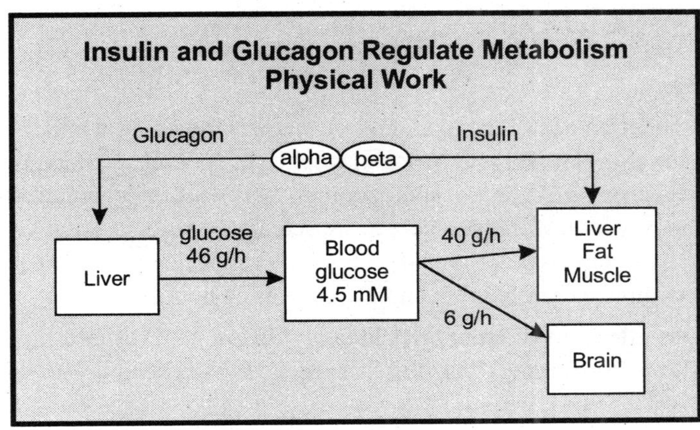

Fig. 8.45 : The Regulatory Action on Metabolism at Physical Work

muscles are dependent upon glucose as a substrate for anaerobic energy production. As we can see from the figure below, despite the huge increase in the amount of glucose taken up from blood by working muscles, the level of blood glucose is not altered. Decreased secretion of insulin and a marked increase in glucagon secretion prompt the liver to break down glycogen and start gluconeogenesis. These actions provide the glucose required to balance glucose uptake by muscles. One might protest, " but we need insulin to activate glucose uptake by the working muscle fibers". The answer to this is that glucose uptake through GLUT4 is activated by glucose consumption as well as insulin in muscle cells. A feedback control mechanism coordinates glucose metabolism and glucose uptake in muscle.

An often ignored point is that insulin inhibits hepatic gluconeogenesis. The reduction in insulin secretion while working is a significant factor in the promotion of gluconeogenesis. The importance of this is obvious when we consider diabetes. Diabetic hyperglycemia is partially due to glucagon-stimulated hepatic glucose synthesis which occurs in spite of the high blood glucose levels that follow the loss of insulin or insulin resistance. Inhibition of hepatic gluconeogenesis with metformin is an important component of treatment for type 2 diabetes.

Note again that the brain takes up glucose at the same rate as that seen under basal conditions.

Extremely hard and prolonged physical work can lead to a pronounced fall in blood sugar. Marathon runners, skiers and others that press their bodies to maximal performance sometimes do collapse before the finish line. Why? The answer lies in the fact that hepatic gluconeogenesis alone cannot provide glucose as quickly as skeletal muscles can remove it from the blood. When glycogen stores are exhausted one of two things must happen: (1) the individual must reduce speed and energy output so that the blood sugar level can remain within normal boundaries or (2) with time a hypoglycemic episode will occur and collapse will follow.

After a Good Meal

One act does alter blood glucose levels; eating a good meal. A normally balanced meal provides about 90 grams of glucose, mostly as polysaccharides. These are usually absorbed over a period of about 120 minutes. The glucose released from food is used as an immediate energy substrate and any excess will be stored as glycogen.

A prerequisite for oxidation or storage of glucose is that it is transported into cells over their plasma membrane. Uptake of larger molecules requires the presence of specific carriers. In the case of glucose we find a family of at least five glucose transport proteins with varying characteristics. The important point here is that the body's largest tissue, skeletal muscle, is dependent upon GLUT4 for uptake of glucose and that this transport must be activated by insulin or muscle activity. In adipose tissue insulin is required for storage of lipids from the diet and for control of lipolysis.

Blood glucose levels after a meal must be held below the renal threshold for recovery of glucose from the glomerular filtrate (8-10 mmol/l). If this is exceeded, glucose is lost to the urine as is seen in diabetes.

So, after a meal insulin secretion is activated, glucagon secretion is minimized, and the liver takes up glucose which is then stored as glycogen to be used to buffer blood glucose at a later time. Insulin also stimulates glucose uptake and glycogen synthesis in muscles. Note that muscle glycogen cannot be released to the circulation. Muscle glycogen is used exclusively as a substrate for muscle activity.

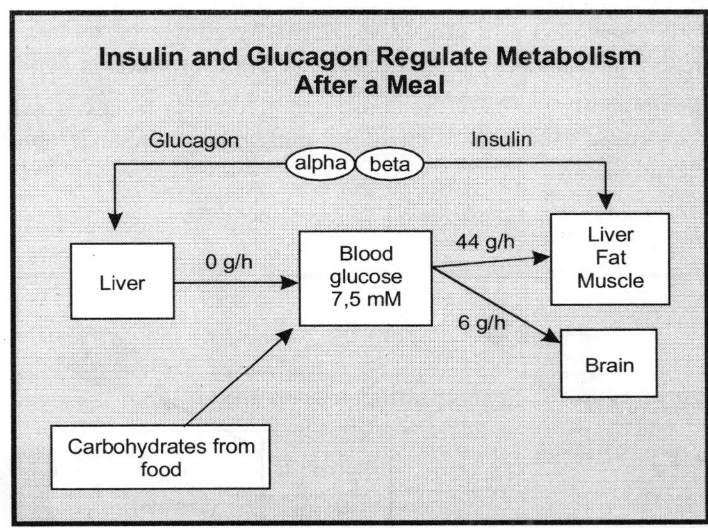

Fig. 8.46 : The Action of Insulin After Meal (and fat in the case of over-nutrition)

Note once more that the brain's glucose uptake is constant.

Tissue Distribution of Glucose after a Meal

If one is to understand the actions of insulin and the pathology that follows insulin deficiency or resistance one must know where this hormone acts. We can gain insight in this by following the fate of glucose after a meal. An average meal contains about 90 grams of glucose. This glucose will either be utilized as an immediate energy substrate for ATP synthesis or will be stored as glycogen or fat.

Let us consider a person who has a good control over his/her weight and where food consumption and the energy used are in balance.

Around 15-18% of the ingested glucose goes to nourish the brain during the

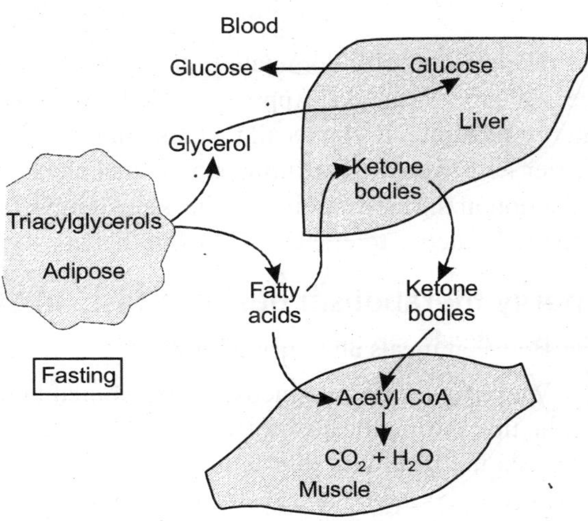

Fig. 8.47 : The Distribution of Insulin ater Meal

absorptive period. Note that there is no storage form of glucose in the CNS; all of the glucose that is taken up is "burned". The brain is, therefore, extremely sensitive to reduced blood glucose levels.

The liver stores excess glucose as glycogen, readying a buffer for blood glucose to meet the coming post-absorptive period. Little glucose is normally converted to fat. Note that over-eating carbohydrates (sucrose and fructose) can and does lead to fat production and storage.

The kidneys take up about 9-10% of the consumed glucose as lactate which is excreted from red blood cells. RBCs lack mitochondria and, therefore, must release the anaerobically oxidized glucose they use as pyruvate and lactate.

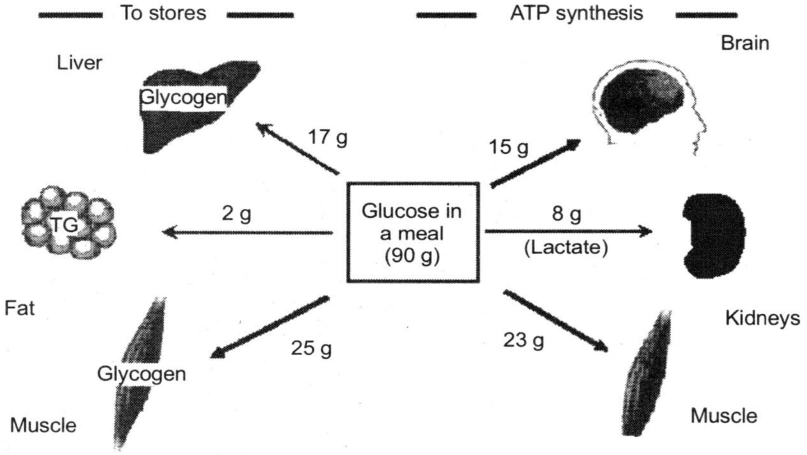

Fig. 8.48 : The Distribution of Glucose after Meal

Skeletal muscles dominate in the fight for blood sugar after a meal, accounting for about 50% of the total glucose uptake. Approximately half of this is stored as glycogen, the rest is used as an immediate energy substrate. If you recall that insulin is needed to activate glucose uptake in muscle (GLUT4 is the carrier here) you will acknowledge that skeletal muscle must be the major target organ for insulin. A reduction in the effect of insulin in skeletal muscle (insulin resistance) is the key mechanism leading to impaired glucose tolerance (IGT) and diabetes type 2.

Energy metabolism during a fast, in Starvation and Uncontrolled Diabetes

Between meals and during a fast:

We have seen how glucose is distributed among the body's various tissues after a meal. Energy production during the absorptive period uses glucose from the diet as substrate. Furthermore, we have seen that the insulin to glucagon ratio is lowered during the post-absorptive period and during a fast. Let us now examine the effects of this alteration in hormone levels upon distribution of substrates for energy metabolism. As pointed out earlier, insulin inhibits lipolysis in adipose tissue while glucagon is a major activator of hormone-sensitive lipase. The change in the insulin/glucagon ratio seen in fasting and between meals activates adipocyte lipolysis. As we can see in the figure to the left, adipose tissue supplies fatty acids to drive aerobic metabolism in muscle, liver and other tissues (but not the brain; fatty acids are not taken up here). Excess fat can be converted to ketone bodies in the liver. The levels of these remain rather low and they are used as energy substrates in muscle. Once more, we see that glycerol arising from lipolysis cannot be reused in fat cells but is circulated to the liver where it enters gluconeogenesis.

Let's sum up. After a meal, fat from food is stored and carbohydrates from the diet are burned. Between meals stored fat is released from fat cells and supplies fuel for most of the body's organs until the next meal. Remember, blood cells and the brain must use glucose as their energy source. We say that "fat spares sugar" so that an even supply of fuel is available to all of the body's tissues. And, remember, this is possible only through the minute-to-minute adjustments in the ratio of insulin to glucagon.

Starvation

We have already seen that gluconeogenesis supplies glucose during fasting and starvation.

Blood Levels of Energy Substrates in Starvation

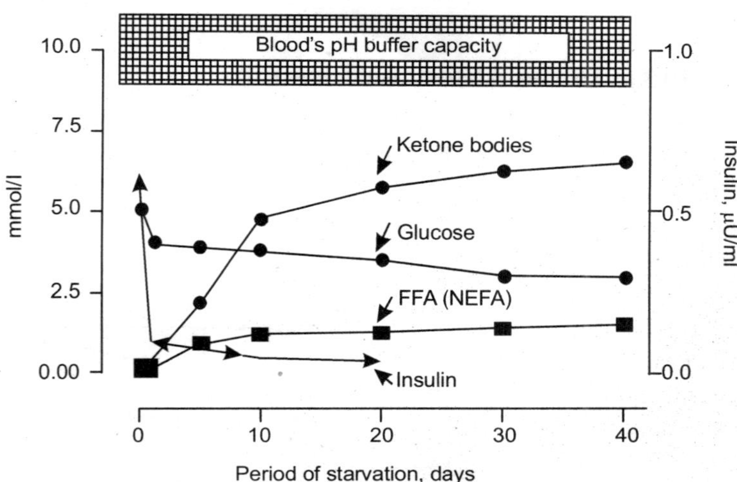

Fig. 8.49 : The Energy Levels Substrates during Starvation

Consider the changes in blood levels of insulin and fuels shown in the next figure. The marked fall in insulin levels seen in starvation leads to an increase in plasma levels of both free fatty acid and especially ketone body levels. Most of the fatty acids liberated from adipose tissue are converted to ketone bodies (acetoacetate and beta-hydroxy butyrate) by the liver during starvation. The level of these "ketone bodies" rises abruptly during the two first weeks of starvation, and then slowly increases. When plasma ketone body levels reach about 5mmol/l they can supply the brain with around 50% of the substrate required for ATP production. Even during starvation about 50% of the brain's energy

Hypothalamic K_{ATP} Channels Regulate Hepatic Gluconeogenesis and Postpranial Blood Glucose Levels.

Blood sugar levels are dependent upon glucose uptake after meals and hepatic release of glucose between meals. The sugar released from the liver comes either from stored glycogen or production of glucose from lactate and amino acids. This production of glucose is largely responsible for stabilization of postprandial blood sugar levels. The hyperglycemia noted in type 2 diabetes partially results from lack of control over hepatic glucose formation due to resistance to insulin. It has recently become clear that part of this insulin effect occurs indirectly through insulin-sensitive receptors in the brain (more precisely, in the hypothalamus).

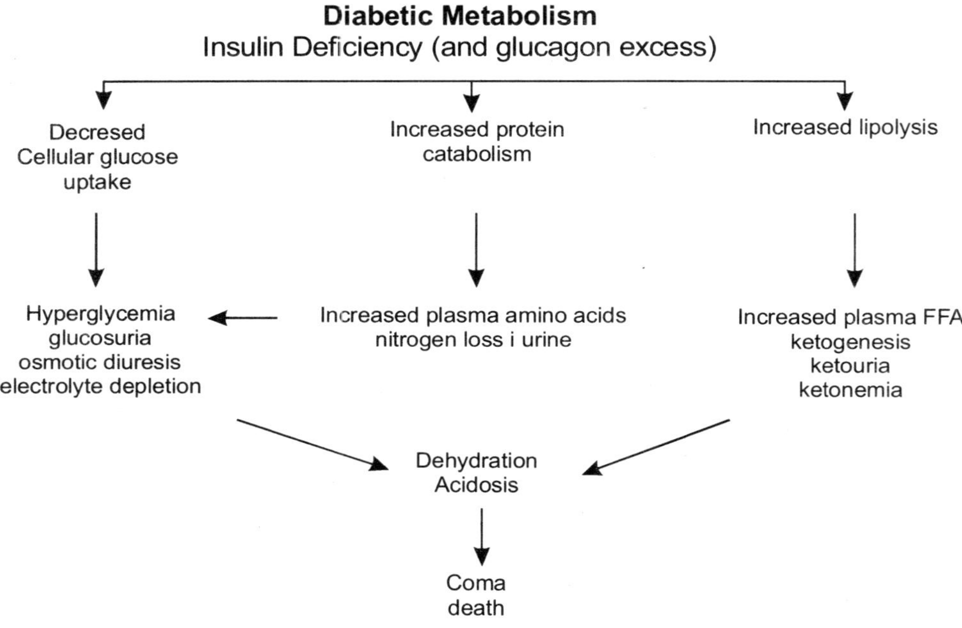

Fig. 8.51 : The Diabetic Metabolism Flow Chart

In a very recent article in Nature, Alessandro Pocai and coauthors presented convincing data that couples insulin-stimulation of hypothalamic K_{ATP} channels with neural control of hepatic gluconeogenesis)). Insulin stimulated opening of hypothalamic K_{ATP} channels results in vagal nerve signaling to the liver and inhibition of gluconeogenesis. This is part of the normal response to meals and following insulin release from the pancreatic ß-cells. Thus, signaling from the brain is one of the important control mechanisms which establish correct "between-meal" blood sugar levels. Hypothalamic insulin resistance and therefore loss of control over hepatic gluconeogenesis may well be one of the important factors involved in development of type 2 diabetes.

AIMASTHÆNIA – THE BLOOD DISEASES

In this chapter, we have dealt with the chemistry and functions of blood, now under this subhead we are going to discuss about the different kinds of diseases formed in the blood.

As this is a subject of biochemistry and not a aimatology or direct medicine so the discussion will be subsequently less, the diseases covered are the Anæmia, types, hemophilia, hæmolysis, Leukemia, Edema and sub acute CO poisoning , diseases of bone marrow.

Before understanding about the disease of the blood we have to understand something about the bone marrow.

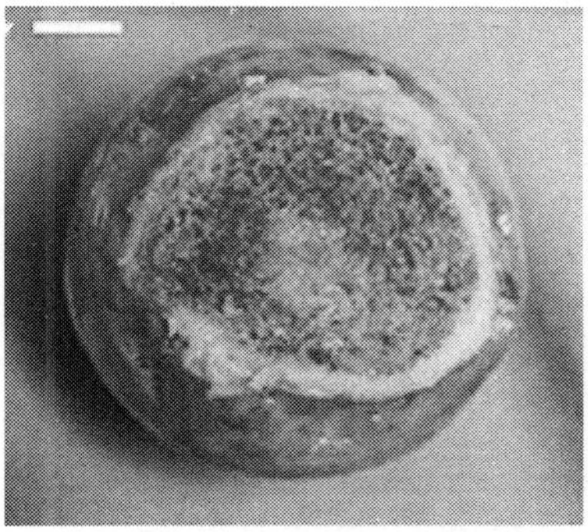

Fig. 8.52 : A femur with a cortex of cortical bone and medulla of trabecular bone showing its red bone marrow and a focus of yellow bone marrow

Bone Marrow

It is the flexible tissue found in the hollow interior of bones. In adults, marrow in large bones produces new blood cells. It constitutes 4% of total body weight, i.e. approximately 2.6 kg (5.7 lbs.) in adults.

There are two types of bone marrow: **red marrow** (consisting mainly of myeloid tissue) and **yellow marrow** (consisting mainly of fat cells). Red blood cells, platelets and most white blood cells arise in red marrow. Both types of bone marrow contain numerous blood vessels and capillaries.

At birth, all bone marrow is red. With age, more and more of it is converted to the yellow type. About half of adult bone marrow is red. yellow marrow is found mainly in the flat bones, such as the hip bone, breast bone, skull, ribs, vertebrae and shoulder blades, and in the cancellous ("spongy") material at the epiphyseal ends of the long bones such as the femur and humerus. Yellow marrow is found in the hollow interior of the middle portion of long bones.

In cases of severe blood periods, the body can convert yellow marrow back to red marrow to increase blood cell production.

Stroma

The *stroma of the bone marrow* is all tissue not directly involved in the primary function of hematopoiesis. The yellow bone marrow belongs here, and makes the majority of the bone marrow stroma, in addition to stromal cells located in the red bone marrow. Yellow bone marrow is found in the Medullary cavity.

Still, the stroma is indirectly involved in hematopoiesis, since it provides the *hematopoietic microenvironment* that facilitates hematopoiesis by the parenchymal cells. For instance, they generate colony stimulating factors, affecting hematopoiesis.

Cells that constitute the bone marrow stroma are:

- fibroblasts (reticular connective tissue)

- macrophages
- adipocytes
- osteoblasts
- osteoclasts
- endothelial cells forming the sinusoids

Macrophages contribute especially to red blood cell production. They deliver iron for hemoglobin-production. The blood vessels constitute a barrier, inhibiting immature blood cells from leaving the bone marrow. Only mature blood cells contain the membrane proteins required to attach to and pass the blood vessel endothelium. Hematopoietic stem cells may also cross the bone marrow barrier, and may thus be harvested from blood.

Stem Cells

The bone marrow stroma contain *mesenchymal stem cells* (also called *marrow stromal cells*). These cells are multipotent stem cells that can differentiate into a variety of cell types. Cell types that MSCs have been shown to differentiate into in vitro or in vivo include osteoblasts, chondrocytes, myocytes, adipocytes, and, as described lately, beta-pancreatic islets cells. They can also transdifferentiate into neuronal cells.

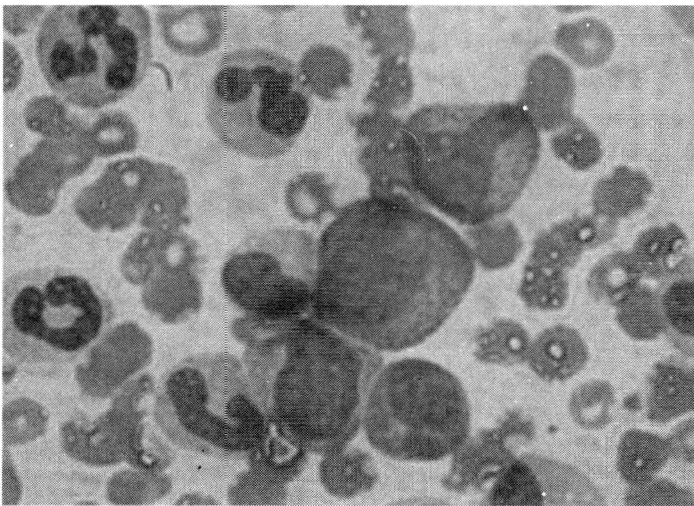

Fig. 8.53 : Hematopoietic precursor cells: promyelocyte in the center, two metamyelocytes next to it and band cells from a bone marrow aspirate

There is biologic compartmentalization in the bone marrow, in that certain cell types tend to aggregate in specific areas. For instance, erythrocytes, macrophages and their precursors tend to gather around blood vessels, while granulocytes gather at the borders of the bone marrow.

Types of Stem Cells

Bone marrow contains three types of stem cells:

- Hematopoietic stem cells give rise to the three classes of blood cells that are found in the circulation: white blood cells (leukocytes), red blood cells (erythrocytes), and platelets (thrombocytes).
- Mesenchymal stem cells are found arrayed around the central sinus in the bone marrow. They have the capability to differentiate into osteoblasts, chondrocytes, myocytes, and many other types of cells. They also function as "gatekeeper" cells of the bone marrow.
- Endothelial stem cells

Diseases Involving the Bone Marrow

The normal bone marrow architecture can be displaced by malignancies or infections such as tuberculosis, leading to a decrease in the production of blood cells and blood platelets. In addition, cancers of the hematologic progenitor cells in the bone marrow can arise; these are the leukemias.

To diagnose diseases involving the bone marrow, a bone marrow aspiration is sometimes performed. This typically involves using a hollow needle to acquire a sample of red bone marrow from the crest of the ilium under general or local anesthesia. The average number of cells in a leg bone is about 440,000,000,000 (440×10^9).

Exposure to radiation or chemotherapy will kill many of the rapidly dividing cells of the bone marrow and will therefore result in a depressed immune system. Many of the symptoms of radiation sickness are due to damage to the bone marrow cells.

Examination

Bone marrow examination is the pathologic analysis of samples of bone marrow obtained by *bone marrow biopsy* and *bone marrow aspiration*. Bone marrow examination is used in the diagnosis of a number of conditions, including leukemia, multiple myeloma, anemia, and pancytopenia. The bone marrow produces the cellular elements of the blood, including platelets, red blood cells and white blood cells. While much information can be gleaned by testing the blood itself (drawn from a vein by phlebotomy), it is sometimes necessary to examine the source of the blood cells in the bone marrow to obtain more information on hematopoiesis; this is the role of bone marrow aspiration and biopsy.

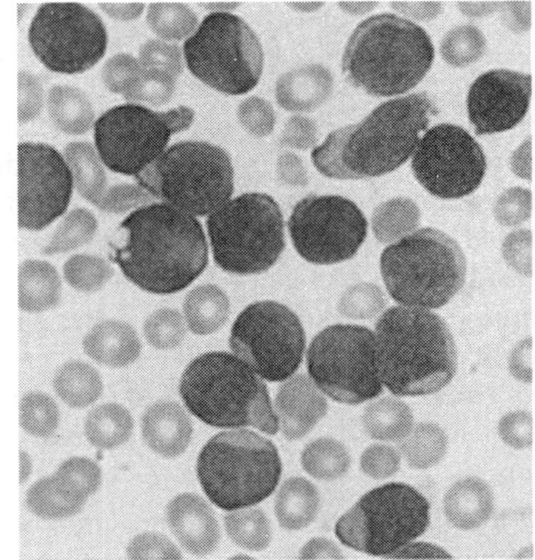

Fig. 8.54 : A Wright's stained bone marrow aspirate smear from a patient with leukemia

Anaemia or Anæmia

From Ancient Greek ναιμια *anaimia*, meaning "lack of blood") is a decrease in normal number of red blood cells (RBCs) or less than the normal quantity of hemoglobin in the blood. However, it can include decreased oxygen-binding ability of each hemoglobin molecule due to deformity or lack in numerical development as in some other types of hemoglobin deficiency.

Since hemoglobin (found inside RBCs) normally carries oxygen from the lungs to the tissues, anemia leads to hypoxia (lack of oxygen) in organs. Since all human cells depend on oxygen for survival, varying degrees of anemia can have a wide range of clinical consequences.

The three main classes of anemia include excessive blood loss (acutely such as a hemorrhage or chronically through low-volume loss), excessive blood cell destruction (hemolysis) or deficient red blood cell production (ineffective hematopoiesis).

Anemia is the most common disorder of the blood. There are several kinds of anemia, produced by a variety of underlying causes. Anemia can be classified in a variety of ways, based on the morphology of RBCs, underlying etiologic mechanisms, and discernible clinical spectra, to mention a few.

There are two major approaches: the "kinetic" approach which involves evaluating production, destruction and loss, and the "morphologic" approach which groups anemia by red blood cell size. The morphologic approach uses a quickly available and cheap lab test as its starting point (the MCV). On the other hand, focusing early on the question of production may allow the clinician more rapidly to expose cases where multiple causes of anemia coexist.

Signs and Symptoms

Anemia goes undetected in many people, and symptoms can be minor or vague. The signs and symptoms can be related to the anemia itself, or the underlying cause.

Most commonly, people with anemia report non-specific symptoms of a feeling of weakness, or fatigue, general malaise and sometimes poor concentration. They may also report shortness of breath, dyspnea, on exertion. In very severe anemia, the body may compensate for the lack of oxygen carrying capability of the blood by increasing cardiac output. The patient may have symptoms related to this, such as palpitations, angina (if preexisting heart disease is present), intermittent claudication of the legs, and symptoms of heart failure.On examination, the signs exhibited may include pallor (pale skin, mucosal linings and nail beds) but this is not a reliable sign. There may be signs of specific causes of anemia, e.g., koilonychia (in iron deficiency), jaundice (when anemia results from abnormal break down of red blood cells—in hemolytic anemia), bone deformities (found in thalassaemia major) or leg ulcers (seen in sickle cell disease).In severe anemia, there may be signs of a hyperdynamic circulation: a fast heart rate (tachycardia), flow murmurs, and cardiac enlargement. There may be signs of heart failure.

Pica, the consumption of non-food based items such as dirt, paper, wax, grass, ice, and hair, may be a symptom of iron deficiency, although it occurs often in those who have normal levels of hemoglobin.

Chronic anemia may result in behavioral disturbances in children as a direct result of impaired neurological development in infants, and reduced scholastic performance in children of school age.

Restless legs syndrome is more common in those with iron deficiency anemia.

Less common symptoms may include swelling of the legs or arms, chronic heartburn, vague bruises, vomiting, increased sweating, and blood in stool.

Diagnosis

Generally, clinicians request complete blood counts in the first batch of blood tests in the diagnosis of an anemia. Apart from reporting the number of red blood cells and the hemoglobin level, the automatic counters also measure the size of the red blood cells by flow cytometry, which is an important tool in distinguishing between the causes of anemia. Examination of a stained blood smear using a microscope can also be helpful, and is sometimes a necessity in regions of the world where automated analysis is less accessible.

In modern counters, four parameters (RBC count, hemoglobin concentration, MCV and RDW) are measured, allowing others (hematocrit, MCH and MCHC) to be calculated, and compared to values adjusted for age and sex. Some counters estimate hematocrit from direct measurements.

WHO's Hemoglobin thresholds used to define anemia (1 g/dL = 0.6206 mmol/L)

Age or gender group	Hb threshold (g/dl)	Hb threshold (mmol/l)
Children (0.5-5.0 yrs)	11.0	6.8
Children (5-12 yrs)	11.5	7.1
Children (12-15 yrs)	12.0	7.4
Women, non-pregnant (>15yrs)	12.0	7.4
Women, pregnant	11.0	6.8
Men (>15yrs)	13.0	8.1

Reticulocyte counts, and the "kinetic" approach to anemia, have become more common than in the past in the large medical centers of the United States and some other wealthy nations, in part because some automatic counters now have the capacity to include reticulocyte counts. A reticulocyte count is a quantitative measure of the bone marrow's production of new red blood cells. The reticulocyte production index is a calculation of the ratio between the level of anemia and the extent to which the reticulocyte count has risen in response. If the degree of anemia is significant, even a "normal" reticulocyte count actually may reflect an inadequate response.

If an automated count is not available, a reticulocyte count can be done manually following special staining of the blood film. In manual examination, activity of the bone marrow can also be gauged qualitatively by subtle changes in the numbers and the morphology of young RBCs by examination under a microscope. Newly formed RBCs are usually slightly larger than older RBCs and show polychromasia. Even where the source of blood loss is obvious, evaluation of erythropoiesis can help assess whether the bone marrow will be able to compensate for the loss, and at what rate.

When the cause is not obvious, clinicians use other tests: ESR, ferritin, serum iron, transferrin, RBC folate level, serum vitamin B_{12}, hemoglobin electrophoresis, renal function tests (e.g. serum creatinine). When the diagnosis remains difficult, a bone marrow examination allows direct examination of the precursors to red cells.

Classification

The "kinetic" approach to anemia yields what many argue is the most clinically relevant classification of anemia. This classification depends on evaluation of several hematological parameters, particularly the blood reticulocyte (precursor of mature RBCs) count. This then yields the classification of defects by decreased RBC production versus increased RBC destruction and/or loss. Clinical signs of loss or destruction include abnormal peripheral blood smear with signs of hemolysis; elevated LDH suggesting cell destruction; or clinical signs of bleeding, such as guiaic-positive stool, radiographic findings, or frank bleeding.

The following is a simplified schematic of this approach:

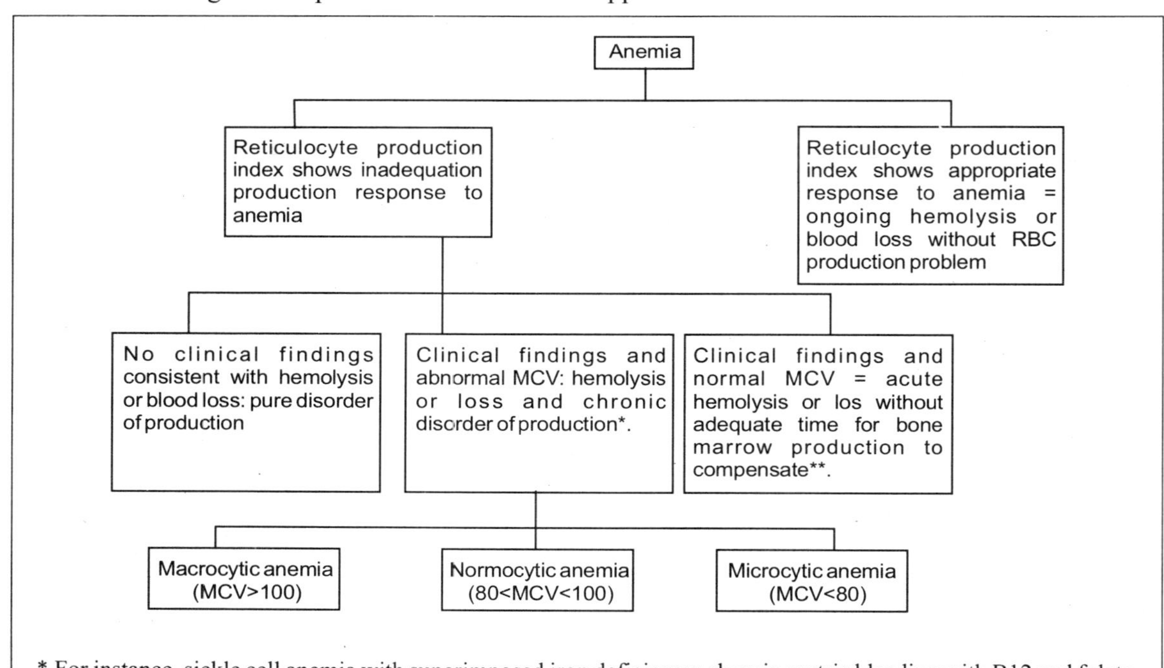

* For instance, sickle cell anemia with superimposed iron deficiency; chronic gastric bleeding with B12 and folate deficiency; and other instances of anemia with more than one cause.

**Confirm by repeating reticulocyte count: ongoing combination of low reticulocyte production index, normal MCV and hemolysis or loss may be seen in bone marrow failure or anemia of chronic disease, with superimposed or related hemolysis or blood loss.

In the morphological approach, anemia is classified by the size of red blood cells; this is either done automatically or on microscopic examination of a peripheral blood smear. The size is reflected in the *mean corpuscular volume* (MCV). If the cells are smaller than normal (under 80 fl), the anemia is said to be *microcytic*; if they are normal size (80-100 fl), *normocytic*; and if they are larger than normal (over 100 fl), the anemia is classified as *macrocytic*. This scheme quickly exposes some of the most common causes of anemia; for instance, a microcytic anemia is often the result of iron deficiency. In clinical workup, the MCV will be one of the first pieces of information available; so even among clinicians who consider the "kinetic" approach more useful philosophically, morphology will remain an important element of classification and diagnosis.

Here is a schematic representation of how to consider anemia with MCV as the starting point:

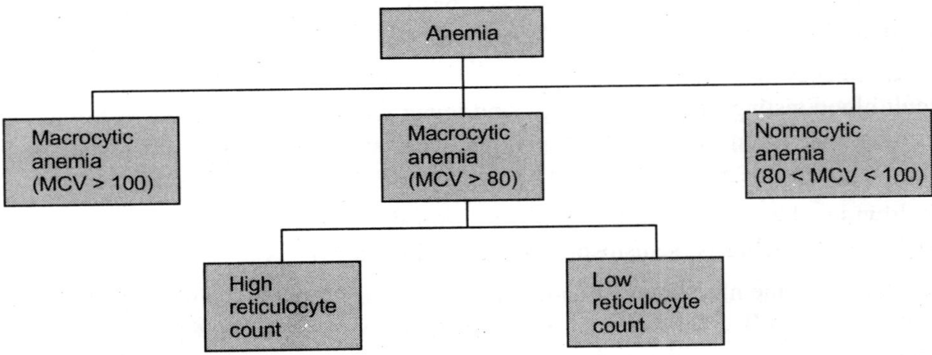

Other characteristics visible on the peripheral smear may provide valuable clues about a more specific diagnosis; for example, abnormal white blood cells may point to a cause in the bone marrow.

Microcytic anemia is primarily a result of hemoglobin synthesis failure/insufficiency, which could be caused by several etiologies:

- Heme synthesis defect
  - Iron deficiency anemia
  - Anemia of chronic disease (more commonly presenting as normocytic anemia)
- Globin synthesis defect
  - alpha-, and beta-thalassemia
  - HbE syndrome
  - HbC syndrome
  - and various other unstable hemoglobin diseases
- Sideroblastic defect
  - Hereditary sideroblastic anemia
  - Acquired sideroblastic anemia, including lead toxicity
  - Reversible sideroblastic anemia

Iron deficiency anemia is the most common type of anemia overall and it has many causes. RBCs often appear hypochromic (paler than usual) and microcytic (smaller than usual) when viewed with a microscope.

- Iron deficiency anemia is caused by insufficient dietary intake or absorption of iron to replace losses from menstruation or losses due to diseases. Iron is an essential part of hemoglobin, and low iron levels result in decreased incorporation of hemoglobin into red blood cells. In the United States, 20% of all women of childbearing age have iron deficiency anemia, compared with only 2% of adult men. The principal cause of iron deficiency anemia in premenopausal women is blood lost during menses. Studies have shown that iron deficiency without anemia causes poor school performance and lower IQ in teenage girls. Iron deficiency is the most prevalent deficiency state on a worldwide basis. Iron deficiency is sometimes the cause of abnormal fissuring of the angular (corner) sections of the lips (angular stomatitis).
- Iron deficiency anemia can also be due to bleeding lesions of the gastrointestinal tract. Faecal occult blood testing, upper endoscopy and lower endoscopy should be performed to identify bleeding lesions. In men and post-menopausal women the chances are higher that bleeding from the gastrointestinal tract could be due to colon polyp or colorectal cancer.
- Worldwide, the most common cause of iron deficiency anemia is parasitic infestation (hookworm, amebiasis, schistosomiasis and whipworm).
- Megaloblastic anemia, the most common cause of macrocytic anemia, is due to a deficiency of either vitamin B_{12}, folic acid (or both). Deficiency in folate and/or vitamin B_{12} can be due either to inadequate intake or insufficient absorption. Folate deficiency normally does not produce neurological symptoms, while B_{12} deficiency does.
  - Pernicious anemia is caused by a lack of intrinsic factor. Intrinsic factor is required to absorb vitamin B_{12} from food. A lack of intrinsic factor may arise from an autoimmune condition targeting the parietal cells (atrophic gastritis) that produce intrinsic factor or against intrinsic factor itself. These lead to poor absorption of vitamin B_{12}.
  - Macrocytic anemia can also be caused by removal of the functional portion of the stomach, such as during gastric bypass surgery, leading to reduced vit B_{12}/folate absorption. Therefore one must always be aware of anemia following this procedure.
- Hypothyroidism
- Alcoholism commonly causes a macrocytosis, although not specifically anemia. Other types of Liver Disease can also cause macrocytosis.
- Methotrexate, zidovudine, and other drugs that inhibit DNA replication.

Macrocytic anemia can be further divided into "megaloblastic anemia" or "non-megaloblastic macrocytic anemia". The cause of megaloblastic anemia is primarily a failure of DNA synthesis with preserved RNA synthesis, which result in restricted cell division of the progenitor cells. The megaloblastic anemias often present with neutrophil hypersegmentation (6-10 lobes). The non-megaloblastic macrocytic anemias have different etiologies (i.e. there is unimpaired DNA globin synthesis) which occur, for example in alcoholism.

In addition to the non-specific symptoms of anemia, specific features of vitamin B_{12} deficiency include peripheral neuropathy and subacute combined degeneration of the cord with resulting balance difficulties from posterior column spinal cord pathology. Other features may include a smooth, red tongue and glossitis.

The treatment for vitamin B_{12}-deficient anemia was first devised by William Murphy who bled dogs to make them anemic and then fed them various substances to see what (if anything) would make them healthy again. He discovered that ingesting large amounts of liver seemed to cure the disease. George Minot and George Whipple then set about to chemically isolate the curative substance and ultimately were able to isolate the vitamin B_{12} from the liver. All three shared the 1934 Nobel Prize in Medicine.

Normocytic anemia occurs when the overall hemoglobin levels are always decreased, but the red blood cell size (Mean corpuscular volume) remains normal. Causes include:

- Acute blood loss
- Anemia of chronic disease
- Aplastic anemia (bone marrow failure)
- Hemolytic anemia

When two causes of anemia act simultaneously, e.g., macrocytic hypochromic, due to hookworm infestation leading to deficiency of both iron and vitamin B_{12} or folic acid or following a blood transfusion more than one abnormality of red cell indices may be seen. Evidence for multiple causes appears with an elevated RBC distribution width (RDW), which suggests a wider-than-normal range of red cell sizes.

Heinz bodies form in the cytoplasm of RBCs and appear like small dark dots under the microscope. There are many causes of Heinz body anemia, and some forms can be drug induced. It is triggered in cats by eating onions or acetaminophen (Tylenol). It can be triggered in dogs by ingesting onions or zinc, and in horses by ingesting dry red maple leaves.

Cause

- Anemia of prematurity occurs in premature infants at 2 to 6 weeks of age and results from diminished erythropoietin response to declining hematocrit levels.
- Aplastic anemia is a condition generally unresponsive to anti-anemia therapies where bone marrow fails to produce enough red blood cells.
- Fanconi anemia is a hereditary disorder or defect featuring aplastic anemia and various other abnormalities.
- Hemolytic anemia causes a separate constellation of symptoms (also featuring jaundice and elevated LDH levels) with numerous potential causes. It can be autoimmune, immune, hereditary or mechanical (e.g. heart surgery). It can result (because of cell fragmentation) in a microcytic

anemia, a normochromic anemia, or (because of premature release of immature red blood cells from the bone marrow), a macrocytic anemia.

- Hereditary spherocytosis is a hereditary defect that results in defects in the RBC cell membrane, causing the erythrocytes to be sequestered and destroyed by the spleen. This leads to a decrease in the number of circulating RBCs and, hence, anemia.
- Sickle-cell anemia, a hereditary disorder, is due to homozygous hemoglobin S genes.
- Warm autoimmune hemolytic anemia is an anemia caused by autoimmune attack against red blood cells, primarily by IgG.
- Cold agglutinin hemolytic anemia is primarily mediated by IgM.
- Pernicious anemia is a form of megaloblastic anemia due to vitamin B_{12} deficiency dependent on impaired absorption of vitamin B_{12}.
- Myelophthisic anemia or Myelophthisis is a severe type of anemia resulting from the replacement of bone marrow by other materials, such as malignant tumors or granulomas.
- Anemia of Pregnancy is anemia that is induced by blood volume expansion experienced in pregnancy.

Anemia diminishes the capability of affected individuals to perform physical activities, a result of one's muscles being forced to depend on anaerobic metabolism. The lack of iron associated with anemia can cause many complications, including hypoxemia, brittle or rigid fingernails, cold intolerance, and possible behavioral disturbances in children. Hypoxemia resulting from anemia can worsen the cardio-pulmonary status of patients with pre-existing chronic pulmonary disease. Cold intolerance occurs in one in five patients with iron deficiency anemia, and becomes visible through numbness and tingling.

Anemia affects 20 percent of all women of childbearing age in the United States. Due to the subtlety of the symptoms, affected women are often unaware that they have this disorder, as they attribute the symptoms to the stresses of their daily lives. Possible problems for the fetus include increased risk of growth retardation, prematurity, intrauterine death, rupture of the amnion and infection.

An expectant woman should be especially aware of the symptoms of anemia, as an adult female loses an average of two milligramsof iron daily. Therefore, she must intake a similar quantity of iron in order to compensate for this loss. Additionally, a woman loses approximately 500 milligramsof iron with each pregnancy, compared with a loss of 4–100 milligramsof iron with each period. Possible consequences include cardiovascular symptoms, reduced physical and mental performance, reduced immune function, fatigue, reduced peripartal blood reserves and increased need for blood transfusion in the postpartum period. This may severely affect the health of the child.

Thalassemia

Thalassemia (British spelling, "thalassaemia") is an inherited autosomal recessive blood disease. In thalassemia, the genetic defect results in reduced rate of synthesis of one of the globin chains that make up hemoglobin. Reduced synthesis of one of the globin chains can cause the formation of abnormal hemoglobin molecules, thus causing anemia, the characteristic presenting symptom of the thalassemias.

Thalassemia is a quantitative problem of too few globins synthesized, whereas sickle-cell anemia (a hemoglobinopathy) is a qualitative problem of synthesis of an incorrectly functioning globin. Thalassemias usually result in underproduction of normal globin proteins, often through mutations in regulatory genes. Hemoglobinopathies imply structural abnormalities in the globin proteins themselves.[1] The two conditions may overlap, however, since some conditions which cause abnormalities in globin proteins (hemoglobinopathy) also affect their production (thalassemia). Thus, some thalassemias are hemoglobinopathies, but most are not. Either or both of these conditions may cause anemia.

The disease is particularly prevalent among Mediterranean people, and this geographical association was responsible for its naming: *Thalassa* is Greek for the sea, *Haema* is Greek for blood. In Europe, the highest concentrations of the disease are found in Greece and in parts of Italy, in particular, Southern Italy and the lower Po valley. The major Mediterranean islands (except the Balearics) such as Sicily, Sardinia, Malta, Corsica, Cyprus and Crete are heavily affected in particular. Other Mediterranean people, as well as those in the vicinity of the Mediterranean, also have high rates of thalassemia, including people from the middle east and North Africa. Far from the Mediterranean, South Asians are also affected, with the world's highest concentration of carriers (18% of the population) being in the Maldives.

Generally, thalassemias are prevalent in populations that evolved in humid climates where malaria was endemic. It affects all races, as thalassemias protected these people from malaria due to the blood cells' easy degradation.

Thalassemias are particularly associated with people of Mediterranean origin, Arabs, and Asians. The Maldives has the highest incidence of Thalassemia in the world with a carrier rate of 18% of the population. The estimated prevalence is 16% in people from Cyprus, 1% in Thailand, and 3-8% in populations from Bangladesh, China, India, Malaysia and Pakistan. There are also prevalences in descendants of people from Latin America and Mediterranean countries (e.g. Greece, Italy, Portugal, Spain, and others). A very low prevalence has been reported from people in Northern Europe (0.1%) and Africa (0.9%), with those in North Africa having the highest prevalence.

Ancient Egyptians also suffered from thalassemia, with as many as 40% of studied pre-dynastic and dynastic mummies found to carry the genetic defect. Today, it is particularly common in populations of indigenous ethnic minorities of Upper Egypt such as the Beja, Hadendoa, Saiddi and also peoples of the Nile Delta, Red Sea Hill Region and especially amongst the Siwans.

Normal haemoglobin is composed of two chains each of α and β globin. Thalassemia patients produce a *deficiency* of either α or β globin, unlike sickle-cell disease which produces a specific mutant form of β globin.

The **thalassemias** are classified according to which chain of the hemoglobin molecule is affected. In α thalassemias, production of the a globin chain is affected, while in β thalassemia production of the β globin chain is affected.

β globin chains are encoded by a single gene on chromosome 11; α globin chains are encoded by two closely linked genes on chromosome 16. Thus in a normal person with two copies of each chromosome, there are **two** loci encoding the β chain, and **four** loci encoding the α chain. Deletion of one of the α loci has a high prevalence in people of African or Asian descent, making them more likely to develop α thalassemias. β thalassemias are common in Africans, but also in Greeks and Italians.

Alpha (α) thalassemias

The α thalassemias involve the genes HBA1and HBA2, inherited in a Mendelian recessive fashion. It is also connected to the deletion of the 16p chromosome. α thalassemias result in decreased alpha-globin production, therefore fewer alpha-globin chains are produced, resulting in an excess of β chains in adults and excess ã chains in newborns. The excess β chains form unstable tetramers (called Hemoglobin H or HbH of 4 beta chains) which have abnormal oxygen dissociation curves.

Beta (β) thalassemias

Beta thalassemias are due to mutations in the HBB gene on chromosome 11, also inherited in an autosomal-recessive fashion. The severity of the disease depends on the nature of the mutation. Mutations are characterized as (β^o or β thalassemia major) if they prevent any formation of β chains (which is the most severe form of beta Thalassemia); they are characterized as (β^+ or β thalassemia intermedia) if they allow some β chain formation to occur. In either case there is a relative excess of α chains, but these do not form tetramers: rather, they bind to the red blood cell membranes, producing membrane damage, and at high concentrations they form toxic aggregates.

Delta (δ) thalassemia

As well as alpha and beta chains being present in hemoglobin about 3% of adult hemoglobin is made of alpha and delta chains. Just as with beta thalassemia, mutations can occur which affect the ability of this gene to produce delta chains.

In Combination with Other Hemoglobinopathies

Thalassemia can co-exist with other hemoglobinopathies. The most common of these are:

- hemoglobin E/thalassemia: common in Cambodia, Thailand, and parts of India; clinically similar to β thalassemia major or thalassemia intermedia.
- hemoglobin S/thalassemia, common in African and Mediterranean populations; clinically similar to sickle cell anemia, with the additional feature of splenomegaly
- hemoglobin C/thalassemia: common in Mediterranean and African populations, hemoglobin C/β^o thalassemia causes a moderately severe hemolytic anemia with splenomegaly; hemoglobin C/β^+ thalassemia produces a milder disease.

α and β thalassemia are often inherited in an autosomal recessive fashion although this is not always the case. Cases of dominantly inherited α and β thalassemias have been reported, the first of which was in an Irish family who had a two deletions of 4 and 11 bp in exon 3 interrupted by an insertion of 5 bp in the β-globin gene. For the autosomal recessive forms of the disease both parents must be carriers in order for a child to be affected. If both parents carry a hemoglobinopathy trait, there is a 25% chance with each pregnancy for an affected child. Genetic counseling and genetic testing is recommended for families that carry a thalassemia trait.

There are an estimated 60-80 million people in the world who carry the beta thalassemia trait alone. This is a very rough estimate and the actual number of thalassemia Major patients is unknown due to the prevalence of thalassemia in less developed countries in the Middle East and Asia where genetic screening resources are limited. Countries such as India, Pakistan and Iran are seeing a large increase of thalassemia patients due to lack of genetic counseling and screening. There is growing concern that thalassemia may become a very serious problem in the next 50 years, one that will burden the world's blood bank supplies and the health system in general. There are an estimated 1000 people living with Thalassemia Major in the United States and an unknown number of carriers. Because of the prevalence of the disease in countries with little knowledge of thalassemia, access to proper treatment and diagnosis can be difficult.

As with other genetically acquired disorders, genetic counseling is recommended.

Thalassemia Major

This a very serious blood disease that begins in early childhood. Children with thalassemia major are normal at birth but become anemic between the age of three months and eighteen months. They become pale, do not sleep well, do not want to eat, and may vomit frequently after feedings. If thalassemia major goes untreated, children usually die between one and eight years of age.

Haemophilia

Haemophilia is a group of hereditary genetic disorders that impair the body's ability to control blood clotting or coagulation, which is used to stop bleeding when a blood vessel is broken. Haemophilia A (clotting factor VIII deficiency) is the most common form of the disorder, occurring at about 1 in 5,000–10,000 male births. Haemophilia B (factor IX deficiency) occurs at about 1 in about 20,000–34,000 male births.

Similarly to most recessive sex-linked, X chromosome disorders, only males exhibit symptoms. This is due to the fact that females have two X chromosomes while males have only one, lacking a 'back up' copy for the defective gene the defective gene becomes manifest more easily in males. Because females have two X chromosomes and because haemophilia is rare, the chance of a female having two defective copies of the gene is very low, thus females are almost exclusively asymptomatic carriers of the disorder. Female carriers may inherit the defective gene from either their mother, father, or it may be a new mutation. Only under rare circumstances do females actually have haemophilia.

Haemophilia lowers blood plasma clotting factor levels of the coagulation factors needed for a normal clotting process. Thus when a blood vessel is injured, a temporary scab does form, but the missing coagulation factors prevent fibrin formation, which is necessary to maintain the blood clot. Thus a haemophiliac does not bleed more intensely than a normal person, but can bleed for a much

plasma. The concentration of potassium inside red blood cells is much higher than in the plasma and so an elevated potassium is usually found in biochemistry tests of hemolysed blood. If as little as 0.5% of the red blood cells are hemolysed the serum will have a visually obvious pinkish colour, due to hemoglobin.

Most causes of *in vitro* hemolysis are related to specimen collection. Difficult collections, unsecure line connections, contamination, and incorrect needle size, as well as improper tube mixing and incorrectly filled tubes are all frequent causes of hemolysis. Excessive suction can cause the red blood cells to be literally smashed on their way through the hypodermic needle owing to turbulence and physical forces. Such hemolysis is more likely to occur when a patient's veins are difficult to find or when they collapse when blood is removed by a syringe or a modern vacuum tube. Experience and proper technique are key for any phlebotomist or nurse to prevent hemolysis. *In vitro* hemolysis can also occur in a blood sample owing to prolonged storage or storage in incorrect conditions (*i.e.* too hot, too cold).

Hemolysis due to Mechanical Blood Processing during Surgery

In some surgical procedures (especially some heart operations) where substantial blood loss is expected, machinery is used for intraoperative blood salvage. A centrifuge process takes blood from the patient, washes the red blood cells with normal saline, and returns them to the patient's blood circulation. Hemolysis may occur if the centrifuge rotates too quickly (generally greater than 500 rpm)—essentially this is hemolysis occurring outside of the body. Unfortunately, increased hemolysis occurs with massive amounts of sudden blood loss, because the process of returning a patient's cells must be done at a correspondingly higher speed to prevent hypotension, pH imbalance, and a number of other hemodynamic and blood level factors.

Hemolytic patterns of the various Gram positive cocci; Streptococci are differentiated by hemolysis of red blood cells on blood agar (BA) plates.

- Alpha hemolysis is shown by a greenish halo around the colony and is the result of hemoglobin oxidation to methemoglobin in red blood cells.
- Beta hemolysis is shown by a clear halo around the colony and is produced by complete hemolysis of the red blood cells.
- Gamma hemolysis is shown as no hemolysis or discoloration of the blood.

Leukæmia—The Blood Cancer

Leukaemia (Leukemia) is a cancer of the blood or bone marrow and is characterized by an abnormal proliferation (production by multiplication) of blood cells, usually white blood cells (leukocytes). Leukemia is a broad term covering a spectrum of diseases. In turn, it is part of the even broader group of diseases called hematological neoplasms.

Classification

Leukemia is clinically and pathologically subdivided into a variety of large groups. The first division is between its *acute* and *chronic* forms:

Severe complications are much more common in severe and moderate haemophiliacs. Complications may be both directly from the disease or from its treatment:

- **Deep internal bleeding**, e.g. deep-muscle bleeding, leading to swelling, numbness or pain of a limb.
- **Joint damage**, potentially with severe pain and even destruction of the joint and development of arthritis.
- **Transfusion transmitted infection** from blood transfusions that are given as treatment.
- **Adverse reactions** to clotting factor treatment, including the development of an immune inhibitor which renders factor replacement less effective.
- **Intracranial hemorrhage**, is a serious medical emergency cause by the buildup of pressure inside the skull. It can cause disorientation, nausea, loss of consciousness, brain damage, and death.

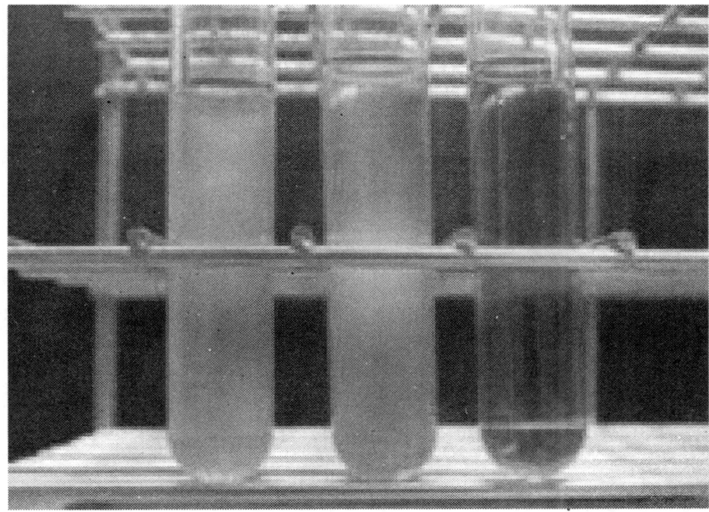

Fig. 8.55 : The Haemolysis in Blood

Subsection 8.6q Haemolysis (or **hemolysis**)—from the Greek **Hemo-**, Greek aima meaning blood, **-lysis**, meaning to break open—is the breaking open of red blood cells and the release of hemoglobin into the surrounding fluid (plasma, *in vivo*).

In vivo Hemolysis

In vivo (inside the body) hemolysis, which can be caused by a large number of conditions, can lead to anemia. Anemias caused by *in vivo* hemolysis are collectively called hemolytic anemias.

In vitro Hemolysis

In vitro (outside the body) hemolysis can be an important unwanted effect in medical tests and can cause inaccurate results, because the contents of hemolysed red blood cells are included with the plasma. The concentration of potassium inside red blood cells is much higher than in the plasma and so an elevated potassium is usually found in biochemistry tests of hemolysed blood. If as little as 0.5% of the red blood cells are hemolysed the serum will have a visually obvious pinkish colour, due to hemoglobin.

Most causes of *in vitro* hemolysis are related to specimen collection. Difficult collections, unsecure line connections, contamination, and incorrect needle size, as well as improper tube mixing and incorrectly filled tubes are all frequent causes of hemolysis. Excessive suction can cause the red blood cells to be literally smashed on their way through the hypodermic needle owing to turbulence and physical forces. Such hemolysis is more likely to occur when a patient's veins are difficult to find or when they collapse when blood is removed by a syringe or a modern vacuum tube. Experience and proper technique are key for any phlebotomist or nurse to prevent hemolysis. *In vitro* hemolysis can also occur in a blood sample owing to prolonged storage or storage in incorrect conditions (*i.e.* too hot, too cold).

Hemolysis due to Mechanical Blood Processing during Surgery

In some surgical procedures (especially some heart operations) where substantial blood loss is expected, machinery is used for intraoperative blood salvage. A centrifuge process takes blood from the patient, washes the red blood cells with normal saline, and returns them to the patient's blood circulation. Hemolysis may occur if the centrifuge rotates too quickly (generally greater than 500 rpm)—essentially this is hemolysis occurring outside of the body. Unfortunately, increased hemolysis occurs with massive amounts of sudden blood loss, because the process of returning a patient's cells must be done at a correspondingly higher speed to prevent hypotension, pH imbalance, and a number of other hemodynamic and blood level factors.

Hemolytic patterns of the various Gram positive cocci; Streptococci are differentiated by hemolysis of red blood cells on blood agar (BA) plates.

- Alpha hemolysis is shown by a greenish halo around the colony and is the result of hemoglobin oxidation to methemoglobin in red blood cells.
- Beta hemolysis is shown by a clear halo around the colony and is produced by complete hemolysis of the red blood cells.
- Gamma hemolysis is shown as no hemolysis or discoloration of the blood.

Leukæmia—The Blood Cancer

Leukaemia (Leukemia) is a cancer of the blood or bone marrow and is characterized by an abnormal proliferation (production by multiplication) of blood cells, usually white blood cells (leukocytes). Leukemia is a broad term covering a spectrum of diseases. In turn, it is part of the even broader group of diseases called hematological neoplasms.

Classification

Leukemia is clinically and pathologically subdivided into a variety of large groups. The first division is between its *acute* and *chronic* forms:

- *Acute leukemia* is characterized by the rapid increase of immature blood cells. This crowding makes the bone marrow unable to produce healthy blood cells. Immediate treatment is required in acute leukemia due to the rapid progression and accumulation of the malignant cells, which then spill over into the bloodstream and spread to other organs of the body. Acute forms of leukemia are the most common forms of leukemia in children.
- *Chronic leukemia* is distinguished by the excessive build up of relatively mature, but still abnormal, white blood cells. Typically taking months or years to progress, the cells are produced at a much higher rate than normal cells, resulting in many abnormal white blood cells in the blood. Whereas acute leukemia must be treated immediately, chronic forms are sometimes monitored for some time before treatment to ensure maximum effectiveness of therapy. Chronic leukemia mostly occurs in older people, but can theoretically occur in any age group.

Additionally, the diseases are subdivided according to which kind of blood cell is affected. This split divides leukemias into lymphoblastic or *lymphocytic leukemias* and myeloid or *myelogenous leukemias*:

- In lymphoblastic or *lymphocytic leukemias*, the cancerous change takes place in a type of marrow cell that normally goes on to form lymphocytes, which are infection-fighting immune system cells. Most lymphocytic leukemias involve a specific subtype of lymphocyte, the B cell.
- In myeloid or *myelogenous leukemias*, the cancerous change takes place in a type of marrow cell that normally goes on to form red blood cells, some other types of white cells, and platelets.

Combining these two classifications provides a total of four main categories:

Four Major Kinds of Leukemia

Cell type	Acute	Chronic
Lymphocytic leukemia (or "lymphoblastic")	Acute lymphoblastic leukemia (ALL)	Chronic lymphocytic leukemia (CLL)
Myelogenous leukemia (also "myeloid" or "nonlymphocytic")	Acute myelogenous leukemia (AML)	Chronic myelogenous leukemia (CML)

Within these main categories, there are typically several subcategories. Finally, hairy cell leukemia and T-cell prolymphocytic leukemia are usually considered to be outside of this classification scheme.

- *Acute lymphoblastic leukemia* (ALL) is the most common type of leukemia in young children. This disease also affects adults, especially those age 65 and older. Standard treatments involve chemotherapy and radiation. The survival rates vary by age: 85% in children and 50% in adults.[1] Subtypes include precursor B acute lymphoblastic leukemia, precursor T acute lymphoblastic leukemia, Burkitt's leukemia, and acute biphenotypic leukemia.
- *Chronic lymphocytic leukemia* (CLL) most often affects adults over the age of 55. It sometimes occurs in younger adults, but it almost never affects children. Two-thirds of affected people are men. The five-year survival rate is 75%. It is incurable, but there are many effective treatments. One subtype is B-cell prolymphocytic leukemia, a more aggressive disease.

- *Acute myelogenous leukemia* (AML) occurs more commonly in adults than in children, and more commonly in men than women. AML is treated with chemotherapy. The five-year survival rate is 40%. Subtypes of AML include acute promyelocytic leukemia, acute myeloblastic leukemia, and acute megakaryoblastic leukemia.
- *Chronic myelogenous leukemia* (CML) occurs mainly in adults. A very small number of children also develop this disease. Treatment is with imatinib (Gleevec) or other drugs. The five-year survival rate is 90%. One subtype is chronic monocytic leukemia.
- *Hairy cell leukemia* (HCL) is sometimes considered a subset of CLL, but does not fit neatly into this pattern. About 80% of affected people are adult men. There are no reported cases in young children. HCL is incurable, but easily treatable. Survival is 96% to 100% at ten years.
- T-cell prolymphocytic leukemia (T-PLL) is a very rare and aggressive leukemia affecting adults; somewhat more men than women are diagnosed with this disease. Despite its overall rarity, it is also the most common type of mature T cell leukemia; nearly all other leukemias involve B cells. It is difficult to treat, and the median survival is measured in months.
- Large granular lymphocytic leukemia may involve either T cell or NK cells; like hairy cell leukemia, which involves solely B cells, it is a rare and indolent (not aggressive) leukemia.

Signs and Symptoms

Damage to the bone marrow, by way of displacing the normal bone marrow cells with higher numbers of immature white blood cells, results in a lack of blood platelets, which are important in the blood clotting process. This means people with leukemia may easily become bruised, bleed excessively, or develop pinprick bleeds (petechiae).

White blood cells, which are involved in fighting pathogens, may be suppressed or dysfunctional. This could cause the patient's immune system to be unable to fight off a simple infection or to start attacking other body cells. Because leukemia prevents the immune system from working normally, some patients experience frequent infection, ranging from infected tonsils, sores in the mouth, or diarrhea to life-threatening pneumonia or opportunistic infections.

Finally, the red blood cell deficiency leads to anemia, which may cause dyspnea and pallor.

Some patients experience other symptoms. These symptoms might include feeling sick, such as having fevers, chills, night sweats and other flu-like symptoms, or feeling fatigued. Some patients experience nausea or a feeling of fullness due to an enlarged liver and spleen; this can result in unintentional weight loss. If the leukemic cells invade the central nervous system, then neurological symptoms (notably headaches) can occur.

All symptoms associated with leukemia can be attributed to other diseases. Consequently, leukemia is always diagnosed through medical tests.

The word *leukemia*, which means 'white blood', is derived from the disease's namesake high white blood cell counts that most leukemia patients have before treatment. The high number of white blood cells are apparent when a blood sample is viewed under a microscope. Frequently, these extra white blood cells are immature or dysfunctional. The excessive number of cells can also interfere with the level of other cells, causing a harmful imbalance in the blood count.

Some leukemia patients do not have high white blood cell counts visible during a regular blood count. This less-common condition is called *aleukemia*. The bone marrow still contains cancerous white blood cells which disrupt the normal production of blood cells. However, the leukemic cells are staying in the marrow instead of entering the bloodstream, where they would be visible in a blood test. For an aleukemic patient, the white blood cell counts in the bloodstream can be normal or low. Aleukemia can occur in any of the four major types of leukemia, and is particularly common in hairy cell leukemia.

Causes

No single known cause for all of the different types of leukemia exists. The different leukemias likely have different causes. Known causes include natural and artificial ionizing radiation, viruses such as Human T-lymphotropic virus, and some chemicals, notably benzene and alkylating chemotherapy agents for previous malignancies. Use of tobacco is associated with a small increase in the risk of developing acute myeloid leukemia in adults. A few cases of maternal-fetal transmission have been reported.

Leukemia, like other cancers, results from somatic mutations in the DNA which activate oncogenes or deactivate tumor suppressor genes, and disrupt the regulation of cell death, differentiation or division. These mutations may occur spontaneously or as a result of exposure to radiation or carcinogenic substances and are likely to be influenced by genetic factors. Cohort and case-control studies have linked exposure to petrochemicals, such as benzene, and hair dyes to the development of some forms of leukemia.

Viruses have also been linked to some forms of leukemia. For example, certain cases of ALL are associated with viral infections by either the human immunodeficiency virus or human T-lymphotropic virus (HTLV-1 and -2, causing adult T-cell leukemia/lymphoma). However, one report suggests exposure to certain germs may offer children limited protection against leukemia.

Some people have a genetic predisposition towards developing leukemia. This predisposition is demonstrated by family histories and twin studies. The affected people may have a single gene or multiple genes in common. In some cases, families tend to develop the same kind of leukemia as other members; in other families, affected people may develop different forms of leukemia or related blood cancers.

In addition to these genetic issues, people with chromosomal abnormalities or certain other genetic conditions have a greater risk of leukemia. For example, people with Down syndrome have a significantly increased risk of developing forms of acute leukemia, and Fanconi anemia is a risk factor for developing acute myeloid leukemia.

Whether non-ionizing radiation causes leukemia has been studied for several decades. The International Agency for Research on Cancer expert working group undertook a detailed review of all data on static and extremely low frequency electromagnetic energy, which occurs naturally and in association with the generation, transmission, and use of electrical power. They concluded that there is limited evidence that high levels of ELF magnetic (but not electric) fields might cause childhood leukemia. Exposure to significant ELF magnetic fields might result in twofold excess risk for leukemia for children exposed to these high levels of magnetic fields. However, the report also says that methodological weaknesses and biases in these studies have likely caused the risk to be overstated. No evidence for a relationship to leukemia or an other form of malignancy in adults has been demonstrated. Since exposure to such levels of ELFs is relatively uncommon, the World Health Organization concludes that ELF exposure, if later proven to be causative, would account for just 100 to 2400 cases worldwide each year, representing 0.2 to 4.95% of the total incidence for that year.

Until the cause or causes of leukemia are found, there is no way to prevent the disease. Even when the causes become known, they may not be readily controllable, such as naturally occurring background radiation, and therefore not especially helpful for prevention purposes.

Edema

Edmma (words from the Greek *edo,* means here and *aima* blood), formerly known as **dropsy** or **hydropsy**, is an abnormal accumulation of fluid beneath the skin or in one or more cavities of the body. Generally, the amount of interstitial fluid is determined by the balance of fluid homeostasis, and increased secretion of fluid into the interstitium or impaired removal of this fluid may cause edema.

Formation

Five factors can contribute to the formation of edema:

1. It may be facilitated by increased hydrostatic pressure or,
2. reduced oncotic pressure within blood vessels;
3. by increased blood vessel wall permeability as in inflammation;
4. by obstruction of fluid clearance via the lymphatic; or,
5. by changes in the water retaining properties of the tissues themselves. Raised hydrostatic pressure often reflects retention of water and sodium by the kidney.

Mechanism

Generation of interstitial fluid is regulated by the forces of the Starling equation. Hydrostatic pressure within blood vessels tends to cause water to filter out into the tissue. This leads to a difference in protein concentration between blood plasma and tissue. As a result the oncotic pressure of the higher level of protein in the plasma tends to suck water back into the blood vessels from the tissue. Starling's equation states that the rate of leakage of fluid is determined by the difference between the two forces and also by the permeability of the vessel wall to water, which determines the rate of flow for a given force imbalance. Most water leakage occurs in capillaries or post capillary venules, which have a semi-permeable membrane wall that allows water to pass more freely than protein. (The protein is said to be reflected and the efficiency of reflection is given by a reflection constant of up to 1.) If the gaps between the cells of the vessel wall open up then permeability to water is increased first, but as the gaps increase in size permeability to protein also increases with a fall in reflection coefficient.

Changes in the variables in Starling's equation can contribute to the formation of edema either by an increase in hydrostatic pressure within the blood vessel, a decrease in the oncotic pressure within the blood vessel or an increase in vessel wall permeability. The latter has two effects. It allows water to flow more freely and it reduces the oncotic pressure difference by allowing protein to leave the vessel more easily.

Generalized Edema

A rise in hydrostatic pressure occurs in cardiac failure. A fall in oncotic pressure occurs in nephrotic syndrome and liver failure. It is commonly taught that these facts explain the occurrence of edema in these conditions. However, it has been known since the 1950s that the situation is more complex and it is still far from completely understood.

Causes of edema which are generalized to the whole body can cause edema in multiple organs and peripherally. For example, severe heart failure can cause pulmonary edema, pleural effusions, ascites and peripheral edema, the last of which effects can also derive from less serious causes.

Although a low plasma oncotic pressure is widely cited for the edema of nephrotic syndrome, most physicians note that the edema may occur before there is any significant loss of protein in the urine or fall in plasma protein level. Fortunately there is another explanation available. Most forms of nephrotic syndrome are due to biochemical and structural changes in the basement membrane of capillaries in the kidney glomerulae, and these changes occur, if to a lesser degree, in the vessels of most other tissues of the body. Thus the resulting increase in permeability that leads to protein in the urine can explain the edema if all other vessels are more permeable as well.

Organ-specific Edema

Edema will occur in specific organs as part of inflammation, as in pharyngitis, tendonitis or pancreatitis, for instance. Certain organs develop edema through tissue specific mechanisms.

Examples of edema in specific organs:

- Cerebral edema is extracellular fluid accumulation in the brain. It can occur in toxic or abnormal metabolic states and conditions such as systemic lupus. It causes drowsiness or loss of consciousness.
- Pulmonary edema occurs when the pressure in blood vessels in the lung is raised because of obstruction to removal of blood via the pulmonary veins. This is usually due to failure of the left ventricle of the heart. It can also occur in altitude sickness or on inhalation of toxic chemicals. Pulmonary edema produces shortness of breath. Pleural effusions may occur when fluid also accumulates in the pleural cavity.
- Edema may also be found in the cornea of the eye with glaucoma, severe conjunctivitis or keratitis or after surgery. It may produce coloured haloes around bright lights.
- Edema surrounding the eyes is called *periorbital edema* or eye puffiness. The periorbital tissues are most noticeably swollen immediately after waking, perhaps due to the gravitational redistribution of fluid in the horizontal position.
- Common appearances of cutaneous edema are observed with mosquito bites, spider bites, bee stings (wheal and flare), and skin contact with certain plants such as Poison Ivy or Western Poison Oak, the latter of which are termed *contact dermatitis*.
- Another cutaneous form of edema is myxedema, which is caused by increased deposition of connective tissue. In myxedema (and a variety of other rarer conditions) edema is due to an increased tendency of the tissue to hold water within its extracellular space. In myxedema this

is because of an increase in hydrophilic carbohydrate-rich molecules (perhaps mostly hyaluronan) deposited in the tissue matrix. Edema forms more easily in dependent areas in the elderly (sitting in chairs at home or on aeroplanes) and this is not well understood. Estrogens alter body weight in part through changes in tissue water content. There may be a variety of poorly understood situations in which transfer of water from tissue matrix to lymphatics is impaired because of changes in the hydrophilicity of the tissue or failure of the 'wicking' function of terminal lymphatic capillaries.

- In the case of human **feet**, the Starling forces are always a long way out of balance, because the variation in hydrostatic pressure in the vessels in the feet as compared to the face is about a metre of water. In severe heart failure the change in central venous pressure is tiny in comparison and cannot explain why edema of the feet develops simply through an effect on capillary leakage. Three other factors may be involved. If the central venous pressure rises to equal that of the thoracic lymph duct then clearance of fluid from the tissue will be impeded (see below). That is to say the edema may actually be caused by a change in output of fluid from the tissue, as much as input to the tissue. Secondly, severe heart failure is one of the most exhausting conditions there is. The sufferers tend to spend what little effort they can make trying to breathe with edematous lungs. They tend to sit up to make breathing easier and their feet hang immobile on the floor. Immobility is perhaps the commonest of all causes of edema, because clearance of fluid via the lymphatics needs muscle action. Thirdly, in severe heart failure endocrine and neural changes alter the way tissues are perfused in ways that are not fully understood.

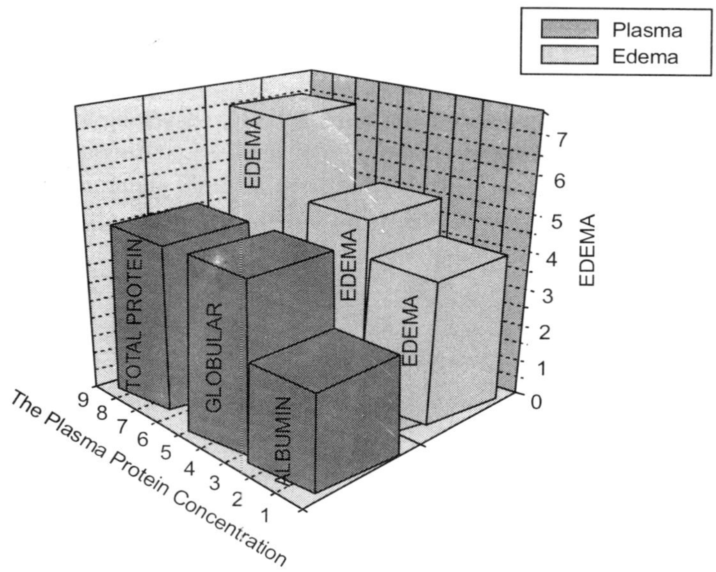

Fig. 8.56 : The Comaprison of the Plasma Protein Concetration and Edema Fromation in Humman Serum

- In lymphedema abnormal removal of interstitial fluid is caused by failure of the lymphatic system. This may be due to obstruction from, for example, pressure from a cancer or enlarged lymph nodes, destruction of lymph vessels by radiotherapy, or infiltration of the lymphatics by infection (such as elephantiasis). It is most commonly due to a failure of the pumping action of muscles due to immobility, most strikingly in conditions such as multiple sclerosis, or paraplegia. Lymphatic return of fluid is also dependent on a pumping action of structures known as lymph hearts. It has been suggested that the edema that occurs in some people following use of aspirin-like cyclo-oxygenase inhibitors such as ibuprofen or indomethacin may be due to inhibition of lymph heart action.

The Subacute CO Poisoning

Suffering from vertigo, crushing chest pain, nausea a dry coughy chills and a headache. On admission to the hospital the patient's blood pressure was 110/90 mm Hg, the pulse was 112 beats/min, and the respirations were 28/min. The patient was admitted *to* the coronary care unit. Arterial blood gas measurements were made while he was receiving oxygen at 4 L/min: oxygen saturation, 83% (calculated 98%); pO_2, 101 mm Hg;pCO_2, 32 mm Hg (0.992 mmol/L); pH, 7.45; HCO), 22.1 mmol/L; and carboxyhæmoglobin, 15.6%. At the same time a blood sample taken from his wife showed 18.1% carboxyhæmoglobin.

The gas furnace in their home was found later to have a rusted-out flue.

1. What is the structure of carboxyhæmoglobin?
2. How is carboxyhæmoglobin determined in blood?
3. What is the effect of CO on the oxygen-carrying capacity of the blood?

1. Structure Carbon monoxide combines with the iron in ferrohæmoglobin and ferro-myoglobin to form a complex very similar in structure to oxyhæmoglobin and oxymy-oglobin Stereochemically the O_2 and CO bind so that the Fe and the C and 0 atoms of CO are not linear.

Whereas this structure for the O_2 complex is normal and produces the optimum binding for O_2, the nonlinear structure for the CO complex is forced on it by a histidine residue that is close to the heme group in the globin chain. CO binds to free iron porphyrin so that the Fe, C, and O atoms are linear.

2. Determination of carboxyhæmoglobin Carboxyhæmoglobin (HbCO) is determined spectrophotometrically, taking advantage of the different spectra for the various hæmoglobin derivatives. Oxyhæmoglobin (HbO_2) (absorption maxima 577 nm and 541 nm) has a similar spectrum in alkaline solution to HbCO (absorption maxima 570 nm and 539 nm). Hæmoglobin has a single broad absorbance band at 555 nm. If weakly alkaline blood is reduced with sodium dithionite (NazSz04), the HbO_4 is converted to hæmoglobin, but HbCO is unaffected. Determination therefore involves diluting a blood sample with ammonium hydroxide and centrifuging it if the blood does not clear. One adds dithionite to an aliquot of this clear solution and measures absorbance at 555 nm and 480 nm against a reference standard containing dithionite in dilute ammonium hydroxide. The ratio of the two absorbancies then is read on a standard curve constructed from known mixtures of hæmoglobin and HbCO.

3. Effects **of carbon monoxide** CO binds 25,000 times as strongly to *free* iron por-phyrin as O_2. Because the CO in hæmoglobin and myoglobin is forced into a less stable nonlinear structure, the binding to hæmoglobin is only 200 times as strong as O_2. Otherwise the toxicity of CO would probably be fatal to smokers, who have 4% to 8% saturation of hæmoglobin with CO. Aside from the competitive binding of the CO, which reduces the amount of O_2 that can be carried to the tissues, the loss of the cooperactivity of O_2 binding and the consequent tighter binding of O_2 to the hæmoglobin means that O2 is not as readily liberated to the tissues. These combined effects produce significant tissue anoxia, resulting in shortness of breath on exertion (at 10 to 20% CO saturation). At 30% to 50% saturation, headache, confusion, and fainting on exertion occur; concentrations of 80% and greater are rapidly fatal.

The half-life of hæmoglobin-bound CO when a person is breathing air is about 5 to 8 hr. CO is released fairly rapidly from hæmoglobin if the patient is receiving O_2, as was done in this case. One must also remember that a patient receiving O_2 when transported to a hospital will have a much reduced HbCO concentration, which may confuse the diagnosis.

The Blood Grouping and Testing Methods

There are different types of Blood testing Methods as because thisi s book of theoretical biochemistry so only the methods of hæmoglobin estimation and blood counting methods has been explained

Here we shall discuss about the latest testing method of hæmoglobin and Blood grouping of blood

Technique

If the patient bleeds freely from a skin puncture,0.05 cc. of blood may be taken with an accurate 0.1 cc. pipette. After wiping off the tip of the pipette the blood is quickly blown out into 2.45 cc. of water in a test-tube and mixed at once. The tube may be set aside (corked) until a convenient time for estimation, or the work may proceed as soon as the blood is fully laked. Add exactly 2.5 cc. of 0.2 N HCI (18 cc. of C.P. acid per liter are close enough). The volume of the mixture is 5 cc., and the dilution of the blood is 1 part in 100. Warm the tube in a water bath at 55-60" (often hot tap water is satisfactory) for 7 minutes or more. This develops the maximum color of the acid hematin (we did not find 1 hour standing at room temperature sufficiently reliable). In the meantime the calorimeter has been set and tested for uniformity of the light on both sides. Put the standard solution in the left hand cup and set at 15 mm. Cool the blood mixture and put it in the other cup. Make several readings, being careful to avoid eye-strain. Determine the temperature of the standard at once by inserting a thermometer in the liquid. No calculation is necessary since the percentage of hemoglobin can be read directly from the table using the average reading and the temperature.

Table of Hæmoglobin Calculation is an abbreviation of the complete table that will be furnished with the solution. An estimation of 100 per cent indicates that the blood contains 13.8 g. of hemoglobinin 100 cc. Normal bloods will, of course, give a distinctly higher estimation. It is more satisfactory to use oxalated venous blood. In this case measure exactly 1 cc. into a 100 cc. flask, containing about 40cc. of water. After laking has occurred add 50 cc. of 0.2 N HCl while mixing, and dilute to the mark. To cut the foam use a drop of alcohol (caprylic alcohol must not be used). Mix well, and heat some of the solution in a test-tube as described above. In cases of marked anemia use double quantity of blood, anddivide the estimated per cent of hemoglobin by 2. If 0.1 cc. is taken for the micro method 2.45 cc. of HCl are to be used instead of 2.5 cc. Preservation of the standard. The solution is kept in a Non-sol or Pyrex flask, corked with a rubber stopper (if a new stopper is used, scour off the powdered material). The standard after being used is stored in a second flask. Before the main stock of standard is exhausted check up the used standard against it by making estimations with both, using the same acid hematin preparation.

Per cent of Hemoglobin Corrected for the Temperature of the Standard, Corresponding to Various Colorimeter Readings

Colorimeter reading	Temperature										
	15.5°	16.5°	17.5°	18.5°	19.5°	20.5°	21.5°	22.5°	23.5°	24.5°	25.5°
8.0	131.5	134.1	136.7	139.5	142.0	144.7	147.4	150.8	154.0	157.0	160.6
8.5	123.7	126.2	128.6	131.1	133.6	136.2	139.2	142.0	144.8	148.0	151.2
9.0	116.9	119.2	121.3	123.8	126.2	128.8	131.4	133.9	136.5	139.1	142.7
9.2	114.5	116.7	119.0	121.2	123.5	125.9	128.3	131.3	133.9	136.9	139.8
9.4	111.9	114.1	116.3	118.5	120.8	123.3	125.8	129.4	131.2	133.9	136.6
9.6	109.4	111.6	113.9	116.1	118.4	120.6	122.8	125.4	128.5	131.1	133.8
9.8	107.2	109.3	111.4	113.6	115.8	118.1	120.5	123.3	126.2	128.6	131.0
10.0	105.0	107.1	109.2	111.5	113.5	115.8	118.2	120.5	123.1	125.6	128.2
10.2	103.0	105.0	107.0	109.1	111.2	113.5	115.9	118.1	120.5	123.1	125.7
10.4	101.0	103.0	105.0	107.0	109.0	111.1	113.2	115.6	118.0	120.6	123.2
10.6	99.0	101.1	103.0	105.0	107.0	109.0	111.0	113.3	116.0	118.4	120.8
10.8	97.4	99.4	101.4	103.3	105.2	107.2	109.3	111.6	113.9	116.2	118.7
11.0	95.8	97.6	99.4	101.2	103.0	105.0	107.0	109.2	111.5	114.0	116.5
11.5	91.3	93.2	95.1	96.9	98.8	100.5	102.5	104.7	106.9	109.1	111.5
12.0	87.8	89.5	91.2	92.9	94.7	96.5	98.4	100.4	102.5	104.6	106.8
12.5	84.2	85.9	87.6	89.2	91.2	92.8	94.5	96.4	98.5	100.5	102.5
13.0	80.8	82.5	84.3	85.9	87.5	89.2	91.0	92.8	94.7	96.8	98.9
13.5	78.0	79.5	81.0	82.6	84.2	85.8	87.5	89.4	91.3	93.2	95.1
14.0	75.2	76.7	78.2	79.7	81.3	82.8	84.3	86.1	88.0	89.9	91.8
15.0	70.2	71.6	73.0	74.4	75.8	77.3	78.6	80.4	82.2	83.9	85.4
16.0	65.8	67.1	68.4	69.7	71.0	72.4	73.8	75.4	77.0	78.6	80.3

If they agree, pour the used standard into the other flask and use again. We have found that with reasonable care no change in the solution is noticed during months of repeated use. Of course, the cup and plunger of the calorimeter must be clean and dry before the standard is put in the cup. If dried salts collect on the stopper and on the mouth of the flask, wipe off with a damp cloth. Advantages of the Method. The technique is quick, simple, and accurate. Special apparatus is not required. The standard is permanent.

The Blood Grouping and Blood Cell Counting

The most common type of grouping is the ABO grouping. Red Blood Cells have a protein coat on their surface which distinguishes them. According to this blood is divided into four groups:

- A (A oligosaccharide is present)
- B (B oligosaccharide is present)
- AB (A and B oligosaccharides are present)
- O (neither A nor B, only their precursor H oligosaccharide present)

There are subtypes under this grouping (listed as A1, A2, A1B or A2B...) some of which are quite rare. Apart from this there is a protein which plays an important part in the grouping of blood. This is called the Rh factor. If this is present, the particular blood type is called positive. If it is absent, it is called negative. Thus we have the following broad categories:

- A1 Negative (A1 -ve)
- A1 Positive (A1 +ve)
- A1B Negative (A1B -ve)
- A1B Positive (A1B +ve)
- A2 Negative (A2 -ve)
- A2 Positive (A2 +ve)
- A2B Negative (A2B -ve)
- A2B Positive (A2B +ve)
- B Negative (B -ve)
- B Positive (B +ve)
- O Negative (O -ve)
- O Positive (O +ve)

Rare Blood Types

A blood type is classified as rare when more than 200 donors have to be screened to find one compatible donor with blood of that type. In the “ABO” system, all blood belongs to one of four major group: A, B, AB, or O. But there are more than two hundred minor blood groups that can complicate blood transfusions. These are known as rare blood types. About one person in 1,000 will inherit a rare blood type. Whereas common blood types are expressed in a letter or two, with maybe a plus or a minus, a fewer number of people express their blood type in an extensive series of letters in addition to their ‘ABO’ type designation. For example, AB +ve, O -ve, and A1 -ve are rare types.

Blood Group Systems

ISBT N°	System name	System symbol	Epitope or carrier, notes	Chromosome
1	2	3	4	5
001	ABO	ABO	Carbohydrate (N-Acetylgalactosamine, galactose). A, B and H antigens mainly elicit IgM antibody reactions, although anti-H is very rare, see the Hh antigen system (Bombay phenotype, ISBT #18).	9
002	MNS	MNS	GPA / GPB (glycophorins A and B). Main antigens M, N, S, s.	4
003	P	P1	Glycolipid. Antigen P1.	22
004	Rh	RH	Protein. C, c, D, E, e antigens (there is no "d" antigen; lowercase "d" indicates the absence of D).	1
005	Lutheran	LU	Protein (member of the immunoglobulin superfamily). Set of 21 antigens	19
006	Kell	KEL	Glycoprotein. K1 can cause hemolytic disease of the newborn (anti-Kell), which can be severe.	7
007	Lewis	LE	Carbohydrate (fucose residue). Main antigens Lea and Leb - associated with tissue ABH antigen secretion.	19
008	Duffy	FY	Protein (chemokine receptor). Main antigens Fya and Fyb. Individuals lacking Duffy antigens altogether are immune to malaria caused by Plasmodium vivax and Plasmodium knowlesi.	1
009	Kidd	JK	Protein (urea transporter). Main antigens Jka and Jkb	18
010	Diego	DI	Glycoprotein (band 3, AE 1, or anion exchange). Positive blood is found only among East Asians and Native Americans.	17
011	Yt or Cartwright	YT	Protein (AChE, acetylcholinesterase).	7
012	XG	XG	Glycoprotein	X
013	Scianna	SC	Glycoprotein	1
014	Dombrock	DO	Glycoprotein (fixed to cell membrane by GPI, or glycosyl-phosphatidyl-inositol)	12
015	Colton	CO	Aquaporin 1. Main antigens Co(a) and Co(b).	7

1	2	3	4	5
016	Landsteiner-Wiener	LW	Protein (member of the immunoglobulin superfamily)	19
017	Chido/ Rodgers	CH/RG	C4A C4B (complement fractions).	6
018	Hh/Bombay	H	Carbohydrate (fucose residue).	19
019	Kx	XK	Glycoprotein.	X
020	Gerbich	GE	GPC / GPD (Glycophorins C and D).	2
021	Cromer	CROM	Glycoprotein (DAF or CD55, regulates complement fractions C3 and C5, attached to the membrane by GPI).	1
022	Knops	KN	Glycoprotein (CR1 or CD35, immune complex receptor)	1
023	Indian	IN	Glycoprotein (CD44 adhesion function?)	11
024	Ok	OK	Glycoprotein (CD147)	19
025	Raph	MER2	Transmembrane glycoprotein.	11
026	JMH	JMH	Protein (fixed to cell membrane by GPI).	6
027	I_i	I	Branched (I) / unbranched (i) polysaccharide	6
028	Globoside	GLOB	Glycolipid. Antigen P.	3
029	GIL	GIL	Aquaporin 3.	9
030	Rh-associated glycoprotein	RHAG	Rh-associated glycoprotein.	6

Determination of Blood Groups[2]

The blood type tests are viewed as one of the important tests and this is done prior to acquiring a blood transfusion as well as to examine the blood type of a pregnant woman. The blood of humans is typed on the red blood cells surface by markers called antigens. The blood type test is done also to see if they are blood relatives. However, the important antigen is the blood group ABO as well as the Rh antigen. The common blood type tests are the Rh and ABO tests. The ABO blood type test indicates that people have four types of blood namely A, AB, B or O. The antigen A shows that you have A blood group and the plasma consists of antibodies to fight against the blood group B type. The B antigen indicates that the person has B blood group. The plasma present here has antibodies to fight against group A blood. In case you have type O blood, the plasma has antibodies that can face and stand against both A and B

* This testing is done with the help of a chemical kit (SPAN).

blood types. Both A and B antigens indicate that the person has group AB blood. In this group the plasma lacks antibodies to fight against type A or B blood. Blood that is received while in transfusion should certainly have the identical antigen as yours. A transfusion having different antigens is regarded as incompatible blood. On the other hand, the incompatible blood will have antibodies and ruin the donor blood cells. Such reaction is immediate and this is known as the transfusion reaction. This reaction may be mild or serious and sometimes may also lead to health. The blood group O negative does not consist of any antigens and hence is referred to as the universal donor and is compatible with all the blood types. The AB positive blood is well-known to be the universal recipient. They are known so because a person with this type of blood can receive any type of blood. Though, the universal donor or the universal recipient are useful in classifying blood during emergency, it is essential to know the blood type tests to avoid transfusion reactions. Minor antigens occurring on the red blood cells also can cause problems. Grave transfusion reactions are very rare nowadays owing to the blood type tests. Rh test checks the blood for R antigen. For instance, if Rh and A antigens are in you, then the blood type is A positive. Similarly, if the blood has B antigen and not Rh antigen, then the blood type is B negative.

Blood Counting

The Blood counting is done by using a hemacytometer (also spelled hemocytometer) is an etched glass chamber with raised sides that will hold a quartz coverslip exactly 0.1 mm above the chamber floor. The counting chamber is etched in a total surface area of 9 mm^2 (see Figure 8.57).

Calculation of concentration is based on the volume underneath the cover slip. One large square has a volume of 0.0001 ml (length × width × height; i.e., 0.1 cm × 0.1 cm × 0.01 cm).

In both methods, the hemacytometer is filled by capillary action - place the pipette filled with a well-suspended mix of cells at the notch at the edge of the hemacytometer and then slowly expel some contents so that the fluidis drawn into the chamber by capillary action.

Staining of cells often facilitates visualization and counting. Either mix cells with an equal volume of trypan blue [0.4% (w/v) tyrypan blue in PBS] to determine live/dead count (dead cells are blue) or kill cells with 10% formalin and then stain with trypan blue or other stain (to improve visualization of all cells.

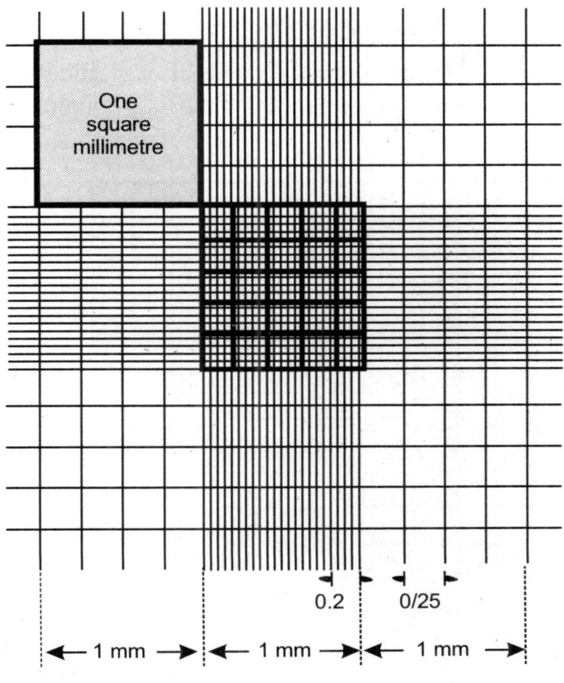

Fig. 8.57 : The Dimensions of a Hemacytometer

Here are two simple methods for counting cells based on the surface area of the hemacytometer used to determine cell count. Other counting schemes are accetable also. The choice of methods depends upon the cell concentration - the accuracy of the procedure depends upon the number of cells counted. When cell concentration is low, one should count more grids.

Method A

Count the number of cells in the 4 outer squares (see the left panel of Figure 8.58).

The cell concentration is calculated as follows:

Cell concentration per milliliter = Total cell count in 4 squares × 2500 × dilution factor

Example: If one counted 450 cells after diluting an aliquot of the cell suspension 1:10, the original cell concentration = 450 × 2500 × 10 = 11,250,000/ml

Method B

Estimate cell concentration by counting 5 squares in the large middle square (see the right panel in Figure 8.58).

The cell concentration is calculated as follows:

Cell concentration per milliliter = Total cell count in 5 squares × 50,000 × dilution factor.

Example: If one counted 45 cells after diluting an aliquot of the cell suspension 1:10, the original cell concentration = 45 × 50,000 × 10 = 22,500,000/ml.

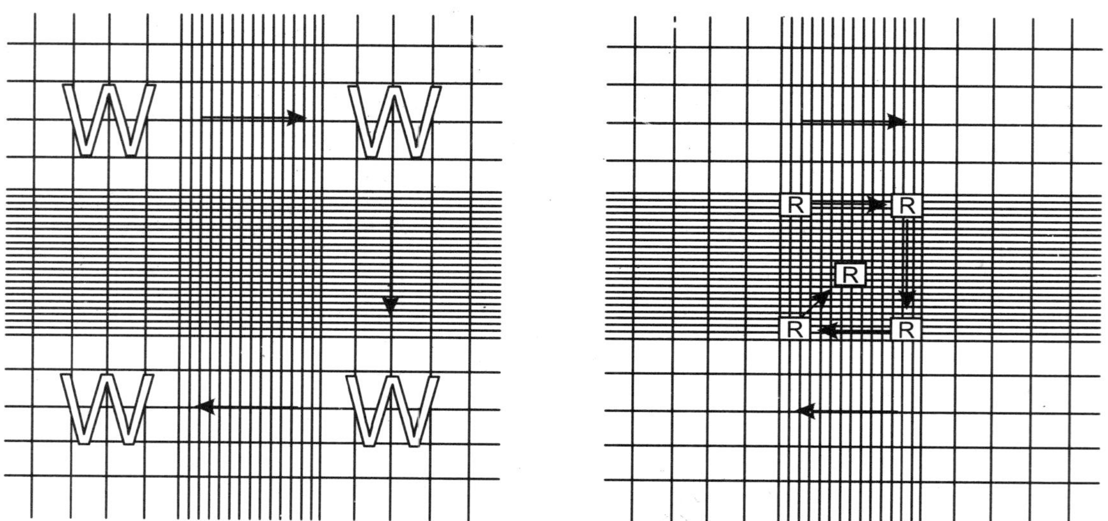

Fig. 8.58 : Counting Procedure for Methods A (left panel) and B (right panel)